Pharmaceutical Medicine and Translational Clinical Research

Pharmaceutical Medicine and Translational Clinical Research

Edited by

Divya Vohora

Gursharan Singh

Academic Press is an imprint of Elsevier
125 London Wall, London EC2Y 5AS, United Kingdom
525 B Street, Suite 1800, San Diego, CA 92101-4495, United States
50 Hampshire Street, 5th Floor, Cambridge, MA 02139, United States
The Boulevard, Langford Lane, Kidlington, Oxford OX5 1GB, United Kingdom

Library of Congress Cataloging-in-Publication Data
A catalog record for this book is available from the Library of Congress

British Library Cataloguing-in-Publication Data
A catalogue record for this book is available from the British Library

ISBN: 978-0-12-802103-3

For Information on all Academic Press publications
visit our website at https://www.elsevier.com/books-and-journals

Publisher: Mica Haley
Acquisition Editor: Kattie Washington
Editorial Project Manager: Tracy Tufaga
Production Project Manager: Anusha Sambamoorthy
Cover Designer: Miles Hitchen

Typeset by MPS Limited, Chennai, India

Contents

SECTION III PHARMACEUTICAL LAW AND ETHICS

SECTION V GENERICS, SUPERGENERICS, BIOLOGICS, BIOSIMILARS, AND BIOBETTERS

SECTION VI MEDICAL SERVICES

SECTION VII PHARMACOVIGILANCE

SECTION VIII DRUG UTILIZATION AND PHARMACOECONOMICS

List of Contributors

Nidhi B. Agarwal
Jamia Hamdard University, New Delhi, India

Geeta O. Bedi
Providentia Research, Gurgaon, Haryana, India

Subodh Bhardwaj
Consultant Biopharmaceuticals, New York, NY, United States

Richa Goyal
QuintilesIMS, Mumbai, Maharashtra, India

Debleena Guin
CSIR-Institute of Genomics and Integrative Biology, New Delhi, India

Rajinder K. Jalali
Sun Pharmaceutical Industries Ltd, Gurgaon, Haryana, India

Manoj Karwa
Auriga Research Pvt. Ltd, New Delhi, India

Bharti Khanna
Pharmalex India Pvt. Ltd, New Delhi, India

Lalit K. Khurana
Sun Pharmaceutical Industries Ltd, Gurgaon, Haryana, India

Ritushree Kukreti
CSIR-Institute of Genomics and Integrative Biology, New Delhi, India

Sunita Narang
Ranbaxy Laboratories Limited, Gurgaon, Haryana, India

Gerfried K.H. Nell
NPC Nell Pharma Connect Ltd, Vienna, Austria; Rokitan Ltd, Vienna, Austria

Sten Olsson
Uppsala Monitoring Centre, Uppsala, Sweden

Mahendra Rai
Tata Consultancy Services, Mumbai, Maharashtra, India

Deepa Rasaily
Sun Pharmaceutical Industries Ltd, New Delhi, India

Nilanjan Saha
Jamia Hamdard University, New Delhi, India

Bindu Sharma
Origiin IP Solutions LLP, Bengaluru, Karnataka, India

Sangeeta Sharma
Institute of Human Behaviour & Allied Sciences, New Delhi, India

Gursharan Singh
Life Sciences, SmartAnalyst India Private Limited, Gurgaon, Haryana, India

Harinder Singh
Sun Pharmaceutical Industries Ltd, Gurgaon, Haryana, India

Romi Singh
Sun Pharmaceutical Industries Ltd, Gurgaon, Haryana, India

Sandeep Sinha
Jamia Hamdard University, New Delhi, India

Ankit Srivastava
Jamia Hamdard University, New Delhi, India; CSIR-Institute of Genomics and Integrative Biology, New Delhi, India

Divya Vohora
Jamia Hamdard University, New Delhi, India

Preeti Vyas
Jamia Hamdard University, New Delhi, India

About the Editors

Prof. Divya Vohora, MPharm, PhD

Dr. Divya Vohora is Professor of Pharmacology and In-Charge of Pharmaceutical Medicine Programme in School of Pharmaceutical Education and Research at Jamia Hamdard, New Delhi, India. She has 20 years of teaching/research experience with more than 100 publications in reputed national and international journals with more than 1500 citations. She has worked as principal investigator for various research projects funded by AICTE, ICMR, UGC, CSIR, and DST, Government of India. She is also coordinating University Grants Commission Special Assistance Programme of Department of Pharmacology at DRS Phase II level. Forty two (MPharm: 31 and PhD: 11) degrees have been awarded under her supervision. She has participated in various national and international conferences/symposia and presented invited lectures/research papers in India and abroad (Australia, France, Malaysia, Poland, Singapore, London, etc.). She was also organizing secretary, coorganizer, chairperson for some of these meetings and workshops. Her major areas of preclinical, clinical, and translational research interests are epilepsy, neurobehavioral research, cognitive functions, histamine H3 receptors, and secondary osteoporosis. For her research contributions, she has been awarded Dr. D. N. Prasad Memorial Award of ICMR, Chandra Kanta Dandiya Prize for best published paper in Pharmacology, DST Fast Track Award for Young Scientists, AICTE Career Award for Young Teachers, etc. She is a member of many learned societies, professional bodies, and expert committees. She is Member, International Advisory Group, British Pharmacological Society, London; Reviewer, National Science Centre, Krakow, Poland; Member, Executive Committee, Indian Society of Pharmacoeconomics and Outcomes Research and on editorial boards (Behavioral Neurology, Current Psychopharmacology), and reviewer for more than 30 journals of repute.

Dr. Gursharan Singh, MBBS, PhD

Dr. Gursharan Singh, MBBS, PhD (Pharmaceutical Medicine), MIPL started his career as Assistant Professor of Pharmacology at Dolphin Institute, Dehradun, India. He then worked as Senior Medical Affairs and Clinical Research Physician at India's largest pharmaceutical MNC, the erstwhile Ranbaxy Labs Limited for almost 8 years prior to joining Gurgaon office of a New York–based Global Life Sciences Consulting Firm, Smart Analyst as Senior Principal, Life Sciences. He has a total of 14 years of experience with almost 12 years of industry experience. His areas of expertise and interest include clinical development strategy, medical affairs, medical writing, new product ideation especially re-innovation including supergenerics and biobetters, orphan drugs, and gene therapy. He has contributed to development of target product profile and robust Phase III study design for both New Chemical Entities as well as re-purposed molecules for many Big Pharma and small biotech clients as well as appropriate BE study design for generic drugs. Throughout his career, he has been involved in various training activities in the field of drug development, approval, and lifecycle management.

Foreword

With the recent advances in drug discovery and development processes, and frequently changing pharmaceutical regulations in the world, Prof. Divya Vohora and Dr. Gursharan Singh present the first edition of "Pharmaceutical Medicine and Translational Clinical Research," which provides an up-to-date information on Pharmaceutical Medicine, a specialty that broadly comprises of the whole lifecycle of a drug with emphasis from the discovery, preclinical testing, pharmaceutical development to details on postmarketing surveillance, and pharmacovigilance.

The book not only covers an overview of the drug discovery and development processes but also incorporates up-to-date information on pharmaceutical regulations in the United States, European Union, and India and also for complementary medicines as well as ethical considerations. While providing comprehensive understanding of various aspects of medicine and regulations governing discovery and development, the book has some topics like pharmacogenomics, pharmacoeconomics, clinical pharmacokinetics, generic drugs and bioequivalence studies, tools for monitoring medicine use, etc. A thorough understanding of these aspects is critical for professionals working in the pharmaceutical industry, drug regulation, contract research organizations, clinical research professional, and students.

The book will be extremely useful for medical and pharmaceutical professionals and also for students pursuing PG diploma, MSc, and PhD, and MSc (Clinical Research) courses as well as for teachers and professionals, including medical doctors, nurse pharmacists involved in clinical practice or clinical research.

I congratulate the authors for bringing this book in the rapidly advancing field of Pharmaceutical Medicine.

18-10-17

Dr. G.N. Singh
Drugs Controller General, India

Preface

Pharmaceutical medicine is the specialty concerned with the discovery, preclinical and clinical development, risk benefit evaluation, regulatory approval, safety monitoring, and medico-marketing of drugs.

In 2014 the Tufts Center for the Study of Drug Development (CSDD) projected the cost of developing a prescription drug that gains market approval at US $2.6 billion with a total lifecycle cost of US $2.9 billion. The increase in costs is attributed to increased complexity of clinical trials to assure regulators of favorable risk benefit and insurers regarding comparative drug effectiveness data to justify the pricing of drugs. Clinical development now requires large number of clinical studies in much larger patient population. To improve the overall efficiency, studies are being conducted across multiple geographies. In view of increasingly global face of drug development, the International Council for Harmonization of Technical Requirements for Pharmaceuticals for Human Use (ICH) has made attempts to bring together the regulatory authorities and pharmaceutical industry to discuss scientific and technical aspects of drug registration. Despite commendable efforts, pharmaceutical regulations continue to vary across geographies. Professionals working in the field of pharmaceutical medicine need to be trained in both the scientific and the regulatory aspects of this discipline.

Today, dedicated organizations and training in the field of pharmaceutical medicine is available in many countries like Australia, Belgium, France, Germany, India, Ireland, Italy, Japan, Netherlands, Singapore, South Africa, Spain, Sweden, Switzerland, United Kingdom, and United States.

The textbook "Pharmaceutical Medicine and Translational Clinical Research" provides a comprehensive review of scientific and regulatory aspects of drug discovery, development, approval, intellectual property, pharmacovigilance, medical affairs, and pharmacoeconomics. Special chapters have been included to cover recent advancement in the field of pharmacogenomics, biosimilars, supergenerics, and biobetters.

The target audience for this book includes students studying pharmaceutical medicine as a specialty or those studying drug development and approval as part of any other course. The book would also be useful for all professionals working in the field of pharmaceutical medicine irrespective of their educational background and roles.

Writing a textbook on such a vast subject requires making balanced and well thought through choices on the content and contributors. We, as editors would like to thank all our expert authors with diverse industry and academic expertise whose contribution and dedication helped us create this work. We also acknowledge the contribution of the team at Elsevier, especially Tracy and Anusha for handling this project patiently and efficiently despite repeated delays due to busy schedule of contributors. Last but not the least, we would like to thank our families for providing us continuous support and encouragement.

The first edition of this book is just a beginning and we invite constructive feedback on the book from the readers to further improve the book in its subsequent editions.

Divya Vohora and Gursharan Singh

SECTION

OVERVIEW OF PHARMACEUTICAL MEDICINE

I

CHAPTER

THE SPECIALTY OF PHARMACEUTICAL MEDICINE 1

Gerfried K.H. Nell[1,2]

[1]NPC Nell Pharma Connect Ltd, Vienna, Austria [2]Rokitan Ltd, Vienna, Austria

1.1 WHAT IS PHARMACEUTICAL MEDICINE?

The content of pharmaceutical medicine may be described according to the definition given by International Federation of Associations of Pharmaceutical Physicians & Pharmaceutical Medicine (IFAPP):

> Pharmaceutical medicine is a medical scientific discipline concerned with the discovery, development, evaluation, registration, monitoring and medical aspects of marketing medicines for the benefit of patients and public health [1].

Regarding the content of this definition of pharmaceutical medicine one may add medicinal components of medical devices, diagnostics (especially companion diagnostics in order to select appropriate and optimal therapies in the context of a patient's genetic content or other molecule or cellular analysis [2]), and dietary supplements which are loosely summarized under the term nutraceuticals (dietary supplements, isolated nutrients, and herbal products [3]). The term *medicines development* is used synonymously with *pharmaceutical medicine* [4] in order to indicate that the biomedical input can be provided not only by medically trained persons but also to a large extent by other life scientists.

Table 1.1. shows the syllabus of Pharmaceutical Medicine as approved by the Faculty of Pharmaceutical Medicine, UK [5], PharmaTrain, an originally European public–private partnership which has been developed into a global standard-setting organization in pharmaceutical medicine [4], and IFAPP, which serves as the international platform of national associations of professionals in pharmaceutical medicine. The content of Pharmaceutical Medicine is outlined in 14 Sections containing a total of 183 topics [5].

It is clear that the sections mentioned comprise topics which cover the whole field of medicine, life sciences, chemical and technological sciences and in addition ethics, technical, mathematical, legal, financial, and economical aspects important for drug development and marketing. This does not mean that the combined knowledge and competencies in this vast area can be summarized as medical discipline. However, drug or medical device development and lifecycle management have essential medical aspects at every stage of the whole process which have to be taken into account because the ultimate

Pharmaceutical Medicine and Translational Clinical Research. DOI: http://dx.doi.org/10.1016/B978-0-12-802103-3.00001-8

Table 1.1 Syllabus for Medicines Development according to PharmaTrain[a]

Training Syllabus for Medicines Development	
Section	**Section Overview**
1.	Discovery of Medicines
2.	Development of Medicines: Planning
3.	Non-Clinical Testing
4.	Pharmaceutical Development
5.	Exploratory Development (Molecule to Proof of Concept)
6.	Confirmatory Development: Strategies
7.	Clinical Trials
8.	Ethics and Legal Issues
9.	Data Management and Statistics
10.	Regulatory Affairs
11.	Drug Safety, Pharmacovigilance and Pharmacoepidemiology
12.	Information, Promotion and Education
13.	Economics of Healthcare
14.	Therapeutics

[a]*http://www.pharmatrain.eu/ [accessed 28.07.15].*

goal of all these efforts is the benefit for the patients, i.e., in selecting and developing new drug candidates. The particular features of the targeted diseases have to be considered, e.g., is the biochemical target specific enough or do other factors also perhaps play a decisive role rendering the therapeutic option less efficacious. Thus, pharmaceutical medicine may be regarded as an umbrella term, covering the medical aspects of medicines development and lifecycle management of medicinal products.

The multidisciplinary character of pharmaceutical medicine is reflected by the specialists working in the field. A specialist in pharmaceutical medicine/drug development is a biomedical professional or physician who is engaged in discovery of medicines, translational medicine, clinical development, regulatory affairs, and support of ethical promotion and safe use of medicines. Drug development scientists are professionals from biomedical sciences working in the aforementioned areas but who do not have a medical degree. A huge group of physicians is engaged in clinical research as physician investigators planning, supervising, monitoring, or conducting clinical trials. These investigators are usually working in academia or other clinical trial sites, in industry, health authorities, or other governmental institutions.

Pharmaceutical medicine is thus a multifaceted interdisciplinary area where a huge number of specialists are working in teams. The professionals involved in medicines development beside physicians are pharmacists, life scientists of several disciplines, chemists, statisticians, information technology specialists, project managers, regulatory specialists, lawyers, communication specialists, financial advisers, and accountants.

A certain depth of knowledge in all areas mentioned in the syllabus is necessary for Pharmaceutical Medicine professionals in order to enable them to put their biomedical input in the process of Medicines Development and Lifecycle Management into perspective, which is one of the major advantages of an education in Pharmaceutical Medicine.

Pharmaceutical physicians are working in pharmaceutical industry, Contract Research Organizations (CROs), Regulatory Agencies and Authorities, and in clinical research located in academia, other hospitals, physicians' offices, and SMOs (site management organizations, i.e., consortia of clinical trial centers).

Pharmaceutical medicine is a discipline which overlaps with almost all other medical specialties and with many other disciplines leading to a working environment which is characterized by teamwork, flexibility, and the attitude to learn from and cooperate with colleagues from a huge area of disciplines.

One particular point deserves attention, which is the relation to clinical pharmacology, which is a recognized medical specialty in many countries either as a standalone specialty or as a subspecialty to other medical specializations. Clinical pharmacology may be defined as follows: "Clinical pharmacology is the scientific study of the actions and modes of action of drugs in the human species and the actions and modes of action of human physiology and metabolism on drugs" [6]. The World Health Organization defines clinical pharmacology as follows: The functions of a clinical pharmacologist are: (1) to improve patient care by promoting the safe and more effective use of drugs; (2) to increase knowledge through research; (3) to pass on knowledge through teaching; (4) to provide services e.g. analyses, drug information and advice in the design of experiments, which clearly points to clinical pharmacology as an academic discipline [7,8].

Obviously there is a huge overlap regarding the content of the two disciplines which originated about 50 years ago out of the need to develop better medicines and to improve their therapeutic use. Both disciplines evolved in parallel and with close exchange of ideas and concepts [9]. However, the main emphasis is different, as can be easily derived from the above-mentioned definitions. The differences can be best described by looking at the sections of the Syllabus for Medicines Development (Table 1.1). Both disciplines center on action, efficacy, and safety of medicinal products in patients. Clinical pharmacology focuses on research regarding actions and mode of actions of drugs in human diseases and research in pharmacokinetics and drug monitoring regarding therapeutic and toxicological questions.

Pharmaceutical medicines comprises the medical aspects of the whole lifecycle of a drug with additional emphasis on discovery and planning of development of drugs, nonclinical testing, pharmaceutical development and—most importantly for all who are working in pharmaceutical industry—information, promotion, and education regarding medicinal products.

It requires the attitude to face and tackle new problems every working day, necessitates corporations with colleges of several disciplines in a team, and makes working rewarding because of the strong feeling of accomplishment based on team work and cooperation. Working in pharmaceutical medicine also confers a global perspective since the principles of medicines development are based on globally accepted scientific standards [10].

1.2 ORIGIN AND DEVELOPMENT OF PHARMACEUTICAL MEDICINE

In principle the history of pharmaceutical medicines dates back several millennia to the times when the administration of extracts of herbs and other natural medicaments became one of the pillars of treatment of pain and other ailments. Over the centuries the preparation of medications from herbal and mineral origin was in the hands of pharmacists and administered to patients based on medical advice. It was one of the cornerstones of caring for the patients besides surgical measures and physician's counseling. Over the centuries, efforts have been made in order to assure quality standards of this kind of medicine, e.g., by exact descriptions of their preparation in the London Pharmacopoeia of 1618 in Europe [10].

During the nineteenth century two events changed the whole situation, namely the emergence of modern life sciences, biology, and biochemistry, and their application, to explore the causes of human illnesses—which led to a scientific approach to medical problems on the one hand, and to the science of chemistry, which allowed production of medicinal products on an industrial scale, on the other hand. The latter process moved drug production from pharmacies to chemical manufacturers. Thus, most present Big Pharma companies trace back its origin to the nineteenth century. At the beginning of industrial drug manufacturing the emphasis was laid on quality in the chemical sense. Therefore the industry was led by chemists and pharmacists.

The scientific approach to the practice of medicine led to an explosion of new areas of medicinal treatment, e.g., general and local anesthetics, antibiotics, analgesics, antiphlogistics, and hypnotics. These achievements were accompanied by a rapid development of statistic methodology which was applied to toxicology, pharmacology, and later to the development of controlled randomized clinical trials. This new armamentarium led to the employment of life scientists and physicians in pharmaceutical companies in order to take care of medical aspects of these new developments [10].

The next important milestone was the focus on safety in the sense of toxicity and adverse effects. It was triggered by unfortunate events which caused worldwide attention. One of the major events demonstrating the need for more safety evaluation was the series of diethylene poisonings in the United States of America in 1937. More than 100 people, mostly children, died because diethylene glycol was used as a solvent for preparing a sulfanilamide elixir without any safety testing. This event resulted in the introduction of The Federal Food, Drug and Cosmetic Act in 1938 which made premarketing notification for new drugs mandatory. The second and probably up till now most important event in the development of safety standards for drugs was the thalidomide disaster. Thalidomide was introduced as a sedative and hypnotic in 1956 in Germany and later on in 46 countries worldwide till 1960. This resulted in about 10,000 babies born with phocomelia and other deformities because at this time no standard test program was performed to detect these risks. The consequence of the thalidomide catastrophe was the reshaping of the regulatory system regarding safety of drugs including the implementation of programs of voluntary adverse drug reaction reporting systems in many countries [11].

Obviously, questions of bioethics are closely interrelated with many decisions in medicines development, and have to be carefully considered. Bioethics came to the forefront of daily work in research and development, especially in clinical research. One of the most important triggers of this development was the egregious violation of ethical principles in the era of National Socialism in Germany during the Second World War, e.g., deadly experiments on prisoners of war. These atrocities led to the Nuremberg Code and then to the Declaration of Helsinki defining the ethical foundation of clinical research [12].

The development of Pharmaceutical Medicine over the last 50 years has thus been shaped by the tremendous progress in medicine and biosciences, by the emergence of regulations especially governing research and development and regulatory affairs (GXP, Good...Practice) and by the increasing relevance of bioethics.

With the increasing number of physicians working in the pharmaceutical industry the need of professional associations, either from the side of the workers in medicines development or, later on, from the side of regulatory bodies, was felt. Since 1957 national societies for pharmaceutical medicine were founded, starting in UK and South Africa (for information please see IFAPP website under members [1]). In 1975 the representatives of 12 national societies (Argentina, Belgium, Brazil, France, Germany, India, Italy, Japan, South Africa, Sweden, The Netherlands, and the UK) convened and created an international umbrella organization, the IFAPP (International Federation of Associations of Pharmaceutical Physicians, today known as the International Federation of Associations of Pharmaceutical Physicians and Pharmaceutical Medicine). The geographical distribution of the founding members shows that it was clear from the start that Pharmaceutical Medicine can be developed on a global scale only [10]. Today IFAPP is an organization with 29 member associations on all continents.

The mission of IFAPP is defined as follows:

The mission of the Federation is to promote pharmaceutical medicine by enhancing the knowledge, expertise, and skills of pharmaceutical physicians and biomedical scientists (all science disciplines involved in the discovery, development, processing, and usage of medicines research) worldwide, thus leading to the availability and appropriate use of medicines for the benefit of patients and society (1, see constitution).

In cooperation between national societies and IFAPP the framework of education and training medicines development was developed during the last 40 years. The first structured training program in pharmaceutical medicine was introduced in 1975 in the UK and since 1978 is jointly organized by the British Association of Pharmaceutical Physicians (BrAPP) and the University of Wales in Cardiff. Following the creation of a diploma examination in Pharmaceutical Medicine, the Royal Colleges of Physicians of the UK set up a Faculty of Pharmaceutical Medicine (FPM) in 1989 as a standard setting body [10]. Based on the legislation of the European Union establishing pan-European postgraduate medical training the FPM developed a postgraduate training program leading to a Certificate of Completion of Training (CCT), which is the prerequisite for recognition of a medical specialty. This procedure was followed also in Ireland and Switzerland. Outside Europe, recognition of pharmaceutical medicine as a medical specialty was granted in Mexico [10].

In about 15 countries inside and outside Europe, courses similar to the course in Cardiff, based on the same syllabus, were introduced. In order to evaluate quality and adherence to the common syllabus of IFAPP and the FPM and to harmonize the content and standards of these courses, IFAPP created the Council for Education in Pharmaceutical Medicine (CEPM) in 2001. The CEPM evaluated more than 10 courses in Pharmaceutical Medicine worldwide and issued Guidance Notes for the Establishment of Structured National Continuing Medical Education/Continuing Professional Development (CME/CPD) Programs for Pharmaceutical Medicine [10].

The activities of the CEPM were transferred to a newly created initiative in 2008. In this year the European project Innovative Medicines Initiative (IMI) was launched. The mission of the Innovative Medicines Initiative "is working to improve health by speeding up the development of, and patient access to, innovative medicines, particularly in areas where there is an unmet medical or social need. It does this by facilitating cooperation between the key players involved in healthcare research, including universities, the pharmaceutical and other industries, small and medium-sized enterprises (SMEs), patient organizations, medicines regulators, learned societies, and professional societies of health care professionals. IMI is a partnership between the European Union (represented by the European Commission) and the European pharmaceutical industry represented by EFPIA, the European Federation of Pharmaceutical Industries and Associations (modified based on [13,14]).

The importance of training and education was recognized from the beginning and four of the first 42 projects developed to address bottlenecks in discovering and developing medicines were devoted to training, one of them PharmaTrain. IFAPP was one of the founder members of PharmaTrain, which comprised 24 universities (most of them running the courses in Pharmaceutical Medicine acknowledged by the CEPM of IFAPP), 13 learned societies/professional organizations, and several other partners, e.g., regulatory authorities on the public side and 15 pharmaceutical companies on the private side. The syllabus, curriculum, and learning outcomes of courses in Pharmaceutical Medicine were revised and agreed upon by PharmaTrain, FPM (UK), and IFAPP. and a set of competencies for professionals in Medicines Development was defined. The curriculum contents for these programs have been harmonized across the participating academic institutions throughout Europe [14]. In addition, training programs for clinical investigators and specialists in regulatory affairs have been created. A training program for obtaining recognition as a specialist in medicines development/pharmaceutical medicine was also established. The quality of these standard setting activities is maintained by the implementation of a comprehensive quality management system. [14].

In 2014 the PharmaTrain project was concluded after having achieved its goals. A substantial part of the membership decided to continue its activities as an independent association called PharmaTrain Federation. This organization adopted a principally global mission since medicines development does follow the same principles scientifically and regarding regulatory affairs worldwide and cooperation with universities outside Europe was already established during the life span of IMI PharmaTrain. PharmaTrain Federation is getting increasing attention as a global standard setting body in all aspects of medicines development/pharmaceutical medicine.

1.3 PRESENT STATE OF THE SPECIALTY OF PHARMACEUTICAL MEDICINE

At present pharmaceutical medicine is an acknowledged medical specialty in a few countries (Ireland, Mexico, Switzerland, UK). In addition, in Switzerland an acknowledgment as expert in pharmaceutical medicine is also available for nonmedics building on similar qualification steps as for the physicians, except in clinical education, of course [10,15].

Education and training in pharmaceutical medicine is based on a theoretical aspect acquiring the requested knowledge base, and supplemented by a practical competency-based training in an individualized program centered on an approved workplace training environment (see e.g., UK, 16). In the UK successful passing of the examination of the theoretical aspect and completion of the training program leads to a Certificate of Completion of Training, the holder of which is eligible to apply for entry in the Specialist Register of the General Medical Council (GMC).

The content of the theoretical part has been defined by PharmaTrain in cooperation with the FPM and IFAPP [5,17]. The content of Pharmaceutical Medicine is organized in 14 sections (Table 1) comprising 183 topics. This is the syllabus of Pharmaceutical Medicine, which is essentially a list of topics comprising a subject, discipline, or specialty field. In order to allow for temporal and geographical flexibility whilst guaranteeing a defined quality standard of the theoretical courses in Pharmaceutical Medicine a modular approach based on the Bologna process has been chosen. The Bologna process defines a framework for the harmonization of academic education in Europe [18]. The ultimate goal is to enable mutual recognition on a modular basis between all courses following the principles of PharmaTrain.

In order to transfer the topics of the syllabus into a modular structure a curriculum has been created which allocates the syllabus content to the 6 modules of the "base course" which serves as the theoretical base of the already existing programs of obtaining the degree of a specialist in Pharmaceutical Medicine. The curriculum is a statement of the aims and the content, experiences, and outcome of a program. It describes the structure and expected methods of learning, teaching, feedback, and supervision. The main goal of the curriculum is to set out what knowledge, skills, attitudes, and behaviors the trainee is expected to acquire. These are the learning outcomes. The learning outcomes are the leading principles for the development of the curriculum and each module. (For details see the respective chapters of [4]).

In addition to the base course in Pharmaceutical Medicine the curriculum of a Master Course in Pharmaceutical Medicine has been established and the course providers also offer elective modules, which can be chosen as building blocks of the master program [4].

PharmaTrain has recognized a series of academic courses as following its standards (see Diploma and Master Courses in [4]). The evaluation is based on the content of the syllabus and the learning outcomes and appropriate organization of teaching and examinations. In addition a set of quality standards has to be followed: In brief the set of quality criteria is based on the following principles outlined in Table 1.2 (for references see quality control 4, [19])

Base and Master courses in Pharmaceutical Medicine are the theoretical part of training in this specialty. In order to obtain recognition as a specialist in this field the required competencies have to be demonstrated. This procedure is based on the emerging trend of transformative learning, which has been deemed as particularly important in the education of health professionals. The main changes involved in education in health related areas as compared to the more static model prevalent up till now are shifting from [1] memorizing facts to search, analysis, and synthesis of information for decision-making, [2] from seeking individual professional credentials to achieving core competencies for effective teamwork in health systems, and [3] from noncritical adoption of educational models to creative adaptation of global resources to address local needs [20]. A competency as building block of education is defined as "an observable ability of any professional, integrating multiple components such as knowledge, skills, values, and attitudes." Because competencies are

Table 1.2 Quality Criteria PharmaTrain[a]

1.	Trainees are supported to acquire the necessary knowledge and skills
2.	Course structures encourage exchange and multidisciplinarity
3.	Facilities, infrastructure, leadership and competences adequate to deliver the approved curriculum
4.	Equality principles
5.	Teaching methods appropriate to the goals of the course
6.	Transparency regarding potential conflicts of interest

[a] *http://www.pharmatrain.eu/ (Accessed 28 July 2015).*

observable they can be measured and assessed which is the basis of granting the specialist recognition in Pharmaceutical Medicine.

Regarding the definition of the usually interchangeable terms competency and competence, it should be mentioned that in this context competency does refer to the skill itself and competence does denote the ability to perform, and the attribute of the performer [20]. Competence and performance are closely related but not synonymous since performance can be influenced also by factors other than the competence of the performer [20].

The core competencies for pharmaceutical physicians and drug development scientists (life scientist other than physicians) have been defined. 7 core competency domains and 60 core competencies within the domains have been identified. Up till now the identification of competencies is based on cognitive aspects mainly. A statement of competence summarizing the competency domains has been prepared (Table 1.3, [20]).

It should be clarified at this point that the concept of pharmaceutical medicine as outlined in this chapter was originally conceived of as a medical specialty. However, in the reality of R&D in the pharmaceutical industry, CROs, and big and small enterprises, there is no distinction between the assignments of physicians and nonphysicians with an educational background in pharmacy, veterinary medicine, biology, biochemistry, etc. The existing specialty recognition in UK and Ireland follows the usual pathway for physicians, and nonphysicians are not eligible. The only exception is Switzerland where there is a recognition for physicians as specialists in Pharmaceutical Medicine in place in the same manner as for other medical specialties whereas, in addition, nonphysicians

Table 1.3 Statement of Competence in Pharmaceutical Medicine[a]

Statement of Competence
The Pharmaceutical Physician/Drug Development Scientist: • Is able to identify unmet therapeutic needs, evaluate the evidence for a new candidate for clinical development and design a Clinical Development Plan for a Target Product Profile. • Is able to design, execute and evaluate exploratory and confirmatory clinical trials and prepare manuscripts or reports for publication and regulatory submissions. • Is able to interpret effectively the regulatory requirements for the clinical development of a new drug through the product lifecycle to ensure its appropriate therapeutic use and proper risk management. • Is able to evaluate the choice, application, and analysis of post-authorization surveillance methods to meet the requirements of national/international agencies for proper information and risk minimization to patients and clinical trial subjects. • Is able to combine the principles of clinical research and business ethics for the conduct of clinical trials and commercial operations within the organization. • Is able to appraise the pharmaceutical business activities in the healthcare environment to ensure that they remain appropriate, ethical and legal to keep the welfare of patients and subjects at the forefront of decision making in the promotion of medicines and design of clinical trials. • Is able to interpret the principles and practices of people management and leadership, using effective communication techniques and interpersonal skills to influence key stakeholders and achieve the scientific and business objectives.

[a] *Silva H, Stonier P, Buhler F, Deslypere JP, Criscuolo D, Nell G, Massud J, Geary S, Kerpel-Fronius S, Koski G, Clemens N, Klingmann I, Kesselring G, van Olden R, Dubois D. Core competencies for pharmaceutical physicians and drug development scientists. Frontiers in Pharmacology 2013; doi: 10.3389/fphar.2013.00105.*

undergoing the same education—with the exception of a period spent in patient care, of course—can obtain the SwAPP (Swiss Association of Pharmaceutical Professionals) diploma in Pharmaceutical Medicine [15].

The PharmaTrain Specialist in Medicines Development (SMD) program is based on the experiences with the Pharmaceutical Medicine Specialty training in UK and in Switzerland [5,15,21]. It is open for physicians and drug development scientists. The training and the recognition is based, supervised, and granted on the PharmaTrain SMD certification process & curriculum path and (see 22 for details) globally, according to the same standards.

The theoretical part has already been described. The practical competency-based training in an individualized program is based on the following principles: the applicant has to provide evidence over a four-year period of gaining practical training and competencies in medicines development in an appropriate institution (pharmaceutical company, CRO, clinical or preclinical research institute, or competent authority) which offers the respective opportunities to gain such experience in medicines development. The practical training must be recorded e.g., in the PharmaTrain Training Record e-portfolio. The progress of the trainee has to be supervised by a named mentor. This qualified mentor has to provide documentation of their training to substantiate their qualification as a mentor. Alternatively training can be performed in recognized training sites. Part of the practical training can be accomplished by attending appropriate courses or other suitable offers outside the workplace.

In addition to their training program trainees should:

- Maintain knowledge and awareness of the need to move on from the scientific and technological challenges in the medicines development industries to addressing industry bottlenecks and resolving health care and societal challenges through prevention and therapy.
- Acquire a thorough understanding of the management and administration of the organizations and the applied business models in the area of medicines development.

On completion of the SMD training the trainees are expected to be competent in all domains of the SMD curriculum.

PharmaTrain has set up a PharmaTrain Certification Board (PCB), which is responsible for the implementation, monitoring, standards, and quality of the SMD certification program. After a positive review of the application consisting of satisfactory documentation of four years of practical training the PCB will issue the PharmaTrain title of a Specialist in Medicines Development.

In principle the SMD award is a global title. At present (2015) it is introduced in Italy. Another country interested in this program is Japan.

Continuing professional development (CPD) is an integral part of lifelong learning for all professionals in order to keep up with the progress in their professional activities. This concept has been introduced into the professional curriculum of physicians in many countries. The principle has also been adopted for Specialists in Pharmaceutical Medicine by e.g., the Faculty of Pharmaceutical Medicine in the UK and the respective bodies in Switzerland [5,15,21]. PharmaTrain is in the process of setting up a system for CPD for PharmaTrain Specialists in Medicines Development for physicians and nonphysicians [22]. Particular features of CPD systems are to take advantage of a variety of learning modalities with an emphasis on self-directed and active learning. Offering such systems recognizing lifelong learning does meet the needs of the members of the national societies who expressed their wish for an efficient, self-directed pathway to CPD repeatedly in surveys [14]. At present, structured CPD systems for specialists in pharmaceutical medicine/medicines development are introduced in a few countries only. Therefore it is hoped that the global recognition by PharmaTrain will help to close this gap.

1.4 STATE OF AFFAIRS IN PHARMACEUTICAL MEDICINE IN SELECTED COUNTRIES

As outlined already, pharmaceutical medicine is a discipline which is based on principles of global validity. However, since it is an evolving discipline the degree of development in various regions differs. For details please consult the website of IFAPP where you may find additional information on the activities of the member associations of the federation. In this section reference is given to several recent publications on the activities of some national associations of pharmaceutical medicine.

1.4.1 HUNGARY

After the Second World War the pharmaceutical companies in Hungary were nationalized and their activities were coordinated centrally according to a national plan. In line with this organization clinical development was also centralized within a national clinical-pharmacological network. A board examination in clinical pharmacology was introduced in 1979. In practice this network may be described as a pharmaceutical medicine network. After the end of the socialist regime the pharmaceutical companies became independent again and increased their medical staff. Based on the above-mentioned network an independent Clinical Management Society was founded and became a member of IFAPP in 2004. In order to train experts in pharmaceutical medicine a course in medicine development was initiated by the Semmelweis University in Budapest based on the principles of PharmaTrain. This course was successfully developed into an international educational network of universities in Central and Eastern Europe and Mediterranean regions under the name "Cooperative European Medicine Development Course" [23].

1.4.2 INDIA

A joint collaborative program for PhD in Pharmaceutical Medicine has been established between Jamia Hamdard (Hamdard University) in New Delhi and Ranbaxy Laboratories (now Sun Pharmaceuticals Ltd) in 1999. It is a unique program illustrating a combination of academic excellence and professional expertise through a joint university/industry effort. The primary goal of the program is to provide students with extensive expertise in the field of pharmaceutical medicine.

The PhD program is divided into two parts: (I) Course work in the area of Pharmaceutical Medicine including hands-on training in carrying out clinical pharmacokinetic studies at Sun Pharmaceuticals Clinical Pharmacology Unit. (II) Research work, either clinical or preclinical, followed by submission of a thesis in a minimum period of two years after completing the course work ([24], and personal communication).

1.4.3 MEXICO

In Mexico a successful and thriving society for specialists in Pharmaceutical Medicine was founded in 1967 (AMEIFAC). The society is active in many ways, publishing books and articles, organizing presentations, workshops, symposia, and forums, and participating in both national and international conferences. In 2000 a university-based course in Pharmaceutical Medicine was initiated based on the IFAPP syllabus. Based on this course a recognition of the specialty of Pharmaceutical Medicine is granted by the national postgraduate institutions of upper education [25].

1.4.4 SWITZERLAND

Education and Training in Pharmaceutical Medicine/Medicines development are governed by Swiss law and a number of respective ordinances and are overseen by the Swiss Institute of Postgraduate Training and Continuing Medical Education. As already explained in the preceding section, Switzerland is the only country with an established pathway for postgraduate education and training and CPD in pharmaceutical medicine for both physicians and drug development scientists. For physicians, the postgraduate training program lasts five years and is divided into two years of clinical work in any specialty, and three years of specific postgraduate training. The practical part of the postgraduate training can be performed at training centers in Clinical Trial Units of university hospitals, affiliates of global pharmaceutical companies, or other institutions working in the field of Pharmaceutical Medicine. The theoretical part of the post-graduate training is based on the PharmaTrain program of the base course in Pharmaceutical Medicine [15,21,26].

1.4.5 UK

UK has always been at the forefront of creating and implementing concepts of education and training in Pharmaceutical Medicine. The concepts introduced and elaborated by PharmaTrain are based to a large extent on the groundbreaking work accomplished by the Faculty of Pharmaceutical Medicine UK, which was founded in 1989 [27]. The PharmaTrain program of education and training in Pharmaceutical Medicine is very similar to the program of the FPM [16]. Up till now this program is open for physicians only. The details of the requirements and the execution of the program are thoroughly explained on the FPM website [16].

According to British requirements FPM has also introduced a program of revalidation for licensed physician specialists in Pharmaceutical Medicine. The purpose of the program is to demonstrate on a regular basis (every five years) that the licensed physicians are up to date and fit for practice. The doctors have to connect to the respective designated body, which is in most cases the FPM. The FPM appoints a responsible officer. The doctor is required to participate in a process of annual appraisal based on a portfolio of supporting information. Based on these appraisals the responsible officer will make a recommendation to the GMC about a doctor's fitness to practice [28].

1.4.6 EMERGING COUNTRIES

Increasing attention is paid to the needs of low income countries. The discussions center on the needs, optimal methods, and practical approaches for extending education and teaching of medicines development, regulation, and clinical research. In these countries the rapidly growing number of patients suffering from noncommunicable diseases has to be treated efficiently which requires that modern drug therapy has to become available more widely and with a shorter time lag. In order to achieve these goals additional experts in medicines development, regulation, and clinical research have to be trained. The programs and concepts developed by PharmaTrain seem suitable for achieving these goals [29].

1.5 CONCLUSIONS AND OUTLOOK

Biopharmaceutical industry is facing many challenges and critiques that the process of medicines development suffers from inefficiency, takes too much time, and has costs that are too high. It is unanimously acknowledged that this inefficiency may be due to a considerable extent to inadequate numbers of appropriately educated and trained professionals in the pharmaceutical industry, CROs, and the clinical research enterprise [30].

The groundwork for a competency based, interdisciplinary education, and training and CPD, has been laid by a number of stakeholders, e.g., the Faculty of Pharmaceutical Medicine in the UK, IFAPP, PharmaTrain, and many other academic and nonacademic institutions worldwide. The implementation, however, suffers from a lack of clear direction and leadership in many regions. Therefore the task for all stakeholders is to implement the proposed programs stepwise on a global level. This rollout has the potential to transform drug development procedures into a more efficient and integrated process, which is expected to result in the availability of better and safer medicines more rapidly to the benefit of patients and society.

REFERENCES

[1] <www.ifapp.org>; Aims and Objectives [accessed 29.03.15].
[2] <http://www.personalizedmedicinecoalition.org/Resources/Personalized_Medicine_101>; [accessed 28.07.15].
[3] <http://en.wikipedia.org/wiki/Nutraceutical#Classification_of_nutraceuticals>; [accessed 28.07.15].
[4] <http://www.pharmatrain.eu/>; [accessed 28.07.15].
[5] <https://www.fpm.org.uk/trainingexams/exams/dippharmmed>; [accessed 29.07.15].
[6] Yates R. A Career in Clinical Pharmacology in Stonier PD (ed.) Careers with the Pharmaceutical Industry, second edition, John Wiley & Sons Ltd, Chichester, UK; pp 47–54.
[7] World Health Organization: Clinical Pharmacology. Scope, Organization, Training. Technical Report Series. Report Nr. 446 (WHO, Geneva, 1970).
[8] Nell G. In: Prostran M, Stanulović M, Marisavljević D, Đurić D, editors. Pharmaceutical medicine and clinical pharmacology: common ground and differences. Hemofarm AD, Beograd/Vršac: Farmaceutska Medicina; 2009. p. 20–5.
[9] Stonier PD, Baber NS. Clinical pharmacology and the Faculty of Pharmaceutical Medicine. Br J Clin Pharmacol 2000;49:523–4.
[10] Stonier PD, Silva H, Lahon H. Pharmaceutical medicine. History, global status, evolution and development. Int J Pharm Med 2007;21(4):253–62.

[11] Rägo L, Santoso B. In: van Boxtel CJ, Santoso B, Edwards IR, editors. Drug regulation: History, present and future. Drugs benefits and risks: international textbook of clinical pharmacology. revised 2nd ed. IOS Press and Uppsala Monitoring Centre; 2008. p. 65–77.

[12] Fletcher AJ. In: Edwards LD, Fletcher AJ, Fox AW, Stonier PD, editors. Introduction to bioethics for pharmaceutical professionals. Principles and practice of pharmaceutical medicine. 2nd ed. John Wiley & Sons, Ltd; 2007. p. 587–94.

[13] <http://www.imi.europa.eu/content/mission>; [accessed 28.07.15].

[14] Silva H, Bühler FR, Maillet B, Maisonneuve H, Miller LA, Negri A, et al. Continuing Medical Education and Professional Development in the European Union. Pharm Med 2012;26:223–33.

[15] <http://www.swapp.ch/education/>; [accessed 30.07.15].

[16] <https://www.fpm.org.uk/trainingexams/pmst/curriculumassessment>; [accessed 30.07.15].

[17] <http://www.pharmatrain.eu/_downloads/Appendix_12_1_PharmaTrain_Syllabus_V1_0_February_2010.pdf>; [accessed 30.07.15].

[18] <https://www.google.at/webhp?sourceid = navclient&hl = de&ie = UTF-8&gws_rd = ssl#hl = de-AT&q = bologna + process + official + website>; [accessed 30.07.15].

[19] Klech H, Brooksbank C, Price S, Verpillat P, Bühler FR, Dubois D, et al. European initiative towards quality standards in education and training for discovery, development and use of medicines. Eur J Pharmaceut Sci 2012;45:515–20.

[20] Silva H, Stonier P, Buhler F, Deslypere JP, Criscuolo D, Nell G, et al. Core competencies for pharmaceutical physicians and drug development scientists. Front Pharmacol 2013. Available from: http://dx.doi.org/10.3389/fphar.2013.00105.

[21] <http://www.sgpm.ch/cgi-bin/index.pl?p = 119&l = en>; [accessed 31.07.15].

[22] <http://www.pharmatrain.eu/_downloads/Appendix_12_8_SMD_Procedural_Document_V1_0.pdf>; [accessed 31.07.15].

[23] Kerpel-Fronius S. The development of pharmaceutical medicine in Hungary. Pharm Med 2013;27:289–95.

[24] <http://jamiahamdard.edu/faculty-of-pharmacy/>; [accessed 31.03.16].

[25] Cohen-Muñoz V, Llópiz-Avilés M, Llorens F, Perniche M, Vargas J. Pharmaceutical Medicine in Mexico. Phar Med 2010;24:211–18.

[26] Traber M, Althaus B. Pharmaceutical medicine in Switzerland. Pharm Med 2010;24:75–81.

[27] Daniels S. Pharmaceuitcal medicine in the UK. Pharm Med 2011;25:1–5.

[28] <https://www.fpm.org.uk/revalidationcpd/revalidation/revalidationoverview>; [accessed 31.07.15].

[29] Kerpel-Fronius S, Rosenkranz B, Allen E, Bass R, Mainard JD, Dodoo A, et al. Education and training for medicines development, regulation, and clinical research in emerging countries. Front Pharmacol 2015. Available from: http://dx.doi.org/10.3389/fphar.2015.00080.

[30] Silva H, Sonstein S, Stonier P, Dubois D, Gladson B, Jones TC, et al. Alignment of competencies to adress inefficiences in medicines development and clinical research: need for inter-professional education. Pharm Med 2015. doi 10.1007/s40290-015-0097-3.

SECTION

DRUG DISCOVERY AND DEVELOPMENT

II

CHAPTER 2

DRUG DISCOVERY AND DEVELOPMENT: AN OVERVIEW

Sandeep Sinha and Divya Vohora
Jamia Hamdard University, New Delhi, India

2.1 INTRODUCTION

Drug discovery is a process which is intended to identify a small synthetic molecule or a large biomolecule for comprehensive evaluation as a potential drug candidate. Broadly, the modern drug discovery process includes identification of disease to be treated and its unmet medical need, selection of a druggable molecular target and its validation, in vitro assay development followed by high throughput screening of compound libraries against the target to identify hits, and hit optimization to generate lead compounds that exhibit adequate potency and selectivity towards the biological target in vitro and which demonstrate efficacy in animal models of disease. Subsequently, the lead compounds are further optimized to improve their efficacy and pharmacokinetics before they advance towards drug development (Fig. 2.1). Drug development process can be segregated into preclinical and clinical development stages (Fig. 2.2). In preclinical development, toxicological and safety pharmacology studies of the candidate are conducted in order to establish the maximum safe concentrations in animals and determine the adverse effect potential of the drug-in-development. Additionally, studies are conducted to finalize cost-effective processes required for manufacturing the candidate drug as well as deciding on its best formulation. If the candidate exhibits sufficient efficacy and safety in preclinical evaluation, permission is sought from drug regulatory agencies to initiate its clinical development wherein the safety and efficacy of the drug candidate is assessed in pilot and pivotal studies.

The discovery and development of innovative drugs is time and cost intensive and currently approximately twelve years and an average of $1.8 billion is required to launch a new drug. Over the years, there is a decreasing trend in the number of innovative drugs obtaining marketing approval. This is a consequence of heightened scrutiny of the safety and efficacy of new drugs by regulatory agencies, which leads to increased costs and prolonged development times. Further, the top management of pharmaceutical industry wants to avert the risks associated with drug discovery and development. Moreover, pressure on national health services due to costs associated with pharmaceuticals has an adverse impact on their pricing. Additionally, patent expirations and their generic substitutions have reduced the profits and subsequent growth of the pharmaceutical industry, resulting in reduced investment in innovative research. A new concern is adverse environmental impact of pharmaceuticals and there are clear directives from governmental agencies to ensure steps to reduce this impact [1]. A direct consequence of the dismal drop in productivity is increased

Pharmaceutical Medicine and Translational Clinical Research. DOI: http://dx.doi.org/10.1016/B978-0-12-802103-3.00002-X

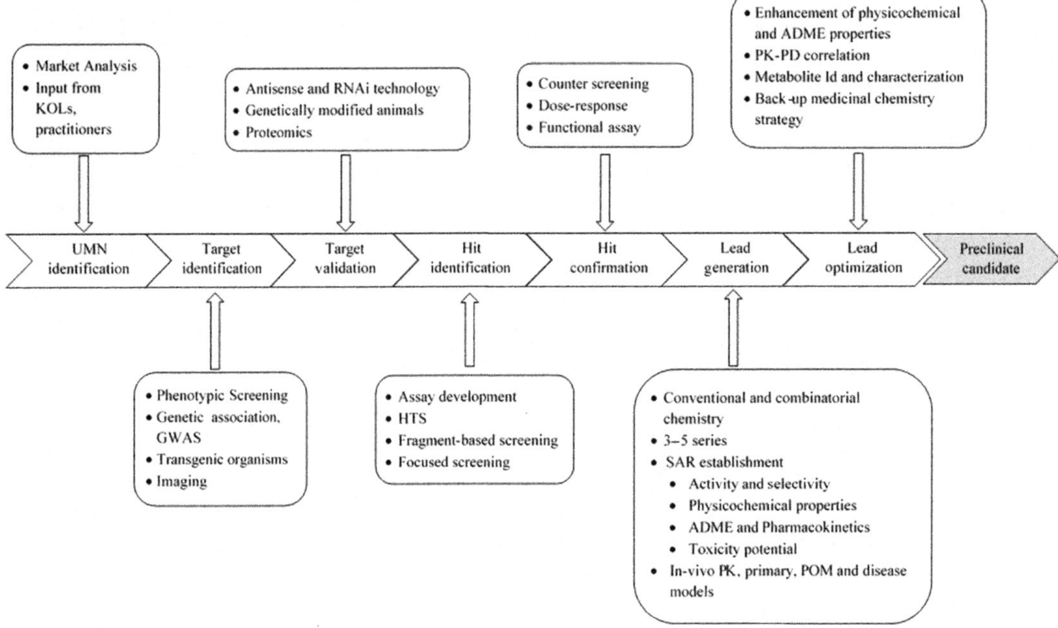

FIGURE 2.1

Overview of drug-discovery process. *UMN*, unmet medical needs; *KOL*, key opinion leader; *SAR*, structure activity relationship; *GWAS*, genome wide association studies; *HTS*, high throughput screening; *POM*, proof of mechanism; *PK*, pharmacokinetics; *PD*, pharmacodynamics.

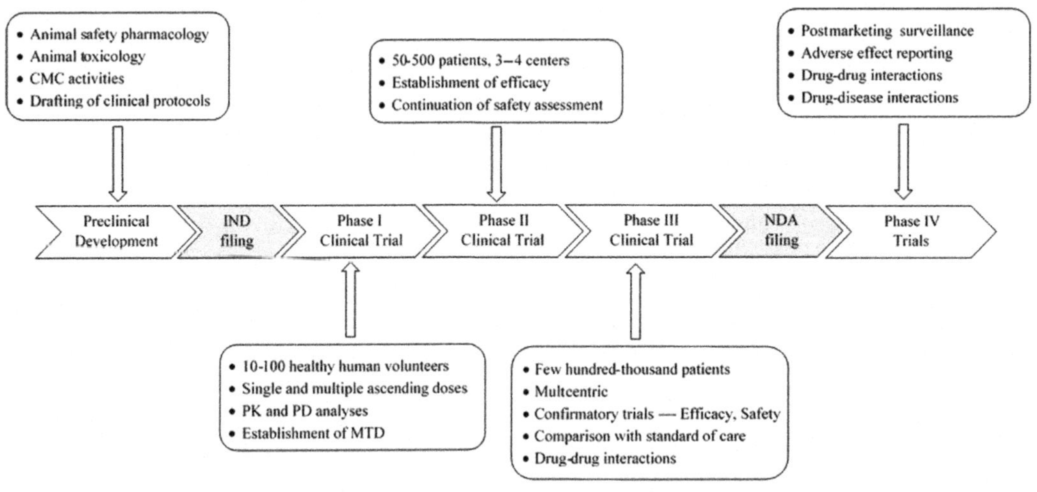

FIGURE 2.2

Overview of drug development process. *CMC*, chemistry, manufacturing and controls; *MTD*, maximum tolerated dose; *IND*, investigated new drug application; *NDA*, new drug application.

mergers and acquisitions seen in pharmaceutical industry, with the primary objective of reducing R&D costs and creating synergies. However, investments in innovative research post-mergers and acquisitions and pipeline advancements have actually decreased and hence industry consolidation to increase productivity is questionable [2].

In spite of the fall in R&D productivity due to the above mentioned concerns, there is still a high unmet need in the therapeutic areas of cancer, Alzheimer's disease, and diabetes, and there is a global emergence of multidrug resistant bacterial infections for which innovative drugs are urgently required.

2.2 IDENTIFICATION OF UNMET MEDICAL NEED

The trigger to initiate a drug discovery program is a medical condition whose treatment is not satisfactorily addressed by currently available treatment modalities. This is referred to as an unmet medical need for that condition. To elaborate, new drugs are needed either to treat a disease for which no other treatment exists or which offer additional advantages over existing treatments like superior therapeutic efficacy, reduced adverse effects, improved compliance, fewer drug-drug interactions, and consequently an overall improvement in the quality of life of a patient. Approaches to identify the unmet medical need include market analysis, inputs from key opinion leaders in a therapeutic area, feedback from medical practitioners and scientific conferences. Additionally, a thorough understanding of disease etiology, epidemiology, available therapeutic options and their shortcomings drive a prudent gap analysis and thereby facilitate shortlisting of medical needs in particular disease condition.

2.3 TARGET IDENTIFICATION

Biologically active compounds, whether they be a small synthetic molecule or a large molecular weight antibody, elicit their activity and thereby a measurable clinical effect by interacting with a naturally existing molecular structure, which in the context of drug discovery is referred to as a "target." These include enzymes, receptors, metabolites, substrates, ion channels, transport proteins, DNA, RNA, and ribosomes [3]. These targets can broadly be classified into established and novel targets. The targets which have been scientifically proven to have well-defined physiological and pathophysiological roles fall in the former category, whereas newly discovered ones whose role is turning out to be clearer with advancing research constitute the latter class. The various approaches to target identification are briefly discussed below.

In the phenotypic screening approach, compounds or antibodies are evaluated in cell-based assays or animal models of disease with an aim to identify compounds which elicit an anticipated change in the phenotype. This may include change in expression of a single or multiple proteins in vitro or obtaining desired pharmacological response in vivo. Subsequent to the identification of an active compound, its molecular target is then determined by genetic approaches like expression cloning techniques, in silico approaches, or chemical proteomic based approaches like affinity chromatography, activity based protein profiling, and label free techniques [4]. To exemplify, in an elegant study conducted by Sandercock et al., [5], single chain variable fragment (scFv) antibodies and designed ankyrin repeat proteins (DARPins) against primary non-small cell lung carcinoma cells were isolated and evaluated for pro-apoptotic and antiproliferative activity against primary cells. The phenotypic changes were detected in an ultra-high content screen by using multiple parametric profiling and subsequently CUB domain containing protein 1 (CDCP1) was identified as a target by employing a cell-surface membrane protein array.

Another process employed for target identification is genetic association study, wherein genetic variants—like single nucleotide polymorphisms (SNPs)—associated with risk for a disease or its progression, are identified. For example, distinct vascular endothelial growth factor (VEGF) polymorphisms lead to predisposition for psoriasis and hence modulation of VEGF signaling pathway might be a potential therapeutic option for the disease [6]. Further, genome wide association studies (GWAS) that explore the entire human genome for a large number of SNPs at a time, with an aim of identifying those variants that occur in most patients with a complex disease, have also aided in target selection as well as repositioning of existing drugs [7]. For example, denosumab, which is indicated for osteoporosis, targets tumor necrosis factor super family member 11 (TNFS11). However, GWAS suggest that the gene for this target is also associated with Crohn's disease and hence denosumab may have potential therapeutic utility in this disorder as well [8]. Transgenic organisms are also employed in target identification. For example, through bacterial artificial chromosome (BAC) transgenics in mice, the putative genomic region linked to a disease phenotype can be identified and further can be narrowed down to the target gene of interest [9]. Molecular and functional imaging techniques are also valuable tools for target identification [10].

2.4 TARGET VALIDATION

Subsequent to identification, a potential drug target needs to undergo the process of validation wherein its function in a disease state is ascertained. There are multiple approaches towards validation and some of them will be discussed here.

Use of antisense technology is a popular route wherein short oligonucleotides (single stranded nucleic acids), complimentary to a specific region of a messenger RNA of interest, are designed. The interaction of the oligonucleotides with the target mRNA results in disruption of translation and subsequently impedes the synthesis of the protein. To exemplify, knockdown of tetrodotoxin-resistant sodium channel $Na_v1.8$ by antisense oligonucleotides, obliterated intrathecal *N*-Methyl-D-Aspartate-induced mechanical hypernociception in rats, thereby highlighting the importance of these ion-channels in pain pathophysiology [11].

An alternative approach is RNA interference (RNAi) technology, wherein silencing of the target gene in a cell or an organism is triggered by introduction of a double stranded RNA (dsRNA) specific to the gene. The long dsRNAs are cleaved by the RNase, Dicer, into small interfering RNAs (siRNAs), which are double stranded fragments of 21–25 nucleotides with some unpaired base pairs at each end. Subsequently, the siRNAs are unwound into two single strands, referred to as a guide and a passenger strand, respectively. The guide strand is incorporated into the RNA interference specificity complex (RISC) and it locates the mRNA possessing the complimentary sequence, resulting in cleavage of the target mRNA and shutting down of the translation machinery. This approach of using long dsRNAs is marred by variability in response, overall decrease in mRNA levels, and expensive design. Alternatively, these issues can be overcome by design and subsequently direct introduction of 21–25 nucleotide long siRNAs into the cellular machinery. The potential utility of the above modality in target validation is exemplified by an experiment conducted by [12], wherein mice injected with siRNA against the chemokine CCR2 and subjected to ischemia-reperfusion, demonstrated lower infarct size and marked decrease in cardiac inflammatory monocytes as compared to control siRNA administered mice, thus demonstrating the role of CCR2 in inflammatory cell trafficking in cardiac tissue.

Employing genetically modified animals for target validation is an appealing methodology as it permits the scrutiny of the phenotypic consequences of gene manipulation. Development of knock-outs, knock-ins, conditional knock-outs, and transgenic animals are instances of genetically modified animals. An animal lacking a particular gene from the embryonic stage is one approach to studying the in vivo functions of diverse genes. For example, contraction to carbamylcholine was virtually abolished in the urinary bladder from muscarinic M_3-receptor knock-out mice, suggesting that contraction was predominantly due to M_3 receptor activation [13]. In gene knock-in animals, the desired gene is inserted into a specific locus in the target genome and can hence be said to be a gain of function mutation. To exemplify, knock-in mice with deficits in the AMPA receptor GluR1 Serine 831 and Serine 845 phosphorylation had a higher threshold and longer latencies to pentylenetetrazole induced seizures in postnatal day 9 as compared to wild type mice thereby supporting the validation of this potential therapeutic target for neonatal seizures [14].

In a conditional knock-out approach the gene of interest is deactivated in the target tissue at a specific time point thereby limiting the risk of embryonic lethality and developmental abnormalities which are often experienced with conventional knock-out models. The utility of this approach in target validation can be exemplified by a study conducted by [15] wherein they show that mice in which GPR88 receptor located on adenosine $A_{2A}R$ neurons were conditionally knocked-out, demonstrated decreased anxiety-like behaviors in light/dark and elevated plus maze tests.

Transgenic animals are also an attractive validation tool wherein a foreign gene is intentionally inserted in their genome. For example, transgenic LXR α mice demonstrated improved myocardial glucose tolerance and reduced cardiac hypertrophy in a mouse model of obesity-induced type 2 diabetes thus supporting use of therapies targeting LXR α for cardiovascular diseases [16].

A limitation of the genetic approach to target validation is that genes generate various isoforms of the protein which may have slightly different functions, and variations in proteins can also be a consequence of post-translational modifications. Hence, an improved and emerging approach to target validation is the proteomics approach which aims on the modulation of the activity of the target protein itself [17].

2.5 HIT IDENTIFICATION AND DEVELOPMENT OF ASSAYS

Following validation of a therapeutic target the next step is identification of 'hits', which in a broad sense may be defined as compounds that elicit the desired activity in a screening assay. For hit identification there exist diverse screening methods of which some will be briefly described. In a high throughput screening (HTS) method, libraries of, in some cases, up to a million drug-like compounds may be directly and quickly evaluated against a protein target in a biochemical assay (wherein requirement for a protein is less) or a cell-based assay, if the target has proven to modulate the cellular activity. This involves complex automation like the use of robotic liquid handler systems. In this methodology, the researchers do not have prior information of the chemotype required for demonstration of activity against the target and as an outcome one or few compounds may be identified with the desired activity in the high affinity range (IC_{50} in μM). Some demerits of HTS include poor coverage of chemical space and difficulty in optimization due to complex nature of hits [18]. Subsequent to identification of hits and subject to accessibility of the three-dimensional structure of the protein, the compound(s) may be co-crystallized with the protein and through X-ray crystallography the structure of the protein-ligand complex is acquired. This aids in determining the structure of the binding site on the protein, thereby providing the information for hit optimization, and thus forms the foundation for structure based drug design.

Another method which is developing popularity is the fragment-based screening approach wherein libraries of small molecule fragments are evaluated at a high concentration and the hits are defined as compounds which although weak in activity show efficient binding. These fragments can be then be used as building blocks for the synthesis of potent and drug-like compounds.

In the focused screening strategy, limited sets of compounds that have demonstrated activity against a specific class of targets (e.g., GPCRs) or structurally similar compounds are evaluated in the assay. This approach also extends to virtual screening, wherein a virtual library of existing compounds may be docked with the three-dimensional protein structures and their activity against the target might be predicted computationally.

The principles of phenotypic screening and its applications in target identification have been previously discussed. The time-honored phenotypic route has been additionally proven to be effective in the discovery of first-in-class drugs. The alternative approach, which is termed as target based screening, involves assessment of the activity of a large number of compounds against a single protein target. This route has been successful in the generation of follow-up drugs, but its main disadvantage with respect to discovery of first-in-class molecules is cross reactivity with several other targets, which cannot be captured in the single protein assay setup [4]. Some examples of drugs discovered through phenotypic screening are: aripiprazole—a conformational/partial receptor agonist; azacitidine—an irreversible enzyme inhibitor; cinacalcet—allosteric activator of receptor; ezetimibe—affects transporter activity; and miglustat—an enzyme inhibitor demonstrating reversible inhibition. On the other hand, drugs discovered through target based screening include: eltrombopag—a non-competitive receptor agonist; imatinib—an enzyme inhibitor that works by stabilizing inactive conformation; mifepristone—a conformational receptor antagonist; orlistat—an irreversible enzyme inhibitor; and raltegravir—an enzyme modulator acting by trapping the conformational state [19].

For most of the hit identification strategies it is crucial to develop biological assays in which compounds are evaluated for their activity. Biological assays may be cell-free or biochemical assays in which human or other mammalian recombinant proteins are employed for evaluation of compounds either for their affinity as in the case of receptors or inhibitory activity for enzymes. Cell-based assays on the other hand are functional assays with a specific read-out, for example intracellular calcium concentration. Factors to be considered for a selection of an assay format include relevance of the assay, its reproducibility, assay quality, cost assessment, effect of compound or its solvent on the assay, and the screening concentrations [20].

2.6 CONFIRMATION OF HITS

In this phase the identified hits are subjected to confirmatory evaluation using the same assay conditions which were employed during hit identification. Further, it is imperative to ascertain that the activity is linked to the anticipated mechanism and is not due to artifacts. Subsequently, frequent or promiscuous hitters are eliminated from further consideration. A procedure of detecting false positives is to employ a counter-screening assay in which hits are evaluated for their activity against an alternative member of the target family under identical assay conditions and if the hit demonstrates similar activity then it is most likely a false positive. Additionally, precipitation of the small molecule as well as aggregate formation can also lead to false positive results and such a counter-screen aids in their exclusion from further attention [21]. Consideration must also be given to identifying compounds that may produce activity based on detrimental mechanisms, as these may lead to toxicity. Establishing dose-response relationships is also essential as an all-or-none response may point to non-specific effects or interaction with any other constituent of the assay condition. Additionally, generation of reliable dose-response curves allows rank ordering of hits through the estimation of half maximal inhibitory concentration (IC_{50}) in case of inhibitors/antagonists and half maximal effective concentration (EC_{50}) for activators/agonists. Experiments examining the nature of binding of the actives at this stage are also advantageous as they filter out compounds showing non-competitive interaction with the target and which will not be the preferred candidates for evaluation in a clinical setting. Finally, the hits may be screened in a secondary cell-based or a functional assay in which the target has been proven to play a role in order to ascertain their efficacy.

During hit-confirmation, medicinal chemists utilize the biological data to rank as well as cluster the hits into groups and gain initial insight into structure-activity relationships (SAR) between members of a group. Additionally, feasibility of chemical synthesis is also scrutinized by the medicinal chemists.

The potency of a hit identified against a target is usually in the range of 1 μM to 5 μM and in the next drug discovery phase chemists aim to enhance its potency and many other features, which will be explained in the following section (Fig. 2.1).

2.7 LEAD GENERATION

Lead generation, also referred to as hit to lead phase, involves optimization of the identified hits from a diverse series to generate lead compounds. Three to five chemical series are typically chosen for lead generation and analogous compounds are evaluated to establish a quantitative SAR towards activity, target selectivity, physicochemical properties, ADME properties, pharmacokinetics, and toxicity potential.

In this stage, compounds synthesis is initiated by medicinal chemists by using various approaches like conventional organic chemistry and combinatorial chemistry. Through the combinatorial chemistry approach large numbers of single compounds or mixtures of compounds can be synthesized in parallel and can be used for the synthesis of both small molecules and peptides. Combinatorial chemistry may be defined as the systematic and repetitive, covalent connection of a set of different building blocks of various structures to one another to yield a large array of diverse molecular entities [22]. The advantage of combinatorial chemistry over the classical approach is faster synthesis of, in some cases, up to a million compounds simultaneously, and hence it aids in rapid as well as efficient discovery of lead compounds. In case of the synthesis of a concoction of compounds, the entire mixture may be subjected to evaluation of its activity followed by identification of its active component(s). If on the other hand no actives are found, then no further attention is given to the mixture. A drawback for a mixture of compounds is interference of one compound with another, and it is prone to generation of false-positive hits.

Screening flow for lead identification comprises of in vitro evaluation in primary/cell-free assays as well as specificity of the compounds for the target. Further, the activity of compounds is also evaluated in known animal orthologs of the target as the compounds have to be evaluated for their efficacy in animal models. Subsequently, data for active compounds is also generated in in vitro functional or cell based assays.

In addition, physicochemical properties of representative compounds from the series being explored are also studied to confirm drug-likeness of the compounds. As the most preferred route for administration of the drug is oral, the new chemical entity in development should observe the Lipinski rule of 5 which asserts that a compound is more likely to be membrane permeable and absorbed by the body if it matches the following criteria [23]:

- Its molecular mass is less than 500 daltons
- Its logP, which is a measure of lipophilicity, is less than 5
- The number of hydrogen bond donors is less than 5
- The number of hydrogen bond acceptors is less than 10

Solubility assessments are additionally conducted as it has a bearing on both in vitro and in vivo assays as well as its absorption from the intestine, and the objective of the medicinal chemists is to obtain compounds having a solubility of >60 μg/mL [24].

Further, the in vitro ADME properties of compounds are also profiled. These include permeability assessment in colon carcinoma (Caco-2) cell line as a model for intestinal absorption [25], metabolic stability evaluation using human liver microsomes to determine the intrinsic clearance [26], cytochrome P450 inhibition and induction to assess whether the compound will have the potential to influence the metabolism of concomitantly administered drugs [27] and plasma protein binding assay which has a bearing on drug distribution and overall pharmacological action [28]).

It is also prudent to assess the toxic potential of compounds in the early stage of drug discovery, and several in vitro assays employing human cell lines have been developed to address this evaluation. These include cytotoxicity assays to investigate the effect of compounds on cell viability [29]; hERG inhibition assay using hERG overexpressing cell lines to predict the QT interval prolongation liability of the compounds under investigation [30]; hepatoxicity assay using a variety of systems like hepatic cell lines, isolated liver cells in suspensions, liver slices, and subcellular fractions [31]; in vitro micronucleus assay [32] to assess the potential for genotoxicities like clastogenic activity (structural aberrations in chromosomes); and aneugenic activity (numerical chromosome aberrations).

Potent and selective compounds having desirable physicochemical and ADME properties are also profiled for their pharmacokinetics in the same animal species in which the efficacy of the compounds have to be evaluated. Compounds having appropriate pharmacokinetics are then evaluated in primary animal models, which may also include proof of mechanism models that demonstrate target engagement. Finally, compounds are screened in animal models of human disease for their efficacy.

2.8 LEAD OPTIMIZATION

The goal of lead optimization is to generate preclinical development candidates by improving the shortcomings of the lead structure by chemical modifications. Generally, the aim is to enhance the physicochemical and ADME properties and minimize the toxicity liabilities so that a potentially safe compound with favorable pharmacokinetics is identified.

It is important to demonstrate a direct correlation between concentrations of the compound in plasma with its pharmacodynamic effect, and such data might be later utilized to predict dosing regimen of the compound. Additionally, it also beneficial to establish dose-linear exposure, as the compounds which do not exhibit such behavior have limited clinical utility, particularly if they have a narrow therapeutic window. Identification and characterization of the metabolites of the compound is also conducted during this stage as metabolites may influence the compound efficacy, may themselves be active, may elicit toxicities, and additionally active metabolites may be considered as new exploratory lead structures for the medicinal chemists.

Medicinal chemistry does not conclude once a preclinical candidate has been identified, as the chemists initiate effort on a back-up strategy with an aim to identify compounds that can substitute for any failures in preclinical and clinical development.

2.9 PRECLINICAL DRUG DEVELOPMENT

Once a preclinical drug candidate is selected, the drug development process begins. The drug is progressed through various studies designed to support its approval by the regulatory bodies to move the candidate into clinical (human) study by submission of an Investigational New Drug (IND) application. The preclinical development program consists of various activities, including safety pharmacology and toxicology studies in animals and other activities related to chemistry, manufacturing, and control (CMC) such as formulation development, stability studies and quality control measures etc. and detailed proposed clinical protocols for initiating clinical studies. For details on preclinical development and safety pharmacology, toxicology, please refer to Chapter 4, on Preclinical Drug Development (Fig. 2.2).

2.10 CRITERIA TO SELECT A CLINICAL CANDIDATE

A compound should demonstrate the following properties for selection as a clinical candidate [33]:

1. *Chemical properties*: It should be a stable molecule whose synthesis is simple and can be scaled up with ease.

2. *Physicochemical properties*: It should observe Lipinski rule of 5 and should have acceptable solubility.
3. *Pharmacological properties*: It should bind with the target site with high affinity, should demonstrate selectivity for its molecular target and should elicit potent functional effect in vitro. Efficacy of the compound should be demonstrable in animal model of human disease.
4. *Pharmacokinetic properties*: It should possess acceptable bioavailability, adequate half-life and proper distribution in animals. Metabolic pathways of the compound should be well characterized and activities of metabolites should be evaluated.
5. *Safety and toxicity potential*: It should be devoid of cardiac toxicity (hERG binding), genotoxicity, and hepatotoxicity, and should demonstrate an acceptable profile for induction and inhibition of cytochrome P450 enzymes. Ultimately it should be devoid of any serious animal toxicity.

2.11 CLINICAL DRUG DEVELOPMENT

If the IND is approved, clinical drug development begins. For details, please refer to Chapter 5, on Target Product Profile and Clinical Development. The general goals of various phases of clinical trials are similar even though the designs of these trials can be substantially different. In general, a phase I trial is conducted to assess safety and tolerability of a drug and is usually conducted on 10–100 healthy volunteers. Both pharmacokinetic (ADME) and pharmacodynamic aspects are monitored. The maximum tolerated dose (MTD) is determined. The trial is generally open-label (nonblinded). Phase II trial is the first study that investigates clinical effectiveness of the drug and hence this is carried out in patients. In this trial, about 50–500 patients receive the investigational new drug mainly to assess efficacy of the drug in patients. However, the trial can have multiple objectives like studying dose-response relationship and determining dosing regimen (optimum dose and frequency of administration, etc.). The safety assessment continues as in Phase I. The trial is generally randomized and controlled and may be single or double blind trial. The majority of clinical candidates fail in this phase due to lack of efficacy or safety issues. Phase III trials confirms the efficacy of investigational drug in a larger population, usually a few hundred to a few thousand participants (patients). The trial is multicentric (conducted at multiple sites) and compares the investigational drug with the best existing treatment or standard of care in that particular disease. The safety is also assessed in a larger pool so that less common adverse events may be detected. They are typically randomized, controlled, double-blind trials with multiple study arms, and are the most expensive and complex trials. If positive results are obtained, all data till date is compiled into a dossier and a New Drug Application (NDA) is filed for regulatory approval to license the drug. Once the drug is marketed, post-marketing surveillance or Phase IV trials begin as additional follow-up studies to detect rare or long-term adverse effects across a much larger population or effects in certain special population, drug-drug or drug-disease interactions, etc. mainly to test the drug in a real world setting. Phase IV studies have huge implications, including altering the labeling of the drug, contraindications, interactions, and even withdrawal of a marketed drug. The Phase IV studies are described in detail in Section VII Pharmacovigilance (see Chapters 26–31).

2.12 OTHER APPROACHES IN DRUG DISCOVERY AND DEVELOPMENT

One of the reasons for fall in the number of innovative drugs is one-drug-one target paradigm and due to the involvement of multiple targets in complex diseases it is prudent to modulate them simultaneously. This can be achieved with either different or a single agents, with the latter being challenging from a medicinal chemistry perspective. However, discovery of new drugs for Alzheimer's disease has shown some promise with this approach [34].

Another alternative approach is repositioning existing drugs for new indications (repurposing). An example in this case is of thalidomide, which was originally approved as a sedative, was later approved for leprosy, and now has been licensed for treatment of multiple myeloma. An advantage of this approach is the bypassing of preclinical safety as well as Phase 1 trials, which saves the industry a lot of time and money [35].

Allosteric modulation of drug targets is another novel approach towards drug discovery, wherein drugs bind at binding sites of the biological target—which are distinct from the active sites. A benefit with this approach is that while active sites might be common in several proteins, the allosteric sites might be unique, which allows for selective targeting and consequently either fewer or target-specific adverse effects. For example, a positive allosteric modulator of M_1 muscarinic receptor, benzylquinolone carboxylic acid (BQCA), has shown efficacy in animal models of schizophrenia [36].

Natural products have been the most valuable sources for small molecules for the treatment of diseases and as leads for drug discovery, but chemical modifications are required to improve their physicochemical properties and to generate derivatives for SAR, which is often challenging. However, new approaches are under development to overcome the bottlenecks to enable a full exploitation of their potential in generating innovative drugs [37].

Pharmaceutical industries are also venturing into discovery and development of biologics or biologically derived medications which have the potential to generate blockbusters in the future. These include monoclonal antibodies, polypeptides, hormones, growth factors, interferons, and interleukins as well as vaccines, and require recombinant DNA process for their production. They work by targeting either a genotype or a protein target. However, they are quite expensive, complex to manufacture, and being mainly proteins they have the potential for immunogenicity. Generally, the target patients are those on whom the conventional therapies do not work or for whom no therapeutic options exist [38].

Crowd sourcing which involves collaboration between pharmaceutical industry and academia is being actively pursued to promote innovation in early drug discovery research typically through the use of internet with the optimism to increase R&D productivity [39].

Network pharmacology is an emerging paradigm in drug discovery, which aims at revealing synergistic interactions between individual drugs administered in combination and thereby determining the group of proteins which are most significant in disease pathophysiology. Subsequently, the goal is to identify molecules which target those proteins [40].

As to which of these strategies would yield the desired results remains to be seen. Science and the long-term approach would be the foundation and the way forward to support innovation. While following these strategies, it is important to get the right attrition at proper times and look for low risk and high pay of drugs.

With respect to clinical drug development, microdosing is an approach which has played a useful role in increasing R&D productivity. Here, a subpharmacologically active dose is administered to humans, and exploratory pharmacokinetics of the parent or its metabolite are studied. This phase is also referred to as Phase 0 and regulatory agencies grant approval for its conduct without a full preclinical safety package. The compounds showing poor pharmacokinetics are dropped from further development [41].

Clinical development is the most expensive stage in drug development and it has been suggested that we should move away from the conventional approach based on different phases towards an integrative view in which one uses adaptive design tools to increase flexibility and maximize use of accumulated knowledge, which could result in achieving the desired goal [42].

2.13 CONCLUSION

Pharmaceutical industry is currently under immense pressure due to rising costs, pricing, and risks associated with drug discovery and development. However, due to high unmet medical needs, the disease treatment will continue to be determined by innovation generated by the industry in collaboration with academic institutions and other modes of public-private partnerships. Intellectual property protection, however, is important in supporting pharmaceutical R&D. New developments in science and technology and other innovative and emerging approaches to improve R&D productivity need to be adopted with a long-term approach in order to be truly supportive of novel drug research.

ACKNOWLEDGMENTS

The authors are grateful to Prof. C. L. Kaul, Former Director, NIPER for reviewing the manuscript and providing substantive and useful inputs.

REFERENCES

[1] Paul MS, Mytelka DS, Dunwiddie CT, Persinger CC, Munos BH, Lindborg SR, et al. How to improve R&D productivity: the pharmaceutical industry's grand challenge. Nat Rev Drug Discov 2010;9:203–14.

[2] LaMattina JL. The impact of mergers on pharmaceutical R&D. Nat Rev Drug Discov 2011;10 (8):559–60.

[3] Imming P, Sinning C, Meyer A. Drugs, their targets and the nature and number of drug targets. Nat Rev Drug Discov 2006;5(10):821–34.

[4] Lee J, Bogyo M. Target deconvolution techniques in modern phenotypic profiling. Curr Opin Chem Biol 2013;17(1):118–26.

[5] Sandercock AM, Rust S, Guillard S, Sachsenmeier KF, Holoweckyj N, Hay C, et al. Identification of anti-tumour biologics using primary tumour models, 3-D phenotypic screening and image-based multi-parametric profiling. Mol Cancer 2015;14(147).

[6] Lee YH, Song GG. Vascular endothelial growth factor gene polymorphisms and psoriasis susceptibility: a meta-analysis. Genet Mol Res 2015;14(4) 14396-40.

[7] Cao C, Moult J. GWAS and drug targets. BMC Genomics 2014;15(Suppl. 4):S5.

[8] Sanseau P, Agarwal P, Barnes MR, Pastinen T, Richards JB, Cardon LR, et al. Use of genome-wide association studies for drug repositioning. Nat Biotechnol 2012;30:317–20.

[9] Snaith MR, Törnell J. The use of transgenic systems in pharmaceutical research. Brief Funct Genomic Proteomic 2002;1(2):119–30.

[10] Willmann JK, van Bruggen N, Dinkelborg LM, Gambhir SS. Molecular imaging in drug development. Nat Rev Drug Discov 2008;7(7):591–607.

[11] Parada CA, Vivancos GG, Tambeli CH, Cunha FQ, Ferreira SH. Activation of presynaptic NMDA receptors coupled to NaV1.8-resistant sodium channel C-fibers causes retrograde mechanical nociceptor sensitization. Proc Natl Acad Sci USA 2003;100(5):2923–8.

[12] Leuschner F, Dutta P, Gorbatov R, Novobrantseva TI, Donahoe JS, Courties G, et al. Therapeutic siRNA silencing in inflammatory monocytes in mice. Nat Biotechnol 2011;29(11):1005–10.

[13] Stengel PW, Yamada M, Wess J, Cohen ML. M(3)-receptor knockout mice: muscarinic receptor function in atria, stomach fundus, urinary bladder, and trachea. Am J Physiol Regul Integr Comp Physiol 2002;282(5):R1443–9.

[14] Rakhade SN, Fitzgerald EF, Klein PM, Zhou C, Sun H, Huganir RL, et al. Glutamate receptor 1 phosphorylation at serine 831 and 845 modulates seizure susceptibility and hippocampal hyperexcitability after early life seizures. J Neurosci December 2012;32(49):17800–12.

[15] Meirsman AC, Robé A, de Kerchove d'Exaerde A, Kieffer BL. GPR88 in A2AR neurons enhances anxiety-like behaviors. eNeuro August 17 2016;3(4).

[16] Cannon MV, Silljé HH, Sijbesma JW, Khan MA, Steffensen KR, van Gilst WH, et al. LXRα improves myocardial glucose tolerance and reduces cardiac hypertrophy in a mouse model of obesity-induced type 2 diabetes. Diabetologia March 2016;59(3):634–43.

[17] Smith C. Drug target validation: hitting the target. Nature 2003;422(6929) 341, 343, 345.

[18] Mashalidis EH, Śledź P, Lang S, Abell C. A three-stage biophysical screening cascade for fragment-based drug discovery. Nat Protoc November 2013;8(11):2309–24.

[19] Swinney DC, Anthony J. How were new medicines discovered? Nat Rev Drug Discov June 24 2011;10 (7):507–19.

[20] Hughes JP, Rees S, Kalindjian SB, Philpott KL. Principles of early drug discovery. Br J Pharmacol March 2011;162(6):1239–49.

[21] Keseru GM, Makara GM. Hit discovery and hit-to-lead approaches. Drug Discov Today. August 2006;11(15-16):741–8.

[22] Pandeya SN, Thakkar D. Combinatorial chemistry: A novel method in drug discovery and its application. Indian J Chem February 2005;44B:335–48.

[23] Leeson P. Drug discovery: chemical beauty contest. Nature January 25 2012;481(7382):455–6.

[24] Kerns EH, Di L, Carter GT. In vitro solubility assays in drug discovery. Curr Drug Metab November 2008;9(9):879–85.

[25] van Breemen RB, Li Y. Caco-2 cell permeability assays to measure drug absorption. Expert Opin Drug Metab Toxicol August 2005;1(2):175–85.

[26] Baranczewski P, Stańczak A, Sundberg K, Svensson R, Wallin A, Jansson J, et al. Introduction to in vitro estimation of metabolic stability and drug interactions of new chemical entities in drug discovery and development. Pharmacol Rep July–August 2006;58(4):453–72.

[27] Yan Z, Caldwell GW. Metabolism profiling, and cytochrome P450 inhibition & induction in drug discovery. Curr Top Med Chem November 2001;1(5):403–25.

[28] Lambrinidis G, Vallianatou T, Tsantili-Kakoulidou A. In vitro, in silico and integrated strategies for the estimation of plasma protein binding. A review. Adv Drug Deliv Rev June 23 2015;86:27–45.

[29] Riss TL, Moravec RA, Niles AL. Cytotoxicity testing: measuring viable cells, dead cells, and detecting mechanism of cell death. Methods Mol Biol 2011;740:103–14.
[30] Pollard CE, Valentin JP, Hammond TG. Strategies to reduce the risk of drug-induced QT interval prolongation: a pharmaceutical company perspective. Br J Pharmacol August 2008;154(7):1538–43.
[31] Gómez-Lechón MJ, Castell JV, Donato MT. The use of hepatocytes to investigate drug toxicity. Methods Mol Biol 2010;640:389–415.
[32] Kirsch-Volders M, Plas G, Elhajouji A, Lukamowicz M, Gonzalez L, Vande Loock K, et al. The in vitro MN assay in 2011: origin and fate, biological significance, protocols, high throughput methodologies and toxicological relevance. Arch Toxicol August 2011;85(8):873–99.
[33] Hefti FF. Requirements for a lead compound to become a clinical candidate. BMC Neurosci December 10 2008;9(Suppl. 3):S7.
[34] Zhang Hong-Yu. One-compound-multiple-targets strategy to combat Alzheimer's disease. FEBS Lett 2005;579:5260–4.
[35] Azvolinsky A. Repurposing existing drugs for new indications. The Scientist January 1 2017.
[36] Grover AK. Use of allosteric targets in the discovery of safer drugs. Med Princ Pract 2013;22:418–26.
[37] Robles O, Romo D. Chemo- and site-selective derivatizations of natural products enabling biological studies. Nat Prod Rep March 2014;31(3):318–34.
[38] Morrow T, Felcone LH. Defining the difference: what makes biologics unique. Biotechnol Healthcare September 2004;1(4):24–9.
[39] Lessl M, Bryans JS, Richards D, Asadullah K. Crowd sourcing in drug discovery. Nat Rev Drug Discov April 2011;10(4):241–2.
[40] Hopkins AL. Network pharmacology: the next paradigm in drug discovery. Nat Chem Biol November 2008;4(11):682–90.
[41] Lappin G, Noveck R, Burt T. Microdosing and drug development: past, present and future. J Expert Opin Drug Metab Toxicol July 2013;9(7):817–34.
[42] Orloff JJ, Stanski D. Innovative approaches to clinical development and trial design. Ann Ist Super Sanita 2011;47(1):8–13.

CHAPTER

PHARMACEUTICAL DEVELOPMENT 3

Harinder Singh, Lalit K. Khurana, and Romi Singh
Sun Pharmaceutical Industries Ltd, Gurgaon, Haryana, India

3.1 INTRODUCTION

Pharmaceutical development is envisioned to design a quality product and a manufacturing process that can consistently deliver the product with its intended performance. A pharmaceutical product should be designed to meet patients' needs. Strategies for product development vary from company to company and from product to product. An applicant might choose either an empirical approach or a more systematic approach to product development, or a combination of both [1]. The knowledge and information acquired from pharmaceutical development studies and manufacturing experience offer scientific understanding to support the establishment of the design space, specifications, and manufacturing controls. Pharmaceutical development establishes that the type of dosage form selected, and the formulation proposed, are suitable for the proposed use.

A more systematic approach to development (also defined as quality by design) can include, for example, incorporation of prior knowledge, results of studies using design of experiments, use of quality risk management, and use of knowledge management throughout the lifecycle of the product [1]. Data obtained from the pharmaceutical development studies form the basis for quality risk management. It is important to know that quality cannot be tested into products, i.e., quality should be built in by design optimization. Changes in formulation and manufacturing processes during its development and lifecycle management should be taken as opportunities to increase knowledge and further support establishment of the design space. Design space is proposed by the applicant and is subject to regulatory assessment and approval. An alteration within the design space is not considered as a change. Deviation out of the design space is considered to be a change and would normally initiate a regulatory postapproval change process. Such a systematic approach can enhance achieving the desired quality of the product and help the regulators to better understand a company's strategy. Product and process understanding can be updated with the knowledge gained over the product lifecycle [1].

Pharmaceutical development studies provide an opportunity to demonstrate a higher degree of understanding of material attributes, manufacturing processes, and their controls. This scientific understanding facilitates establishment of an expanded design space. At a minimum, those aspects of drug substances, excipients, container closure systems, and manufacturing processes which are critical to the quality of the product, should be determined and control strategies justified. Critical formulation attributes and process parameters should be identified through an assessment of the

Pharmaceutical Medicine and Translational Clinical Research. DOI: http://dx.doi.org/10.1016/B978-0-12-802103-3.00003-1

extent to which their variation can have impact on the quality of the drug product [2]. In these situations, opportunities exist to develop more flexible regulatory approaches, for example, to facilitate:

- Risk-based regulatory decisions (reviews and inspections).
- Manufacturing process improvements, within the approved design space described in the dossier, without further regulatory review.
- Reduction of postapproval submissions.
- Real-time quality control, leading to a reduction of end-product release testing.

The design and conduct of pharmaceutical development studies should be consistent with their intended scientific purpose. A greater understanding of the product and its manufacturing process can create a basis for more flexible regulatory approaches. The degree of regulatory flexibility is predicated on the level of relevant scientific knowledge provided in the registration application. It is the knowledge gained and submitted to the authorities, and not the volume of data collected, that forms the basis for science- and risk-based submissions and regulatory evaluations. Nevertheless, appropriate data demonstrating that this knowledge is based on sound scientific principles should be presented with each application [1].

Pharmaceutical development includes the following elements:

- Defining the quality target product profile (QTPP) as it relates to quality, safety, and efficacy, considering e.g., the route of administration, dosage form, bioavailability, strength, and stability.
- Identifying potential critical quality attributes (CQAs) of the drug product, so that those product characteristics having an impact on product quality can be studied and controlled.
- Determining the critical quality attributes of the drug substance, excipients, etc., and selecting the type and amount of excipients to deliver drug product of the desired quality.
- Selecting an appropriate manufacturing process.
- Defining a control strategy.

An enhanced, quality by design approach to product development would additionally include the following elements:

- A systematic evaluation, understanding and refining of the formulation and manufacturing process, including:
- Identifying (through, e.g., prior knowledge, experimentation, and risk assessment) the material attributes and process parameters that can have an effect on product CQAs. Determining the functional relationships that link material attributes and process parameters to product CQAs.
- Using the enhanced product and process understanding in combination with quality risk management to establish an appropriate control strategy, which can, for example, include a proposal for a design space(s) and/or real-time release testing.

As a result, this more systematic approach could facilitate continual improvement and innovation throughout the product lifecycle.

3.2 ELEMENTS OF PHARMACEUTICAL DEVELOPMENT

This section elaborates on the possible approaches to gaining a more systematic, enhanced understanding of the product and process under development [1].

3.2.1 QUALITY TARGET PRODUCT PROFILE

The quality target product profile relates to quality, safety, and efficacy of the pharmaceutical product. It forms the basis of design for the development of the product [3]. Considerations for the quality target product profile could include:

- Intended use in clinical setting, route of administration, dosage form, delivery systems.
- Dosage strength(s).
- Container closure system.
- Therapeutic moiety release or delivery and attributes affecting pharmacokinetic characteristics (e.g., dissolution, aerodynamic performance) appropriate to the drug product dosage form being developed.
- Drug product quality criteria (e.g., sterility, purity, stability, and drug release) appropriate for the intended marketed product.

3.2.2 CRITICAL QUALITY ATTRIBUTES (CQA)

A CQA is a physical, chemical, biological, or microbiological property or characteristic that should be within an appropriate limit, range, or distribution to ensure the desired product quality. CQAs are generally associated with the drug substance, excipients, intermediates (in-process materials), and drug product.

CQAs of solid oral dosage forms are typically those aspects affecting product purity, strength, drug release, and stability. CQAs for other delivery systems can additionally include more product specific aspects, such as aerodynamic properties for inhaled products, sterility for parenterals, and adhesion properties for transdermal patches. For drug substances, raw materials, and intermediates, the CQAs can additionally include those properties (e.g., particle size distribution, bulk density) that affect drug product CQAs.

Potential drug product CQAs derived from the quality target product profile and/or prior knowledge are used to guide the product and process development. The list of potential CQAs can be modified when the formulation and manufacturing process are selected and as product knowledge and process understanding increase. Quality risk management can be used to prioritize the list of potential CQAs for subsequent evaluation. Relevant CQAs can be identified by an iterative process of quality risk management and experimentation that assesses the extent to which their variation can have an impact on the quality of the drug product [2].

3.2.3 RISK ASSESSMENT: LINKING MATERIAL ATTRIBUTES AND PROCESS PARAMETERS TO DRUG PRODUCT CQAS

Risk assessment is a valuable science-based process used in quality risk management that can aid in identifying which material attributes and process parameters potentially have an effect on product CQAs. Risk assessment is typically performed early in the pharmaceutical development process and is repeated as more information becomes available and greater knowledge is obtained.

Risk assessment tools can be used to identify and rank parameters (e.g., process, equipment, input materials) with potential to have an impact on product quality, based on prior knowledge and initial experimental data. The initial list of potential parameters can be quite extensive, but can be modified and prioritized by further studies (e.g., through a combination of design of experiments, mechanistic models). The list can be refined further through experimentation to determine the significance of individual variables and potential interactions. Once the significant parameters are identified, they can be further studied (e.g., through a combination of design of experiments, mathematical models, or studies that lead to mechanistic understanding) to achieve a higher level of process understanding [4].

3.2.4 PREFORMULATION

Preformulation studies involve (1) selection of the drug candidate itself; (2) selection of formulation components, (3) API & drug product manufacturing processes; (4) determination of the most appropriate container closure system; (5) development of analytical methods; (6) assignment of API retest periods; (7) the synthetic route of the API; (8) toxicological strategy. Preformulation studies strengthen the scientific foundation of the guidance, provide regulatory relief, and conserve resources in the drug development and evaluation process, improve public safety standards, enhance product quality, facilitate the implementation of new technologies, facilitate policy development and regulatory [5].

3.2.4.1 Drug substance

The physicochemical and biological properties of the drug substance that can influence the performance of the drug product and its manufacturability, or were specifically designed into the drug substance (e.g., solid state properties), should be identified. Examples of physicochemical and biological properties that should be examined as appropriate include solubility, water content, particle size, crystal properties, biological activity, and permeability. These properties could be interrelated and, when appropriate, should be considered in combination.

To evaluate the potential effect of drug substance physicochemical properties on the performance of the drug product, studies on drug product might be warranted. The knowledge gained from the studies investigating the potential effect of drug substance properties on drug product performance can be used, as appropriate, to justify elements of the drug substance specification.

The compatibility of the drug substance with excipients should be evaluated. For products that contain more than one drug substance, the compatibility of the drug substances with each other should also be evaluated [2].

Bulk characteristics of API, like particle size and surface area, polymorphism, crystallinity, hygroscopicity, flow properties & bulk density, compressibility, drug-excipient compatibility, electrostatic charge, osmolarity, rheology, and wettability, are evaluated [5]. Solubility analysis of API involves aqueous solubility, Solubilization, partition coefficient, thermal effect, common ion effect, dissolution [5].

Compatibility Studies of drug and excipients involves: Drug: active part of dosages form and mainly responsible for therapeutic value. Excipients: substances which are included along with drugs being formulated in a dosage form so as to impart specific qualities to them [6].

Different analytical techniques used to detect drug-excipients compatibility involve thermal methods of analysis (Differential Scanning Calorimetric, Differential Thermal Analysis, FT-IR Spectroscopy and Diffuse Reflectance Spectroscopy). Chromatography involves Self Interactive

Chromatography, Thin Layer Chromatography, High Pressure Liquid Chromatography, and Miscellaneous involves-Radiolabel Techniques Vapor Pressure, Fluorescence Spectroscopy [6].

Analytical methods provide required data for a given analytical problem, the required sensitivity, the required accuracy, the required range of analysis, The required precision, i.e., the minimum requirements, which essentially are the specifications of the method for the intended purpose—to be able to analyze the desired analyte in different matrices with surety and certainty. The following steps are common to most types of projects: Method development plane definition, background information gathering, laboratory method development, generation of test procedure, methods validation protocol definition, laboratory methods validation, validated test method generation, validation report [7].

3.2.4.2 Excipients

The excipients chosen, their concentration, and the characteristics that can influence the drug product performance (e.g., stability, bioavailability) or manufacturability should be discussed relative to the respective function of each excipient. This should include all substances used in the manufacture of the drug product, whether they appear in the finished product or not (e.g., processing aids). Compatibility of excipients with other excipients, where relevant (for example, combination of preservatives in a dual preservative system), should be established. The ability of excipients (e.g., antioxidants, penetration enhancers, disintegrants, release controlling agents) to provide their intended functionality and to perform throughout the intended drug product shelf life should also be demonstrated. The information on excipient performance can be used, as appropriate, to justify the choice and quality attributes of the excipient and to support the justification of the drug product specification. Information to support the safety of excipients, when appropriate, should be cross-referenced [2].

3.2.4.3 Overages

In general, use of an overage of a drug substance to compensate for degradation during manufacture or a product's shelf life, or to extend shelf life, is discouraged. Any overages in the manufacture of the drug product, whether they appear in the final formulated product or not, should be justified considering the safety and efficacy of the product. Information should be provided on the (1) amount of overage, (2) reason for the overage (e.g., to compensate for expected and documented manufacturing losses), and (3) justification for the amount of overage. The overage should be included in the amount of drug substance listed in the batch formula [2].

3.2.4.4 Physicochemical and Biological Properties

The physicochemical and biological properties relevant to the safety, performance, or manufacturability of the drug product should be identified and discussed. This includes the physiological implications of drug substance and formulation attributes. Studies could include, for example, the development of a test for respirable fraction of an inhaled product. Similarly, information supporting the selection of dissolution vs. disintegration testing (or other means to ensure drug release) and the development and suitability of the chosen test could be provided in this section [2].

3.2.5 DESIGN SPACE

The relationship between the process inputs (material attributes and process parameters) and the critical quality attributes can be described in the design space.

1. *Selection of Variables*

   The risk assessment and process development experiments described can lead to an understanding of the linkage and effect of process parameters and material attributes on product CQAs and also help identify the variables and their ranges within which consistent quality can be achieved. These process parameters and material attributes can thus be selected for inclusion in the design space.

   A description should be provided in the application of the process parameters and material attributes considered for the design space, those that were included, and their effect on product quality. The rationale for inclusion in the design space should be presented. In some cases, it is helpful to provide also the rationale as to why some parameters were excluded. Knowledge gained from studies should be described in the submission. Process parameters and material attributes that were not varied through development should be highlighted [1].

2. *Describing a Design Space in a Submission*

   A design space can be described in terms of ranges of material attributes and process parameters, or through more complex mathematical relationships. It is possible to describe a design space as a time dependent function (e.g., temperature and pressure cycle of a lyophilization cycle), or as a combination of variables such as components of a multivariate model. Scaling factors can also be included if the design space is intended to span multiple operational scales. Analysis of historical data can contribute to the establishment of a design space. Regardless of how a design space is developed, it is expected that operation within the design space will result in a product meeting the defined quality [1].

3. *Unit Operation Design Space(s)*

   The applicant can choose to establish independent design spaces for one or more unit operations, or to establish a single design space that spans multiple operations. While a separate design space for each unit operation is often simpler to develop, a design space that spans the entire process can provide more operational flexibility. For example, in the case of a drug product that undergoes degradation in solution before lyophilization, the design space to control the extent of degradation (e.g., concentration, time, temperature) could be expressed for each unit operation or as a sum over all unit operations [1].

4. *Relationship of Design Space to Scale and Equipment*

   When describing a design space, the applicant should consider the type of operational flexibility desired. A design space can be developed at any scale. The applicant should justify the relevance of a design space developed at small or pilot scale to the proposed production scale manufacturing process and discuss the potential risks in the scale-up operation.

   If the applicant proposes the design space to be applicable to multiple operational scales, the design space should be described in terms of relevant scale-independent parameters. For example, if a product was determined to be shear sensitive in a mixing operation, the design space could include shear rate, rather than agitation rate. Dimensionless numbers and/or models for scaling can be included as part of the design space description [1].

5. *Design Space Versus Proven Acceptable Ranges*
A combination of proven acceptable ranges does not constitute a design space. However, proven acceptable ranges based on univariate experimentation can provide useful knowledge about the process [1].
6. *Design Space and Edge of Failure*
It can be helpful to determine the edge of failure for process parameters or material attributes, beyond which the relevant quality attributes cannot be met. However, determining the edge of failure or demonstrating failure modes are not essential parts of establishing a design space [1].

3.2.6 FORMULATION DEVELOPMENT

A summary describing the development of the formulation, including identification of those attributes that are critical to the quality of the drug product—taking into consideration intended usage and route of administration—should be provided. Information from formal experimental designs can be useful in identifying critical or interacting variables that might be important to ensure the quality of the drug product.

The summary should highlight the evolution of the formulation design from initial concept up to the final design. This summary should also take into consideration the choice of drug product components (e.g., the properties of the drug substance, excipients, container closure system, any relevant dosing device), the manufacturing process, and, if appropriate, knowledge gained from the development of similar drug product(s) [7].

A summary of formulations used in clinical safety and efficacy and in any relevant bioavailability or bioequivalence studies should be provided. Any changes between the proposed commercial formulation and those formulations used in pivotal clinical batches and primary stability batches should be clearly described and the rationale for the changes provided.

Information from comparative in vitro studies (e.g., dissolution) or comparative in vivo studies (e.g., bioequivalence) that links clinical formulations to the proposed commercial formulation should be summarized and a cross-reference to the studies (with study numbers) should be provided. Where attempts have been made to establish an in vitro/in vivo correlation, the results of those studies and a cross-reference to the studies (with study numbers) are provided. A successful correlation can assist in the selection of appropriate dissolution acceptance criteria and can potentially reduce the need for further bioequivalence studies following changes to the product or its manufacturing process.

Any special design features of the drug product (e.g., tablet score line, overfill, anti- counterfeiting measure as it affects the drug product) should be identified and a rationale provided for their use [2].

3.2.6.1 Prototype development

Prototype development is a stage in the new product development process. The idea is a descriptive statement that can be written or only verbalized. The idea is refined into a product concept that includes consumer benefits and features of the product. The concept is developed into a prototype, i.e., a working model or preliminary version of the product. After several iterations the prototype is perfected into the final product [8].

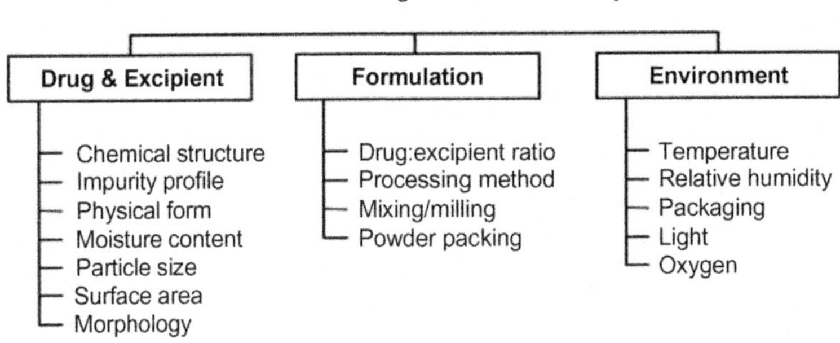

FIGURE 3.1

Factors affecting formulation stability [8].

Phases of prototype development are:

1. Identify the basic requirements of the product.
2. Develop a design meeting these requirements and implement it.
3. Have users experiment with the prototype.
4. Revise the prototype, thereby redefining and completing the requirements [8].

Lab Stability Studies are initiated (Fig. 3.1)
Importance of stability

- Extensive chemical degradation—a substantial loss of potency.
- Degradation products may result in adverse events or be unsafe.
- Instability may cause undesired change in performance, i.e., dissolution/bioavailability.
- Substantial changes in physical appearance of the dosage form causing product failures requirement for approval by regulatory agencies [8].

3.2.7 CONTROL STRATEGY

A control strategy is designed to ensure that a product of required quality will be produced consistently. The elements of the control strategy should describe and justify how in-process controls and the controls of input materials (drug substance and excipients), intermediates (in-process materials), container closure systems, and drug products contribute to the final product quality. These controls should be based on product, formulation, and process understanding and should include, at a minimum, control of the critical process parameters and material attributes [1].

A comprehensive pharmaceutical development approach will generate process and product understanding and identify sources of variability. Sources of variability that can have an impact on product quality should be identified, appropriately understood, and subsequently controlled. Understanding sources of variability and their impact on downstream processes or processing, in-process materials, and drug product quality can provide an opportunity to shift controls upstream and minimize the need for end-product testing. Product and process understanding, in combination with quality risk management, will support the control of the process such that the variability (e.g., of raw materials) can be compensated for in an adaptable manner to deliver consistent product quality [1].

This process understanding can enable an alternative manufacturing paradigm where the variability of input materials could be less tightly constrained. Instead, it can be possible to design an adaptive process step (a step that is responsive to the input materials) with appropriate process control to ensure consistent product quality.

Enhanced understanding of product performance can justify the use of alternative approaches to determine that the material is meeting its quality attributes. The use of such alternatives could support real time release testing. For example, disintegration could serve as a surrogate for dissolution for fast-disintegrating solid forms with highly soluble drug substances. Unit dose uniformity performed in-process (e.g., using weight variation coupled with near infrared (NIR) assay) can enable real-time release testing and provide an increased level of quality assurance compared to the traditional end-product testing using compendial content uniformity standards. Real-time release testing can replace end-product testing, but does not replace the review and quality control steps called for under GMP to release the batch [2].

A control strategy can include, but is not limited to, the following:

- Control of input material attributes (e.g., drug substance, excipients, primary packaging materials) based on an understanding of their impact on processability or product quality.
- Product specification(s).
- Controls for unit operations that have an impact on downstream processing or product quality (e.g., the impact of drying on degradation, particle size distribution of the granulate on dissolution).
- In-process or real-time release testing in lieu of end-product testing (e.g., measurement and control of CQAs during processing).
- A monitoring program (e.g., full product testing at regular intervals) for verifying multivariate prediction models.

A control strategy can include different elements. For example, one element of the control strategy could rely on end-product testing, whereas another could depend on real-time release testing. The rationale for using these alternative approaches should be described in the submission.

3.2.8 SCALEUP AND STABILITY STUDY OF BATCH

As the development of a new drug product progresses, the batch sizes manufactured generally increase. Current International Conference on Harmonization (ICH) guidelines state that data from stability studies should be provided on at least three primary batches of the drug product. The primary batches should be of the same formulation and packaged in the same container-closure system as proposed for the marketed formulation [9].

Laboratory-scale batches are produced at the research and early development laboratory stage. They may be of very small size (e.g., 100–1000 times less than production scale). Laboratory-scale batches may be used to support formulation and packaging development. Laboratory-scale batches can also be analyzed to assist in the evaluation and definition of critical quality attributes (CQAs) [9].

Pilot-scale batches may be used in the process-development or optimization stage. For oral solid dosage forms, this size should generally be 10% of production scale or 100,000 units, whichever is greater. The choice of pilot scale is often difficult for the project team as members must balance parameters such as anticipated product volumes [10].

Process optimization is the discipline of adjusting a process so as to optimize some specified set of parameters without violating some constraint. The most common goals are minimizing cost, maximizing throughput, and/or efficiency. This is one of the major quantitative tools in industrial decision-making. When optimizing a process, the goal is to maximize one or more of the process specifications, while keeping all others within their constraints [11].

3.2.9 ANALYTICAL METHOD VALIDATION

The steps of methods development and method validation depend upon the type of method being developed. However, the following steps are common to most types of projects: Method development plane, background information gathering, laboratory method development, generation of test procedure, methods validation protocol definition, laboratory methods validation, validated test method generation, validation report. Pharmaceutical analytical methods are categorized into five general types: Identification tests, potency assays, impurity tests: quantitative, impurity tests: limit, specific tests. [11,12]

Validation requirements depend upon the type of test method, including:

- *Specificity*: ability to measure desired analyte in a complex mixture.
- *Accuracy*: agreement between measured and real value.
- *Linearity*: proportionality of measured value to concentration.
- *Precision*: agreement between a series of measurements.
- *Range*: concentration interval where method is precise, accurate, and linear.
- *Detection limit*: lowest amount of analyte that can be detected.
- *Quantitation limit*: lowest amount of analyte that can be measured.
- *Robustness*: reproducibility under normal but variable laboratory conditions [11,12].

3.2.10 DOCUMENTATION

3.2.10.1 IPR and regulatory clearance

Regulator is usually a government or professional body that 'regulates'—meaning makes up thc rules for certain types of businesses. Getting regulatory clearance can apply to many things. For example, if researcher wanted to become a Financial Advisor then researcher employer would need to submit an application, probably to the Financial Services Authority, to seek approval along with a list of researcher relevant experience and qualifications [13].

3.2.10.2 Technology transfer dossier

Technology transfer is the practice of transferring scientific findings from one organization to another for further development. Technology transfer is both integral and critical to the drug discovery and development process for new medicinal product (Fig. 3.2). This process is important for

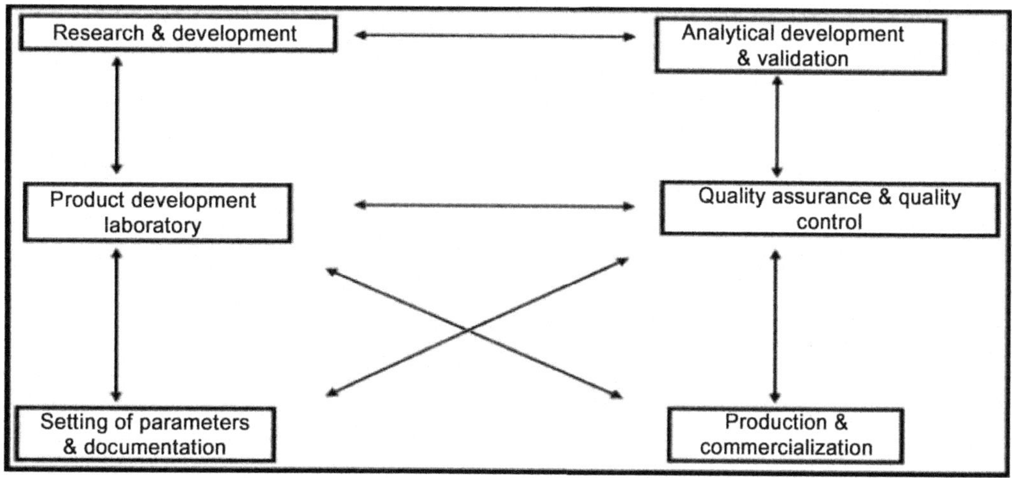

FIGURE 3.2

Process of Technology transfer.

to elucidate necessary information for technology transfer from R & D to product development laboratory and for development of existing products to the production for commercialization [13,14].

3.2.10.3 Technology transfer from R & D to production

R & D provides technology transfer dossier (TTD) document to product development laboratory, which contains all the information of formulation and drug product as given below: [13,14]

3.2.10.4 Master formula card (MFC) includes

Product name along with its strength, generic name, MFC number, page number, effective date, shelf life, and market.

3.2.10.5 Master packaging card

Gives information about packaging type, material used for packaging, stability profile of packaging, and shelf life of packaging.

3.2.10.6 Master formula

Describes formulation order and manufacturing instructions. Formulation order and manufacturing instructions give an idea of process order, environment conditions required, and manufacturing instructions for dosage form development.

3.2.10.7 Specifications and standard test procedure (STPs)

Helps to know active ingredients and excipients profile, in process parameters and specifications, product release specification, and finished product details.

3.2.11 EXHIBIT BATCH MANUFACTURING

3.2.11.1 Validation batch

Validation is a process of establishing documentary evidence demonstrating that a procedure, process, or activity carried out in production or testing maintains the desired level of compliance at all stages similarly; the activity of qualifying systems and equipment is divided into a number of subsections including the following:

- Design qualification (DQ), Component qualification (CQ), Installation qualification (IQ), and Operational qualification (OQ), Performance qualification (PQ). [15]

3.2.11.2 The validation report

A written report should be available after completion of the validation. If found to be acceptable, it should be approved and authorized. The report should include at least the following: Title and objective of study, reference to protocol, details of material, equipment, programs and cycles used, details of procedures and test methods, results (compared with acceptance criteria), and recommendations on the limit and criteria to be applied on a future basis. [15]

3.2.11.3 Pivotal BE Studies

This batch shall be submitted to Regulatory Authority:

- Shall be charged for both accelerated and long term stability testing.
- Biostudy shall be done on the same. [16]

3.2.11.4 Dossier Compilation and filling

Dossier compiling, documentation, and presentation in a compliant format is very involving and yet an integral part of the process of registration of any pharmaceutical product in India as well as international markets. This is why companies look for ways to increase efficiency and cut time and costs in the dossier submission process. We understand the challenges of Pharmaceutical Dossier submission and registration, involving repeated interaction between the submitting pharmaceutical firm and the regulatory authority. We incorporate Quality by Design into your CTD and ACTD requirements and create the optimal Modules for dossier filling in different regulated, semi-regulated, and non-regulated countries. Dossiers can be prepared in the Common Technical Document Format (CTD) [16].

3.2.12 PRODUCT LIFECYCLE MANAGEMENT AND CONTINUAL IMPROVEMENT

Throughout the product lifecycle, companies have opportunities to evaluate innovative approaches to improve product quality.

Process performance can be monitored to ensure that it is working as anticipated to deliver product quality attributes as predicted by the design space. This monitoring could include trend analysis of the manufacturing process as additional experience is gained during routine manufacture. For certain design spaces using mathematical models, periodic maintenance could be useful to ensure the model's performance. The model maintenance is an example of activity that can be managed within a company's own internal quality system provided the design space is unchanged.

Expansion, reduction, or redefinition of the design space could be desired upon gaining additional process knowledge. Change of design space is subject to regional requirements [2].

3.3 CONCLUSION

Companies are constantly designing and developing new products, and improving their product development processes [PDPs]. Companies have difficulty designing or selecting the process. This research used a series of case studies to define the components that distinguish product development processes from each other, providing companies with a more analytical method of comparison. This product development process design is a directly applicable for research contribution, and provides companies with a framework for efficiently designing product development processes that suit specific project needs. Recording product development process into reviews and iterations can be helpful in product design and development. Having good ideas for new products is not enough; to successfully bring those products to market, companies also need to design product development processes.

REFERENCES

[1] Guidance for Industry Q8(R2) Pharmaceutical Development. U.S. Department of Health and Human Services, Food and Drug Administration Center for Drug Evaluation and Research (CDER); November 2009.

[2] DiFeo TJ. Drug product development: a technical review of chemistry, manufacturing, and controls information for the support of pharmaceutical compound licensing activities. Drug Dev Industrial Pharmacy 2003;29(9):939–58.

[3] <https://amcrasto.wordpress.com/2014/02/04/qtpp-the-quality-target-product-profile/>; [accessed 11.08.15].

[4] <http://www.ema.europa.eu/docs/en_GB/document_library/Presentation/2011/10/WC500115824.pdf>; [accessed 11.08.15].

[5] Aulton ME. Pharmaceutics-the science of dosage form design. 2nd ed.; 1988, 214859, 113.

[6] WHO Expert Committee on Specifications for Pharmaceutical Preparations. Guidelines for stability testing of pharmaceutical products containing well established drug substances in conventional dosage forms. Thirty-fourth report. Geneva, World Health Organization; 1996, Annex 5 (WHO Technical Report Series, No. 863).

[7] <https://www.tga.gov.au/book/4-development-pharmaceutics-and-formulation>; [accessed 0709.15].

[8] Jatto E, Okhamafe AO. An overview of pharmaceutical validation and process controls in drug development. Trop J Pharmaceut Res 2002;1(2):115–22.

[9] International Conferences on Harmonisation. ICH Q1A (R2): stability testing of new drug substances and products, <http://www.ich.org/LOB/media/MEDIA419.pdf>.

[10] Guidance for industry dissolution testing of immediate release solid oral dosage forms U.S. Department of Health and Human Services, Food and Drug Administration Center for Drug Evaluation and Research (CDER) August 1997, 88.

[11] Long CP, McQuaid J. Strategic approaches to process optimization and scale-up. Pharmaceut Tech-Process Opt Scale Up 2010;34(Issue 9):1–44.

[12] Oona Mcpolin.Validations of analytical methods for pharmaceutical analysis publisher. Mourne Training service. ISBN 978-0-9561528, 1–7.

[13] Singh A, Aggarwal G. Technology transfer in pharmaceutical industry a discussion. Int J Pharma Bio Sci 2010;13:1–5.

[14] George P, Millili. Scale up & technology transfer as a part of pharmaceutical quality systems. Pharmaceut Qual Syst 201114–16.

[15] Breaux J, Jones K, Boulas P. Understanding and implementing efficient analytical methods development and validation. Pharmaceut Technol Anal Chem Testing 2000;65(169):8–13.

[16] Guidance for Industry. Waiver of in vivo bioavailability and bioequivalence studies for immediate-release solid oral dosage forms based on a biopharmaceutics classification system. U.S. Department of Health and Human Services, Food and Drug Administration, Center for Drug Evaluation and Research (CDER), August 2000.

CHAPTER

PRECLINICAL DRUG DEVELOPMENT

4

Gursharan Singh
Life Sciences, SmartAnalyst India Private Limited, Gurgaon, Haryana, India

4.1 INTRODUCTION

In the earlier chapter on Drug Discovery, the process of selection of optimized leads and identifying the preclinical candidate was described. Most companies work on 7–10 optimized leads, before notifying a preclinical development candidate along with a back-up candidate. Preclinical development candidates generally need to meet the criteria given in the Table 4.1 below [1].

Once the Preclinical Candidate is notified, the preclinical development program consists of various activities, given in Table 4.2 below, which act as a bridge between drug discovery and the initiation of first in human studies [2].

While the data from pivotal GLP safety pharmacology and toxicology studies is a critical component of IND application, many studies are done prior to IND, enabling preclinical development. An adequate quantity of drug substance is required for various preclinical activities. Non-GMP material in the milligram to gram range is required for early formulation activities, non-good laboratory practice (non-GLP) in vivo efficacy studies, pharmacokinetic/metabolism studies, dose range-finding studies, and experimental toxicology studies. The data from these studies is used to define the dose, route, and frequency required for subsequent studies. Higher quantity of GMP material (in kilo grams range) is required for the IND-enabling GLP toxicology studies [2].

Activities related to API, formulation, analytical and bioanalytical methods, and GMP manufacturing have been described in detail in the chapter on Pharmaceutical Development. The focus of this chapter is on pharmacokinetic/metabolism studies and safety pharmacology and toxicology studies.

4.2 PHARMACOKINETIC AND METABOLISM STUDIES

A comprehensive knowledge of the absorption, distribution, metabolism, and elimination of a compound is required to determine the dose levels and frequency of administration for the safety pharmacology and toxicology studies. This knowledge also helps in interpretation of the results of toxicology studies for determining the first in man dose [3].

Early PK studies are aimed at determining the absorption and excretion pathways for the drug. Later studies focus on the differences in the extent of tissue distribution and the metabolites

Pharmaceutical Medicine and Translational Clinical Research. DOI: http://dx.doi.org/10.1016/B978-0-12-802103-3.00004-3

Table 4.1 Considerations for Preclinical Candidate Notification

S. No.	Consideration
1.	Acceptable PK (with a validated bioanalytical method)
2.	Demonstrated in vivo efficacy/activity
3.	Acceptable safety margin (toxicity in rodents or dogs when appropriate)
4.	Feasibility of GMP manufacture
5.	Acceptable drug interaction profile

Source: Strovel J, Sittampalam S, Coussens NP, et al. Early Drug Discovery and Development Guidelines: For Academic Researchers, Collaborators, and Start-up Companies. 2012 May 1 [Updated 2016 Jul 1]. In: Sittampalam GS, Coussens NP, Nelson H, et al., editors. Assay Guidance Manual [Internet]. Bethesda (MD): Eli Lilly & Company and the National Center for Advancing Translational Sciences; 2004. Available from: http://www.ncbi.nlm.nih.gov/books/NBK92015/.

Table 4.2 Preclinical Development Activities

S. No.	Activity
1	API preparation
2	Formulation studies
3	Analytical and bioanalytical methods
4	Pharmacokinetic and metabolism studies
5	GMP manufacturing
6	Safety pharmacology and toxicology studies

Sources: Steinmetz KL, Spack EG. The basics of preclinical drug development for neurodegenerative disease indications. BMC Neurol 2009; 9(Suppl 1):S2.

between human and animal species, so that appropriate species can be selected for the nonclinical toxicology program. Metabolism studies are typically conducted using in vitro methods for exposing hepatic microsomes, cytosolic fractions, hepatocyte cultures, etc., derived from different species. Apart from generating drug metabolite profiles, metabolism studies are also used to evaluate the potential for drug-drug interactions and cytochrome P450 inhibition [2].

21 CFR Sec. 312.23 (IND content and format) (8)(i) requires that information on the absorption, distribution, metabolism, and excretion of the drug in animal species, if known, be included in the IND application [4].

Further, section 7.3.5 (b) of ICH GCP Guidelines requires that information on pharmacokinetics and biological transformation and disposition of the investigational product in all species studied should be included in the investigator's brochure. The description should include information on absorption and the local and systemic bioavailability of the investigational product and its metabolites, and their relationship to the pharmacological effects [5].

4.3 SAFETY PHARMACOLOGY STUDIES

Section 7.3.5 (a) of ICH GCP Guidelines requires that data on both primary pharmacodynamics as well as safety pharmacology be included in the investigator's brochure [5].

Primary pharmacodynamic studies are aimed at evaluating the mechanism of action and/or effects of a compound in relation to its desired therapeutic target. Such studies are generally conducted during the discovery phase of drug development and have been described in previous chapters. These studies help in determining the appropriate dose for both nonclinical and clinical studies [6].

4.3.1 INTRODUCTION AND OBJECTIVES

In safety pharmacology studies, the compound is administered to provide exposure in the therapeutic range and above. These studies are done to identify undesirable pharmacodynamic (PD) properties of the compound that may have relevance to its human safety, to evaluate adverse PD and/or pathophysiological effects of a compound observed in toxicology and/or clinical studies, and to understand the mechanism of the observed and/or suspected effects [7,8].

4.3.2 ROUTE/DOSE AND DURATION

The most appropriate route of drug administration for the conduct of safety pharmacology studies is the intended clinical route. Dose is selected so as to ensure that exposure to the parent compound and its major metabolites are greater than that achieved in humans. These studies usually involve single-dose administration. However, when PD effects occur only after a certain duration of treatment, or when safety concerns arise from repeat dose nonclinical studies or clinical studies, appropriate duration is selected for the studies [7,8].

4.3.3 TYPE OF SAFETY PHARMACOLOGY STUDIES

A minimal set of safety pharmacology studies are required to be completed for all drugs. These include evaluation of basic parameters pertaining to the vital organs, i.e., the central nervous system, cardiovascular system, and respiratory system. These are referred to as essential safety pharmacology studies, or the core-battery [7,8].

When potential adverse effects raise concern for human safety, these are explored in follow-up or supplemental studies. These may be based on:

1. Pharmacological properties or chemical class of the test substance.
2. Safety pharmacology core battery, clinical trials, pharmacovigilance, experimental in vitro or in vivo studies, or from literature reports.

Follow-up studies are aimed at gaining in-depth understanding of the effect of the drug on vital functions. Supplemental studies evaluate the potential adverse PD effects on organ system functions not addressed by core battery or repeated dose toxicity studies [7,8].

The Table 4.3 below describes the parameters evaluated as part of various safety pharmacology studies.

In addition to above, the ICH Safety Guideline S7B provides guidance on the non-clinical evaluation of the potential of QT interval prolongation. The QT interval of the electrocardiogram is a measure of the duration of ventricular depolarization and repolarization. Prolongation of QT interval results in an increased risk of ventricular tachyarrhythmia, including torsade de pointes.

Table 4.3 Safety Pharmacology Studies in Drug Development

S. No.	Organ System	Essential Safety Pharmacology (Core Battery)	Follow-up Safety Pharmacology	Supplemental Safety Pharmacology
1	CNS	Motor activity, behavioral changes, coordination, sensory/motor reflex responses, and body temperature are evaluated. Functional observation battery, modified Irwin's, etc., are used for assessment.	Behavioral pharmacology, learning and memory, ligand-specific binding, neurochemistry, visual, auditory and/or electrophysiology assessments are done.	NA
2	CVS	Blood pressure, heart rate, and electrocardiogram are evaluated. In vivo, in vitro and/or ex vivo evaluations (including repolarization and conductance) are done.	Cardiac output, ventricular contractility, vascular resistance, the effects of endogenous and/or exogenous substances on the cardiovascular responses are evaluated.	NA
3	RS	Respiratory rate and other measures of respiratory function (e.g., tidal volume or hemoglobin oxygen saturation etc.) are evaluated.	Airway resistance, compliance, pulmonary arterial pressure, blood gases, blood pH, etc. are evaluated.	NA
4	Renal/Urinary	NA	NA	Urinary volume, specific gravity, osmolality, pH, fluid/electrolyte balance, proteins, cytology, and blood chemistry analysis such as blood urea nitrogen, creatinine, and plasma proteins are done.
5	ANS	NA	NA	Parameters evaluated include binding to receptors relevant for the ANS, functional responses to agonists or antagonists in vivo or in vitro, direct stimulation of autonomic nerves and measurement of cardiovascular responses, baroreflex testing, and heart rate variability.
6	GI	NA	NA	Parameters evaluated include gastric secretion, gastrointestinal injury potential, bile secretion, transit time in vivo, ileal contraction in vitro, gastric pH measurement, and pooling.
7	Others	NA	NA	Immune, endocrine, skeletal muscle

CNS, *Central Nervous System;* CVS, *Cardiovascular System;* RS, *Respiratory System;* ANS, *Autonomic Nervous System;* GI, *Gastro-intestinal.*

Sources: ICH Safety Guideline S 7A. Safety Pharmacology Studies for Human Pharmaceuticals, http://www.ich.org/products/guidelines/safety/article/safety-guidelines.html; 2000 [accessed 7.08.16]; Schedule Y (Subs. G.S.R. 32(E), dt, 20.1.2005), Drugs and cosmetics rules, 1945, http://cdsco.nic.in/html/D&C_Rules_Schedule_Y.pdf; 2005 [accessed 7.08.16]; Mattsson JL, Spencer PJ, Albee RR. A performance standard for clinical and Functional observational battery examinations of rats. J Am Coll Toxicol 1996; 15, 239; Irwin S. Comprehensive observational assessment: 1a. A systematic, quantitative procedure for assessing the behavioral and physiologic state of the mouse. Psychopharmacologia (Berl.) 1968; 13, 222–257 [9,10].

Table 4.4 Safety Pharmacology Studies Not Required
1. Locally applied agents (e.g., dermal or ocular) where the pharmacology is well characterized, and systemic exposure or distribution is low.
2. Prior to the first administration in humans of cytotoxic agents (except first in class) for treatment of end-stage cancer patients.
3. For biotechnology-derived products (except first in class) that achieve highly specific receptor targeting. Here, it is often sufficient to evaluate safety pharmacology endpoints as a part of toxicology and/or pharmacodynamic studies.
4. A new salt having similar pharmacokinetics and pharmacodynamics.
Sources: ICH Safety Guideline S 7A. Safety Pharmacology Studies for Human Pharmaceuticals, http://www.ich.org/products/guidelines/safety/article/safety-guidelines.html; 2000 [accessed 7.08.16]; Schedule Y (Subs. G.S.R. 32(E), dt, 20.1.2005), Drugs and cosmetics rules, 1945, http://cdsco.nic.in/html/D&C_Rules_Schedule_Y.pdf; 2005 [accessed 7.08.16].

In vitro I_{Kr} and in vivo QT assays are considered complementary approaches and the guideline recommends that both assay types should be conducted [11].

4.3.4 TIMING OF SAFETY PHARMACOLOGY STUDIES

Not all safety pharmacology studies are required to be completed at the time of IND filing. Prior to first in man studies, core battery & any follow-up or supplemental studies deemed appropriate, based on a cause for concern, are required to be completed. Information from toxicology studies can result in reduction or elimination of separate safety pharmacology studies. Based on observed or suspected adverse effects in animals and humans, additional studies may be required to be conducted during clinical development [7,8].

In general, safety pharmacology effects on all systems are required to be assessed prior to product approval. Appropriate justification has to be provided if certain studies have not been conducted [7,8].

The Table 4.4 below describes the circumstances where safety pharmacology studies are not required to be conducted.

4.4 TOXICOLOGY STUDIES

The two critical questions to be addressed prior to human testing are:

1. Is the compound safe enough for human administration?
2. What should be the maximum safe starting dose?

Though significant progress has been made in the field of toxicology, even today a variety of animal studies done in rodent/non-rodent species are considered the primary method of addressing these questions [2].

Table 4.5 Toxicology Studies in Drug Development

Systemic Toxicology Studies	**Local Toxicity Studies**	
Single Dose Toxicity Studies Dose-ranging studies Repeated Dose/Chronic Toxicity Studies	Dermal toxicity study Photo-allergy or dermal photo-toxicity Vaginal Toxicity Test	Rectal Tolerance Test Ocular toxicity studies Inhalation toxicity studies
Reproductive Toxicity Studies		*Allergenicity & Hypersensitivity Studies*
Male Fertility Studies	*Female Reproductive Developmental & Toxicity Studies* Segment I/Segment II/Segment III	*Mutagenicity/Genotoxicity Studies* *Carcinogenicity Studies* *Immunotoxicity Studies*

Sources: IND content and format (312.23), Part 312, Title 21, Subchapter B Electronic Code of Federal Regulations, http://www.ecfr.gov/cgi-bin/text-idx?SID=892d1c8a72af225b1fa8d57ae4da1067&mc=true&node=se21.5.312_123&rgn=div8; [accessed 21.08.16]; ICH Efficacy Guideline E6 (1996). Guideline for Good Clinical Practice, http://www.ich.org/fileadmin/Public_Web_Site/ICH_Products/Guidelines/Efficacy/E6/E6_R1_Guideline.pdf; [accessed 21.08.16]; ICH Multidisciplinary Guideline M3 (2009). Guidance on nonclinical safety studies for the conduct of human clinical trials and marketing authorization for pharmaceuticals, http://www.ich.org/fileadmin/Public_Web_Site/ICH_Products/Guidelines/Multidisciplinary/M3_R2/Step4/M3_R2__Guideline.pdf; [accessed 7.08.16]; Schedule Y (Subs. G.S.R. 32(E), dt, 20.1.2005). Drugs and cosmetics rules, 1945, http://cdsco.nic.in/html/D&C_Rules_Schedule_Y.pdf; 2005 [accessed 7.08.16].

21 CFR Sec. 312.23 (IND content and format) (8)(ii) and Section 7.3.5 (c) of ICH GCP Guidelines require that information on the toxicological effects of the drug in animals and in vitro be included in the IND application and investigator brochure respectively [4,5].

The Table 4.5 provides a list of animal studies required to be conducted as part of IND toxicology package.

4.4.1 SYSTEMIC TOXICOLOGY STUDIES

The Table 4.6 provides a list of systemic toxicology studies along with the animal species.

4.4.1.1 Single Dose Toxicity Studies/Dose Ranging Studies

The purpose of these studies is to generate data on acute toxicity. These studies are carried out using the intravenous route as well as the intended clinical route of administration. Animals are observed for 14 days after the drug administration for signs of intoxication, effect on body weight, gross pathological changes, and histopathology of grossly affected organs. Minimum lethal dose (MLD), maximum tolerated dose (MTD), and the target organ of toxicity (if possible) is determined. LD-50 is no more required to be an intended endpoint in these studies, though certain regulatory agencies still insist on this data [6,8].

It may be noted that dose-escalation studies or short-duration dose-ranging studies in the general toxicity test species can also provide information on MTD and target organ of toxicity as above and can therefore serve as an alternative to these studies. However, in certain situations like

Table 4.6 Systemic Toxicology Studies

S. No.	Type of Study	Animal Species	Number of Animals Used and Dose/Duration of Drug Treatment
1	Single Dose Toxicity Studies	2 rodent species (mice and rats)	5 animals of either sex, each exposed to at least four graded doses of the test substance over 24 h
2	Dose-ranging studies	Two mammalian species [rodent (preferably rat); one nonrodent]	5 animals of either sex for rodent studies, each exposed to at least four graded doses (with highest dose as MTD of single dose study) given daily for 10 consecutive days One male and one female for nonrodent studies; starting dose 3 to 5 times the extrapolated effective dose or MTD (whichever is less), and dose escalated/lowered every third day stepwise. Well tolerated dose given daily for 10 consecutive days
3	Repeated Dose/ Chronic Toxicity Studies	Two mammalian species (one nonrodent)	For 14–28 days studies: 6–10/sex/group for rodent studies and 2–3/sex/group for nonrodent studies at three graded dose levels plus a control group For 90–180 days studies: 15–30/sex/group for rodent studies and 4–6/sex/group for nonrodent studies at three graded dose levels plus a control group

MTD, *Maximum tolerated dose.*
Sources: IND content and format (312.23), Part 312, Title 21, Subchapter B Electronic Code of Federal Regulations, http://www.ecfr.gov/cgi-bin/text-idx?SID=892d1c8a72af225b1fa8d57ae4da1067&mc=true&node=se21.5.312_123&rgn=div8; [accessed 21.08.16]; ICH Efficacy Guideline E6 (1996). Guideline for Good Clinical Practice, http://www.ich.org/fileadmin/Public_Web_Site/ICH_Products/Guidelines/Efficacy/E6/E6_R1_Guideline.pdf; [accessed 21.08.16]; ICH Multidisciplinary Guideline M3 (2009). Guidance on nonclinical safety studies for the conduct of human clinical trials and marketing authorization for pharmaceuticals, http://www.ich.org/fileadmin/Public_Web_Site/ICH_Products/Guidelines/Multidisciplinary/M3_R2/Step4/M3_R2__Guideline.pdf; [accessed 7.08.16]; Schedule Y (Subs. G.S.R. 32(E), dt, 20.1.2005). Drugs and cosmetics rules, 1945, http://cdsco.nic.in/html/D&C_Rules_Schedule_Y.pdf; 2005 [accessed 7.08.16].

microdose trials, where data on dose-escalation or dose ranging studies is not needed, appropriate GLP compliant single dose toxicity studies are required to be conducted [6,8].

Information on the acute toxicity provides insights into the possible impact of human overdose and therefore this data should generally be available prior to conduct of Phase III studies. However, in certain indications like depression, pain, and dementia, where patients are at risk of overdosing, this data should be available prior to the conduct of phase II studies [6,8].

4.4.1.2 Repeated Dose/Chronic Toxicity Studies

The purpose of repeated-dose studies is to identify safe levels of drug (no observed adverse effect levels) following continuous exposure of the animals. The No Observed Adverse Effect Level (NOAEL) determined in the most sensitive animal species gives critical information for calculation of first in man dose. In general, the duration of the repeated dose toxicity studies should be equal to or exceed the duration of the human clinical trials up to the maximum recommended duration of the repeated-dose toxicity studies. In exceptional circumstances where meaningful therapeutic benefit has been shown, trials can be extended beyond the duration of the repeated-dose toxicity studies on a case-by case basis [6,8].

The highest dose evaluated in repeated dose toxicity studies should produce observable toxicity, the lowest dose should not cause observable toxicity, but should be comparable to or a multiple of the intended therapeutic dose in humans, and the intermediate dose should be placed logarithmically between the other two doses [6,8].

Parameters evaluated in repeated dose toxicity studies include signs of intoxication (general appearance, activity, and behavior etc.), body weight, food intake, biochemical and hematological measurements, urine analysis, organ weights, gross and microscopic study of viscera and tissues. Electrocardiogram and fundus examination are also done in the non-rodent species [6,8].

Generally, dose limits of 1000 mg/kg/day for rodents and non-rodents are recommended for single dose and repeated dose toxicity studies. Tables 4.7 and 4.8 present the generally recommended duration of repeated-dose toxicity studies to support the conduct of clinical trials and marketing authorization. For further details and exceptions to these recommendations, readers can refer the ICH Guidance on "Nonclinical Safety Studies for The Conduct of Human Clinical Trials and Marketing Authorization for Pharmaceuticals" [6].

Table 4.7 Recommended Duration of Repeated-Dose Toxicity Studies to Support the Conduct of Clinical Trials

Maximum Duration of Clinical Trial	Recommended Minimum Duration of Repeated-Dose Toxicity Studies to Support Clinical Trials
Up to 2 weeks	2 weeks
Between 2 weeks and 6 months	Same as clinical trial
>6 months	6 months (rodents) and 9 months (non-rodents)

Source: ICH Multidisciplinary Guideline M3 (2009). Guidance on nonclinical safety studies for the conduct of human clinical trials and marketing authorization for pharmaceuticals, http://www.ich.org/fileadmin/Public_Web_Site/ICH_Products/Guidelines/Multidisciplinary/M3_R2/Step4/M3_R2__Guideline.pdf; [accessed 7.08.16]; ICH Safety Guideline S 4. Duration of chronic toxicity testing in animals (rodent and non rodent toxicity testing), http://www.ich.org/products/guidelines/safety/article/safety-guidelines.html; 1998 [accesses 7.08.16] [12].

Table 4.8 Recommended Duration of Repeated-Dose Toxicity Studies to Support Marketing

Duration of Indicated Treatment	Recommended Minimum Duration of Repeated-Dose Toxicity Studies to Support Marketing
Up to 2 weeks	1 month
>2 weeks to 1 month	3 months
>1 month to 3 months	6 months
>3 months	6 months (rodents) and 9 months (nonrodents)

Source: ICH Multidisciplinary Guideline M3 (2009). Guidance on nonclinical safety studies for the conduct of human clinical trials and marketing authorization for pharmaceuticals, http://www.ich.org/fileadmin/Public_Web_Site/ICH_Products/Guidelines/Multidisciplinary/M3_R2/Step4/M3_R2__Guideline.pdf; [accessed 7.08.16]; ICH Safety Guideline S 4. Duration of chronic toxicity testing in animals (rodent and non rodent toxicity testing), http://www.ich.org/products/guidelines/safety/article/safety-guidelines.html; 1998 [accesses 7.08.16].

In India, the recommended duration of repeated-dose toxicity studies differs from the ICH recommendations. Unlike ICH recommendations, where the duration of studies to support marketing is longer than duration of studies to support clinical trial, the Schedule Y of Indian Drugs and Cosmetic Act provides a common recommendation both for supporting clinical trials and marketing. As such, in general, Indian regulations require longer duration of repeated dose toxicity studies to support clinical trials of equivalent duration compared to ICH regions. Thus, for a 3-week clinical study, while ICH regions require a 3-week repeated dose toxicity study, Indian regulations require a 12-week study. However, to support marketing for a product proposed to be administered for a 3-week period, ICH requires a 3-month repeated dose toxicity study, while Indian regulations require a 12-week study [6,8]. These regional differences need to be considered while preparing for a Global IND filing.

4.4.2 REPRODUCTION AND TERATOLOGY STUDIES

Reproduction toxicity studies are aimed at studying the impact of the drug on reproduction. The observations in these studies are continued through one complete lifecycle, i.e., from conception in one generation through to conception in the next generation. The ICH Guidance "Detection of toxicity to reproduction for medicinal products & toxicity to male fertility S5 (r2)" dated 24 June 1993 divides this lifecycle into the following stages [13]:

1. Premating to conception
2. Conception to implantation
3. Implantation to closure of the hard palate
4. Closure of the hard palate to the end of pregnancy
5. Birth to weaning
6. Weaning to sexual maturity

Broadly, three phases of the reproductive process are evaluated during reproductive studies. The fertility studies evaluate the effect on the fertility and the early implantation stages of embryogenesis. The teratogenicity studies evaluate the effect on fetal organ development. The perinatal studies evaluate the effect on late gestation, parturition, and lactation. Behavioral and neurodevelopmental assessments in the offspring are also made in perinatal studies [8,13].

While male fertility studies are required to be conducted for all drugs, female reproductive developmental and toxicity studies are required to be conducted for drugs proposed to be studied or used in women of child-bearing age. For including permanently sterilized or postmenopausal women in clinical trials, reproduction toxicity studies are not generally required if the relevant repeated-dose toxicity studies (including an evaluation of the female reproductive organs) have been conducted [6,8,13]. Table 4.9 briefly describes the various reproductive toxicity studies.

On 27 March 2015, a concept paper "S5(R3): Detection of Toxicity to Reproduction for Medicinal Products & Toxicity to Male Fertility dated 9 February 2015" was endorsed by the ICH Steering Committee. The paper proposed revision of the existing 20-year old guidance document on reproductive toxicity.

Table 4.9 Reproductive Toxicity Studies

S. No.	Type of Study	Animal Species	Number of Animals Used and Dose/Duration of Drug Treatment
1	Male fertility studies	One rodent species (preferably rat)	6 adult male animals/group for 3 dose groups (based on 14–28 days study) dosed for 28–70 days before pairing with female animals; drug treatment of the male animals continued during pairing.
2	**Female Reproductive Developmental & Toxicity Studies**		
	Segment I/female fertility studies	Albino rats (preferred) over mice Segment II study should include albino rabbits[a] also as a second test species.	15 adult males and females/group for 3 dose groups (the highest dose should usually be the MTD obtained from previous systemic toxicity studies). Dosing initiated prior to mating (28 and 14 days in advance in case of males and females respectively). Drug treatment continued during mating and, subsequently, during the gestation period for female animals.
	Segment II/ teratogenicity studies		20 pregnant rats (or mice) and 12 rabbits, on each of the three-dose level as for segment I. Drug administered throughout the period of organogenesis.
	Segment III/ perinatal studies		Atleast 4 groups (including control), each consisting of 15 dams used. Drug administered throughout the last trimester of pregnancy (from day 15 of gestation) and then the dose that causes low fetal loss continued throughout lactation and weaning. One male and one female from each litter of Fl generation (total 15 males and 15 females in each group) selected at weaning and treated with vehicle or test substance (at the dose levels described above) throughout their periods of growth to sexual maturity, pairing, gestation, parturition, and lactation.

[a]On the occasion, when the test article is not compatible with the rabbit (e.g., antibiotics) the Segment II data in the mouse may be substituted.

Source: ICH Multidisciplinary Guideline M3 (2009). Guidance on nonclinical safety studies for the conduct of human clinical trials and marketing authorization for pharmaceuticals, http://www.ich.org/fileadmin/Public_Web_Site/ICH_Products/Guidelines/Multidisciplinary/M3_R2/Step4/M3_R2__Guideline.pdf; [accessed 7.08.16]; Schedule Y (Subs. G.S.R. 32(E), dt, 20.1.2005). Drugs and cosmetics rules, 1945, http://cdsco.nic.in/html/D&C_Rules_Schedule_Y.pdf; 2005 [accessed 7.08.16]; ICH Safety Guideline S 5. Detection of toxicity to reproduction for medicinal products & toxicity to male fertility, http://www.ich.org/products/guidelines/safety/article/safety-guidelines.html; 2005 [accessed 7.07.16].

4.4.3 LOCAL TOXICITY STUDIES

ICH guidance M3 states that it is preferable to evaluate local tolerance by the intended clinical route as part of the general toxicity studies. However, the approach to such studies differs in the

Table 4.10 Local Toxicity Studies

S. No.	Type of Toxicology Study	Animal Species	Comments
1.	Dermal toxicity study	Rabbit and rat	Period of application varies from 7 to 90 days
2.	Photo-allergy or dermal photo-toxicity	Guinea pig	Armstrong/Harber Test in guinea pig. This test is done if the drug or a metabolite is related to an agent causing photosensitivity or the nature of action suggests such a potential (e.g., drugs to be used in treatment of leucoderma).
3.	Vaginal toxicity test	Rabbit or dog	Period of application varies from 7 to 30 days
4.	Rectal tolerance test	Rabbit or dog	Period of application varies from 7 to 30 days
5.	Ocular toxicity studies	Two species including albino rabbit	Albino rabbit is used as it has a large conjunctival sac. Period of application extends up to a maximum of 90 days
6.	Inhalation toxicity studies	One rodent and one nonrodent	Acute, subacute, and chronic toxicity studies performed for intended duration of human exposure.

Source: Schedule Y (Subs. G.S.R. 32(E), dt, 20.1.2005). Drugs and cosmetics rules, 1945, http://cdsco.nic.in/html/D&C_Rules_Schedule_Y.pdf; 2005 [accessed 7.08.16].

various regions. These studies are required to be conducted when the new drug is proposed to be administered by some special route (other than oral) in humans. Depending on the route of administration, the drug is applied to an appropriate site to determine local effects in a suitable species [6,8]. Table 4.10 lists the various local toxicity studies and the species used.

For products meant for intravenous or intramuscular or subcutaneous or intradermal injection, the sites of injection in systemic toxicity studies are required to be specially examined both grossly and microscopically [6,8].

4.4.4 ALLERGENICITY/HYPERSENSITIVITY STUDIES

Standard tests include guinea pig maximization test (GPMT) and local lymph node assay (LLNA) in mouse. Anyone of the two may be done [8].

4.4.5 MUTAGENICITY/GENOTOXICITY STUDIES

Genotoxicity tests are relatively inexpensive in vitro and in vivo tests conducted early in the drug development to enable hazard identification with respect to damage to DNA and its fixation, which may manifest later in studies of carcinogenicity or teratogenicity. However, it may be noted that not all carcinogenic or teratogenic effects are due to genetic alterations and not all genetic alterations will translate into human risks [6,8,14].

Table 4.11 Standard battery for Genotoxicity Testing

	Option 1	Option 2
Bacterial Cell	**A Test for Gene Mutation in Bacteria**	
Mammalian cell	A cytogenetic test for chromosomal damage (the in vitro metaphase chromosome aberration test or in vitro micronucleus test) OR An in vitro mouse lymphoma *tk* gene mutation assay An in vivo test for genotoxicity, generally a test for chromosomal damage using rodent hematopoietic cells, either for micronuclei or for chromosomal aberrations in metaphase cells	An in vivo assessment of genotoxicity with two tissues, usually an assay for micronuclei using rodent hematopoietic cells and a second in vivo assay

Source: ICH Safety Guideline S2. Guidance on genotoxicity testing and data interpretation for pharmaceuticals intended for human use, http://www.ich.org/products/guidelines/safety/article/safety-guidelines.html; 2011 [accessed 7.08.16].

In a mini-review (Brambilla et al., 2013) involving compilation of genotoxicity and carcinogenicity information of bronchodilators and antiasthma drugs, caffeine was reported to give positive responses in several genotoxicity assays as well as tested positive in carcinogenicity studies done in both mice and rats. Formoterol, orciprenaline, salbutamol, salmeterol, zafirlukast, and zileuton were reported to give negative responses in genotoxicity assays but tested positive in carcinogenicity studies done in mice and/or rats. Theophylline was reported to test positive in several genotoxicity assays but tested negative in carcinogenicity studies [15].

The bacterial reverse gene mutation (Ames) test has been shown to detect the majority of relevant genotoxic rodent and human carcinogens. However, ICH S2 guidance on genotoxicity testing recommends a battery approach including assessment in mammalian cells as assessment done in bacterial cells alone is not capable of detecting all genotoxic mechanisms relevant in tumorigenesis. Further, in vivo test(s) are included in the test battery because some agents are mutagenic only in vivo and this testing also accounts for absorption, distribution, metabolism, and excretion. Table 4.11 provides the two options for the standard battery which are considered equally appropriate [14].

For compounds that test negative in the standard battery, no additional tests are needed. However, for compounds that test positive in the standard test battery, additional tests may be required depending on their therapeutic use [14].

In case of antibiotics which are themselves toxic to bacteria, the Ames test should be carried out at lower concentrations and any one of the in vitro mammalian cell assays are recommended to be done [14].

4.4.6 CARCINOGENICITY STUDIES

Carcinogenicity Studies are aimed at identifying a tumorigenic potential in animals and to assess the relevant risk in humans. These are required for pharmaceuticals whose expected clinical use is

Table 4.12 Factors to be Considered to Determine Concern for Carcinogenic Potential
Previous demonstration of carcinogenic potential in product class that is relevant to humans
Structure-activity relationship suggesting carcinogenic risk
Evidence of preneoplastic lesions in repeated dose toxicity studies
Long-term tissue retention of parent compound or metabolite(s) resulting in local tissue reactions or other pathophysiological responses

Source: Schedule Y (Subs. G.S.R. 32(E), dt, 20.1.2005), Drugs and cosmetics rules, 1945, http://cdsco.nic.in/html/D&C_Rules_Schedule_Y.pdf; 2005 [accessed 7.08.16]; ICH Safety Guideline S1A. Guideline on the need for carcinogenicity studies of pharmaceuticals, http://www.ich.org/products/guidelines/safety/article/safety-guidelines.html; 1995 [7.08.16].

continuous for at least 6 months or which are intended to be used frequently in an intermittent manner in the treatment of chronic or recurrent conditions (e.g., allergic rhinitis, depression, and anxiety). These are also required for pharmaceuticals with a significant cause of concern for carcinogenic risk (Table 4.12) [8,16].

The choice of species for carcinogenicity studies is recommended to be based on various considerations like pharmacology, repeated-dose toxicology, metabolism, toxicokinetics, and route of administration. In the absence of clear evidence favoring one species, the studies are generally recommended to be done in rats through the intended clinical route. Mice may be employed only with proper scientific justification. Generally, the period of dosing should be 24 months for rats and 18 months for mice. ICH Guideline S1C provides recommendations on appropriate dose for carcinogenicity studies. The guideline recommends that in most cases (where the maximum recommended human dose does not exceed 500 mg/day), the dose for carcinogenicity studies be limited to 1500 mg/kg/day [8,17,18].

4.4.6.1 Carcinogenicity Studies Not Required

Unequivocally, genotoxic compounds, in the absence of other data, are presumed to be trans-species carcinogens, implying a hazard to humans, and carcinogenicity studies are generally not required. However, if such a drug is intended for long-term treatment, a chronic toxicity study (up to one year) may be required to detect early tumorigenic potential. Also, for pharmaceuticals intended to treat diseases where the life expectancy is less than 2–3 years (e.g., in the case of cancers and the treatment is unlikely to prolong survival), carcinogenicity studies may not be required. Carcinogenicity studies are not generally needed for endogenous substances given as replacement therapy [8,16].

4.4.7 IMMUNOTOXICITY

Immunotoxicity refers to unintended immunosuppression or enhancement. The ICH S8 guidance recommends that immunotoxicity potential should be evaluated based on standard toxicity studies and additional immunotoxicity studies if required based on evidence such as immune-related signals from standard toxicity studies [6,19].

4.4.8 PHOTOSAFETY EVALUATION

As briefly mentioned in Table 4.10, the decision to conduct photosafety studies is based on a variety of factors like the photochemical properties of the substance, the known risk of the phototoxic potential of chemically related compounds, tissue distribution, and clinical or nonclinical data suggesting phototoxicity. In the case that photosafety studies suggest a phototoxic or photocarcinogenic risk, appropriate protective measures need to be taken during outpatient clinical studies [6].

4.5 TOXICOKINETICS AND TISSUE DISTRIBUTION STUDIES

4.5.1 TOXICOKINETICS

Toxicokinetics is the generation of pharmacokinetic data as a part of various toxicity studies in order to assess systemic exposure. The measurement of peak and total exposure in these studies helps to determine the relationship between the toxicological effects and the exposure. Compared to the actual dose administered in the toxicology studies, the data on exposure is more relevant for comparing effects in animals and man. This is because the pharmacokinetics of a drug varies extensively between the species [8,20].

These studies generally include repeated-dose toxicity studies, reproductive, and carcinogenicity studies. Since, single dose studies are usually done before a bioanalytical method has been developed, toxicokinetic monitoring cannot be integrated in these studies. Data on systemic exposure becomes particularly important in cases of negative results of in vivo genotoxicity studies [20].

The main purpose of toxicokinetics is to describe the systemic exposure achieved in animals and its relationship to dose level and the time course of the toxicity study. The data relates the exposure achieved in toxicity studies to toxicological findings and contributes to the assessment of the relevance of these findings to human safety. It provides information on linear/non-linear pharmacokinetics, accumulation, and whether the effects are related to Cmax (peak concentration) or total exposure (AUC). Toxicokinetic data helps to determine the appropriate species, study design, and treatment regimen in subsequent non-clinical toxicity studies. Toxicokinetic information also helps in evaluating the impact of a proposed change in the clinical route of administration [8,20].

Normally, in the case of large animals, blood samples for the generation of toxicokinetic data are collected from main study itself. However, in the case of smaller species satellite groups may be required. ICH Guidance S3A provides detailed recommendations on toxicokinetic assessment [20].

4.5.2 TISSUE DISTRIBUTION STUDIES

Tissue distribution studies in animal models help to determine the distribution and accumulation of the compound and/or metabolites, especially in relation to potential sites of action. In general, single dose tissue distribution studies provide sufficient information for designing toxicology and pharmacology studies and for interpreting the results of these experiments [8,21].

The repeated dose tissue distribution studies may be most appropriate for compounds which have an apparently long half-life in the tissues compared to plasma; incomplete elimination as

Table 4.13 Toxicology Data Requirements to Support Clinical Trials

Clinical Trial Phase	Systemic Toxicology Studies[a]	Male Fertility Studies[b]	Female Reproductive and Developmental Toxicity Studies[c]	Genotoxicity Studies[d]	Carcinogenicity Studies[e]
Phase I	Yes	No[f]	No	In vitro	No
Phase II	Yes	No[f]	Segment II	In vitro and in vivo	No
Phase III	Yes	Yes	Segment I, II, III	In vitro and in vivo	Yes
Marketing	Yes	Yes	Segment I, II, III	In vitro and in vivo	Yes

[a]*Refer Table 4.7 for duration of systemic toxicology studies to support clinical trials.*
[b]*Men can be included in Phase I and II trials before the conduct of the male fertility study since an evaluation of the male reproductive organs is performed in the repeated-dose toxicity studies.*
[c]*Regional differences; In the US, Segment II studies can be deferred until before Phase III for women of child-bearing potential using precautions to prevent pregnancy in clinical trials. In all ICH regions, the Segment III studies are required to be submitted for marketing approval. However, in India Segment III studies need to be completed prior to Phase III for drugs to be given to pregnant or nursing mothers for long periods or where there are indications of possible adverse effects on fetal development.*
[d]*An assay for gene mutation is generally considered sufficient to support single dose studies. To support multiple dose studies, additional assessment(s) in mammalian system(s) is/are required to be completed.*
[e]*Generally required to be conducted to support the marketing authorization application. However, where there is a significant concern for carcinogenic risk, studies need to be completed to support Phase III clinical trials. For pharmaceuticals intended to treat certain serious diseases, these studies can be completed even after the grant of marketing authorization.*
[f]*In India, male fertility studies are required to be completed before Phase I trials.*
Source: ICH Multidisciplinary Guideline M3 (2009). Guidance on nonclinical safety studies for the conduct of human clinical trials and marketing authorization for pharmaceuticals, http://www.ich.org/fileadmin/Public_Web_Site/ICH_Products/Guidelines/Multidisciplinary/M3_R2/Step4/M3_R2__Guideline.pdf; [accessed 7.08.16]; Schedule Y (Subs. G.S.R. 32(E), dt, 20.1.2005). Drugs and cosmetics rules, 1945, http://cdsco.nic.in/html/D&C_Rules_Schedule_Y.pdf; 2005 [accessed 7.08.16].

evident by higher steady state levels of compound/metabolite or unanticipated organ toxicity. The repeated dose tissue distribution studies are also considered appropriate for drugs intended for site-specific targeted delivery design [8,21].

4.5.3 TIMING OF TOXICOLOGY STUDIES

The timing of toxicology studies in the development program depends on the data required to support a particular phase of clinical study (Table 4.13). Despite efforts being made to ensure harmonization across the globe, there are regional differences which need to be considered while planning for a global development program.

4.6 GOOD LABORATORY PRACTICES

Pivotal safety pharmacology and toxicology studies need to comply with the norms of Good Laboratory Practice (GLP). In the United States, 21 CFR part 58 deals with GLP studies for non-clinical studies. GLP standards require that the studies should be performed by appropriately trained and qualified personnel using properly calibrated and standardized equipment of adequate size and capacity. The studies should be done as per written protocols and standard operating procedures. Any amendments/deviations should be documented. The whole process should be retrospectively verifiable. The GLP norms also specify requirements for the testing facilities; test and control articles, as well as records and reports. The documentation records, raw data, and specimens are required to be retained for a period which depends on whether the application for research or market permit of the compound for which the studies were conducted was submitted or not.

In the case that the application was submitted, the period-length also depends on whether it was approved or not. In the case that the results were not submitted, the records are to be retained for a period of at least two years following the date on which the study was completed, terminated, or discontinued. In the case that the results were submitted but the application was not approved, the records are to be retained for a period of at least 5 years following the date on which the results were submitted. In the case that the application is approved, the records are to be retained for a period of at least 2 years following the date on which the application was approved [8,22].

REFERENCES

[1] Strovel J, Sittampalam S, Coussens NP, et al. Early drug discovery and development guidelines: for academic researchers, collaborators, and start-up companies. 2012 May 1 [Updated 2016 Jul 1]. In: Sittampalam GS, Coussens NP, Nelson H, et al., editors. Assay Guidance Manual [Internet]. Bethesda (MD): Eli Lilly & Company and the National Center for Advancing Translational Sciences; 2004. Available from: http://www.ncbi.nlm.nih.gov/books/NBK92015/.

[2] Steinmetz and Spack. The basics of preclinical drug development for neurodegenerative disease indications. BMC Neurol 2009;9(Suppl 1):S2.

[3] ICH Safety Guideline S 3B. Pharmacokinetics: guidance for repeated dose tissue distribution studies, <http://www.ich.org/products/guidelines/safety/article/safety-guidelines.html>; 1994 [accessed 7.08.16].

[4] IND content and format (312.23). Part 312, Title 21, Subchapter B Electronic Code of Federal Regulations, <http://www.ecfr.gov/cgi-bin/text-idx?SID=892d1c8a72af225b1fa8d57ae4da1067&mc=true&node=se21.5.312_123&rgn=div8>; [accessed 21.08.16].

[5] ICH Efficacy Guideline E6. Guideline for Good Clinical Practice, <http://www.ich.org/fileadmin/Public_Web_Site/ICH_Products/Guidelines/Efficacy/E6/E6_R1_Guideline.pdf>; 1996 [accessed 21.08.16].

[6] ICH Multidisciplinary Guideline M3. Guidance on nonclinical safety studies for the conduct of human clinical trials and marketing authorization for pharmaceuticals, <http://www.ich.org/fileadmin/Public_Web_Site/ICH_Products/Guidelines/Multidisciplinary/M3_R2/Step4/M3_R2__Guideline.pdf>; 2009 [accessed 7.08.16].

[7] ICH Safety Guideline S 7A. Safety pharmacology studies for human pharmaceuticals, <http://www.ich.org/products/guidelines/safety/article/safety-guidelines.html>; 2000 [accessed 7.08.16].

[8] Schedule Y. (Subs. G.S.R. 32{E), dt, 20.1.2005), Drugs and cosmetics rules, 1945, <http://cdsco.nic.in/html/D&C_Rules_Schedule_Y.pdf> [accessed 7.08.16].

[9] Mattsson JL, Spencer PJ, Albee RR. A performance standard for clinical and functional observational battery examinations of rats. J Am Colloids Toxicol 1996;15:239.

[10] Irwin S. Comprehensive observational assessment: 1a. A systematic, quantitative procedure for assessing the behavioural and physiologic state of the mouse. Psychopharmacologia (Berl.) 1968;13:222–57.

[11] ICH Safety Guideline S 7B. The non-clinical evaluation of the potential for delayed ventricular repolarization (QT interval prolongation) by human pharmaceuticals, <http://www.ich.org/products/guidelines/safety/article/safety-guidelines.html>; 2005 [accessed 7.08.16].

[12] ICH Safety Guideline S 4. Duration of chronic toxicity testing in animals (rodent and non rodent toxicity testing), <http://www.ich.org/products/guidelines/safety/article/safety-guidelines.html>; 1998 [accessed 7.08.16].

[13] ICH Safety Guideline S 5. Detection of toxicity to reproduction for medicinal products & toxicity to male fertility, <http://www.ich.org/products/guidelines/safety/article/safety-guidelines.html>; 2005 [accessed 7.08.16].

[14] ICH Safety Guideline S2. Guidance on genotoxicity testing and data interpretation for pharmaceuticals intended for human use, <http://www.ich.org/products/guidelines/safety/article/safety-guidelines.html>; 2011 [accessed 7.08.16].

[15] Brambilla G, Mattioli F, Robbiano L, Martelli A. Genotoxicity and carcinogenicity studies of bronchodilators and antiasthma drugs. Basic Clin Pharmacol Toxicol 2013;112:302–13. Available from: http://dx.doi.org/10.1111/bcpt.12054.

[16] ICH Safety Guideline S1A. Guideline on the need for carcinogenicity studies of pharmaceuticals, <http://www.ich.org/products/guidelines/safety/article/safety-guidelines.html>; 1995 [accessed 7.08.16].

[17] ICH Safety Guideline S1B. Testing for carcinogenicity of pharmaceuticals, <http://www.ich.org/products/guidelines/safety/article/safety-guidelines.html>; 1995 [accessed 7.08.16].

[18] ICH Safety Guideline S1C. Dose selection for carcinogenicity studies of pharmaceuticals, <http://www.ich.org/products/guidelines/safety/article/safety-guidelines.html>; 2008 [accessed 7.08.16].

[19] ICH Safety Guideline S8. Immunotoxicity studies for human pharmaceuticals, <http://www.ich.org/products/guidelines/safety/article/safety-guidelines.html>; 2005 [accessed 7.08.16].

[20] ICH Safety Guideline S3A. Note for guidance on toxicokinetics: the assessment of systemic exposure in toxicity studies, <http://www.ich.org/products/guidelines/safety/article/safety-guidelines.html>; 1994 [accessed 7.08.16].

[21] ICH Safety Guideline S3B. Pharmacokinetics: guidance for repeated dose tissue distribution studies, http://www.ich.org/products/guidelines/safety/article/safety-guidelines.html>; 1994 [accessed 7.08.16].

[22] Good Laboratory Practice for Nonclinical Laboratory Studies. Part 58, Title 21, Subchapter a electronic code of federal regulations, <http://www.ecfr.gov/cgi-bin/text-idx?SID=f353aaaebf34ec68957b4d005b45a7b7&mc=true&node=pt21.1.58&rgn=div5>; [accessed 21.08.16].

CHAPTER

TARGET PRODUCT PROFILE AND CLINICAL DEVELOPMENT PLAN

5

Gursharan Singh
Life Sciences, SmartAnalyst India Private Limited, Gurgaon, Haryana, India

TARGET PRODUCT PROFILE

5.1 INTRODUCTION

As the name suggests, "Target Product Profile," also known as the TPP, refers to the targeted or intended profile a pharmaceutical/biotechnology product or technology. The profile provides the product's desired characteristics or features defining the value proposition, and key differentiators aimed at providing a competitive advantage to the product [1–5].

Stephen R. Covey in his self-help book titled "The 7 Habits of Highly Effective People" published in 1989, identified "Begin with the end in mind" as one of the seven habits of effective people. Applying this learning to product development, the TPP is a dynamic multidisciplinary strategic development process tool embodying the notion of beginning with the goal in mind. In the pharmaceutical industry, the term "Target Product Profile" was first suggested by a Clinical Development Working Group in 1997 that was comprised of representatives from the US FDA and the pharmaceutical sponsors [1–5].

The ultimate goal of a drug development program is to obtain regulatory approval with an optimal label (prescribing information), which not only establishes the safety and efficacy of the drug but also distinguishes the drug from other competitor's, thereby maximizing the use of the product in the target population. The TPP summarizes the intended drug label specifying the labeling concepts and then describes the drug development program in the context of prescribing information/labeling goals. The TPP provides a summary of the studies required to be completed and the outcomes required to be demonstrated to support the labeling concepts or product claims. The TPP can be used throughout the drug development process from pre-IND phase to postmarketing phase [1–5].

5.2 THE TPP PROCESS

The development of TPP requires cross-functional discussions between various stakeholders. These include members from departments like Pharmacology, Metabolism & Pharmacokinetics, Safety Pharmacology & Toxicology, Clinical Pharmacology, Clinical Development, Medical Affairs, Marketing, Regulatory, Market Access, Intellectual Property, Strategy, etc.

Pharmaceutical Medicine and Translational Clinical Research. DOI: http://dx.doi.org/10.1016/B978-0-12-802103-3.00005-5

Table 5.1 Key Sections for Regulatory and Development Team TPP

Boxed warning
Indications and usage
Dosage and administration
Dosage forms and strengths
Contraindications
Warnings and precautions
Adverse reactions
Drug interactions
Use in specific populations
Drug abuse and dependence
Overdosage
Description
Clinical pharmacology
Nonclinical toxicology
Clinical studies

Source: *FDA. Guidance for Industry "Labeling for Human Prescription Drug and Biological Products – Implementing the PLR Content and Format Requirements", http://www.fda.gov/downloads/drugs/guidances/ucm075082.pdf; 2013 [accessed 26.01.17] [6].*

The first step is to review the current and emerging disease landscape to identify the current and possible future Standard of Care (SoC). Potential key areas of differentiation are then identified based on unmet needs assessment and gap analysis. The unmet needs may pertain to efficacy, safety, route of administration, or dosing frequency. Depending on the known pharmacological profile of the new drug, the possible value proposition is defined. The value proposition of the product determines the potential place in therapy for the product as well as the potential economic value. The development plan (non clinical/clinical studies) of the product is determined based on the development program of the standard of care with additional studies/considerations based on the differentiation offered by the product [1–5].

5.3 REGULATORY PERSPECTIVE

While TPP is a part of proprietary IND file, the submission of TPP is voluntary and not mandatory. As the TPP summarizes the intended label, the TPP intended for regulatory communications is best organized according to key sections in a drug label (Table 5.1) [1].

The decision regarding the sections of the label to be included in the TPP depends on the stage of drug development as well as the purpose of the meeting for which the TPP is used. The US FDA recommends the following templates for the TPP [1]:

Target Product Profile: Product X				
Milestone (Meeting or Submission)	**Date**	***TPP Submitted? Y/N**	**TPP Version Date**	**TPP Discussed? Y/N**
Pre-IND IND Submission EOP1 EOP2A EOP2/ Pre-Phase 3 Pre-NDA/BLA Other (specify)				
Target			*Annotations*	
[This area includes the intended labeling language]			[This area includes summary information of studies (planned or completed) to support the target including protocol number, serial number, submission date, etc.]	
Comments [This area includes additional information for better clarity]				
Source*: FDA. Guidance for industry and review staff target product profile—a strategic development process tool, http://www.fda.gov/downloads/drugs/guidancecomplianceregulatoryinformation/guidances/ucm080593.pdf; 2007 [accessed 26.01.17[1].*				

Apart from labeling concepts, the TPP provides the opportunity for a constructive dialogue with the regulators with regards to proposed promotional claims and/or presentations for use in product's promotional materials along similar lines as above for labeling concepts [1].

Thus, the target labeling language as well as the proposed promotional claims in the TPP are supported by the evidence generated from nonclinical/clinical studies. The text from label and promotional claims, as well as the nonclinical/clinical studies conducted as part of development program of the standard of care, serve as a reference. Though not required by the US FDA, this information for the standard of care can also be captured in the TPP to provide a ready reference.

The TPP serves as a format aimed at providing an efficient communication between the regulators and the sponsors to minimize the risk of late stage failure in the drug development process. The TPP helps in streamlining discussion with US FDA review staff by distinguishing the TPP entries that have been previously discussed from those pending discussions [1].

For constructive formal milestone meetings with the sponsors, the US FDA requires a briefing document to serve as an information package. The TPP included in a briefing document can help to provide a structure for presenting medical and scientific information relevant in the context of the overall development goal. Based on this information, the review staff can provide a constructive feedback for successful drug development [1].

However, it may be noted that even when submitted, there is no legal obligation on the sponsor to pursue all stated goals in the TPP or submit a draft label identical to the TPP. Similarly, even if the US FDA has provided concordance to the TPP, it does not mandate it to approve the identical language in the final label [1].

5.4 COMMERCIAL PERSPECTIVE

Apart from serving as a strategic planning tool for communicating with regulatory authorities for drug approval, the TPP also serves as a business communication tool for discussions with investors, partners, employees, physicians, and payors [1,3].

The business/commercial team TPPs are usually abbreviated versions of the regulatory TPP but also contain additional information necessary for the assessment of a product's commercial potential. In general, such TPPs include information on the target indication (including patient segment), value proposition/differentiation, dosage and route of administration, study design, clinical efficacy (primary and secondary endpoints and outcomes on each endpoints), and safety (contra-indications and adverse reactions). Though there are no standard guidelines for such TPPs, these TPPs usually have multiple profiles like minimal profile, base profile, and optimal profile. Apart from the above information, these TPPs may also contain information on pricing and reimbursement, patents/exclusivities, and product valuations (Table 5.2) [2,7].

The minimal profile represents the minimum requirement not just to gain regulatory approval but also to be at least on a par with and not inferior to the future standard of care. Benchmarking with regards to efficacy endpoints, route of administration/dosing and clinical efficacy and safety outcomes for the minimal profile is based on data of approved/pipeline products. The optimal profile represents the best case which fulfills the requirements (clinically and economically meaningful differentiation) of all stakeholders (physicians and payors) so as to maximize its usage in the target indication. A market research exercise involving Key Opinion Leaders (KOLs) and payors is

Table 5.2 Contents of a Commercial Team TPP

Target Product Profile: Product X				
Value-proposition Target indication Target patient population Product description Dosage and route of administration				
Study design				
Efficacy	Primary endpoint Secondary endpoints			
Safety	Contraindications Adverse reactions			
Pricing and reimbursement Patents and exclusivities Patient share Product valuation (rNPV)				

rNPV, *Risk adjusted Net Product Value.*

usually undertaken to define the optimal profile on various parameters. While the minimal and optimal profiles represent the two extreme ends, the likely profile of the product which falls between these two is referred to as the base profile. This is based on the most likely outcomes derived from extrapolation of available pharmacology, efficacy, or safety data. The commercial team TPP is tested with prescribers for the potential patient share and this information helps in estimating the net present value (NPV) of the product [4,7].

5.5 QUALITY TARGET PRODUCT PROFILE

While the focus of this chapter is on clinical, regulatory, and commercial target product profiles, it may be noted that to achieve the desired safety and efficacy profile, a systematic approach to pharmaceutical development is being adopted that begins with predefined objectives and emphasizes product and process understanding and process control, based on sound science and quality risk management. This approach is known as "Quality by Design (QbD)." The Quality Target Product Profile (QTPP) provides a prospective summary of the quality characteristics of a drug product that will ensure the desired quality, taking into account safety and efficacy of the drug product [8]. Table 5.3 provides a list of elements included in the QTPP.

CLINICAL DEVELOPMENT PLAN

5.6 CLINICAL DEVELOPMENT

The whole process of clinical development of a new drug is aimed at finding out the indication, dose range, and schedule at which the drug is both safe and effective for specified use in a patient population. To fulfill these requirements, an ordered program of clinical trials, each with its own

Table 5.3 Elements of a Quality Target Product Profile

S. No.	QTPP Element
1.	Dosage form (e.g., tablet, capsule)
2.	Dosage design (e.g., immediate release)
3.	Dosage strength
4.	Route of administration
5.	Method of administration (e.g., with or without food)
6.	Pharmacokinetics
7.	Stability
8.	Drug product quality attributes (e.g., physical attribute, identification, assay, dissolution)
9.	Container closure system

Source: Maguire J, Peng D. How to identify critical quality attributes and critical process parameters, http://pqri.org/wp-content/uploads/2015/10/01-How-to-identify-CQA-CPP-CMA-Final.pdf; 2015 [26.01.17] [9].

specific objectives, is specified in the clinical development plan (CDP). The early studies in clinical development are small short-term trials to provide information on safety and tolerability, pharmacodynamics, and pharmacokinetics, to choose an appropriate dosage range and administration schedule for initial exploratory therapeutic trials. Later larger and more definitive confirmatory studies are done in a diverse patient population [10].

5.7 PHASES OF CLINICAL DEVELOPMENT

Drug development is a step-wise procedure often described as comprising of four phases (Phase I–IV). The Phase I studies are mostly human pharmacology studies, the Phase II studies are mostly therapeutic exploratory studies, the Phase III studies are mostly therapeutic confirmatory studies, and the Phase IV studies are comprised of variety of therapeutic use studies. However, it may be noted that these phases do not imply a fixed order of studies and the types of study are not synonymous with the phases of development. Thus, human pharmacology studies may be done during Phase II or Phase III of development also but may still be labeled as Phase I studies. Table 5.4 describes the various types of studies [11].

5.8 STUDY ENDPOINTS OR VARIABLES

5.8.1 PRIMARY AND SECONDARY ENDPOINTS

The primary endpoint or variable is the one capable of providing a valid and reliable measure of the most clinically relevant benefits in the target patient population. In most clinical studies only one primary endpoint is selected. The primary variable is usually an efficacy variable, but safety/tolerability and measurements relating to quality of life and health economics are other potential primary variables. Sample size calculation is done based on the primary endpoint. Secondary endpoints or variables are either supportive measurements related to the primary objective or measurements of effects related to the secondary objectives. Though theoretically there can be large number of secondary variables, the number should be limited [10].

5.8.2 STUDY ENDPOINTS: ADDITIONAL CONSIDERATIONS

5.8.2.1 Composite endpoints

In cases where multiple measurements are associated with the primary objective, the same can be integrated into a single composite variable, using a predefined algorithm. The benefit of this approach is that it addresses the multiplicity problem without requiring adjustment to the type I error [10].

5.8.2.2 Surrogate endpoints

ICH E9 defines a surrogate variable as a variable that provides an indirect measurement of effect in situations where direct measurement of clinical effect is not feasible or practical. Surrogate variables are used in a number of indications but there are two major challenges with their use. Firstly,

Table 5.4 Types of Studies in Various Phases of Development

Phase of Development	Types of Studies	Objectives	Duration (Months)
Phase I	Mostly human pharmacology but sometimes therapeutic exploratory	– Initial assessment of safety and tolerability through single and multiple dose studies – Define Pharmacokinetics and Pharmacodynamics – Explore drug metabolism and drug interactions – Early estimate of activity	20
Phase II	Mostly therapeutic exploratory but also include human pharmacology and therapeutic confirmatory	– Explore the following – Efficacy for the targeted indication in narrow patient populations using surrogate or pharmacological endpoints or clinical measures – Dosage (including dose response), endpoints, therapeutic regimens and target populations for subsequent studies	30
Phase III	Mostly therapeutic confirmatory but also human pharmacology and sometimes therapeutic exploratory	– Confirm efficacy and Safety through large comparative well controlled studies – Establish favorable benefit/risk relationship to support licensing – Confirm dose-response relationship	30
Phase IV	Mostly therapeutic use but sometimes human pharmacology, therapeutic exploratory and therapeutic confirmatory	– Optimize drug use in general or special patient populations – Identify additional adverse reactions – Refine dosing recommendation – Health Economic Outcomes Research	Varies depending on type of study

Source: ICH Efficacy Guideline E8 (1997). General Considerations for Clinical Trials, accessed online at http://www.ich.org/fileadmin/Public_Web_Site/ICH_Products/Guidelines/Efficacy/E8/Step4/E8_Guideline.pdf on 26th January, 2017.

they may not be able to truly predict the clinical outcome of interest. Secondly, quantitative measurement of clinical benefit to assess the benefit/risk relationship may not be possible [10].

5.8.2.3 Multiple primary endpoints

In clinical studies with more than one primary endpoint, it needs to be clarified in advance whether an impact on any of the endpoints, some minimum number of them, or all of them, would be considered necessary to achieve the trial objectives. The effect on the type I error and the method of controlling it should be described in the protocol [10].

5.8.2.4 Global impression/assessment

ICH 9 defines the Global Assessment Variable as a single variable—usually a scale of ordered categorical ratings—which integrates objective variables and the investigator's overall impression about the state or change in state of a subject. In neurology, psychiatry, and certain other therapeutic areas, such variables are used to measure the overall safety, overall efficacy, and/or overall usefulness of a treatment. Global Impression Variables should generally not be used as a primary variable but when there are specific efficacy and safety outcomes they should separately be used as additional primary variables [10].

5.9 BIAS AND TECHNIQUES TO AVOID BIAS

ICH E 9 provides the following definition of bias.

"The systematic tendency of any factors associated with the design, conduct, analysis, and evaluation of the results of a clinical trial to make the estimate of a treatment effect deviate from its true value." Bias resulting from deviations in conduct is referred to as 'operational' bias while that resulting due to other causes is referred to as 'statistical' bias [10].

Blinding and randomization are the most commonly used techniques to avoid bias.

Blinding or masking refers to preventing the identification of the treatments to limit the occurrence of conscious and unconscious bias in the conduct and interpretation of a clinical study. Blinding is done because knowledge of treatment may influence the recruitment and allocation of subjects, their subsequent care, the attitudes of subjects to the treatments, the assessment of end-points, the handling of withdrawals, the exclusion of data from analysis, etc. Randomization is aimed at introducing a deliberate element of chance into the assignment of treatments to subjects in a clinical study. Randomization helps to produce treatment groups with similar distributions of known and unknown prognostic factors [10].

In a double-blind trial, neither the subject nor any of the investigator or sponsor staff involved in the treatment or evaluations of the subjects are aware of the treatment received. The randomization schedule documents the random allocation of treatments to subjects. While the double-blind randomized trial is considered the optimal approach, if it is not feasible, then the single-blind option should be considered. However, in some cases only an open-label trial is practically or ethically possible. For single-blind and open-label trials, a centralized randomization method, such as telephone randomization, may be used, and the clinical assessments should be made by medical staff who are not involved in treating the subjects and who remain blind to treatment [10].

5.10 DESIGN CONFIGURATIONS

The three most commonly used design configurations in clinical studies are the parallel group, crossover, and factorial design as described in Table 5.5.

Table 5.5 Design Configurations in Clinical Studies

S. No.	Design Configuration	Description	Comments
1	Parallel group design	Subjects are randomized to one of two or more arms, each arm being allocated a different treatment.	These treatments will include the investigational product at one or more doses (X mg, 2X mg etc.), and one or more control treatments, such as placebo (Plc) and/or an active comparator (Y)
2	Crossover design	Each subject is randomized to a sequence of two or more treatments	In the 2 × 2 crossover design each subject receives each of two treatments (X and Y) in randomized order with two possible sequences (XY and YX) in two successive treatment periods (P1 & P2), often separated by a washout period. Since each subject acts as his own control for treatment comparisons, this approach reduces the sample size needed to achieve a specific power
3	Factorial Design	Two or more treatments are evaluated simultaneously through the use of varying combinations of the treatments	In the 2 × 2 factorial design subjects are randomly allocated to one of the four possible combinations of two treatments, say X and Y. X alone; Y alone; X and Y combined; Neither. This helps examine the interaction between X and Y.

Source: *ICH Efficacy Guideline E9. Statistical principles for clinical trials, http://www.ich.org/fileadmin/Public_Web_Site/ICH_Products/Guidelines/Efficacy/E9/Step4/E9_Guideline.pdf; 1998 [26.01.17].*

5.11 CHOICE OF CONTROL GROUP

The control group in clinical trial makes it possible to differentiate patient outcomes caused by the test drug from those caused by other factors, such as the natural progression of the disease, observer or patient expectations, etc. [12].

Control groups in clinical trials are classified based on the type of treatment used and whether the control group and test groups are chosen from the same population and treated concurrently or the control population is separate from the population (external) treated in the trial. Based on these criteria, there are a total of five different types of control groups as described in Table 5.6 [12].

5.12 TRIAL DESIGN FOR ACTIVE CONTROLLED TRIALS

The objective of an active controlled study may be either to demonstrate the superiority, noninferiority, or equivalence of the new product with the active comparator [10,13].

Table 5.6 Control Groups in Clinical Trials

S. No.	Control Group	Description	Comments
Concurrent Control			
1	Placebo control	Subjects randomly assigned to a test treatment or to an identical-appearing treatment that does not contain the test drug.	Generally double-blind randomized trials Controls for all potential influences on the course of the disease other than those due to the pharmacologic action of the test drug. The new treatment and placebo may be added to a common standard therapy
2	No treatment control	Subjects randomly assigned to test treatment or to no treatment.	Major drawback is that the investigator cannot be blinded in this approach.
3	Dose-response	Subjects randomly assigned to one of several fixed-dose groups.	Usually double-blind trials. May include a placebo (zero dose) and/or active control.
4	Active (Positive)	Subjects randomly assigned to the test treatment or to an active control treatment.	Usually double-blind trials. May be aimed to demonstrate noninferiority, equivalence, or superiority.
External Control			
5	External (Historic Control)	Group of patients external to the study.	May involve a group of patients treated at an earlier time (historical control) or treated during the same time period but in another setting. Also includes baseline controlled studies, in which subjects' status on therapy is compared with status before therapy.

Source: ICH Efficacy Guideline E10. Choice of control group and related issues in clinical trials, http://www.ich.org/fileadmin/Public_Web_Site/ICH_Products/Guidelines/Efficacy/E10/Step4/E10_Guideline.pdf; 2000 [accessed 26.01.17].

5.12.1 SUPERIORITY TRIAL

The aim of superiority trial is to demonstrate a statistically significant difference between treatments. The null hypothesis (Ho) is that the beneficial response to the test drug (T) is not greater than the response to the active comparator (AC); the alternative hypothesis (Ha) is that the response to the test drug is greater than AC. Thus:

$$\text{Ho:T} \leq \text{AC; T-AC} \leq 0$$

$$\text{Ha:T} > \text{AC; T-AC} > 0$$

In most cases, the test for a treatment effect corresponds to showing that the lower bound of the two-sided 95% confidence interval (equivalent to the lower bound of a one-sided 97.5% confidence interval) for T-AC is >0. This is equivalent to showing that the two means (effect of test drug vs active comparator) are statistically significantly different at the 5% level ($P < 0.05$) [10,13].

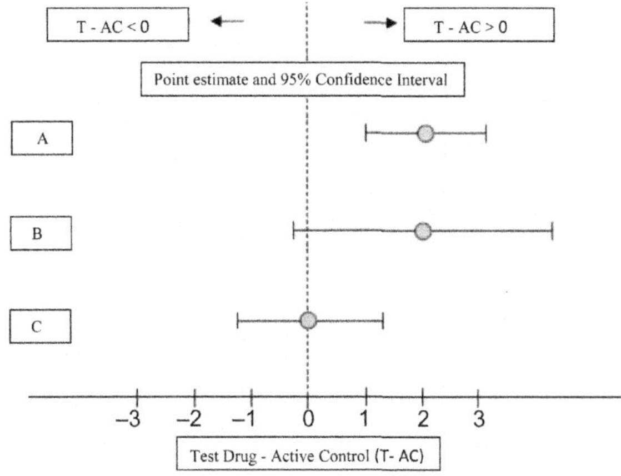

FIGURE 5.1

95% confidence interval in superiority trials. A. 95% CI lower bound is >0 (Test Drug is Superior to Active Control). B. 95% CI lower bound is < 0, even though the point estimate >0 (Test Drug is Not Superior to Active Control). C. 95% CI lower bound is <0 and the point estimate is 0 (Test Drug is Not Superior to Active Control)

The Fig. 5.1 below shows the possible scenarios pertaining to difference in the effect of test drug and active control.

However, in certain cases, the statistics of interest may not be the difference between means but may be the ratio of geometric means. Further, it may be noted that in superiority trials, the observed statistically significant difference may or may not be clinically relevant and determining clinical relevance is a matter of judgment [10,13].

5.12.2 EQUIVALENCE TRIAL

The aim of an equivalence trial is to demonstrate the absence of a clinically meaningful difference between treatments. This requires choosing a margin of clinical equivalence (Δ) which represents the largest difference that is considered to be clinically acceptable. For the two treatments to be considered clinically equivalent, the two sided 95% confidence interval should lie entirely within the interval $-\Delta$ and $+\Delta$. The equivalence approach is commonly used for bioequivalence studies, and for the pharmacokinetic parameters used (AUC and Cmax), 90% confidence interval has become the accepted regulatory standard. For topical formulations, where pharmacokinetic-based bioequivalence approach is not possible, clinical bioequivalence trials based on two-sided 95% confidence interval may be done [10,13].

5.12.3 NONINFERIORITY TRIALS

The noninferiority trials are aimed at demonstrating that a new treatment is no less effective than an existing treatment. Again, the noninferiority margin (Δ) which represents the largest difference that is considered to be clinically acceptable is chosen. For noninferiority, the two sided 95% confidence interval should lie entirely to the right of $-\Delta$ and $+\Delta$ [10,13,14,15].

Thus, common to both equivalence and noninferiority trials is the choice in advance of Δ as well as the appropriate confidence interval. However, no such Δ is needed for superiority trials. While superiority, equivalence, or noninferiority needs to be defined in advance in protocol, switching from noninferiority to superiority or superiority to noninferiority may be feasible under certain circumstances discussed in detail in the EMA document CPMP/EWP/422/99 [10,13].

5.13 SAMPLE SIZE

To provide a reliable answer to the questions being addressed by the clinical study, an adequate number of subjects need to be enrolled. This number is usually determined by the primary objective of the trial. The following factors are considered in calculation of sample size: a primary variable, the test statistic, the null hypothesis, the alternative hypothesis, the probability of erroneously rejecting the null hypothesis (the type I error), and the probability of erroneously failing to reject the null hypothesis (the type II error), as well as the approach to dealing with treatment withdrawals and protocol violations. Conventionally, the probability of type I error is set at 5% or less and the probability of type II error is conventionally set at 10% to 20%. The details of sample size calculation are covered in separate chapters covering biostatistics and medical statistics [10].

5.14 ADAPTIVE DESIGNS IN CLINICAL DEVELOPMENT

Adaptive design study includes a prospectively planned opportunity to use accumulating data to decide on how to modify aspects of the trial without undermining the validity and integrity of the trial [16].

The purpose of adaptive design is to make the study more efficient with fewer patients and shorter duration, more likely to demonstrate an effect of the drug (if it exists), or more informative in terms of dose response relationship. Table 5.7 provides a list of potential modifications for adaptive design clinical trials [16].

For details, the readers can refer to "The US FDA Guidance for Industry—Adaptive Design Clinical Trials for Drugs and Biologics, February 2010."

5.15 CLINICAL DEVELOPMENT PLAN

The Clinical Development Plan is the blueprint of the entire clinical research strategy of a drug, which defines the critical path for the clinical program, including development, assessment, and decision points, and the project resource (personnel and budget) estimates. The purpose of the CDP is to provide an efficient marketing application with the minimum studies needed for global registration and approval. The clinical development plan is comprised of two components: a strategic plan and an operational plan. The strategic plan includes the overall direction while the operational plan outlines the execution. The CDP is a dynamic document that evolves over time. Three phases

Table 5.7 Potential Study Design Modifications for Adaptive Design Clinical Trials

- Study duration: Early termination (futility, early rejection)
- Sample size
- Treatment allocation ratios
- Treatment arms
- Concomitant treatments
- Randomization procedure
- Study endpoints
- Test hypotheses (noninferiority vs superiority)
- Study population (e.g., inclusion/exclusion criteria)
- Statistical analysis (e.g., combine trials/treatment phases)

Source: *FDA. The US FDA Guidance for industry - adaptive design clinical trials for drugs and biologics, http://www.fda.gov/downloads/drugs/guidances/ucm201790.pdf; 2010 [accessed 26.01.17].*

of CDP have been defined, including the preclinical phase, the early development phase, and the late development phase [17,18].

The preclinical phase or initial CDP includes the background for the development of the compound including the pharmacology (PK & PD), safety pharmacology/toxicology data from animal studies, the scientific rationale for drug development in the proposed indication, along with the target population. The initial CDP is driven by the preclinical product champion and is developed approximately 1 year prior to the preparation of the Investigational New Drug (IND) application. The purpose is to provide adequate data to gain approval for conduct of first in man studies [17].

The early development phase begins just prior to the conduct of first in man studies and ends 9–12 months before the phase IIB dose-response trial. The early development plan is driven by someone with expertise in clinical pharmacology who provides outline for Phase I and Phase IIA (proof of concept) studies including the countries where these studies would be conducted, the study design, and the biomarkers used. Safety and proof of concept are the key considerations during this phase [17].

The late development phase begins with patient enrollment in the phase IIB, dose-response trial. The late development plan is driven by a global clinical leader who provides outline for Phase III studies, including the countries where these studies would be conducted. The key considerations are availability of experienced investigators, patients, regulatory environment, clinical trial costs, monitoring resources, and availability of comparator drugs that will meet the TPP requirements. The clinical operations plan is built based on the strategic elements contained in the late development plan [17]. Table 5.8 provides the general content of a CDP.

Table 5.9 provides a template for outline of clinical studies included in the clinical development plan.

5.16 COMPETENCIES REQUIRED FOR CDP

As would be obvious from the content of a CDP, developing a robust CDP requires weighing and balancing medical, scientific, regulatory, and marketing opinions. Table 5.10 provides the list of knowledge, skills, and abilities required to successfully develop a CDP.

Table 5.8 Content of a Clinical Development Plan

	Target Product Profile	
1.	Commercial rationale	Disease epidemiology
		Current standard of care
		Evolving disease landscape (competition)
		Unmet medical needs
		Target markets and market size
2.	Scientific Rationale	Mechanism of action (in vitro pharmacology) and target population
		Animal pharmacology (PK & PD)
		Safety pharmacology
		Toxicology
3.	Clinical Strategy	Outline of clinical studies (Refer Table 5.9)
		Timeframe for the study in Gantt Chart
		Go-No-Go Criteria based on the TPP
		Clinical supplies, project resources, manpower, and development costs
		Protocol synopsis for each study
4.	Strategic Planning	Go/no go decision points
		Risk assessment and contingency plan
5.	Regulatory considerations	Pre-IND meeting planning and IND filing
		Regulatory communications and meetings
		Regulatory trends and intelligence
	List of experts/thought leaders	

Source: *Evens R, editor. R&D planning and governance. Drug and biological development from molecule to product and beyond ISBN 978-0-387-69094-0, XI; 2007, pp. 33–65; Biostrategics Consulting Limited. The Clinical Development Plan Guide, http://www.biostrategics.com/pdf/The-Clinical-Development-Plan-Guide.pdf; 2011 [accessed 26.01.17].*

Table 5.9 Outline of Clinical Studies

Phase	Trial #	Objectives	Comparator	# of Subjects	Site (country)	Duration (months)	End points	Budget estimate

Source: *European vaccine initiative early phase clinical development plan template, www.euvaccine.eu/sites/default/.../2-EVI_EPCDP_template_120203.doc; [accessed 26.01.17].*

Table 5.10 Competencies Required for CDP

- Research design and statistics
- Clinical research and investigational drug development
- Clinical pharmacology—pharmacokinetics & pharmacodynamics
- Animal toxicology
- Clinical therapeutics
- FDA/EU/ICH guidelines/pharmaceutical regulations
- Project planning and scheduling
- Use of PK/PD modeling/simulations and computer-assisted trial design
- 'Right-sizing' trials and alternative statistical designs (e.g., futility analyses, adaptive designs)

5.17 THE CDP PROCESS

The clinical development plan follows a hypothetical drug label/Target Product Profile. The target counties of interest are then identified. Next step is to determine the type of clinical studies required in each phase to meet the criteria stated in the target product profile. The regulatory guidance/guidelines as well as the NDA Reviews—FDA (US) and EPARs (EU)—of similar products provide valuable information for deciding the type of studies. For indications, where no product has been approved till date and no regulatory guidance exists, inputs from experts/leaders in the field play an important role in making this decision. Once the types of clinical studies have been decided, the duration of each clinical study and their timings with regards to each other (sequential, overlapping, or concurrent) are worked upon along with the cost of each study. The information on study timelines helps in planning backwards from target approval date [17,18].

REFERENCES

[1] FDA. Guidance for industry and review staff target product profile—a strategic development process tool, <http://www.fda.gov/downloads/drugs/guidancecomplianceregulatoryinformation/guidances/ucm080593.pdf>; 2007 [accessed 26.01.17].

[2] Premier Research. Beginning with the end in mind: Using a Target Product Profile (TPP) to guide strategic medical device and diagnostic development, <http://265qkwxlfwp2ll1ls3vq8qdb.wpengine.netdna-cdn.com/wp-content/uploads/2015/02/Target-Product-Profile-White-Paper-Final.pdf>; 2015 [accessed 26.01.17].

[3] MaRS. Defining your target product profile: therapeutics, <https://www.marsdd.com/wp-content/uploads/2011/02/Defining-Your-Target-Product-Profile-Therapeutics-WorkbookGuide.pdf>; 2011 [accessed 26.01.17].

[4] XOMA. The target product profile: the target product profile: converting discoveries into companies, <http://qb3.org/sites/qb3.org/files/QB3Podcast20120316_2_0.pdf>; 2012 [accessed 26.01.17].

[5] Tebbey PW, Rink C. Target product profile: a renaissance for its definition and use. J Med Market 2009;9:301–7.

[6] FDA. Guidance for industry "Labeling for Human Prescription Drug and Biological Products – Implementing the PLR Content and Format Requirements", <http://www.fda.gov/downloads/drugs/guidances/ucm075082.pdf>; 2013 [accessed 26.01.17].

[7] Aurentz K, Thunecke M. Revitalizing portfolio decision-making at Merck Serono S.A. – Geneva. J Commer Biotechnol 2010;17:24–36.

[8] ICH Quality Guideline Q8. Pharmaceutical development, <http://www.ich.org/fileadmin/Public_Web_Site/ICH_Products/Guidelines/Quality/Q8_R1/Step4/Q8_R2_Guideline.pdf>; 2009 [accessed 26.01.17].

[9] Maguire & Peng. How to Identify Critical Quality Attributes and Critical Process Parameters, <http://pqri.org/wp-content/uploads/2015/10/01-How-to-identify-CQA-CPP-CMA-Final.pdf>; 2015 [accessed 26.01.17].

[10] ICH Efficacy Guideline E9. Statistical principles for clinical trials, <http://www.ich.org/fileadmin/Public_Web_Site/ICH_Products/Guidelines/Efficacy/E9/Step4/E9_Guideline.pdf>; 1998 [accessed 26.01.17].

[11] ICH Efficacy Guideline E8. General considerations for clinical trials, <http://www.ich.org/fileadmin/Public_Web_Site/ICH_Products/Guidelines/Efficacy/E8/Step4/E8_Guideline.pdf>; 1997 [accessed 26.01.17].

[12] ICH Efficacy Guideline E10. Choice of control group and related issues in clinical trials, <http://www.ich.org/fileadmin/Public_Web_Site/ICH_Products/Guidelines/Efficacy/E10/Step4/E10_Guideline.pdf>; 2000 [accessed 26.01.17].

[13] EMA. Points to consider on switching between superiority and non-inferiority, <http://www.ema.europa.eu/docs/en_GB/document_library/Scientific_guideline/2009/09/WC500003658.pdf>; 2000 [accesses 26.01.17].

[14] FDA. Guidance for industry non-inferiority clinical trials to establish effectiveness, <http://www.fda.gov/downloads/drugs/guidancecomplianceregulatoryinformation/guidances/ucm202140.pdf>; 2016 [accessed 26.01.17].

[15] EMA. Guideline on the choice of the non-inferiority margin, <http://www.ema.europa.eu/docs/en_GB/document_library/Scientific_guideline/2009/09/WC500003636.pdf>; 2005 [accessed 26.01.17].

[16] FDA. The US FDA guidance for industry - adaptive design clinical trials for drugs and biologics, <http://www.fda.gov/downloads/drugs/guidances/ucm201790.pdf>; 2010 [accessed 26.01.17].

[17] Evens, R, editor. R&D Planning and governance. Drug and biological development from molecule to product and beyond ISBN 978-0-387-69094-0, XI; 2007, pp. 33–65.

[18] Biostrategics Consulting Limited. The clinical development plan guide, <http://www.biostrategics.com/pdf/The-Clinical-Development-Plan-Guide.pdf>; 2011 [accessed 26.01.17].

[19] European vaccine initiative early phase clinical development plan template, <www.euvaccine.eu/sites/default/.../2-EVI_EPCDP_template_120203.doc>; [accessed 26.01.17].

CHAPTER

CLINICAL PHARMACOKINETICS AND DRUG INTERACTIONS

Nilanjan Saha
Jamia Hamdard University, New Delhi, India

The basic principle of modern pharmacology relates drug effects to its concentration at the receptor site. However, receptor sites are generally inaccessible for practical purposes and drug concentrations in biological fluids that can be easily collected are used to correlate with drug effect and for modeling. Typically, plasma drug concentrations are most frequently measured and are used as surrogate to the drug concentrations at the receptor site. Apart from this, plasma concentration maintains equilibrium with drug concentration in other tissues of the body. As a result, plasma is considered as the central compartment, while other tissues are understood to be peripheral compartments (Figs. 6.1–6.3).

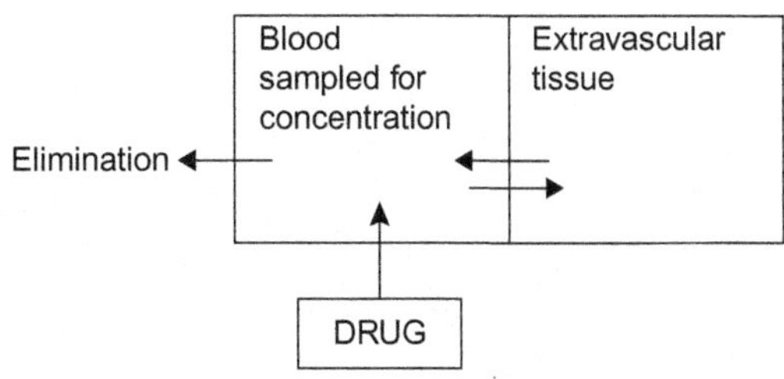

FIGURE 6.1

Drug handling by the body. Arrows depict movement of drug.

Pharmaceutical Medicine and Translational Clinical Research. DOI: http://dx.doi.org/10.1016/B978-0-12-802103-3.00006-7

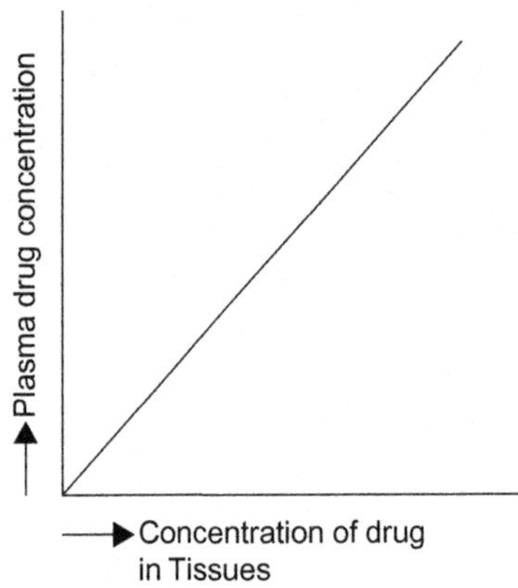

FIGURE 6.2

Relationship between plasma and tissue concentration are directly proportional.

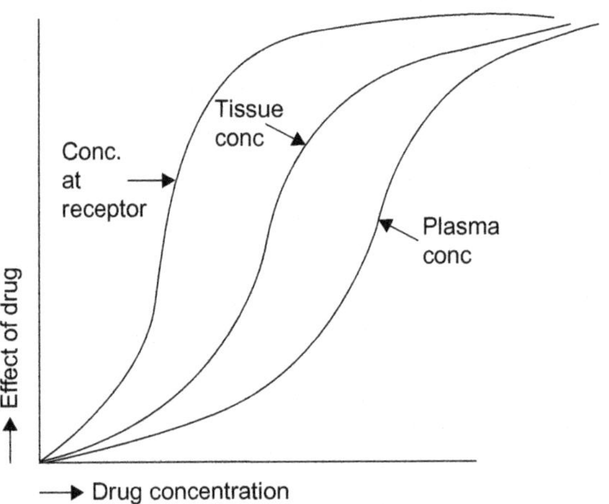

FIGURE 6.3

Drug concentration in various tissues and relation to drug effects.

6.1 PHYSIOLOGICAL PROCESSES

6.1.1 ABSORPTION

Absorption is the process of drug molecules reaching the systemic circulation from the site of administration. Drug substances can be administered through the gastrointestinal tract, lung, skin, mucus membrane in oral cavity, nasal cavity, or vagina, and the subcutaneous or muscular tissue, from where they may be absorbed. The gastrointestinal tract is the most common site of drug absorption and the process of absorption is affected by the physico-chemical characteristics of the drug, gastrointestinal motility, and drug concentration at the site of absorption. PK parameters like bioavailability, the maximum concentration achieved in serum/plasma (C_{max}), the time to attain maximum concentration (T_{max}) and the absorption rate constant (K_a) are dependent on the rate and extent of absorption. The process of absorption can be bypassed if an aqueous solution of the drug is introduced into the blood stream.

6.1.2 DISTRIBUTION

Distribution describes the reversible transfer of drug from the systemic circulation to the extravascular tissues within the body. The distribution of a drug is influenced by lipid solubility of the drug, the concentration in plasma and extravascular tissues, and binding to plasma and transport proteins and tissues. Distribution ensures that the drug reaches the site of action in adequate amount to produce effect. However, the nature of equilibrium attained between plasma and the extravascular tissues determines whether effective concentrations of the drugs are achieved at the receptor site or site of action of the drug.

A drug used in the central nervous system disorder must reach the neuronal tissue. For this the drug must penetrate the blood-brain barrier. Similar conceptual barriers may exist as placenta and blood-testicular barrier. Volume of distribution is the parameter that exemplifies this process.

6.1.3 METABOLISM [1]

Metabolism is the process of irreversible transformation of administered drugs into smaller molecules or metabolites. Most drugs are metabolized in the liver. Other organs like lungs, skin, gastrointestinal tract mucosa, blood, etc. may metabolize drugs. Metabolism serves as the first defense mechanism against harmful chemicals or xenobiotics. The objective of this process is to ensure conversion of drug molecules into more hydrophilic, inactive compounds. Half-life ($t_{1/2}$) reflects the rate of decay of the concentration of the drug in the body and metabolism may contribute to the process.

Metabolism in the liver occurs in two stages: Phase I pathways in liver microsomes where the drug is altered or broken down into intermediate metabolites, and Phase II reaction pathways in the liver cells where the drug or the intermediate metabolites are conjugated. Most of the Phase I reactions occur in the liver microsomes catalyzed by a group of enzymes called the cytochrome P450 system. This enzyme system contributes significantly to drug metabolism and commonly catalyzes chemical reaction pathways like aromatic hydroxylation, aliphatic hydroxylation, oxidative *N*-dealkylation, oxidative *O*-dealkylation, oxidation, reduction, hydrolysis, etc. These reactions may be adequate to make a drug hydrophilic, facilitating elimination through the kidneys.

Conjugation in Phase II reactions involves glucuronidation, sulfation, amino acid conjugation, acetylation, methylations, or glutathione conjugation to further facilitate elimination. Factors that influence drug metabolism including route of administration, dose, genetic makeup, concomitant diseases, and metabolic activity.

6.1.3.1 Cytochrome P450 system

Cytochrome P450 (CYP) enzymes are a group of enzymes encoded by P450 genes and are expressed as membrane bound proteins mostly found in the endoplasmic reticulum of the liver.

CYP enzymes function as monoxygenases and effect oxidation by transfer of one oxygen atom through a number of steps.

The CYP enzymes are grouped into families, designated by a digit after CYP, and assuring 40% homology in amino acid sequence of the enzymes; and sub-families, designated by an alphabet after the digit following CYP and characterizing 60% homology in the amino acid sequence of the enzymes. The last number in the name of an enzyme represents the sequence of discovery. Thus CYP3A4 is a cytochrome P450 enzyme belonging to the family 3 and sub-family A. Several CYP drug-metabolizing enzymes are controlled by more than one variant gene. This is called polymorphism and can explain the difference in efficacy, toxicity, and metabolism between individuals.

Of the 30 CYP enzymes, only 6 have a major role in drug metabolism. These are CYP1A2, CYP2C9, CYP2C19, CYP2D6, CYP2E1, and CYP3A4. Some of the drugs metabolized by these CYP enzymes are presented in Table 6.1 and representative drugs which can induce or inhibit the functions of these CYP are listed in Table 6.2. Inducers increase the activity of CYP and accelerate the metabolism of the drugs handled by the respective enzymes leading to reduced effectiveness. On the contrary, inhibitors of CYP reduce the metabolism of the drugs and may lead to increased efficacy or toxicity.

Table 6.1 Some Common CYP450 Substrate Drugs [2]

Enzyme	Substrate Drugs
CYP1A2	Amitriptyline, caffeine, clozapine, haloperidol, olanzapine, ondansetron, theophylline
CYP2C9	Amitriptyline, diclofenac, fluoxetine, ibuprofen, losartan, phenytoin
CYP2C19	Amitriptyline, citalopram, diazepam, omeprazole
CYP2D6	Amitriptyline, codeine, clozapine, fluoxetine, haloperidol, metoclopramide, metoprolol, olanzapine, ondansetron, risperidone, sertraline
CYP2E1	Paracetamol, caffeine, chlorzoxazone, dextromethorphan, ethanol, theophylline
CYP3A4	Alprazolam, amitriptyline, bupropion, caffeine, carbamazepine, clarithromycin, codeine, cyclosporine, dextromethorphan, dexamethasone, diazepam, donepezil, erythromycin, estradiol, fluoxetine, lansoprazole, loratadine, midazolam, nifedipine, omeprazole, progesterone, rifampicin, sertraline, sildenafil, tacrolimus, testosterone, theophylline

Table 6.2 Some Known CYP450 Inducer and Inhibitor Drugs [2]

Enzyme	Inducers	Inhibitors
CYP1A2	Cigarette smoke, charbroiled foods, carbamazepine, phenobarbitone	Ciprofloxacin, fluoxetine, nefazodone
CYP2C9	Rifampicin, carbamazepine, ethanol, phenytoin	Fluoxetine, fluconazole, ritonavir, sulphamethoxazole
CYP2C19	Rifampicin	Fluoxetine, ritonavir, ticlopidine
CYP2D6	Pregnancy	Cimetidine, fluoxetine, haloperidol, ritonavir, sertraline, ticlopidine
CYP2E1	Ethanol, isoniazid	Cimetidine
CYP3A4	Carbamazepine, rifampicin, phenobarbitone, phenytoin	Clarithromycin, erythromycin, grapefruit juice, fluoxetineKetoconazole, ritonavir

6.1.3.2 Factors affecting drug metabolism

Hereditary of genetic differences among human beings or groups of people can lead to excessive and/or prolonged or reduced effect of the desirable or undesirable effect of a drug. This is true in cases of therapeutic use as well as toxic overdosage with drugs.

Genetic polymorphism of CYP enzymes may produce the following:

1. Extensive metabolizers—designated as normal population.
2. Poor metabolizers—people who inherit inactive genes and have deficient enzyme activity.
3. Ultra-extensive Metabolisers—people who have amplified gene expression and demonstrate enhanced enzyme expression.

CYP2D6 and CYP2C19 demonstrate genetic polymorphism and lead to variable effects of drugs metabolized by these enzymes.

Environmental factors, namely diet, alcohol, smoking, pollutants, and concomitant drug use, may also affect drug metabolism. Grapefruit juice as a dietary component may affect drug metabolism by inhibiting CYP3A4. Several other ingredients of neutraceuticals, e.g., ginko biloba and St John's wort, have been implicated in alterations of drug metabolism, and possibilities of drug interactions must be evaluated.

6.1.4 EXCRETION

Excretion is the process of elimination of the drug and/or metabolites from the body. Most of the drugs in the body are eliminated in the urine. Substances with low lipid solubility such as polar metabolites are excreted efficiently. Other tissues which could eliminate drugs are lungs, skin, gastrointestinal tract, etc.

Alveoli in the lungs have a large surface area for absorption and elimination of inhalational anesthetics. Heavy metals are excreted by the skin and its appendages while many drugs are eliminated in the feces. Drugs excreted in the mother's milk may be instrumental in unwanted ingestion of drugs by a newborn.

Clearance (CL) is the measure of the rate of elimination of a drug.

6.2 PK PROFILE AND PK PARAMETERS

Intravenous administration of drugs is followed by rapid distribution to the extravascular tissues without any absorption process in action. For other routes of administration, the absorption process has to be considered. As the drug reaches systemic circulation during the absorption process, distribution, metabolism, and excretion are initiated and continue till all the drug is eliminated from the body. Drug concentration will be seen to increase in biological samples drawn from the systemic circulation when the amount of drug absorbed exceeds the amount of drug that is distributed into the extravascular tissues and the drug that is metabolized and/or excreted during this period. On the contrary, the plasma drug concentration is on the decline when the amount of drug that is distributed, metabolized, and/or excreted exceeds the amount of drug that is absorbed.

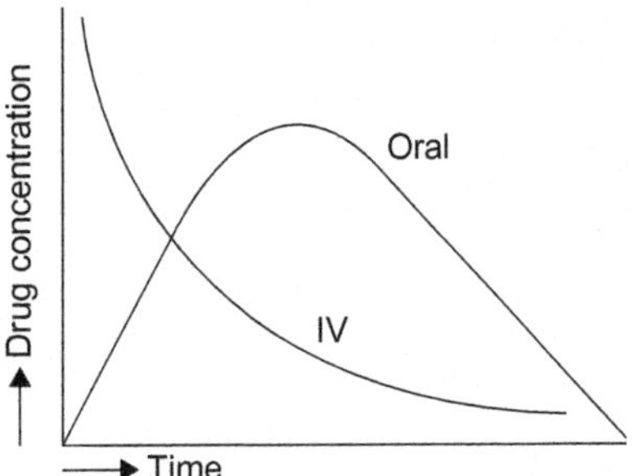

FIGURE 6.4

Reproduction of PK profile for intravenous and oral doses.

It is noteworthy that though distribution is a reversible process, absorption and elimination (metabolism and/or excretion) are irreversible processes. Each of the processes can be described by a rate constant and the rates of the processes determine the concentration of a drug at any point in time. The plot of drug concentration in the plasma versus time is represented as the pharmacokinetic profile (Fig. 6.4).

The processes involved in the elimination of drugs from the body follow first order kinetics, i.e., a fixed proportion of drug present in the body are eliminated per unit time. This implies that the elimination rate increases while elimination rate decreases when drug concentration decreases. Few drugs follow zero order elimination when only a constant amount of drug is eliminated per unit time, for example ethanol and aspirin at high concentrations.

6.2.1 PK PROFILE: EXTRAPOLATION OF PLASMA CONCENTRATIONS [3]

If Cp is the plasma concentration, "t" stands for time and "k" represents the constant for first order elimination for a drug administered intravenously as bolus, then the relationship described can be written as:

$$\mathrm{d}Cp/\mathrm{d}t = -k \cdot Cp \tag{6.1}$$

Integration of this Eq. (6.1) gives

$$Cp_{(t)} = Cp_{(0)} \cdot e^{-kt} \tag{6.2}$$

When $Cp_{(t)}$ is the plasma concentration at time t and $Cp_{(0)}$ is the initial plasma concentration at time "zero." Very often $Cp_{(0)}$ cannot be measured as the assumption is that the intravenous administration as bolus and subsequent distribution process happens instantaneously at time 0 is hypothetical. However, a plasma concentration measured immediately after completion of the bolus injection, assuming that the distribution process is rapid, is considered for calculations and $Cp_{(0)}$ can be obtained from the extrapolation:

$$\log Cp_{(t)} = \log Cp_{(0)} - kt/2.303 \tag{6.3}$$

A semi-logarithmic plot of log $Cp_{(t)}$ versus time will be linear. The slope of plot will correspond to the elimination rate constant.

6.2.2 PHARMACOKINETIC PARAMETERS [3–5]

Mathematical modeling of drug concentrations obtained in biological matrix obtained at different intervals after drug administration results in the generation of PK parameters.

6.2.2.1 Area under the curve (AUC)

Estimation of AUC may be determined by various methods. AUC is commonly estimated by the trapezoidal rule where the area between the curve and the axis is considered as series of smaller trapezoid and the areas of all the trapezoids are added to obtain the total AUC.

$AUC_{0->t}$ represents the area under the plot of drug concentrations versus time curve, from time of drug administration (time "0") to time "t." In most cases "t" is understood to be the last experimental point when a biological sample had been collected and evaluated. Mathematically, this may be expressed as:

$$AUC_{0->t} = \int_0^t c\delta t \quad (6.4)$$

AUC $_{0->\alpha}$ represents the area under the plot of drug concentration versus time curve from time 0 (as above) till the time the concentration becomes zero. This involves calculation of the area of the triangle whose base is represented by the last measured concentration at time "t" and the apex of the triangle is the extrapolated point where the curve meets the time axis. Thus AUC $_{0->\alpha}$ can be expressed as follows:

$$AUC_{0->\alpha} = \int_0^\infty c\delta t = AUC_{0->t} + AUC_{t->\alpha}\ (\text{area of } \Delta) \quad (6.5)$$

6.2.2.1.1 Practical insights

1. The area of the triangle is dependent on how well the slope of the curve has been defined and how close to linearity is the curve. It is important to examine the number of points used to calculate the slope, and the correlation coefficient of the points used to calculate the slope, which is designated as k_{el}. In an ideal case, the part of the curve utilized for extrapolation is almost linear with the coefficient of correlation approaching unity.
2. AUC represents the amount of drug exposure that is related to the action of the drug and is important for the patient. Thus, AUC $_{0->t}$ and AUC $_{0->\alpha}$ are part of the bioequivalence criteria for generic products. Most regulatory authorities expect both these parameters to be calculated for the newly developed formulation (test or generic) and innovator (reference) formulations. The 90% Confidence Intervals of the ratio of test to reference formulations need to be calculated.
3. AUC $_{0->t}$ is also used to evaluate relative bioavailability of a test formulation as compared to an innovator or reference formulation. The reference formulation could be intravenous administration of a New Chemical Entity (NCE) and the AUC $_{0->t}$ of an oral formulation as compared to the intravenous administration, which indicates the absolute bioavailability of the NCE. Any altered release of an innovative formulation of a drug may be compared against the immediate release or the already marketed formulation used in therapy. The AUC $_{0->t}$ could be compared to evaluate the acceptability of the new formulation.
4. $AUC_{0\to T}$ represents the area under the plot of the drug concentration versus the time curve from time "0" to the end of the dosing interval. This is usually calculated and reported for long

acting drugs. For example, following administration of a single dose of amlodipine, the biological samples may be collected up to 72 hours for adequate PK characterization and AUC $_{0->72}$ can be calculated. However, for amlodipine, which is administered once a day for its antihypertensive action, AUC $_{0->24}$ is clinically relevant for making patient related decisions.

5. AUC calculations show considerable variability from one experimental setting to another. Thus AUC values should be compared when data is generated as part of the same protocol.

6.2.2.2 Area under the first moment curve (AUMC)

AUMC is the total area under the first moment curve. The first moment curve is prepared when concentration x time is plotted versus time. AUMC can be mathematically expressed as:

$$\mathrm{AUMC} = \int_0^{\infty} \mathrm{c}\cdot t\delta t \tag{6.6}$$

Knowledge about AUC and AUMC allows further calculation and analysis of drug characteristics. Mean residence time (MRT) which is the average time the drug stays in the body as measured in the plasma and can be calculated by the formula:

$$\mathrm{MRT} = \frac{\mathrm{AUMC}}{\mathrm{AUC}} \tag{6.7}$$

In addition, the Apparent Elimination Rate Constant (kel′) and Apparent Volume of Distribution at Steady State (Vss) can be calculated by the following equations:

$$\mathrm{kel}' = \frac{1}{\mathrm{MRT}};\ \mathrm{Vss} = \frac{\mathrm{Dose}}{\mathrm{AUC}}\cdot\frac{\mathrm{AUMC}}{\mathrm{AUC}} \tag{6.8}$$

6.2.2.3 Half-life

Half-life ($t_{1/2}$) of a drug is the time required for the concentration of the drug to be reduced by half. Putting $Cp = ½\ C_0$ in Eq. (6.2),

$$t_{1/2} = \frac{0.693}{k} \tag{6.9}$$

6.2.2.3.1 Practical insights

1. Half-life is a quantitative parameter and can be used to predict the duration for which the drug will be available in the body after administration. It also provides an estimate of how early or late the drug could be eliminated from the body.
2. Half-life is used to determine the dosing interval of a drug. It is one of the parameters that are considered to determine the dosing interval of a drug. A common practice is to attempt to administer the drug at intervals more or less equal to one half-life.
3. The half-life of a drug is affected by the experimental conditions existing during the PK profiling. Factors like frequency of collection of biological samples in the elimination phase of the drug and the sensitivity of the bioanalytical assay are important determinants of how close the calculated half-life is to the real half-life.

Table 6.3 Amount of Drug Eliminated After Passage of Multiple Half-lives

No.of Half-lives	% of Drug Remaining in the Body	Cumulative % of Drug Eliminated
1	50	50
2	25	75
3	12.5	87.5
4	6.25	93.75
5	3.125	96.875
7	0.78125	99.21875
10	~0.098	>99.9

a. Typically during the drug development path, the sensitivity of the assay may change and the half-life of a drug may not be exactly replicated (Table 6.3).

6.2.2.4 Volume of distribution

Following intravenous administration of a drug, the initial plasma concentration C_0 is calculated by extrapolation backwards, the apparent volume of distribution can be calculated by the following equation:

$$V_D = \frac{\text{IV Dose}}{C_0} \tag{6.10}$$

where C_0 is the initial plasma concentration (calculated).

6.2.2.4.1 Practical insights

1. Physiologically, drugs are unevenly distributed amongst various tissues in the body, even after equilibrium is attained. The apparent V_D provides an estimate of volume of distribution, assuming that in the tissues where the drug reaches it is the same concentration as in the blood at the time of intravenous administration.
2. V_D may or may not correspond to any physiological volume of the body.
3. V_D is related to the protein binding capacity of the drug. When the drug has high plasma protein binding, the V_D is small and concentration of the drug in the blood is high. The plasma protein bound drug acts as a reservoir and continuously equilibrates with the unbound drug in the plasma.
4. Basic drugs have the propensity to bind to extra-vascular tissue proteins; the plasma concentrations of such drugs are low and the V_D high. To achieve the therapeutic concentrations in the blood, administration of a loading dose is considered when early effects are required, e.g., chloroquine, digoxin.
5. V_D is independent of drug concentration (C_0 is an extrapolated value) and thus is a characteristic of the drug.

6. V_D is a hypothetical volume and not a true anatomical or physiological compartment. However, some basic assumptions may be drawn based on knowledge of certain details of the human body. For a 60 kg man, the actual body fluids/structures are as follows:

Body Fluid/Structure	Actual Volume (L)
Blood	7
Plasma	4
Whole body	42

Drugs that have a distribution volume of 4 L or less are thought to be confined to plasma. If the volume is between 4 and 7 L, the drug may be distributed in blood (plasma + erythrocytes). If the V_D is larger than 42 L, the drug may be distributed to all tissues in the body.

6.2.2.5 Clearance

Clearance of a drug may be described as its metabolism and/or excretion from the body by concerned organs or organs systems. The most common organs involved are the liver and the kidneys. Clearance is expressed as the volume of blood which is cleared of the drug per unit of time.

Thus, if C_a represents the concentration of the drug in the artery supplying the organ of clearance, C_v is the venous drug concentration from the same organ, and Q is the blood flow rate to the organ, then:

$$\text{Extraction Ratio} = \frac{C_a - C_v}{C_a} \tag{6.11}$$

$$\text{Clearance (CL)} = Q \times \frac{C_a - C_v}{C_a} \tag{6.12}$$

$$\begin{aligned}\text{Alternatively, clearance (CL) (mL/min)} &= \frac{[L^3]}{[T]} = \frac{[M][L^3]}{[M][T]} = \frac{[M]}{[M][L^{-3}][T]} = \frac{[M]}{[ML^{-3}][T]} \\ &= \frac{\text{Dose}}{\text{Conc} \times \text{Time}} = \frac{\text{Dose}}{\text{AUC}}\end{aligned} \tag{6.13}$$

6.2.2.5.1 Practical insights

1. Organ clearance affected by the kidney may be exponentially evaluated using this formula after intravenous administration of a drug. Following an oral dose, clearance cannot be estimated using the expression Dose/AUC, as the dose administered may not be absorbed completely to reach the systemic circulation.
2. The total body clearance is the sum of individual organ clearances which are involved in elimination of the drug. For drugs which are completely metabolized, renal clearance may be considered as negligible.
3. Hepatic clearance can be predicted or extrapolated from in vitro metabolism data.

6.2.2.6 Bioavailability

Bioavailability (F) of a drug is the rate and extent of absorption of a drug into the body. Thus the fraction of the dose of a drug that reaches the systemic circulation is the bioavailability of the drug.

The bioavailability of an intravenously administered drug is 100% as no absorption process is involved. Following all extravascular administration of the drug including oral administration, absorption occurs and bioavailability is expressed as:

$$F = \frac{\text{AUC } (E_v)}{\text{Dose } (E_v)} \times \frac{\text{Dose (IV)}}{\text{AUC (IV)}} \tag{6.14}$$

Where "IV" represents parameters after intravenous administration and "E_v" represents observations after administration by any other route.

6.2.2.6.1 Practical insights

1. On repetitive administration of a drug, the bioavailability of each dose is important to ensure adequate concentration for effectiveness of the drug. Thus, for treatment of chronic conditions, the extent of absorption is more critical than the rate of absorption.
2. The rate of absorption is important for drugs used for relief of acute symptoms, such as nausea and pain, and the therapy is to be designed to deliver effective concentration of the drug as early as possible.
3. As already discussed, absorption and elimination processes frequently occur simultaneously. In general, absorption is a more rapid process as compared to elimination, though exception may exist. However, if the absorption rate is slow, then the plasma concentration of drugs is a function of both rate and elimination rate. In such situations, the relationship for an orally administered drug may be defined as:

$$C_p = F \times \frac{D}{V_d} \times \frac{K_a\,(e^{-K_t} - e^{-K_{at}})}{(K_a - K)} \tag{6.15}$$

   Where F is the bioavailability of the drug, D is the dose of the drug, V_d is the apparent volume of distribution, K_a is the absorption rate constant, and K is the elimination rate constant—both of first order.

   This relationship assumes the whole body as a single unit and the drug is distributed within the body at the same concentration as in the plasma. The derivation of the relationship is not in the scope of this discussion and the reader is referred to the article by Dhillon & Gill.
4. Several PK software packages are commercially available which calculate the rate constants using the concentration time data. When $K_a \gg K$, then:

$$C_p = F \times \frac{D}{V_d} \times \frac{K_a \times e^{-K_t}}{(K_a - K)} \tag{6.16}$$

   as $e^{-K_{at}}$ is negligible. The description of various symbols remains the same as described in Eq. (6.14).

6.2.2.7 Two compartment model

Following administration of a drug, either intravenously, when no absorption process is involved, or by any other route, the drug reaches systemic circulation after absorption and the distribution phase initiates almost immediately. During this distributive phase, the drug concentrations in the plasma may decline faster as compared to the phase after equilibrium is achieved and the concentration achieved in certain tissues may be more than that in the plasma. This indicates that some tissue is acting as a reservoir of the drug and as the drug is eliminated from the central compartments or plasma, a new equilibrium is established with drug being released from the tissues. In such cases, data cannot be expressed by simple exponential relationships. Complex mathematical models need to be utilized to determine PK relationships. PK data analysis using complex mathematical tools is called Compartmental Pharmacokinetics. The body is envisaged as consisting of a number of compartments, though there may not be any anatomical meaning of each compartment. The rate of transfer of drugs between compartments is assumed to follow first order kinetics Fig. 6.5.

Plasma concentration in this case can be expressed by the equation:

$$C_p = Ae^{-\alpha t} + Be^{-\beta t} \tag{6.16}$$

where α and β are the slopes of the two parts of the concentration time curve and A and B are the intercepts obtained by extrapolation of the linear parts of the curve (see figure 5). The inter-relationship between the rate constants and these mathematical expressions are as follows:

$$K_{21} = \frac{A\beta + B\alpha}{A + B} \tag{6.17}$$

$$K_{12} = \alpha + \beta - K_{21} - \text{Kel} \tag{6.18}$$

$$\text{Kel} = \frac{\alpha\beta}{K_{21}} \tag{6.19}$$

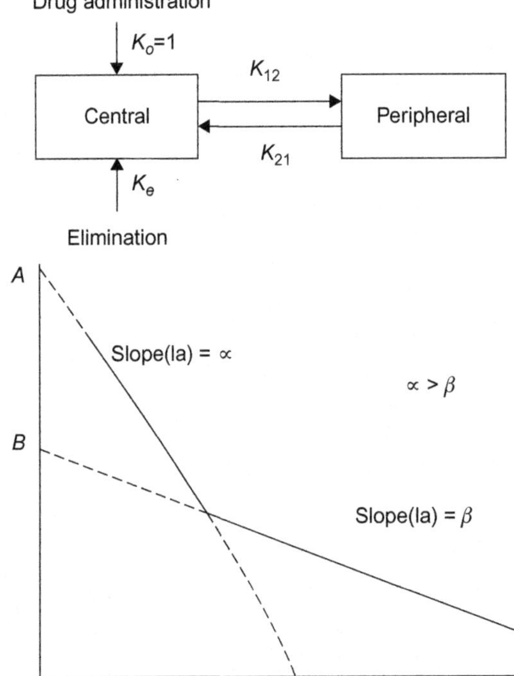

FIGURE 6.5

Intravenous administration of the drug that fits in the two compartment model.

The derivations of these equations are beyond the scope of this discussion. However, using these constants, other PK parameters can also be expressed as:

$$\mathrm{AUC} = \frac{A}{\alpha} + \frac{B}{\beta} \tag{6.20}$$

$$\mathrm{CL} = \frac{\text{Dose IV}}{\mathrm{AUC}} = \frac{\text{Dose IV}}{\frac{A}{\alpha} + \frac{B}{\beta}} \tag{6.21}$$

$$\mathrm{AUMC} = \frac{A}{\alpha^2} + \frac{B}{\beta^2} \tag{6.22}$$

6.2.2.8 Noncompartmental analysis

Without fitting PK data into any specific compartment model, PK parameters can be determined, assuming the drug is distributed uniformly in the body (whole body as one compartment) and following linear kinetics. This analysis is based on the theory of Statistical Moments where AUC under the concentration-time curve is called zero moment, while the area under the concentration-time product and time curve is considered the first moment (AUMC). See discussion on AUC, AUMC, and MRT in Section 3.2.2. Noncompartmental analysis does not adequately characterize nonlinear kinetics.

6.2.2.9 Enzyme kinetics

Systemic clearance of a drug that is metabolized and eliminated from the body is dependent on the hepatic blood flow and on the "intrinsic clearance" by the liver. Intrinsic clearance is solely a measure of enzyme activity in the liver and the rate of metabolism of a drug is proportional to it and the concentration of the drug at the site.

Rate of metabolism is usually defined by the Michaelis-Menten enzymes kinetics relationship and can be expressed as:

$$V_0 = \frac{\mathrm{V}_{\max} \cdot C_s}{K_m + C_s} \tag{6.23}$$

where V_0 is the rate of metabolism, C_s is the drug concentration at the enzyme site, and K_m is the Michaelis-Menten constant. K_m also represents the substrate concentration at half the maximum velocity of the reaction $V_{\max}$.

Under linear conditions, when C_s is much smaller than K_m, $\mathrm{CL_{internal}} = V_0/C_s = V_{\max}/K_m$, thus $\mathrm{CL_{internal}}$ is independent of drug concentration.

6.2.2.9.1 Practical insights

1. Disease states which affect the blood flow to the liver and co-administration of drugs which induce or inhibit CYP enzymes influence the systemic clearance of drugs by the liver.
2. Using scaling techniques, through the experimentally derived in vitro $\mathrm{CL_{internal}}$ using Michaelis-Menten constants, the in vivo clearance can be predicted. The scaling factors that need to be considered are related to miscrosomes and hepatocyte numbers per unit weight of liver tissue, weight of liver, and blood flow to the liver.

6.3 PHARMACOKINETICS IN SPECIAL POPULATION & SPECIAL SITUATIONS

6.3.1 PHARMACOKINETICS AT EXTREMES OF AGE

6.3.1.1 Children [6]

Children as compared to adults have reduced gastric acid secretion. This may increase the bioavailability of acid-labile drugs like penicillin and decreases bioavailability of weakly acidic drugs (like phenobarbitone). There is also reduced bile salt formation in children, which decreases bioavailability of lipophilic drugs like diazepam. Reduced gastric emptying and intestinal motility has also been documented in infants of less than 3 months and these increase the time a drug may take to reach therapeutic concentrations. Drug-metabolizing enzymes present in the intestines of young infants are another cause of reduced drug absorption.

Pain and tissue damage precludes intra-muscular injections in children and are generally avoided, but, when needed, water-soluble drugs in a small volume of injection may be considered. Transdermal absorption is enhanced in neonates and young infants, as the skin is thin. Skin lesions like ulcers and burns increase absorption at any age. Transrectal drug administration is considered only as an alternative to the intravenous route when venous access is not available. Thus rectal diazepam may be administered for treatment of status epilepticus. The venous drainage of the rectum differs in the upper two-thirds (portal circulation) and lower one-third (systemic circulation) and thus the placement of the transrectal formulation is crucial.

Decreased protein binding in neonates is observed due to competitive binding by bilirubin and free fatty acids, which are present in greater concentrations in neonates and infants. This results in increased free drug concentrations and greater drug availability at receptor sites, leading to increased efficacy as well as a higher incidence of adverse effects. Because a higher percentage of body weight in younger children is water, higher doses per kg of body weight are required to be administered for water-soluble drugs.

Drug metabolism and elimination mature with age and most drugs have plasma half-lives 2 to 3 times longer in neonates than in adults. Phase I CYP enzyme activity is low in neonates, but progressively increases to attain adult rates by late puberty. However, for some drugs like barbiturates and phenytoin, adult rates may be attained within 2–4 weeks postnatally. Phase II reaction rates vary considerably by substrate.

Renal blood flow is low at birth and reaches adult levels by 1 year of age. Similarly, GFR is low at birth and increases progressively to adult levels by 6 months of age. Thus all of the factors that affect renal excretion of drugs mature in the first 2 years of life.

6.3.1.2 Elderly [7]

With aging, there are changes in multiple body functions and body composition; however, only some changes are clinically relevant. There is slowed gastric emptying and an increase in gastric pH along with decrease in small-bowel surface area, but these changes tend to have no bearing on absorption of drugs from the gastrointestinal tract. The metabolism and excretion of many drugs decrease, thus requiring the doses of some drugs to be adjusted.

With age, body fat increases, whilst muscle mass and total body water decrease. Increased fat increases the volume of distribution for highly lipophilic drugs like diazepam and increase their elimination half-lives.

CYP enzyme system-mediated hepatic metabolism of drugs decreases with age, in many cases clearance typically decreases 30%–40% in Phase I reactions. Drug maintenance doses may be required to be decreased, though individual dose adjustment is required. First-pass metabolism is also affected by aging, decreasing proportionately with increasing age after 40 years. Thus, for a given oral dose, the elderly may have higher circulating levels for drugs like nitrates, nifedipine, etc.

The other most important age-related change is decreased renal elimination of drugs, though the decrease varies significantly from person to person. Serum creatinine levels remain within normal limits despite a decrease in glomerular filtration rate (GFR) as the elderly produce less creatinine. Decreases in tubular function also parallel the decrease in glomerular function with age. Creatinine clearance is used to guide drug dosing and drugs predominantly excreted by the kidneys should be administered either at lower daily doses or at less frequent intervals.

6.3.2 PHARMACOKINETICS AND PREGNANCY [8]

Pregnancy ushers in several physiologic changes in many organ systems. Some of these changes are secondary to hormonal changes in pregnancy, while others occur to support the anatomical and physiological changes in the mother and her developing fetus. These include increased maternal fat and total body water, decreased plasma protein concentrations, especially albumin by dilution, increased blood volume and cardiac output and decreased blood pressure, and increased blood flow to the kidneys. There is also delayed gastric emptying and gastrointestinal motility along with altered activity of hepatic drug metabolizing enzymes. However, many of these changes affect the PK drugs simultaneously in a manner which counter-balances each other's effect, resulting in a negligible net effect in most cases.

Clinically, increased body fat and total body water results in a larger volume of distribution, and a higher initial and maintenance dose of drugs may be required to be administered to attain therapeutic plasma concentrations. In contrast, decrease in serum albumin concentrations result in higher free levels of highly albumin bound drugs, leading to increased efficacy and toxicity. Midazolam and phenytoin are examples of medications primarily bound to albumin.

CYP enzyme activity increases in pregnancy, especially the Phase I reactions and some Phase II reactions, resulting in increased clearance of drugs, requiring dose adjustment.

Both renal blood flow and GFR increase in pregnancy and elimination rates of renally-excreted drugs increase, producing shorter half-lives. For example, the clearance of lithium, which used to treat bipolar disorder, is doubled during the third trimester of pregnancy compared with the non-pregnant state, leading to drug concentrations lower than that required for optimum effect.

6.3.3 PHARMACOKINETICS AND DISEASE CONDITIONS [9–13]

Edema is frequently observed in chronic heart failure, decompensated liver cirrhosis, and chronic renal failure, and in patients with hypoalbuminemia. Frusemide pharmacokinetics has been extensively researched, as it is a commonly used drug for symptomatic relief of edema. The rate of absorption of frusemide is reduced while the extent of absorption remains unchanged in patients with edema.

Edema causes an increase in the volume of distribution, resulting in lower concentrations of a drug. However, plasma protein binding is likely to be less than in the normal population, owing to hypoalbuminemia and the proportion of the unbound component is greater as compared to the bound component. Theophylline has higher bioavailability in patients with oedema, with a significantly higher Cmax in patients with hepatic cirrhosis and CHF than in healthy volunteers.

An altered volume of distribution of digoxin in thyroid dysfunction has been reported—it is increased in hyperthyroidism and decreased in hypothyroidism.

Liver disease can modify the kinetics of drugs metabolized in the liver. Research suggests that some CYP enzymes are more affected than others by liver disease. The capacity to metabolize drugs depends on hepatic blood flow and liver enzyme activity, both of which can be altered in liver disease. In addition, the collection of toxic metabolic intermediates of endogenous substances in liver failure can affect the binding of drugs to plasma proteins. Drugs having a significant first-pass effect have significantly altered PK when hepatic blood flow is abnormal, however, the PK of drugs with a low first-pass are more affected by hepatic failure rather than liver blood flow changes. Dosage adjustment is required for drugs with altered PK, but it is not easy to quantify the changes required in drug dosage and administration. At present there is no one satisfactory test that quantitatively assesses impaired liver function and the requirement to adjust drug dosages.

Increased hepatic metabolism in hyperthyroidism and reduced metabolism in hypothyroidism have been observed, leading to altered catabolism of vitamin K-dependent clotting factors. This results in an increased dose requirement of oral anticoagulants in hypothyroidism and a decreased dose in hyperthyroidism in the absence of any alteration in the metabolism of the anticoagulant drugs themselves. The antithyroid drug carbimazole has a shorter half-life in hyperthyroid patients than in normal controls, and explains the need for frequent doses while thyroid function is abnormal, and the maintenance dose is once daily when the disease is under control.

Drugs with a narrow therapeutic range that are eliminated by more than 80% metabolism in the liver should be avoided in patients with significant liver disease—e.g., morphine and theophylline. Other drugs with a wider therapeutic range but undergoing extensive hepatic metabolism must be used with increased dosing interval or lower total dose. Only if more than 90% of a drug is excreted unchanged in the urine is hepatic impairment unlikely to result in the accumulation of the drug or produce toxicity. Thus, ascertaining the relative contribution of liver and kidneys in elimination of a drug is key to appropriate use of the drug in impaired organ states. Renal excretion of the drug is a combination of three processes: glomerular filtration, tubular secretion, and tubular reabsorption and dosage adjustment based on GFR and other markers of renal function.

6.3.4 THERAPEUTIC DRUG MONITORING [14]

Therapeutic drug monitoring is the determination of drug concentrations in the biological matrix of patients, the interpretation of which is utilized to evolve drug therapy that is safe and efficacious. Thus, therapeutic drug monitoring (TDM) can result in rapid optimization of therapy which otherwise would require several dose changes using only clinical leads. In conjunction with monitoring of clinical effects produced by the drug, utilization of TDM leads to rapid optimization of drug therapy.

TDM is useful when the following criteria are met:

1. A good correlation exists between the pharmacologic response and plasma concentration. An increase in drug concentration in the biological matrix, for example plasma, is related to an increase in efficacy and/or toxicity related to the drug.
2. Drug concentrations cannot be predicted from a given dose, as result of inter-individual variability.
3. The drug has a narrow therapeutic index (i.e., the therapeutic concentration is close to the toxic concentration).
4. The pharmacological effects of the drug cannot be monitored easily (e.g., monitoring blood pressure for antihypertensives) or the adverse effects cannot be easily differentiated from lack of efficacy of a drug.

TDM is useful for antiepileptics like valproate sodium, carbamazepine, phenytoin, mood stabilizer lithium, antibacterial gentamicin, and antiasthma drug theophylline. However, TDM is a resource intensive procedure and must be used optimally in the following scenarios:

1. Assessment of patient compliance during chronic therapy.
2. To identify drug interaction when new concomitant medication(s) is/are added to a narrow therapeutic index drug.
3. Rapid optimization of therapy.
4. Sometimes for change of brands or to identify batch-to-batch variation in drugs with a very narrow therapeutic window.

TDM is planned with assessment of a minimum number of samples for estimation of drug concentration. The method of analysis must be robust, standardized, and validated, and less time consuming, to allow quick reporting of results. Samples are frequently collected before administration of the next dose of drug for analysis of trough concentration.

6.3.5 PHARMACOKINETICS OF HERBAL PREPARATIONS [15,16]

There is a popular belief that natural medicines are safer than synthetic drugs. This has led to an exponential increase in the use of herbal medicine products (HMP). However, reliable technical data on HMPs is not easily available. In most cases, effective and uniform regulatory system that ensures the safety and efficacy of HMPs is lacking.

In recent years, data on evaluation of efficacy and toxicity of some HMPs have become available. Advances in analytical technology have led to discovery of many new active constituents and have proposed new active ingredients. Thus, elucidating the pharmacological basis for the efficacy of HMPs is a constant challenge. The question of bioavailability of HMP to assess the systemic availability of these "active" compounds after administration has been of much interest and intense research.

Elucidation of metabolic pathways (yielding potentially new active compounds), and the assessment of elimination routes and their kinetics, are also critical as sufficient data may not be available. Availability of these data will provide an important link between bioanalytical assays, concentration of active constituents, and clinical effects. A better understanding of the pharmacokinetics and bioavailability of active constituents of HMPs can also help in rationalizing drug regimens.

Often, conflicting results of studies involving HMPs are reported. Large batch-to-batch variation in the composition and quality of HMPs may explain the conflicting results and gaps in understanding about these products. Most of these inconsistencies with HMPs, when compared with synthetic drugs, present additional challenges, which include:

1. Uniformly accepted methods to establish the quality of HMPs.
2. Methods to monitor batch-to-batch variation.
3. Environmental effects and contamination.
4. Counterfeit practices.
5. Seasonal and geographical variation.

Thus, often great variation is reported in patient response to the use of the same HMP. The results from many of the published studies therefore may be of little value, since the identity, purity, quality, strength, and composition of the supplements is not always confirmed.

6.4 PHARMACOKINETICS IN DRUG DEVELOPMENT

Candidate drugs fail during their development process owing to inadequate efficacy or increased general or organ toxicities. Compounds that achieve adequate concentrations in the systemic circulation, and at the site of action, for a reasonable duration of time that is relevant in the treatment of the concerned disease without producing unacceptable toxicity, have the potential to be converted successfully into drugs. Thus, early pharmacokinetic data on the development compounds can indicate whether the new chemical entity can be successfully developed into a drug.

6.4.1 PRECLINICAL PHARMACOKINETICS [17]

Comprehensive preclinical pharmacokinetic data ensures the selection of the right compounds as potential drug candidates. Both in vitro and in vivo test methodologies are used during the generation of preclinical data. While data is generated rapidly in in vitro systems, the tests systems in most cases do not represent the real physiological environment. Multiple developmental lead compounds can be subjected to in vivo screens to ascertain and confirm the absorption of the compounds across physiological barriers. Candidates that are absorbed into the systemic circulation and possess a favorable in vivo pharmacokinetic profile are further subjected to:

1. Plasma protein binding studies in rat and human plasma.
2. Metabolic stability using rat and human microsomes in vitro.
3. Potential of CYP-450 inhibition and induction using rat and human recombinant CYP-450 enzymes.

It is to be appreciated that the unbound drug is pharmacologically active and the assessment of the bound fraction by the estimation of plasma protein binding of a compound is another important parameter that is evaluated in vitro. Elucidation of the relationship between in vitro intrinsic clearance and in vivo clearance in multiple species is one of the major objectives of the preclinical studies. The relationship between clearance in rat and human liver microsomes and after intravenous administration in rats may be utilized to predict the clearance in humans. The metabolic stability profile provides both qualitative and quantitative comparison of metabolism of the compound in human and animal models in vitro. This data is used to select competing molecules for development. In vitro ADME studies also helps in identifying the right model for toxicity studies.

Extensive first-pass and/or systemic metabolism may reduce the systemic exposure and demonstrates a short half-life of a compound. Several strategies such as modification of structure to reduce lipophilicity and/or blocking of metabolic soft spots and use of enzyme inhibitors have been used to obtain more metabolically stable chemical entities. In contrast, several active metabolites of marketed drugs have been developed as drugs with a better efficacy, safety, and pharmacokinetics profile—the lead of these have been obtained through the process of metabolism. Thus the liability of metabolic instability has been exploited as a tool for discovering better drugs.

The identification of drug metabolizing enzymes involved in the major metabolic pathways of a compound helps in predicting the probable drug–drug interactions in humans. Compounds with more than one metabolic pathway have less likelihood of clinically significant drug interactions. In vitro CYP450 inhibition and induction screens are used to evaluate the potential of compound for drug-drug interactions and the most prone candidates may either be discarded or carried forward with caution.

Toxicity study data is major decisive element in deciding whether a new chemical entity will progress to clinical development. Toxicokinetics is an integral part of toxicity studies and is used to assess the exposure of the new chemical in animals in toxicity studies and correlate the drug levels in blood and various tissues with the toxicological findings.

Although in vitro assays and in vivo screening in animal models are utilized extensively, both approaches have their own limitations. Candidates selected on the basis of preclinical pharmacokinetic and toxicokinetic data may or may not exhibit the desired target PK profile in humans. The real clinical pharmacokinetic data is obtained only after the compound enters Phase-1 clinical trial. The recognition of human micro dosing (Phase 0 clinical trial) by several regulatory agencies may ensure that human PK data is obtained early in the preclinical stage.

6.4.2 PHASE 0 CLINICAL TRIALS

Traditional phase 1 clinical trial programs could be preceded by early human screening studies with sub-pharmacological single doses or microdoses of one or several lead candidates. A microdose is defined as less than 1/100th of the dose of a test substance calculated (based on animal data) to yield a pharmacologic effect of the test substance with a maximum dose of 100 micrograms. This allows human pharmacokinetic and pharmacodynamic data to be generated before the extensive regulatory preclinical studies are completed. Microdoses doses are not expected to have clinically significant toxic potential, so early human screening studies may be supported by limited nonclinical safety data. If appropriately designed, early generation of human pharmacokinetic data in Phase 0 clinical trials may lead to safer and more efficacious doses of novel drugs, reduce attrition in clinical trials, and facilitate more economical drug development.

6.4.3 PHARMACOKINETIC-PHARMACODYNAMIC (PK-PD) MODELING [18]

PK-PD modeling is a mathematical tool defining the pharmacokinetic parameters and pharmacodynamic effects using a statistical relationship. It is used throughout the drug development process allowing researchers to improve the decision-making process, and indirectly saves on cost and time of research programs.

Pharmacokinetics contribute concentrations of the drug achieved over a period of time after administration of a particular dose, pharmacodynamics describes the drug effectiveness measured at the same time and at corresponding concentrations. The variability in response from individual to individual and/or variability demonstrated by the same individual on multiple administration of taking the same drug are accounted for statistically.

PK-PD modeling contributes to selection of lead compound, prediction of potency, and possible clinical doses during preclinical development. During the clinical development, prediction of doses, dosage regimens, and favored designs for the next phase of clinical trials—factors that lead to variability of drug effects in a patient population—are assessed using PK-PD models.

Different models are utilized depending on the setting in which the PK-PD data is generated, either in steady state or nonsteady state conditions.

6.4.3.1 Steady state PK-PD models

The steady state models assume that the concentration of the drug is constant at the site of drug action and the drug effects do not vary over time.

Quantal Effect Model defines the relationship between drug concentration and drug effect which is quantal in nature (response or no response) and can be used to estimate the concentration of a drug, with 50% probability of no response. This model may be useful in the clinical setting.

The Linear Model presumes that the drug concentrations are directly proportional to the drug effects like blood pressure, blood glucose, etc. This model is based on simple regression and presumes no effect in absence of the drug, however, the maximum effect cannot be predicted using this model.

The Log Linear Model is relevant when the log concentration is directly proportional to the drug effects, and has been successfully employed to predict the effects of beta blockers, anticoagulants, etc.

The Emax Model is based on classical drug receptor interaction theory and is popularly used to predict effects related to drug concentrations. E_0 is presumed to be the baseline effect when drug concentration is absent and the maximum effect can be predicted (Emax).

The Sigmoid Emax Model is derived from generalization of the basic Emax model which allows the addition of a best fit factor, or sigmoidicity.

Other models of PK-PD correlation have been described, however they are beyond the scope of this chapter.

6.4.3.2 Nonsteady state PK-PD models

Drug effect and plasma concentrations plotted chronologically may demonstrate presence of hysteresis loop instead of a plain sigmoid concentration effect curve similar to a dose response curve. A counter-clockwise hysteresis may be obtained due to a lag time in producing pharmacodynamic effect or an indirect mechanism of action like inhibition or stimulation of synthesis or degradation

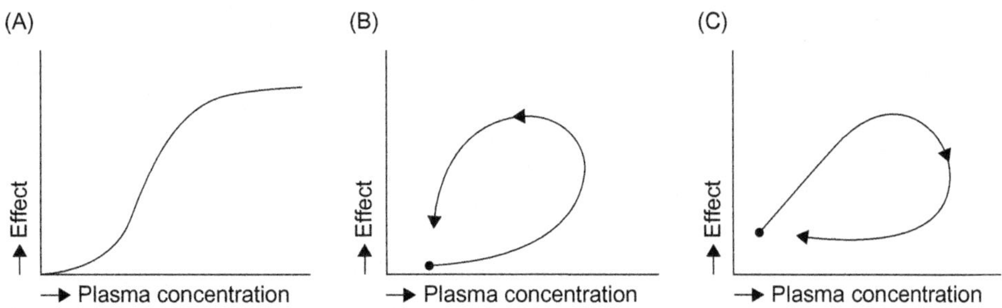

FIGURE 6.6

Effect–plasma concentration plots under different circumstances. (A) A sigmoid curve is obtained at steady state conditions. (B) A counter-clockwise hysteresis in drug effect and plasma concentration. (C) A clockwise hysteresis in drug effect and plasma concentration.

of endogenous products. Interaction of active metabolite(s) of a drug also produces counter-clockwise hysteresis. In contrast, clockwise hysteresis is observed when the drug effect wanes ahead of drug concentration and may be caused by development of tolerance, desensitization of receptors, or physiological counter-regulatory phenomenon (Fig. 6.6).

6.4.4 PHASE I, II, AND III CLINICAL TRIALS [19]

Each phase of the drug development process is designed to accrue the necessary information to assess the probability of success of conversion of a new chemical entity into a drug.

6.4.4.1 Phase I clinical trial

Pharmacokinetic profiling is done in study subjects (healthy human subjects or patients) after administration of single and multiple doses during a Phase I clinical trial program. Biological sample collection schedules may be kept flexible to generate optimum data. Typically data is generated in small cohorts of subjects at each dose and an attempt is made to correlate with the drug effect if measurable. Apart from calculating pharmacokinetic parameters like Cmax, Tmax, AUC, Kel, $T_{1/2}$, and clearance, linearity of drug exposure after administration of single escalating doses is of paramount interest. During administration of multiple escalating doses, possibility of accumulation of drug and time thereof is assessed. The pharmacokinetic data generated in this phase assists in the selection of doses in Phase II clinical trials.

6.4.4.2 Phase II clinical trial

During the phase II clinical trials program, pharmacokinetic data is generated in patients and the characteristics of the drug are confirmed and the correlation between drug concentration and therapeutic effect is evaluated. The data generated is also utilized to identify factors that lead to variability in the effect of the drug (identification of covariates) and propose a POP-PK (population based pharmacokinetics) model.

6.4.4.3 Phase III clinical trial and population pharmacokinetics (POP-PK)

In Phase III, pharmacokinetic data is generated using sparse sampling methodology and POP-PK methods to derive conclusions from the drug concentrations measured. POP-PK is the study of the extent, sources, and correlates of variability in the pharmacokinetics of a drug in a patient population. Age, gender, body weight, coexisting disease, and concomitant medications may alter the pharmacokinetics of a drug. The extent to which these factors produce variability is evaluated by POP-PK. Thus POP-PK evaluates the variation in drug concentrations and key pharmacokinetic parameters like bioavailability and clearance in a given patient population. A quantitative estimation of contribution of variability by identified factors is also identified. Thus, POP-PK in the drug development process helps identify differences in safety and efficacy among population subgroups.

During this phase of development, focused studies are undertaken to evaluate drug–drug interaction potential and the tolerability and pharmacokinetics of the drug in special populations having compromised hepatic and renal functions.

6.4.5 BIOEQUIVALENCE STUDIES

Bioequivalence studies are conducted frequently and data thus generated is utilized to approve generic medicinal products. These are clinical pharmacokinetic studies that assess the in vivo biological equivalence of two drug products containing the same drug ingredient in comparable quantity and which produce comparable rate and extent of exposure of the active ingredients in the experimental subjects. Very often, healthy human volunteers are used as a model to evaluate bioequivalence of products, though select groups of patients are also used when the risk of administration of the medication precludes the use of healthy subjects.

6.5 DRUG INTERACTIONS [20]

When unexpected effects in the form of loss of efficacy or increased toxicity are observed in one or more drugs during simultaneous or concomitant use, a drug–drug interaction (DDI) must be suspected.

6.5.1 PHARMACODYNAMIC DDI

Concomitant us of drugs acting at the same receptor site or on the same organ system may produce pharmacodynamic drug interactions, e.g., use of aspirin and heparin potentiate action on platelets or buprenorphine and methadone act at the same receptors to produce antagonism in pain relief.

6.5.2 PHARMACOKINETIC DDI

Pharmacokinetic interaction alterations in the effects of concomitantly used drugs are accompanied by changes in serum drug concentrations with interference in all the processes from absorption up to excretion.

6.5.2.1 DDI during drug absorption

Several factors may influence the absorption of a drug at the absorption site, most commonly through the gastrointestinal mucosa. Changes in gastric pH (H_2 antagonists like ranitidine, Proton pump inhibitors like pantoprazole inhibit cefpodoxime absorption), gastric motility (motility enhancers like domperidone or antimuscarinics like scopolamine affect drugs like theophylline and aspirin) and formation of complexes in the gut (antacids) may result in altered (increased or decreased) absorption of the interacting drugs (reduce absorption of ciprofloxacin). The severity of the effect of DDIs mainly depends on pharmacodynamic characteristics of the involved drug (e.g., narrow therapeutic range) and the extent of alternation in the absorption process.

P-glycoprotein present in the small and large intestines reduces absorption of drugs, thereby providing a protective mechanism. P-gp reduces the diffusion of drugs across membranes and its inhibition significantly improves the absorption of drugs with low bioavailability. The clinical effects of inhibition of P-gp are significant when drugs with a low therapeutic index (theophylline, some anticancer drugs) are concomitantly used with macrolides (e.g., erythromycin, clarithromycin) and PPIs (e.g., omeprazole or esomeprazole).

6.5.2.2 DDI during drug distribution

Drugs also bind to plasma proteins like albumin, alpha-1 glycoprotein and lipoproteins, while being transported to the tissues. Acidic drugs are bind more to albumin while basic drugs bind more to alpha-1 glycoprotein and/or lipoproteins. The degree of plasma protein binding, expressed by the percentage ratio of bound drug concentration/free drug concentration, varies greatly among drugs and predicts its potential to be involved in a DDI. Drugs which are more than 90% bound are considered as highly bound drugs. In contrast, drugs which bind poorly to plasma proteins have bound to free drug ratios of less than 20%. Drugs that are highly plasma protein bound are potentially more likely to be displaced by another drug with a greater affinity for the same binding site. From a practical clinical view, such displacement causes symptoms or toxicities when the displaced drug is associated with a reduced volume of distribution, a narrow therapeutic index, and is characterized by a faster onset of the effect. A typical pharmacological displacement can be observed when warfarin and diclofenac are co-administered.

6.5.2.3 DDI during drug metabolism

The competitive inhibition of CYP enzymes occurs when inhibitor and substrate compete for the same binding site on the enzyme, resulting in a DDI. In this type of interaction, the inhibition is reversible. It depends on the relative concentrations of substrate and the inhibitor. Some of the inhibitors of CYP3A4 that act by this mechanism of inhibition include azole antifungal agent ketoconazole. In the noncompetitive mechanism, the inhibitor and substrate do not compete for the same site on the CYP enzyme. When a ligand binds to the allosteric site, the conformational changes of the active site occur, leading to reduced binding of the substrate. Many drugs are noncompetitive inhibitors of CYP enzymes, like omeprazole and lansoprazole, and cimetidine. The inhibition is reversed when new enzymes are synthesized after the inhibitor drug is withdrawn.

Drug interactions involving enzyme induction are less common than inhibition-based drug interactions but are clinically important. Environmental pollutants as well as lipophilic drugs can result in induction of CYP enzymes. The increased synthesis of CYP enzyme proteins induced by one drug speed up the metabolism and clearance of the other drug in DDI. The most commons enzyme inducers are rifampicin, phenytoin, and carbamazepine.

6.5.2.4 DDI during drug excretion by kidney

The kidney is responsible for the elimination of most drugs and their metabolites. DDI may occur at the level of active tubular secretion, where two or more drugs use the same transport system. NSAIDs frequently cause appearance of toxic effects of methotrexate when the renal excretion of the antiproliferative drug is blocked. However, similar competition between other pairs of drugs can be exploited for therapeutic purposes, e.g., probenecid can increase the serum concentration of beta-lactams, delaying their renal excretion and thus saving in terms of dosage.

DDI can also occur during tubular reabsorption. Many drugs, when they are in an ionized form in the urine, pass by diffusion in tubular cells. The changes in urinary pH, pharmacologically induced, influence the state of ionization of certain drugs and may therefore affect the reabsorption from the renal tubule.

6.5.3 DESIGN OF DDI STUDIES [21]

The selection of a study design depends on a number of factors for both the substrate and interacting drug, including mechanism of DDI, duration of use of the interacting drugs, safety considerations and knowledge of CYP enzyme inhibition, and/or inhibition.

6.5.3.1 Study population

Most commonly healthy subjects are used for evaluation of DDI potential. Safety considerations, however, may preclude the use of healthy subjects in studies of certain drugs.

6.5.3.2 Route of administration

For an investigational agent, the route of administration generally should be the one planned for clinical use.

6.5.3.3 Dose selection

The doses of the substrate and interacting drug used in studies should maximize the possibility of demonstrating an interaction. For this reason, the maximum planned or approved dose and shortest dosing interval of the interacting drug (as inhibitors or inducers) should be used.

6.5.3.4 Endpoints

Changes in pharmacokinetic parameters are used to assess the clinical importance of drug–drug interactions. Interpretation of findings (i.e., deciding whether a given effect is clinically important) depends on a good understanding of dose/concentration and concentration/response relationships for both desirable and undesirable drug effects in the general population or in specific populations.

6.5.3.5 Pharmacokinetic endpoints

PK parameters such as AUC, Cmax, time to Cmax (Tmax), and others as appropriate, should be obtained in every study. Calculation of pharmacokinetic parameters such as clearance, volumes of distribution, and half-lives may help in the interpretation of the results of the trial. In cases of chronic administration of drugs, these parameters must be measured for the inhibitor or inducer as well, notably where the study is intended to assess possible changes in the disposition of both study drugs. Additional measures may help in steady state studies (e.g., trough concentration) to demonstrate that dosing strategies were adequate to achieve near steady state before and during the interaction.

6.5.3.6 Statistical considerations and sample size

The goal of the DDI study is to ascertain whether there is any increase or decrease in exposure to the substrate in the presence of the interacting drug. Results of such studies should be reported as 90% confidence intervals about the geometric mean ratio of the observed pharmacokinetic measures Cmax and AUC with $(S+I)$ and without the interacting drug (S alone). Confidence intervals provide an estimate of the distribution of the observed systemic exposure measure ratio of $(S+I)$ versus (S alone) and convey a probability of the magnitude of the interaction. In contrast, tests of significance are not appropriate because small, consistent systemic exposure differences can be statistically significant but not clinically relevant.

When a drug–drug interaction of potential importance is clearly present, specific recommendations regarding the clinical significance of the interaction based on what is known about the interacting drugs form the basis for making recommendations of management of similar situations in the clinical setting.

6.6 CONCLUSION

A drug is considered to have a favorable PK profile when most of the following characteristics are fulfilled:

1. Long duration of exposure in the body after administration.
2. Adequately absorbed or good bioavailability.
3. Long and optimum half-life so the drug can be administered once daily orally, or more infrequently by parenteral route.
4. Optimum distribution to various body tissues with adequate concentrations at the site of action.
5. Optimum plasma protein binding, not too high and not too low.
6. Simple or few metabolism steps with minimum variability and no potential for CYP enzyme induction or inhibition.
7. Less variability in PK, leading to predictable dose.
8. Not accumulated in specific tissues like retina, kidney, etc.
9. Adequate clearance within a reasonable time.
10. Optimum Cmax and Tmax depending on the therapeutic concentration and clinical condition being either acute or chronic.

The importance of plasma drug concentration data is based on the fact that response of a drug is closely related to its concentration at the site of action. For some drugs, the plasma concentration range that is safe and effective in treating specific diseases in patients is well documented. Below the therapeutic range, there is a sub-optimal effect while toxic effects may occur above it. The free drug concentration is more reliable than the total plasma concentration for therapeutic drug monitoring. The free amount of drug in plasma and in tissue and the tissue-bound drug amount remain unchanged under steady state conditions. However, for most drugs, no absolute boundaries divide sub-therapeutic, therapeutic, and toxic drug concentrations and these concentrations overlap due to variability in individual patient response. This inter-patient variability may be caused by one or more of the various pharmacokinetic processes and disease factors such variations in drug absorption, distribution, disease state, and drug interactions.

REFERENCES

[1] Greenblatt DJ, von Moltke LL, Harmatz JS, et al. Pharmacokinetics, pharmacodynamics, and drug disposition. In: Davis KL, Charney D, Coyle JT, et al., editors. Neuropsychopharmacology: The Fifth Generation of Progress. Philadelphia: Lippincott, Williams & Wilkins; 2002. p. 507–24.

[2] Gunaratna C. Drug metabolism and pharmacokinetics in drug discovery: a primer for bioanalytical chemists. Part I. Curr Sep 2000;19(1):17–23.

[3] Gunaratna C. Drug metabolism and pharmacokinetics in drug discovery: a primer for bioanalytical chemists. Part II. Curr Sep 2001;19(3):87–92.

[4] Dhillon S, Gill K. Basic Pharmacokinetics. In: Edis Dhillon S, Kostrzewski A, editors. Clinical Pharmacokinetics. Pharmaceutical Press, RPS Publishing; 2006. p. 1–43. Chapter 1.

[5] Toutain PL, Bousquet-Me ́lou A. Volumes of distribution. J Vet Pharmacol Therap 2004;27:441–53.

[6] Batchelor HK, Marriott JF. Pediatric pharmacokinetics: key considerations. Br J Clin Pharmacol 2015;79:395–404.

[7] Mangoni AA, Jackson SHD. Age related changes in pharmacokinetics and pharmacodynamics: basic principles and practical applications. Br J Clin Pharmacol 2003;57:6–14.

[8] Costantine MM. Physiologic and pharmacokinetic changes in pregnancy. Frontiers in Pharmacology 2014;5:1–5.

[9] Keller F, Maiga M, Neumayer HH, Lode H, Distler A. Pharmacokinetic effects of altered plasma protein binding of drugs in renal disease. Eur J Drug Metab Pharmacokinet 1984;9:275–82.

[10] Rodighiero V. Effects of liver disease on pharmacokinetics. An update. Clin Pharmacokinet 1999;37:399–431.

[11] Shenfield GM. Influence of thyroid dysfunction on drug pharmacokinetics. Clin Pharmacokinet 1981;6:275–97.

[12] Verbeeck RK, Musuamba FT. Pharmacokinetics and dosage adjustment in patients with renal dysfunction. Eur J Clin Pharmacol 2009;65:757–73.

[13] Vrhovac B, Sarapa N, Bakran I, Huic M, Macolic-Sarinic V, Francetic, et al. Pharmacokinetic changes in patients with oedema. Clin Pharmacokinet 1995;28:405–18.

[14] Ashavaid TF, Dheraj AJ. Therapeutic drug monitoring: a review. Ind J Clin Biochem 1999;14:91–4.

[15] Kaliappan I, Kammalla AK, Ramasamy MK, Agrawal A, Dubey GP. Emerging need of pharmacokinetics in Ayurvedic system of medicine. Int J Res Ayurveda Pharm 2013;4(5):647.

[16] Obiageri OO. Pharmacokinetics and drug interactions of herbal medicines: a missing critical step in the phytomedicine/drug development process, readings. In: Ayman Noreddin, editor. Advanced pharmacokinetics-theory, methods and applications, Chapter 7; 2012, pp. 129–156.

[17] Gopu D, Gomathi P. Pharmacokinetic/pharmacodynamic (PK/PD) modelling: an investigational tool for drug development. Int J Pharm Pharmaceut Sci 2012;4(Suppl 3):30–7.

[18] Chien JY, Friedrich S, Heathman MA, de Alwis DP, Sinha V. Pharmacokinetics/pharmacodynamics and the stages of drug development: role of modelling and simulation. AAPS J 2005;7:E544–59.

[19] Jang GR, Harris RZ, Lau DT. Pharmacokinetics and its role in small molecule drug discovery research. Med Res Rev 2001;21(5):382–96.

[20] Palleria C, Di Paolo A, Giofrè C, Caglioti C, Leuzzi G, Siniscalchi A, et al. Pharmacokinetic drug-drug interaction and their implication in clinical management. J Res Med Sci 2013;18(7):601–10.

[21] USFDA. Guidance for industry drug interaction studies—study design, data analysis, implications for dosing, and labeling recommendations; 2012.

CHAPTER

PHARMACOGENOMICS: AN EVOLUTION TOWARDS CLINICAL PRACTICE

7

Ankit Srivastava[1,2], Debleena Guin[2], Ritushree Kukreti[2], and Divya Vohora[1]

[1]*Jamia Hamdard University, New Delhi, India*

[2]*CSIR-Institute of Genomics and Integrative Biology, New Delhi, India*

7.1 INTRODUCTION

Every foreign substance, including drugs, is toxic to the human body. But when it comes to benefit to risk ratio, we usually take the help of a medication, in the case of disease treatment, where benefit is much higher than the risk raised by the disease as well as the drug. Advancement in the healthcare sector has led to a significant improvement in disease management. The evolvement of more potent and target specific drugs, being more effective and having reduced adverse effects, has dramatically enhanced patients' quality of life. For receiving the optimum benefits of these drugs, physicians prescribe a drug dose regimen based on various clinical parameters, such as disease severity, patient's age, gender, weight, etc. Based on clinical studies and considering all parameters, an empiric drug dose ("population average dose") is predicted which provides minimum effective dose and maximum tolerated dose and is called the therapeutic window of a particular drug. Most, but not all, individuals show a good response to such empiric drug administration. Some inter-individual variation exists varying from no response to good response to adverse drug reactions or hypersensitivity. For instance, the same dose of an antidepressant drug given to a group of major depressive disorder (MDD) patients, having similar clinical presentation, may lead to non-responsiveness or even suicidal ideation in some patients while some may classify it as remitter after completing the drug therapy. In the same way, flucloxacillin, an antimicrobial agent which may cause drug-induced liver injury (DILI) in some patients [1], in others remains normal after the drug administration. What accounts for such a differential drug response? Genetics may answer such questions. Genetic variations may play a critical role in variable drug response. In the case of antidepressant therapy, there is involvement of *CYP2D6* gene, which causes rapid metabolism of the drug in some patients and leads the MDD condition to severely worsen the situation—i.e., suicidal ideation. Further, the case of DILI development is explained by Daly et al [1] carriers of HLA-B*5701 genotypes are more prone to the specified side effects compared to the individuals where it is absent.

Pharmaceutical Medicine and Translational Clinical Research. DOI: http://dx.doi.org/10.1016/B978-0-12-802103-3.00007-9

The interaction between genes and drugs is termed as Pharmacogenomics, i.e., how genetic makeup of an individual affects the pharmacology [pharmacokinetics (PK) and pharmacodynamics (PD)] of any drug. In general, when a drug enters the human body, it has to pass through different barriers to the target tissue, where it shows its pharmacological effect and thereafter it has to get metabolized and eliminated out of the body. All these processes are controlled by proteins which in turns are managed by gene expressions. These genes can be classified under two categories: pharmacokinetic genes and pharmacodynamic genes. Pharmacokinetic genes majorly involve drug transporters and drug metabolizing enzymes that play roles in absorption, distribution, metabolism, and elimination of the drug, while pharmacodynamic genes mainly include drug receptors/targets and further downstream genes. For the best possible therapeutic effect with minimum side effects, physicians need to prescribe the best possible dose of a drug concerning its pharmacokinetic and pharmacodynamic profile. The drug may interact with "on target" or "off target" genes, which will lead to the desired therapeutic effects and the undesired adverse drug effects respectively. An "on target" gene may become "off target" with little variation in genomic sequences in some individuals of a population. And here comes the science of Pharmacogenomics, which determines how genomic variation of an individual can alter the pharmacokinetic and pharmacodynamic responses of a drug or a class of drugs. On the basis of this prediction, a physician may prescribe a drug and a dose in order to increase therapeutic efficacy or to reduce unwanted adverse drug effects—or both—in that particular individual.

The term *Pharmacogenetics* was first coined by Friedrich Vogel in 1959, and since then this field of science has continuously grown, with a boom after 1990 with the advancement in molecular biology techniques and availability of advanced genetic biotechnologies (Fig. 7.1 shows the history and development of pharmacogenomics). The launch of Human Genome Project in 1990 was a major step in the field of personalized medicine. This project was jointly run by the National Institute of Health (NIH) and the Department of Energy's Office of Health and Environmental Research (OHER), with the objective of sequencing the whole human genome. The final results were published in 2003 and were made available in the public domain. In 1999, The SNP Consortium (TSC) was established as a collaboration of multiple pharmaceutical companies and institutions with an aim to produce a public resource of SNPs present in the human genome. Finally, 1.8 million SNPs were identified and made available publically. With the availability of such a human genome map, there came the development of genome-wide association studies (GWAS). In pharmacogenomics, we mainly stick to the effects of genetic variation on the traits of drug response. For studying drug response today, researchers are using many high throughput technologies like DNA microarrays, RNA expression studies, Protein arrays, DNA sequencing, etc. In 2008, the "1000 Genomes Project" was launched to sequence the genomes of at least one thousand humans from different ethnic groups. In 2012, they reported the sequencing results of 1092 individuals. This project further made pharmacogenomic studies like exome sequencing or whole genome sequencing less expensive and more realistic.

Currently, physicians are already starting, in some places, genetic testing before any diagnosis or prescribing of drugs, but this concept of personalized medicine needs to spread even more. Pharmacogenomics is still in the budding phase but because of its tremendous importance in increasing drug efficacy and reducing adverse effects on a personal basis, it will surely develop and become more commonplace in day to day clinical practice. (Fig. 7.2).

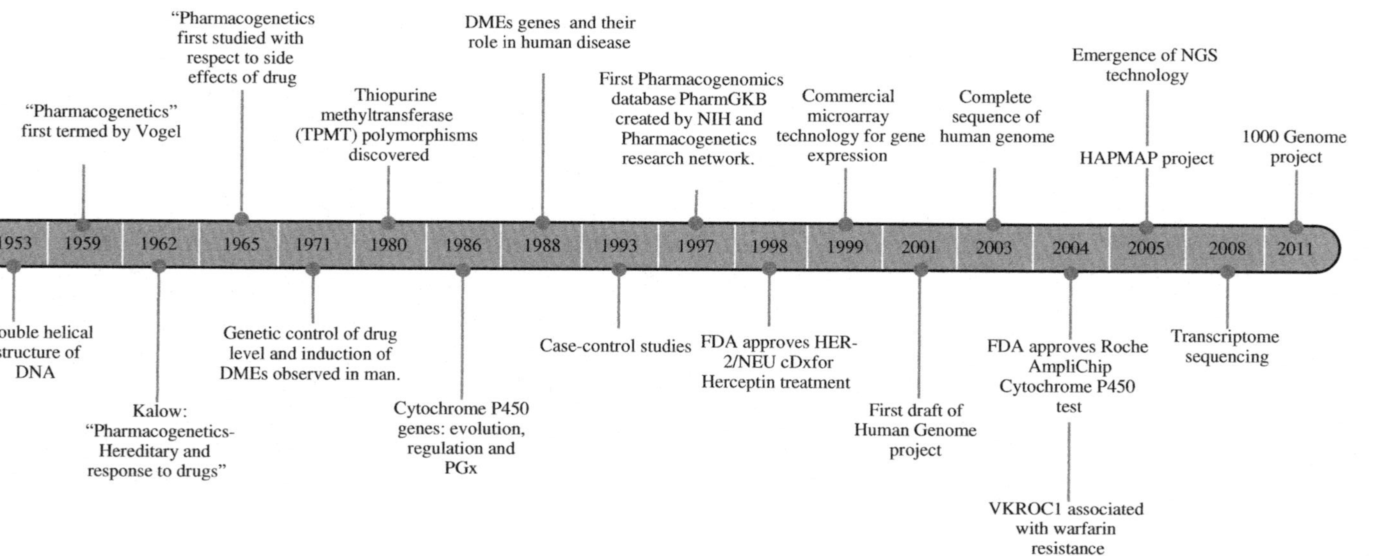

FIGURE 7.1

Major milestones in pharmacogenomics.

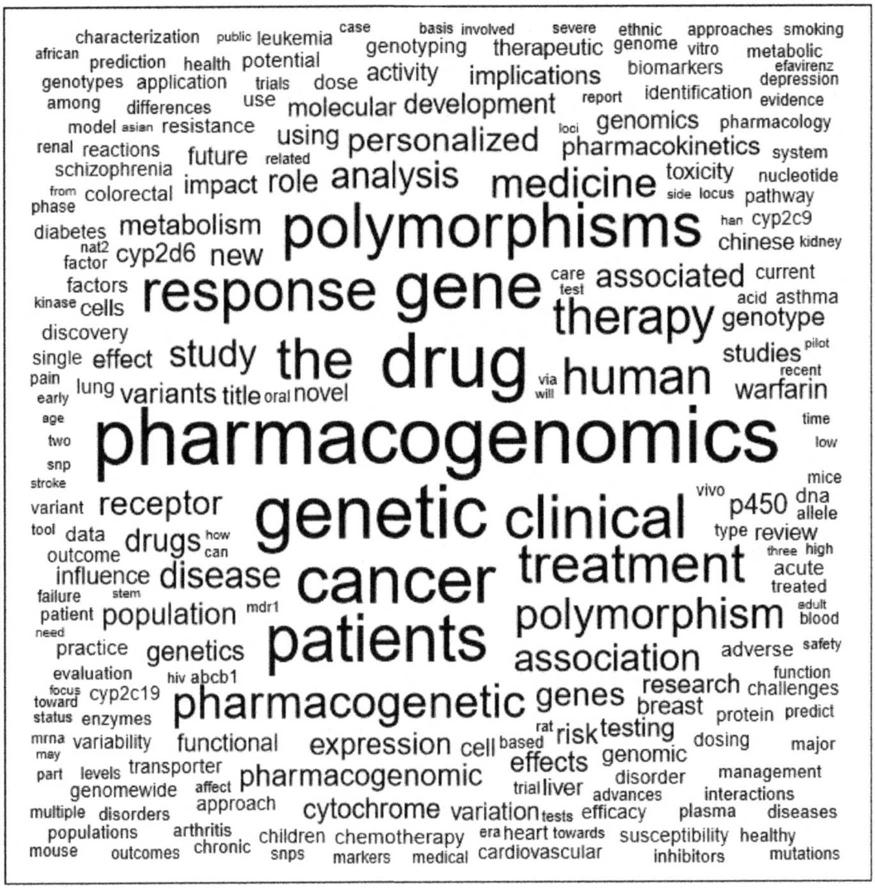

FIGURE 7.2

World cloud of Pharmacogenomics (PGx).

7.2 PK-PD AND PHARMACOGENOMICS

Pharmacokinetic (PK) is the action of the body on the drug and it plays a role in the drug disposition. Pharmacodynamics (PD) deals with the therapeutic actions of the drug on the body physiology. PK and PD are the two pillars of pharmacogenomics, on the basis of which one can predict drug response, dose regimen, and eventually can maximize therapeutic efficacy with minimum adverse drug reactions. PK profile of a drug can be estimated by considering four factors: (1) the extent of absorption of a drug in systemic system from the site of administration, called bioavailability; (2) distribution pattern of the drug to different tissues; (3) kinetics of metabolism of drug; (4) rate and route of elimination of the drug from body. Genetic variability that causes variations in PK profile may involve genes from phase I metabolizing enzymes [2], phase II metabolizing enzymes [3], and absorptive and efflux transporters [4]. Phase I metabolizing enzymes mainly involve the CYP450 enzyme system and are responsible for oxidation, reduction, hydrolysis, and cyclization reactions of the substrate. Phase II metabolizing enzymes are responsible for conjugation reactions which make substrate more polar and hence facilitates their elimination from the body. Transporters play their role during absorption of the drug through the gastro-intestinal wall and in efflux of drugs from the cells to the extracellular fluid. Polymorphism in drug metabolizing enzyme (DME) and transporter genes may lead to either very low or very high concentrations of the drug in various individuals despite them having been given the same dose of that drug. Based on genetic polymorphism in DMEs, individuals can be classified in four different phenotypes: poor metabolizers (PMs) (patients with little or no functional activity for a selected CYP enzyme), intermediate metabolizers (IMs), extensive metabolizers (EMs), and ultra rapid metabolizers (UMs) (patients with very high activity for a selected CYP enzyme).

In addition to drug metabolism for elimination from the body, CYPs are also responsible for activation of many pro-drugs, for example cyclophosphamide is activated by liver cytochrome P450 enzymes to the active metabolite 4-hydroxy cyclophosphamide, and codeine is converted into morphine by CYP2D6. There is direct and inverse relationship between metabolism rate and drug dose requirement in the case of active drug and pro-drug administration respectively, i.e., a high dose of active drug and a low dose of pro-drug will be required for the UMs when compared to that of PMs.

There are total 57 CYP enzymes and 12 out of 57 are responsible for almost 93% of all drug metabolism, with CYP1A2, CYP2D6, CYP2C9, CYP2C19, and CYP3A4 accounting for 60% [5]. The CYP enzymes can be divided into two classes [6] with respect to the level of penetrance for inter-individual variability for differential drug pharmacokinetics: *Class I:* CYP1A1, CYP1A2, CYP2E1, and CYP3A4—they are without important functional polymorphism. *Class II:* CYP2A6, CYP2B6, CYP2C9, CYP2C19, and CYP2D6—they are highly polymorphic and are important for the metabolism of drugs. The highly polymorphic CYP2C9, CYP2C19, and CYP2D6 isozymes are responsible for metabolizing a large portion of routinely prescribed drugs and contribute significantly to adverse drug reactions and therapeutic failures. Table 7.1 shows examples of some FDA-approved drugs with pharmacogenomic information. CYP2D6 is a very well studied in area of psychiatric, pulmonary, and cardiovascular disorders. Some studies have shown the role of CYP2D6 in tamoxifen's (breast cancer drug) therapeutic outcome. CYP2D6 is known to metabolize tamoxifen to its active metabolite, endoxifen that possess approximately 100 times higher affinity for the target receptor than the parent molecule [7]. Stearns et al in 2003 have shown that extensive metabolizers of CYP2D6 have higher endoxifen plasma levels when compared to those having CYP2D6 lower activity and they validated this result using paroxetine (CYP2D6 inhibitor). They found a decrease in endoxifen levels in patients receiving paroxetine as a co-medication to tamoxifen compared to patients who did not use any CYP2D6 inhibitor [8].

Transporters also play a very significant role in therapeutic outcome of drugs by maintaining their proper concentration at target tissue. Endoxifen is a substrate of ABCB1 efflux transporter. Iusuf et al. and Teft et al. have shown that brain concentrations of endoxifen were nearly 20-fold higher in ABCB1 mutant (deficient) mice compared to wild-type animals. [9,10]. Grover et al. have shown the involvement of ABCC2 in controlling seizures in women and they suggest that it could be possibly because of altered interaction of lower expression of ABCC2 with antiepileptic drugs which may help in maintaining proper concentration of drug at the target tissue [11].

Table 7.1 Example of Some FDA-Approved Drugs With Pharmacogenomic Information About Pharmacokinetic and Pharmacodynamic Genes

S. No.	Gene	Gene Polymorphism	Drug	Affect of Polymorphism
Pharmacokinetic genes				
1	CYP2C9	Poor metabolizers	Celecoxib	Multiple fold increase in systemic drug exposure and slow drug dispositioning [13]
		Poor metabolizers	Flurbiprofen	Most of the rare alleles leads to reduced activity of enzyme and thus a higher level of flurbiprofen is observed [14]
		Poor metabolizers	Warfarin	Poor metabolizer requires lower loading/maintenance dose and may lead to bleeding complications [15]
2	CYP2C19	CYP2C19*2 (poor metabolizer)	Citalopram	Patients with decreased activity of enzyme are less tolerant to citalopram when compared to normal ones [16]
		Intermediate or poor metabolizer	Voriconazole	Therapeutic drug monitoring is very essential as it has a very narrow therapeutic window and poor metabolizer may show ADRs [17]
		Intermediate or poor metabolizer	Omeprazole	Standard dose of Omeprazole will lead to higher drug exposure and will improve therapeutic efficacy [18]
		Intermediate or poor metabolizer	Clopidogrel	Reduced platelet inhibition that may lead to increased adverse cardiovascular events
3	CYP2D6	Ultra-rapid metabolizers	Codeine	Increased formation of morphine which may lead to toxicity [19]
		Ultra-rapid metabolizers	Eliglustat	Ultrarapid metabolizers may not achieve adequate concentration for proper therapeutic effect
Pharmacodynamic genes				
4	VKORC1	1639G > A	Warfarin	As patients carrying this polymorphism are more sensitive towards warfarin, they require low doses of it [20]
5	ESR1	Lower expression	Tamoxifen	Low expression of ESR1 is associated with tamoxifen resistance in estrogen receptor positive breast cancer [12]

Once the drug has reached its target, role of pharmacodynamic genes comes into play. Receptors and downstream genes are mainly responsible for any PD variations in drug response. A large number of drug target polymorphisms have been found till now, but in the field of pharmacogenomics, drug response variability based on pharmacodynamic variability is comparatively less explored than pharmacokinetic variations. This may be attributed to nonspecificity of drug development because of less knowledge of pathophysiology of a particular disease. FDA has approved some genes labeling on the drugs like the role of ESR1 in the case of tamoxifen use in breast cancer. Kim et al. had also shown that a low level of ESR1 might be a determinant for tamoxifen resistance in estrogen receptor positive breast cancer [12]. There is very promising role of VKORC1 in the case of warfarin treatment. The VKORC1 gene encodes the vitamin K epoxide reductase enzyme, the drug target of warfarin. Patients carrying 1639G > A polymorphism of the VKORC1 gene are more sensitive towards warfarin and require lower doses. Table 7.1 shows examples of some FDA-approved drugs with pharmacogenomic information about pharmacodynamic genes.

7.3 METHODOLOGIES USED FOR PHARMACOGENOMICS STUDIES

7.3.1 IDENTIFICATION OF GENETIC VARIANTS

The human genome project and subsequently the international HAPMAP project revealed an enormous amount of genotype data open publicly for the researchers. This opened new prospective of research to decode the complexities of the human system and disease occurrence. Pharmacogenomics was one such emerging field that exploited this information in order to screen patients for genetic variations and derive their association with their response to drugs. Technologies developed which could identify both known functional variations as well as novel variations. Several types of genetic variation are found in our genome, the most common of them are the single nucleotide polymorphisms (SNPs). Other variations include copy number variations (CNVs), insertion and deletions (InDels), transversion, translocation at the chromosomal level, and so on. Given the large number of variations present in the human genome, selection of a genotyping method with highest efficacy, productivity, and generating the fastest result is a crucial step in pharmacogenomics. The type of polymorphism to be genotyped, technical expertise required, cost, turn-around time, choosing the technique which can genotype a large number of polymorphisms at a time—with least chances of error and minimal background noise—would determine the choice of a genotyping technique. A genotyping protocol generally is divided into allele-specific products formation (allele discrimination) and detection (allele detection) procedures to identify the products. The subsequent sections focus on the most widely used genotyping techniques.

1. *Sanger Sequencing*: Most often considered as the "gold standard" of genotyping technique, Sanger sequencing has been used for variant identification ever since its inception in 1977 by Frederick Sanger. Also known as the chain termination method, this technique requires dideoxynucleotides (similar to deoxynucleotide, except that they lack a hydroxyl (OH) group at the $3'$ carbon position. A primer oligo anneal to the single stranded DNA template. In every reaction mixture, the DNA polymerase catalyzes to incorporate a deoxynucleotide or dideoxyucleotide into the growing complementary DNA strands. Deoxynucleotides are present at a higher concentration than the dideoxynucleotides in the reaction mixture. If in any case the dideoxynucleotide is incorporated into the growing DNA strand, the chain elongation terminates, since dideoxynucleotide lacks a $3'$OH group to form a phosphodiester bond. This results in chain termination and results in a mixture of DNA fragments of different lengths. All these fragments have common $5'$ end and variable $3'$ end of different sizes. The fragments can then be sized on capillary gel electrophoresis and the length of the fragments can be determined by their position in the gel. In many automated sequencing systems, the primers or the

dideoxynucleotides are labeled with a fluorescent dye. A laser identifies the fragment fluorescence and output is recorded as peaks in chromatogram. For genotyping purposes this output chromatogram can be used to determine if the genotyped individual has a polymorphic or wild type allele at a given site.

2. *PCR-RFLP*: This is one of the traditional methods which discriminates between polymorphic and wild type allele. A restriction enzyme is mixed with the PCR product. The restriction enzyme recognizes either the wild type allele or the polymorphic one present at the restriction site and digests the DNA at that site. Gel electrophoresis then recognizes the DNA fragments based on its size and an individual's genotype is visually assigned based on the digested fragment patterns. Not every allele to be genotyped may lie at a restriction site, hence it is a low-throughput technique not ideal for all systems.
3. *TaqMan*: This assay is based on the real-time PCR technique, where two fluorescent probes are used for allele discrimination, one complementary to the wild type allele and the other to the polymorphic one. These probes are labeled with a reporter dye at the 5′ end and a quencher dye at the 3′ end. The quencher dye neutralizes the reporter dye and emits no fluorescence when the probes are intact. In the PCR reaction, the primer and the fluorescent labeled probe anneal to the template DNA strand. DNA polymerase catalyzes the addition of the deoxynucleotides at the 3′ end of the primer. On chain elongation, when the polymerase encounters the labeled probe that is bound to the DNA strand it cleaves the probe at the 5′ end. This disintegrates the reporter dye and quencher dye, thereby emitting fluorescence. This fluorescence is measured by several commercially available softwares and assigns a genotype. This is a highly assessed assay for genotyping as it requires low sample processing time and reduced expertise. But the high fixed cost for the fluorescent labeled probes designed for each polymorphism is a major limitation [21].
4. *Pyrosequencing*: This is an automated genotyping technique. Prior to genotyping the DNA sequencing of interest, having the polymorphism is amplified by PCR. A sequencing primer is hybridized to the DNA strand. DNA polymerase adds nucleotides in an order on the DNA strand. When a nucleotide is added to the template DNA, a pyrophosphate molecule is released. The release of this pyrophosphate further reacts with ATP sulphurylase, present in the reaction mixture, and produces ATP. The ATP thus produced then reacts with luciferase enzyme and D-luciferin and produces visible light. This light is then captured on the output called a pyrogram. The unbound nucleotide molecules are degraded by an enzyme called apyrase. The higher the intensity of light, the larger the number of nucleotides incorporated into the strand. The software measures the visible light and automatically assigns a genotype for the variant. A number of different polymorphisms can be genotyped in a single run. However, high equipment acquisition cost and labor cost for post PCR processing limit the use of this technique to all laboratories [22].
5. *Mass Spectrometry*: This genotyping technique is especially beneficial when an investigator is looking into a small number of SNPs. Based on the matrix-assisted laser desorption-ionization time of flight (MALDI-TOF) mass spectrometry detects alleles based on mass only. This has an excellent precision and high-throughput abilities. Along with mass spectrometry for allele detection, it is coupled with single primer extension or allele specific hybridization for allele discrimination. For e.g., the Sequenom's iPLEX SNP genotyping uses primer extension where primers are annealed upstream of the polymorphism and are extended to yield products

different in molecular weight. The different masses of the products are detected by MALDI-TOF, and genotypes are assigned accordingly [23].

6. *Genome-wide association study (GWAS)*: This is typically a high-throughput technique in which thousands of SNPs in the entire genome are genotyped in the diseased versus the control group. Comparing the allele frequencies of the genotype depicts the association of the genotype with a possible disease risk. This approach exploits the HAPMAP and the 1000 genome data to identify the allele frequencies associated most often with the disease phenotype. The basis of genome-wide studies is the common disease common variant hypothesis, which says a certain number of common genetic variants are associated with common diseases. When combined, the effect of the variants are largely increased showing the disease risk. In this technique, the disease causing genomic loci are identified across the entire genome and statistically significant association is calculated with respect to a phenotype. The closest gene associated with the common variants represents the most probable candidate gene. Even though GWAS has successfully identified a few definitive risk alleles for common disease, these findings do not completely solve the question of heritability of complex diseases. Several reasons can be behind this problem. Genetic and non-genetic factors contribute to familial aggregation of disease, GWAS identifies genetic loci, which is other than the protein coding regions, making it difficult to identify the disease-causing variants. The GWAS also does not cover all of the genome. Lastly, the rare allele associations are underrepresented by GWAS [24,25].
7. *Next Generation Sequencing (NGS)*: The most comprehensive and widely used approach to identifing genetic variants associated with diseases is performed on several next generation sequencing platforms available commercially. In this process, either the entire genome is sequenced or the specific targeted regions on a large number of samples, to identify all the plausible variants associated with a disease phenotype. A number of platforms are available to be used for this approach, each allowing differences in the scale of sequencing being performed. These technologies have the advantage of unbiased SNP discovery, detecting the less common variants, and large-scale information is retrieved. Other than SNPs, CNVs and structural variants can also be identified by this technology. Often in the major association studies, platforms identify "tag" SNP in a large number of nearby SNPs. This phenomenon is due to linkage disequilibrium, where structures of these "haploblocks" are specific to particular populations. This complicates the understanding of the effect of a given variant in a population. It may be because of a hidden interaction with another genetic factor or other environmental factors. Hence, the SNPs identified by these technologies predicts the "associated" variants rather the causative ones. This demands high-end experimental analysis on genes and variants linked with a disease phenotype. Currently there are many platforms available for NGS, like Roche 454 FLX, Life Technologies SOLiD and 5500xl sequencer and Illumina HiSeq, Illumina MiSeq with different working principle. In Roche 454 FLX, Genomic DNA is fragmented and adapter sequences are ligated onto fragmented DNA mixed with agarose beads. On each of these agarose beads, millions of amplified DNA fragments are generated by emulsion PCR. These beads are deposited into Picotiter wells where sequencing of the entire genome would be performed in numerous picoliter sized wells simultaneously by Pyrosequencing. As discussed above Pyrosequencing reactions are carried out similarly. At each elongation step, the sequence is detected by measuring light emission. This method allows for the amplification of up to 1000 nucleotide size sequences. This technology allows to read 400 nucleotide bases [26,27]. Similar to Roche's the Life technologies, SOLiD sequencing technology

involves the preparation of a sequencing library. DNA is fragmented into smaller pieces and adapter and primer sequences are ligated onto the fragments. The DNA fragments are deposited onto agarose beads and the fragments are enriched during ePCR. The 3′ ends of the amplified fragments are covalently modified to allow for attachment to the glass slide. Following the 3′ modification, the beads are deposited onto glass slides. Sequencing by ligation occurs with the binding of a sequencing primer to the DNA fragment with fluorescently labeled di-bases being ligated to the primer. The specificity of the di-base probe is achieved through the interrogation of the 1st and 2nd base of each ligation reaction. Multiple cycles of ligation, detection, and cleavage are performed, with the number of cycles determining the overall read length. Following the ligation and detection cycles, the extension product is removed and the template is reset with a primer complementary to the $n-1$ position of a second round of ligation cycles. In total, five rounds of primer resets are completed allowing for virtually every base to be interrogated in two independent ligation reactions by two different primers allowing for up to a 99.99% accuracy to be achieved. [28,29] The Illumina HiSeq and MiSeq represent the latest advancement of the sequencing technology which follow the sequence by synthesis (SBS) chemistry. Genomic DNA is fragmented randomly and adapters are ligated to both ends. Bridge amplification is used to create clusters of DNA strands and fluorescently labeled 3′-OH blocked nucleotides are added to the flow cell with DNA polymerase. These technologies sequence millions of fragments by a proprietary terminator based method that detects single bases which are incorporated into the growing strands. The unused nucleotides and DNA polymerase are washed away, and the reaction is imaged. The process is then repeated for another round of nucleotide incorporation. The end result of this assay is highly accurate enabling robust base calling across the genome.

8. *Second and Third Generation sequencing*: These platforms include Helicos, Pacific Biosciences, and Complete Genomics. Helicos provides the True single molecule sequencing (tSMS) which follows the sequence by synthesis method overcoming the DNA amplification step by analyzing single molecule of DNA, thereby significantly reducing the time required. Pacific Biosciences exploits the Single Molecule Real Time (SMRT) using uninterrupted template-directed synthesis, sequencing on a single DNA molecule. Complete Genomics is considered to be the third generation of the sequencing technologies. In this technology sequencing is done by DNA nanoarrays, and the comparison of thousands of sequenced genomes can be performed. These technologies mark the exponential growth of targeted medical resequencing and indicate a vital role pharmacogenomic research. [30] (Table 7.2).

Genotyping technologies had been largely limited to research activities only, however the US Food and drug administration (FDA) has taken up a few of such techniques to be used as pharmacogenomic diagnostic tests in clinical practices. One of the first and most widely used diagnostic tests is the Roche AmpliChip Cytochrome P450 test approved in 2004. This test is based on the Affymetrix DNA microarray gene chip technology to identify a patients CYP2D6 & CYP2C19 genotypes from DNA extracted from whole blood samples. The genotype of the patient is compiled by a software and predicts the respective drug metabolizing enzyme phenotype. The reproducibility of this assay has been reported to be 99.99%. Subsequently many other tests for different diseases and drugs were approved by the FDA for clinical practices. A list of these tests can be found at: http://www.fda.gov/MedicalDevices/ProductsandMedicalProcedures/InVitroDiagnostics/ucm330711.htm. A few of them have been summed up as under: (Table 7.3) (Fig. 7.3).

Table 7.2 Contrasting Features Between Different Genotyping Techniques

S. No.	Genotyping Method	Allele Discrimination Technique	Allele Detection Technique	Cost of genotyping/ SNP	Throughput	Advantages	Disadvantages
1	Sanger Sequencing	Chain termination	Gel electrophoresis	>$4 per SNP	Low	Lowest error rate. Long read length (~750 bp).	Time consuming. Large amount of data per run.
2	PCR-RFLP	Restriction endonucleases	Gel electrophoresis	> $4 per SNP	Low	Low equipment cost.	Long sample processing time. Limited multiplexing. User assigned genotype call, high error chances.
3	Taqman	Allele specific hybridization	Fluorescence	~$1 per SNP	Medium to high	Time and cost effective. Genotype call by automated softwares.	Limited multiplexing. High cost for fluorescent-labeled probes.
4	Pyro-sequencing	Primer extension	Visible light	$1-$4 per SNP	Medium	Sequence information for region around SNP. High sensitivity. Genotype call by automated softwares.	Limited multiplexing. Difficulty in identifying long stretches of same nucleotide regions.
5	Mass spectrometry	Several	Molecular weight	~$1 per SNP	Medium to High	High sensitivity. High multiplexing capability.	High equipment cost. Technical expertise required. Labor intensive processing.

(*Continued*)

Table 7.2 Contrasting Features Between Different Genotyping Techniques *Continued*

S. No.	Genotyping Method	Allele Discrimination Technique	Allele Detection Technique	Cost of genotyping/SNP	Throughput	Advantages	Disadvantages
6	GWAS	Several	Several	~$1 per SNP	High	Discover novel candidate genes/variants.	Large study population required.
						Provides ancestry of subjects.	Requires replication.
						Discovers structural variants.	Detects association and not causation.
7	NGS	Several	Fluorescence	~$1 per SNP	High	Cost and time effective.	Analytical expertise required, especially in alignment of sequence.
						High sensitivity.	Large amount of data.
						High coverage and multiplexing capabilities.	Does not detect structural variants.
						Large amount of data.	Difficulty in interpretation of clinical relevance.
8	2nd & 3rd Generation sequencing	Several	Several	$600 per smart cell	High	Cost and time effective.	Labor intensive.
						High sensitivity.	Large amount of data.
						High coverage and multiplexing capabilities.	High Bioinformatics expertise required.
						Large amount of data.	

Table 7.3 Drug Metabolizing Enzymes Genotyping Tests Approved by FDA for Clinical Practices

S. No.	Trade Name	Manufacturer	Measurand	Type of Test
1	xTAG CYP2D6 Kit v3	Luminex Molecular Diagnostics, Inc.	CYP2D6 genotype	Multiplex PCR followed by multiplex allele specific primer extension. Allele detected by flow cytometry.
2	Spartan RX CYP2C19 Test System	Spartan Bioscience, Inc.	CYP450 2C19 *2, *3, *17	PCR.
3	Verigene CYP2C19 Nucleic Acid Test	Nanosphere, Inc.	CYP2C19	Genotyping microarray.
4	INFINITI CYP2C19 Assay	AutoGenomics, Inc	CYP2C19	Genotyping microarray.
5	Invader UGT1A1 Molecular Assay	Third Wave Technologies Inc.	UGT1A1	Genetic test for single nucleotide polymorphism detection.
6	Roche AmpliChip CYP450 microarray	Roche Molecular Systems, Inc.	CYP2C19, CYP2D6	PCR amplification of purified DNA; fragmentation and labeling of the amplified products; hybridization of the amplified products to a microarray and staining of the bound products; scanning of the microarray.
7	eSensor Warfarin Sensitivity Saliva Test	GenMark Diagnostics	CYP2C9, VKORC1	Qualitative genetic test for single nucleotide polymorphism detection.
8	eQ-PCR LC Warfarin Genotyping kit	TrimGen Corporation	CYP2C9*2 & *3, VKORC 1 (-1639G > A)	Genotyping Realtime PCR.
9	eSensor Warfarin Sensitivity Test and XT-8 Instrument	Osmetech Molecular Diagnostics	CYP2C9, VKORC1	Sandwich hybridization test.
10	Gentris Rapid Genotyping Assay - CYP2C9 & VKORCI	ParagonDx, LLC	CYP2C9 & VKORC1	Qualitative genetic test for single nucleotide polymorphism detection.
11	INFINITI 2C9 & VKORC1 Multiplex Assay for Warfarin	AutoGenomics, Inc.	CYP2C9 & VKORC1	Genotyping microarray.
12	Verigene Warfarin Metabolism Nucleic Acid Test and Verigene System	Nanosphere, Inc.	CYP2C9 & VKORC1	Qualitative genetic test for single nucleotide polymorphism detection.

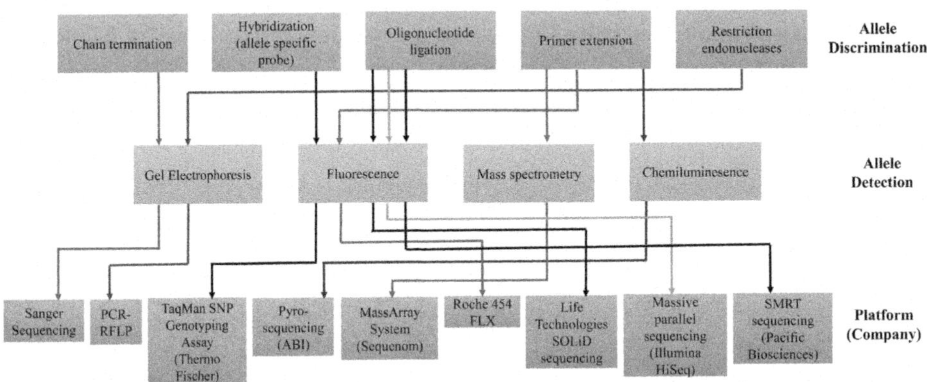

FIGURE 7.3

Comparing the principles of some of the genotyping techniques.

7.3.2 EXPRESSION METHODS

Revealing the details of the human genome also unveiled the central dogma of biology but complicated the understanding of disease occurrence. It is not only the genomic DNA or its variants that might be associated with a phenotype but the regulation of the gene matters equally. Using expression profiles of patients with a diseased condition, one can identify the expression of RNA to find which gene is expressed more/less in diseased phenotype. Elucidating the expression profile may also help us determine the molecular response of a drug. With the growing research in this area, several technological advancements have also come up. The most traditional of them were RT-PCR, Western blot, Fluorescent in situ hybridization (FISH), then came the DNA microarray technology and now the most advanced is RNA sequencing. A few of these techniques have been summed up as below:

1. *RT-PCR*: Reverse transcriptase-PCR is one of the most sensitive and accurate techniques for measuring gene expression. This technique clones the expressed genes by reverse transcriptase activity. cDNA molecules from the RNA of interest are prepared which is further amplified by normal PCR. Subsequent techniques like northern blotting or western blotting are performed to confirm the expression levels [31].
2. *DNA Microarray*: DNA microarray technology has changed the scenario of pharmacogenomic research to address the research questions. The jump from candidate gene approach to genome wide studies had been largely contributed by the microarray technology. DNA microarrays are a collection of probes bound in a grid-like pattern to a solid surface. The bound probes are short oligonucleotides designed to bind to specific regions of DNA. For pharmacogenomic purposes, there are several such gene chips available. DNA microarrays are used to quantitate expression profile of several mRNA which can be used to compare the level of gene transcription in clinical conditions to identify diagnostic or prognostic biomarkers, classify diseases, and monitor the response to therapy—and also to understand the underlying molecular mechanisms involved in the genesis of disease processes. The protocol includes extraction of mRNA from

specific tissues, which is further reversed-transcribed and labeled with a fluorescent dye, and hybridized on the array. Hybridization and washes are repeatedly performed to avoid cross-reaction between other genes. The fluorescent labeled image is then detected using laser-induced fluorescent imaging which quantifies as directly proportional to the amount of mRNA with complementary sequence present in the sample analyzed. Such techniques do not generate an absolute level of expression of a gene, however, it is used to compare the expression profiles between the diseased phenotype and the control [32].

3. *RNA-sequencing (RNA Seq)*: RNA sequencing is the most recently developed transcriptome profiling technology developed by Snyder M. and group that uses deep sequencing principle. A population of RNA molecules are converted to a library of small cDNA fragments, each attached with adapter molecules at one or both ends. The reads can be sequenced by any high-throughput sequencing technology available. After sequencing, the resulting reads are then aligned to a reference genome or reference transcripts or assembled de novo without any mapping sequence. This produces a large-scale transcription map that consists of both transcriptional structure and expression levels of a gene. RNA-Seq is advantageous over several other technologies since it is not limited to detecting transcripts which correspond to an existing genomic sequence. RNA-Seq can also resolve the boundaries of a transcript to a single base level. RNA-Seq has very low background noise due to DNA sequences, unlike microarray technology. The RNA-Seq output presents high reproducibility with both technical and biological replicates. Keeping in mind all these advantages, RNA-Seq is the first sequencing-based method for transcriptome profiling by a high-throughput technique at a much lesser cost [33].

Since studying the molecular effect of a drug treatment to understand the in vivo physiological response in a human subject is unethical, established experimental validation systems on cell lines have been developed to generate gene expression profiles. Tissue specific cell lines are treated with drug molecules to compare the expression profiles in order to identify metabolite-protein interactions that cause or suppress the disease phenotype. Thus, cell lines can be used as surrogates for human subjects and cellular response is considered as a proxy for the in vivo response based on a drug treatment. Similar studies can also substantiate toxicological response of drugs identifying the cellular expression levels in genes due to drug toxicity. (Table 7.4).

Table 7.4 A Few Examples of the Web Resources Used Most Commonly in Pharmacogenomic Research

S. No.	Name	Web Address	Content
Pharmacogenomic databases			
1	PharmGKB	http://www.pharmgkb.org/	Curation of genes/variants, associations between variants and drugs, drug-centered pathways, genotype correlation with pharmacogenomic information [34].
2	The Drug Gene Interaction Database	http://dgidb.genome.wustl.edu/	A database and web interface for identifying known and potential drug-gene relationships [35].

(Continued)

Table 7.4 A Few Examples of the Web Resources Used Most Commonly in Pharmacogenomic Research *Continued*

S. No.	Name	Web Address	Content
3	Side Effect Resource (SIDER 2)	http://sideeffects.embl.de/	It contains information on marketed medicines and their recorded adverse drug reactions [36].
4	Search Tool for the Retrieval of Interacting Genes/Proteins (STRING)	http://string.embl.de/	A database of known and predicted protein interactions [37].
5	Search Tool for Interaction of Chemicals (STITCH)	http://stitch.embl.de/	A resource to explore known and predicted interactions of chemicals and proteins [38].
6	The Comparative Toxicogenomics Database	http://ctdbase.org/	Drugs and their association with genes/variants/pathways/interactions [39,40,41].
Freely available online pharmacogenomic tools			
1	WebGestalt	http://bioinfo.vanderbilt.edu/webgestalt/	Functional genomic, proteomic, and large-scale genetic studies from which large number of gene lists (e.g., differentially expressed gene sets, coexpressed gene sets, etc.) are continuously generated [42].
2	SIFT	http://sift.jcvi.org/	An amino acid substitution affects protein function. SIFT prediction is based on the degree of conservation of amino acid residues in sequence alignments derived from closely related sequences, collected through PSI-BLAST [43].
3	DAVID Bioinformatics Resources	https://david.ncifcrf.gov/	A comprehensive set of functional annotation tools for investigators to understand biological meaning behind large lists of genes [44,45].
4	HaploReg	http://www.broadinstitute.org/mammals/haploreg/haploreg.php	A tool for exploring annotations of the noncoding genome at variants on haplotype blocks, such as candidate regulatory SNPs at disease-associated loci [46].
5	TRANSFAC	http://www.biobase-international.com/gene-regulation	A BIOBASE knowledge-base containing published data on eukaryotic transcription factors and miRNAs, their experimentally-proven binding sites, and regulated genes [47].

7.4 APPLICATIONS OF PHARMACOGENOMICS

So far in this chapter we have discussed the various techniques implemented to obtain the genomic information in response to drugs. Once such information suffices the knowledge base, such as PharmGKB, it can be exploited for further clinical application. This will allow significant characterization of patients' response to an administered drug in a varied population group. Therefore, an effective prescribing of drugs can be concluded. Pharmacogenomics can play a good role in identifying potential drug target. The major applications of pharmacogenomics can be summarized as I) In drug discovery and II) In clinical prescribing. These two are further detailed below.

1. *In drug discovery*: In pharmacogenomics research, identifying the candidate genes and the pathways involved gives new potential drug target candidates. From these drug targets, cheminformatics techniques can recognize the lead molecules which can be carried forward for drug screening and animal studies for efficacy and toxicity profiling. If the lead molecule surpasses the tests, it can be taken ahead for clinical trials, if not it can be discontinued well before the trials. In this way, pharmacogenomics directs a judicious path for drug discovery with reasonable economics for the pharmaceutical industries. Basic efficacy and relative safety profiling of the drug can also be carried out successfully. In the clinical trial phase which engages human subjects, admission of patients based on the response can be used to limit the cohort. Another application of pharmacogenomics can be repurposing of drugs already in the market for better clinical indications. Techniques like molecular docking and simulation can be exploited to discover novel functions of the already discovered drugs.
2. *In clinical prescribing*: One of the foremost applications of this science is with respect to clinical prescribing. Pharmacogenomics enables a personalized approach in clinical decision-making, considering the genetics of the patients and other clinical factors. The genetic variants of a patient determines the fate of the drug within the patient body, which further regulates whether the therapeutic effect is achieved or not, or whether it causes adverse drug reaction. To this, pharmacogenomics can suggest what dosage or alternate therapy is to be administered for optimum effect.
   a. *Dosage*: Once the therapy is decided for the best treatment outcome, the next important factor is choosing a dosage. Traditionally, the dosage is determined based on gender, age, body mass index, and other physiological factors. However, it has been already established with more than a decade of research that genomic details of the patient also plays an equally important role. One of the most elaborately studied drugs in this respect is warfarin. The International Warfarin Pharmacogenetic Consortium ascertains that variants in CYP2C9 and VKROC1 genes help in accurately defining the dosage [48]. An advantage of using such an approach is to minimize the time and overdosing which may lead to increased risk of thromboembolic events in case of warfarin. Therefore, such an approach may broaden effective treatment outcome.
   b. *Efficacy and response*: Polymorphisms in drug metabolizing enzymes and drug transporter altered the pharmacokinetic fate of the drug, which in turn determines if the therapeutic efficacy of the drug would be achieved by the administered dosage of the drug or not. CYP2C19, is a major metabolizing enzyme of the drug, clopidogrel. CYP2C19*2 genetic variant is associated with impaired clopidogrel metabolism in healthy volunteers [49]. Patients with this poor metabolizing phenotype may be administered with an alternate therapy.
   c. *ADRs*: The best therapy is the one which has the least or no adverse effects in the patient. Several drugs have been withdrawn from the clinical trials because of high toxicity profiles. Addressing this problem at the clinical trial phase can prove to be highly economical and safe. On profiling the patients based on the genomic variants, the toxicity of drugs can be controlled. The drugs already approved in markets like anti-arrythmics causing QT elongation leading to ventricular tachyarrhythmias as an adverse effect. In screening the polymorphisms in genes, inter-individual difference in the drug metabolizing enzymes can be addressed in the patients [50] (Fig. 7.4).

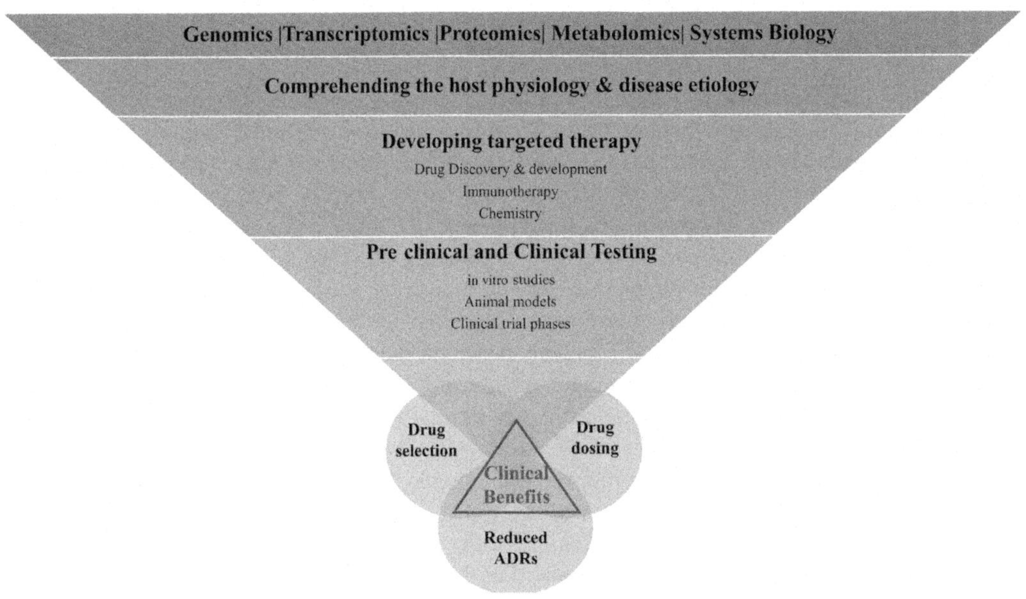

FIGURE 7.4

Application of pharmacogenomics.

7.5 CONCLUSION

Pharmacogenomics is the field of science which deals with the interactions between genetics and drug response. There is huge advancement in genotyping technologies which facilitates research by providing a platform for high throughput studies. By combining high throughput data with online tools and databases, we can find promising biomarkers or targets for personalized therapy and future drug development. As we know, genetics plays a very key role in the regulation of pharmacokinetic and pharmacodynamic responses; its application in clinical practice is extremely promising. As of now, various tests and biomarkers are available, which allows a clinician to decide whether a particular drug shall be given to a patient or not, but at the same time it is not very common to use it in clinical practice because of its high cost and little awareness. Pharmacogenomics is still in the budding phase but to increase drug efficacy and reducing adverse effects on a personal basis, it will develop in future and get a common place in day-to-day clinical practice.

REFERENCES

[1] Daly AK, Donaldson PT, Bhatnagar P, Shen Y, Pe'er I, Floratos A, et al. HLA-B*5701 genotype is a major determinant of drug-induced liver injury due to flucloxacillin. Nat Genet 2009;41(7):816–19.

[2] McGraw J, Waller D. Cytochrome P450 variations in different ethnic populations. Expert Opin Drug Metab Toxicol 2012;8:371–82.

[3] Grover S, Kukreti R. Functional Genetic Polymorphisms from Phase-II Drug Metabolizing Enzymes. CNS Neuroscience Ther 2012;18:705–6.

[4] Deenen MJ, Cats A, Beijnen JH, Schellens JH. Pharmacogenetic variability in drug transport and phase I anticancer drug metabolism part 2. Oncologist 2011;16:820–34.

[5] Preissner SC, Hoffmann MF, Preissner R, Dunkel M, Gewiess A, Preissner S. Polymorphic cytochrome P450 enzymes (CYPs) and their role in personalized therapy. PLoS One 2013;8(12):e82562.

[6] Rodriguez-Antona C, Ingelman-Sundberg M. Cytochrome P450 pharmacogenetics and cancer. Oncogene 2006;25:1679–91.

[7] Brauch H, Schroth W, Eichelbaum M, Schwab M, Harbeck N. Clinical relevance of CYP2D6 genetics for tamoxifen response in breast cancer. Breast Care (Basel) 2008;3:43–50.

[8] Stearns V, Johnson MD, Rae JM, Morocho A, Novielli A, Bhargava P, et al. Active tamoxifen metabolite plasma concentrations after coadministration of tamoxifen and the selective serotonin reuptake inhibitor paroxetine. J Natl Cancer Inst 2003;95:1758–64.

[9] Iusuf D, Teunissen SF, Wagenaar E, Rosing H, Beijnen JH, Schinkel AH. P-glycoprotein (ABCB1) transports the primary active tamoxifen metabolites endoxifen and 4-hydroxytamoxifen and restricts their brain penetration. J Pharmacol Exp Ther 2011;337:710–17.

[10] Teft WA, Mansell SE, Kim RB. Endoxifen, the active metabolite of tamoxifen, is a substrate of the efflux transporter P-glycoprotein (multidrug resistance 1). Drug Metab Dispos 2011;39:558–62.

[11] Grover S, Gourie-Devi M, Bala K, Sharma S, Kukreti R. Genetic association analysis of transporters identifies ABCC2 loci for seizure control in women with epilepsy on first-line antiepileptic drugs. Pharmacogenet Genomics 2012;22(6):447–65.

[12] Kim C, Tang G, Pogue-Geile KL, Costantino JP, Baehner FL, Baker J, et al. Estrogen receptor (ESR1) mRNA expression and benefit from tamoxifen in the treatment and prevention of estrogen receptor-positive breast cancer. J Clin Oncol 2011;29(31):4160–7.

[13] Liu R, Gong C, Tao L, Yang W, Zheng X, Ma P, et al. Influence of genetic polymorphisms on the pharmacokinetics of celecoxib and its two main metabolites in healthy Chinese subjects. Eur J Pharm Sci 2015;79:13–19.

[14] Wang L, Bao SH, Pan PP, Xia MM, Chen MC, Liang BQ, et al. Effect of CYP2C9 genetic polymorphism on the metabolism of flurbiprofen in vitro. Drug Dev Ind Pharm 2015;41(8):1363–7.

[15] Aithal GP, Day CP, Kesteven PJ, Daly AK. Association of polymorphisms in the cytochrome P450 CYP2C9 with warfarin dose requirement and risk of bleeding complications. Lancet 1999;353(9154): 717–19.

[16] Mrazek DA, Biernacka JM, O'Kane DJ, Black JL, Cunningham JM, Drews MS, et al. CYP2C19 variation and citalopram response. Pharmacogenet Genomics 2011;21(1):1–9.

[17] Kim SH, Yim DS, Choi SM, Kwon JC, Han S, Lee DG, et al. Voriconazole-related severe adverse events: clinical application of therapeutic drug monitoring in Korean patients. Int J Infect Dis 2011;15(11): e753–8.

[18] Tang HL, Li Y, Hu YF, Xie HG, Zhai SD. Effects of CYP2C19 loss-of-function variants on the eradication of H. pylori infection in patients treated with proton pump inhibitor-based triple therapy regimens: a meta-analysis of randomized clinical trials. PLoS One 2013;8(4):e62162.

[19] Dean L. Codeine therapy and CYP2D6 Genotype. Medical genetics summaries. Created: September 20, 2012; Last Update: March 8, 2016.

[20] Lee SC, Ng SS, Oldenburg J, Chong PY, Rost S, Guo JY, et al. Interethnic variability of warfarin maintenance requirement is explained by VKORC1 genotype in an Asian population. Clin Pharmacol Ther 2006;79(3):197–205.

[21] Shen GQ, Abdullah KG, Wang QK. The TaqMan method for SNP genotyping. Methods Mol Biol 2009;578:293–306.

[22] Ronaghi M. Pyrosequencing sheds light on DNA sequencing. Genome Res 2001;11(1):3–11.

[23] Gabriel S, Ziaugra L, Tabbaa D. SNP Genotyping Using the Sequenom MassARRAY iPLEX Platform. Curr Protoc Hum Genet 2009 2.12.1-2.12.18.

[24] Norrgard K. Genetic variation and disease: GWAS. Nat Educ 2008;1(1):87.

[25] Welter D, MacArthur J, Morales J, Burdett T, Hall P, Junkins H, et al. The NHGRI GWAS Catalog, a curated resource of SNP-trait associations. Nucleic Acids Research 2014;42(Database issue):D1001–6.

[26] Voelkerding KV, Dames SA, Durtschi JD. Next-generation sequencing: from basic research to diagnostics. Clin Chem. 2009;55(4):641–58.

[27] Wheeler DA, Srinivasan M, Egholm M, Shen Y, Chen L, McGuire A, et al. The complete genome of an individual by massively parallel DNA sequencing. Nature. 2008;452(7189):872–6.

[28] McKernan KJ, Peckham HE, Costa GL, McLaughlin SF, Fu Y, Tsung EF, et al. Sequence and structural variation in a human genome uncovered by short-read, massively parallel ligation sequencing using two-base encoding. Genome Res. 2009;19(9):1527–41.

[29] Valouev A, Ichikawa J, Tonthat T, Stuart J, Ranade S, Peckham H, et al. A high-resolution, nucleosome position map of C. elegans reveals a lack of universal sequence-dictated positioning. Genome Res. 2008;18(7): 1051–63.

[30] Schadt EE, Turner S, Kasarskis A. A window into third-generation sequencing. Hum Mol Genet 2010;19(R2):R227–40.

[31] Derveaux S, Vandesompele J, Hellemans J. How to do successful gene expression analysis using real-time PCR. Methods. 2010;50(4):227–30.

[32] Tarca AL, Romero R, Draghici S. Analysis of microarray experiments of gene expression profiling. Am J Obstet Gynecol 2006;195(2):373–88.

[33] Wang Z, Gerstein M, Snyder M. RNA-Seq: a revolutionary tool for transcriptomics. Nat Rev Genet 2009;10(1):57–63.

[34] Hewett M, Oliver DE, Rubin DL, Easton KL, Stuart JM, Altman RB, et al. PharmGKB: the Pharmacogenetics Knowledge Base. Nucleic Acids Res 2002;30(1):163–5.

[35] Wagner AH, Coffman AC, Ainscough BJ, Spies NC, Skidmore ZL, Campbell KM, et al. DGIdb 2.0: mining clinically relevant drug-gene interactions. Nucleic Acids Res 2016;44(D1):D1036–44.

[36] Kuhn M, Letunic I, Jensen LJ, Bork P. The SIDER database of drugs and side effects. Nucleic Acids Res 2016;44(D1):D1075–9.

[37] Franceschini A, Szklarczyk D, Frankild S, Kuhn M, Simonovic M, Roth A, et al. STRING v9.1: protein-protein interaction networks, with increased coverage and integration. Nucleic Acids Res 2013;41 (Database issue):D808–15.

[38] Kuhn M, von Mering C, Campillos M, Jensen LJ, Bork P. STITCH: interaction networks of chemicals and proteins. Nucleic Acids Res 2008;36(Database issue):D684–8.

[39] Mattingly CJ, Rosenstein MC, Colby GT, Forrest Jr JN, Boyer JL. The Comparative Toxicogenomics Database (CTD): a resource for comparative toxicological studies. J Exp Zool A Comp Exp Biol 2006;305(9):689–92.

[40] Mattingly CJ, Rosenstein MC, Davis AP, Colby GT, Forrest Jr JN, Boyer JL. The comparative toxicogenomics database: a cross-species resource for building chemical-gene interaction networks. Toxicol Sci. 2006;92(2):587–95.

[41] Mattingly CJ, Colby GT, Forrest JN, Boyer JL. The Comparative Toxicogenomics Database (CTD). Environ Health Perspect 2003;111(6):793–5.

[42] Zhang B, Kirov S, Snoddy J. WebGestalt: an integrated system for exploring gene sets in various biological contexts. Nucleic Acids Res 2005;33(Web Server issue):W741–8.

[43] Ng PC, Henikoff S. Predicting the effects of amino acid substitutions on protein function. Annu Rev Genomics Hum Genet 2006;7:61–80.

[44] Huang da W, Sherman BT, Lempicki RA. Systematic and integrative analysis of large gene lists using DAVID bioinformatics resources. Nat Protoc 2009;4(1):44–57.

[45] Huang da W, Sherman BT, Lempicki RA. Bioinformatics enrichment tools: paths toward the comprehensive functional analysis of large gene lists. Nucleic Acids Res 2009;37(1):1–13.

[46] Ward LD, Kellis M. HaploReg: a resource for exploring chromatin states, conservation, and regulatory motif alterations within sets of genetically linked variants. Nucleic Acids Res 2012;40(Database issue): D930–4.

[47] Wingender E, Chen X, Hehl R, Karas H, Liebich I, Matys V, et al. TRANSFAC: an integrated system for gene expression regulation. Nucleic Acids Res 2000;28(1):316–19.

[48] Drozda K, Wong S, Patel SR, Bress AP, Nutescu EA, Kittles RA, et al. Poor warfarin dose prediction with pharmacogenetic algorithms that exclude genotypes important for African Americans. Pharmacogenet Genomics. 2015;25(2):73–81.

[49] Tang XF, Wang J, Zhang JH, Meng XM, Xu B, Qiao SB, et al. Effect of the CYP2C19 2 and 3 genotypes, ABCB1 C3435T and PON1 Q192R alleles on the pharmacodynamics and adverse clinical events of clopidogrel in Chinese people after percutaneous coronary intervention. Eur J Clin Pharmacol 2013;69(5):1103–12.

[50] Potkin SG, Preskorn S, Hochfeld M, Meng X. A thorough QTc study of 3 doses of iloperidone including metabolic inhibition via CYP2D6 and/or CYP3A4 and a comparison to quetiapine and ziprasidone. J Clin Psychopharmacol 2013;33(1):3–10.

CHAPTER

CLINICAL RESEARCH QUALITY ASSURANCE AND AUDITS

Geeta O. Bedi
Providentia Research, Gurgaon, Haryana, India

8.1 DEFINITION

Audit A systematic and independent examination of trial-related activities and documents to determine whether the evaluated trial-related activities were conducted, and whether the data were recorded, analyzed and accurately reported according to the protocol, sponsor's Standard Operating Procedures (SOPs), good clinical practice (GCP), and the applicable regulatory requirement(s). (ICH GCP)

Clinical trial Any investigation in human subjects intended to discover or verify the clinical, pharmacological, and/or other pharmacodynamics effects of one or more investigational medicinal product(s), and/or to identify any adverse reactions to one or more investigational medicinal product(s), and/or to study absorption, distribution, metabolism, and excretion of one or more investigational medicinal product(s), with the object of ascertaining its (their) safety and/or efficacy.

Sponsor An individual, company, institution, or organization which takes responsibility for the initiation, management, and/or financing of a clinical trial.

Quality assurance (QA) All those planned and systematic actions that are established to ensure that the trial is performed and the data are generated, documented (recorded), and reported in compliance with Good Clinical Practice (GCP) and the applicable regulatory requirement(s). (ICH E6)

Quality control (QC) The operational techniques and activities undertaken within the quality assurance system to verify that the requirements for the quality of the trial-related activities have been fulfilled. (ICH GCP)

8.2 WHAT IS AN AUDIT?

The word "audit" comes from the Latin word audire, meaning "to hear."

Auditing originated over 2000 years ago to present publically oral accounts of the handling of funds before a responsible official (an auditor).

In general, auditing involves gathering and evaluating evidence with an unbiased attitude of mind in a structured and ordered manner. Independence of auditors is of great importance as without independence an audit is virtually worthless.

Audits can be:

1. External audits
2. Internal audits

Planning an audit has two phases—audit strategy development and audit program design.

Pharmaceutical Medicine and Translational Clinical Research. DOI: http://dx.doi.org/10.1016/B978-0-12-802103-3.00009-2

8.3 CLINICAL TRIALS AND GOOD CLINICAL PRACTICE (GCP)

Clinical trials are studies that are intended to discover or verify the effects of one or more investigational medicine. The regulation of clinical trials aims to ensure that the rights, safety, and wellbeing of trial subjects are protected, and the results of clinical trials are credible. Clinical trial data is included in clinical study reports that form a large part of the application dossiers submitted by pharmaceutical companies applying for a marketing authorization.

8.3.1 GOOD CLINICAL PRACTICE (GCP)

Good clinical practice (GCP) is an international ethical and scientific quality standard for designing, recording, and reporting trials that involve the participation of human subjects. Compliance with this standard provides public assurance that the rights, safety, and wellbeing of trial subjects are protected and that clinical trial data are credible.

Good clinical practice (GCP) is an international quality standard that is provided by the International Conference on Harmonisation of Technical Requirements for Registration of Pharmaceuticals for Human Use (ICH), an international body that defines a set of standards which governments can then transpose into regulations for clinical trials involving human subjects. High standards are required in terms of comprehensive documentation for the clinical protocol, record keeping, training, and facilities, including computer software. Quality assurance and audits ensure that these standards are achieved.

The ICH GCP Guideline was originally adopted around 20 years ago and still provides an excellent standard for the conduct of clinical trials in humans. Since that time, the scale, complexity, and cost of trials have increased. In particular, there is an increasing use of electronic systems to record data.

The protection of clinical trial subjects is consistent with the principles set out in the Declaration of Helsinki statement of ethical principles developed by the World Medical Association.

Requirements for the conduct of clinical trials in the European Union (EU), including GCP and good manufacturing practice (GMP) and GCP or GMP audits, are implemented in:

- The "Clinical Trial Directive" (Directive 2001/20/EC)
- The "GCP Directive" (Directive 2005/28/EC)

Regardless of where they are conducted, all clinical trials included in applications for marketing authorization in the EEA must be in accordance with:

- Directive 2001/83/EC Annex I, as amended by Directive 2003/63/EC
- The ethical standards of the Clinical Trials Directive (Directive 2001/20/EC)

The sponsor is ultimately responsible for the quality and the integrity of the trial data.

8.4 QUALITY ASSURANCE AND GCP

The key elements of the quality system include:

- Documented procedures and validated methods being developed, implemented, and kept up-to-date

- Documentation system that preserves and allows for the retrieval of any information/documentation (quality records/essential documents) to show actions taken, decisions made and results
- Appropriate training of sponsor personnel as well as of the personnel in affiliates, at the Contract Research Organisations (CROs), vendors, or other service providers, and at trial sites/ Reflection paper on risk-based quality management in clinical trials EMA/269011/2013
- Validation of computerized systems
- Quality control, for example monitoring of trial sites and central technical facilities onsite and/or by using centralized monitoring techniques
- Quality assurance—including internal and external audits performed by independent auditors

ICH GCP requires that the sponsor implements and maintains systems for quality assurance and quality control; similarly the Article 2 of the GCP Directive 2005/28/EC requires the implementation of procedures necessary to secure the quality of every aspect of the trial. The aim of these quality management procedures is to provide assurance that the rights, safety, and wellbeing of trial subjects are protected, and that the results of the clinical trials are credible. The same requirements apply to Contract Research Organisations (CROs), vendors, or other service providers to whom the sponsor has delegated any trial-related duties and functions of the sponsor.

The selection of the required type and number of audits is based on regulatory risk assessment strategies taking the following into consideration:

- Scientific importance of the clinical study within a regulatory submission (i.e., pivotal, first in the new therapeutic, important for key labeling, etc.)
- Compliance history of the auditee
- Complexity of the auditee

Audit program development generation of an audit master schedule for all projects outlining the type and number of audits to be conducted, the assigned auditor(s), the location of audit, number of audits to be conducted and the proposed timelines. Following types of audits are included in the audit program:

- Clinical Investigator Site Audits (routine GCP investigator site audits as well as "for cause" audits) to determine clinical investigator, sponsor, and monitoring compliance with the protocol, regulatory requirements, ICH GCP guidelines, and applicable procedural documents. To verify the integrity and accuracy of the data, and to ensure the rights and safety of the subjects.
  - Site selection criteria includes but is not limited to: subject accrual-high and/or fast enrolling sites
  - Known compliance concerns identified
  - Clinical study site and/or monitoring factors, e.g., new to research, frequent changes in personnel, geographical location
  - Initiation of audits is generally planned after approximately 20% to 50% of subjects have been enrolled and monitored
  - For cause and focus audits are generally performed due to concern raised by clinical research or other responsible personnel regarding the conduct of the study
  - For Phase I studies, periodic qualification of Phase I clinics once every one to three years

- System audits conduced to verify internal processes, procedures to determine whether they are correctly documented, implemented, and compliant with regulatory and project specific requirements.
- Service provider audits include: contract research organizations, vendors providing different services, e.g., monitoring, medical communications, biostatistics, laboratories for bioanalytical and pharmacokinetic testing, central laboratories, investigational drug distribution and randomization, clinical data management. Generally qualification audits of service providers is performed prior to contracting to assess that quality and regulatory compliance requirements are met. On study service provider audits are also done to evaluate the implemented structure and process.
- Database audits and regulatory documentation audits are also conducted for submissions.
- Clinical bioanalytical and core laboratory audits are conducted to determine if the bioanalytical methods have been validated, to ensure that written procedures have been established and are relevant to the operation of the laboratory, and that laboratory personnel are qualified through training and/or experience to perform assigned tasks.

Audit sampling is an important aspect of audit procedures which involves examination of less than 100% items from a defined data. Statistical sampling is advantageous compared to judgmental sampling as it is unbiased and permits quantification of sampling risk. For investigator sites, the minimum number of sites to be audited, sampling criteria of square root (Sqrt) of the lot/population/size (N) + 1.

8.5 OBJECTIVE OF AUDIT

- The objective of a clinical trial audit, which is independent of and separate from routine monitoring or quality control functions, should be to evaluate trial conduct and compliance with the protocol, SOPs, GCP, and the applicable regulatory requirements.

8.6 SCOPE OF A CLINICAL QUALITY ASSURANCE AUDIT

The scope of the clinical audit should be identified and agreed prior to the audit in a formal plan/document, in accordance with company procedures. Areas covered in the scope are usually general/administrative, organization and personnel (including as appropriate, organization, staff qualification and training, Quality assurance and quality control, archives), SOPs, facility, clinical study management/monitoring, pharmacovigilance/safety, investigational product management, data management, biostatistics, medical writing, etc.

8.7 HOW IS THE AUDIT CONDUCTED?

There could be two different approaches:

- System audit
- Specific clinical trial audit

The sponsor should ensure that the auditing of clinical trials/systems is conducted in accordance with the sponsor's written procedures on what to audit, how to audit, the frequency of audits, and the form and content of audit reports.

The sponsor's audit plan and procedures for a trial audit should be guided by the importance of the trial to submissions to regulatory authorities, the number of subjects in the trial, the type and complexity of the trial, the level of risks to the trial subjects, and any identified problem(s).

Audits can be conducted at any time during a clinical trial. Generally, onsite audits are performed after the first patient or healthy subject has been included and the first monitoring visit has taken place.

The trial site is normally informed of the intention to perform an audit approximately 4 to 6 weeks before the planned deadline, generally first by telephone, and then shortly after this in writing. The letter to the auditee should include the date of the audit, the names of the auditors and any accompanying experts, an overview of the topics, and a general timeframe.

Before the start and end of the audit an opening/closing meeting must take place between the auditor(s) and the auditee(s).

The purpose of an opening meeting is to:

- Introduce the auditor(s) and provide an opportunity for introductions of auditee(s)
- Explain the scope and objectives of the audit and expectations from the audit
- Provide a short summary of the methods and procedures to be used to conduct the audit
- Confirm that the resources, documents, and facilities needed by the auditor(s) are available.
- Confirm the time and date for the closing meeting and any interim meetings.

The audit activities should be detailed on the audit plan. Nevertheless, during the audit, the auditor(s) may adjust the plan to ensure the audit objectives are achieved. Sufficient information to fulfill the audit objective(s) should be collected through examination of relevant documents with direct access, interviews, and observation of activities, equipment, and conditions in the audited areas. If access to records or copying of documents is refused for any reason or there is any withholding of documents or denial of access to areas to which the auditor has legal access, these refusals should be documented and included in the audit observations. A closing meeting should be conducted to discuss the main findings and clarify doubts, if any.

The following items should be reviewed in a sponsor/Contract Research Organisation (CRO) system audit:

1. *General and administrative*: Company/site details, composition, business history, range of services, areas of expertise, therapeutic areas, geographic locations, certifications/accreditations, procedures for sub contracting work, regulatory history.
2. *Organization and personnel*:
   a. Auditor covers overall general organization, current and historical organization charts, number of staff and current workload, procedure for security of sponsor information, confidentiality and privacy statement/policy/agreement, use of contract/temporary staff, staff experience, handover of responsibilities, etc. The aim is to evaluate if the sponsor/

CRO has a well-established organization for clinical research activities and has a sufficient number of properly qualified and trained personnel for each area.

b. Staff qualification and training: Selection of staff/personnel for a project/study, training records (project/study training, SOP trainings), GCP training.

c. Quality assurance (QA) and Quality Control (QC): This covers whether the organization has the QA unit. If so, staff organization, experience, and workload is discussed. Scope and range of audits conducted, scheduling of audits, audit planning, conduct and reporting, follow up, handling of significant findings, involvement during regulatory inspections, etc.

3. Standard operating procedures (SOPs): This covers the presence of SOPs, SOP on SOP, version control, revision frequency, handling of SOP deviations/serious non compliances, availability of SOPs to staff, etc.
4. Facility: Access control, after hours access, physical security, general organization, cleanliness, fire and water protection (alarm system, etc.), back-up power supply/generator, etc. Archive facility—access, location, security monitoring/fire alarm, type of fire suppression, location of study files, retention period, document retrieval timeframe, offsite storage facility if any.
5. Project management: Adequacy of staff experience and workload, adequate resource allocation for project, number of current projects for project manager, project team meetings, documentation of responsibilities, communications between sponsor/investigator/contract staff/service provider/IRB, etc.
6. Clinical study management and monitoring: SOPs for clinical trial management and monitoring, selection process of investigators, procedures for study start up activities, e.g., CRF development, coordination of clinical supplies/investigational product, contracts/agreements with service providers, monitoring plan, training materials, monitoring visits, drug accountability, IRB notifications, etc.
7. Safety reporting: Adverse events, serious adverse events, reporting timeframes, internal review of safety information, reporting SAEs to the sponsor/IEC/IRB and as applicable to regulatory authorities, procedures for SAE follow-up, code break procedures, reconciliation of SAE, investigator brochure updates.
8. Investigational Product (IP) Management: Review of procedures for the IP receipt, intermediate storage, shipment to investigator site, temperature control, accounting of returned IP, destruction of IP if applicable, overall IP accountability, expiration date monitoring, generation of code-breaks and rapid identification of IP during medical emergency, recall of defective IP, retention samples, etc.
9. Computerized systems: Disaster recovery plans/business continuity plan, qualifications of hardware/software, formal change control process, environmental logs of the computer room, backups, recovery process, etc.
10. Data management and Biostatistics: Type of the data management system to be used (hardware/software), if the system is validated/non validated/CFR Part 11 compliant, Data collection and Processing (CRF storage and protection, CRF status, query tracking and status, SEA, protocol deviations, data management plans, database setup and verification, listings (generation review and verification), DCF reconciliation process, process for locking of database(s), changes to database after lock, generation of statistical analysis plan, validation of

analyses and process documentation (e.g., how tables are checked/audited), validation of SAS programs (programming plan, documentation maintained, version control, etc.)

11. Medical writing: Review of SOPs/training of medical writing group, Clinical study report (CSR) compilation, internal review, QA/QC procedures, review and approval by sponsor, distribution and archival process of CSR.

A full clinical trial audit should include:

- The trial documents—protocol, information and consent form, blank CRF, and the trial report
- The trial procedures
- The presence, completeness and accuracy of essential documents in the trial master file
- The case report forms and source documents
- The trial database and statistical analysis
- A written report of the audit findings for the investigator and other relevant staff, and
- An audit certificate for the trial master file.

Auditors shall adapt the common audit checklist and incorporate local requirements as well. Auditors shall create local corrective and preventive action (CAPA) plans to deal with the findings and follow up with the team.

The use of computerized systems in clinical trials has greatly increased as new technologies have been developed over the years. An addendum to the ICH GCP principles makes it clear that clinical trial information—irrespective of the type of media used—has to be recorded, handled, and stored in a way that allows its accurate reporting, interpretation, and verification. In addition, the systems used to assure the quality of every aspect of the trial should focus on subject protection and the reliability of trial results.

8.8 QUALIFIED AUDITORS

The sponsor should establish an auditing department with qualified auditors so as to ensure the proper conduct of audits as part of implementing Quality Assurance.

The sponsor should only appoint appropriate individual(s) as auditor(s) based on consideration of their education/training, business experience, and ability. Auditor(s) should have knowledge about necessary laws and regulations, GCP, relevant guidelines, the Declaration of Helsinki, clinical and pharmaceutical knowledge, SOPs, computerized system validation, etc. Auditor(s) should also have specific skills e.g., analytical capability, communication, language, writing, etc.

Each auditor's qualification should be documented to verify that they are a suitable person to properly conduct audits, e.g., records of education/training and work experience.

Auditors should be independent of the activities performed In case the unit does not have its own independent auditor, the sponsor or a contract organization can conduct the audit.

Ideally auditors should be a life sciences or a similar/allied field graduate/postgraduate.

For contract auditors CVs, confidentiality agreements, and master service agreements need to be documented.

To maintain knowledge of current audit standards, strategies, and compliance objectives, QA staff are required to monitor requirements of global and local regulatory environments including guidance documents, government inspection reports, enforcement actions, and industry standards, and to maintain oversight of trends and metrics.

8.9 RISK BASED QUALITY MANAGEMENT

The basic idea of risk-based quality management is the identification of the risks on a continuous basis for risk-bearing activities throughout the design, conduct, evaluation, and reporting of clinical trials. The process should start at the time of protocol design so mitigation can be built into the protocol and other trial-related documents (e.g., monitoring plan). In addition to the mitigation of identified risks, opportunities to introduce beneficial and proportionate adaptation of conventional practices regarding the management, monitoring, and conduct of the trial should also be identified. Risk-based quality management is a systematic process put in place to identify, assess, control, communicate, and review the risks associated with the clinical trial during its lifecycle.

ICH Q92 provides references to various tools that can be used to assist in the risk management process, in particular for risk assessment. Application of risk-based quality management approaches to clinical trials can facilitate better and more informed decision making and better utilization of the available resources. Risk management should be appropriately documented and integrated within existing quality systems. It is the responsibility of all involved parties to contribute to the delivery of an effective risk-based quality management system.

Considering the highly competitive audit environment, the risk-based approach is considered for planning the audits resulting in achieving the maximum level of efficient, cost-effective audits.

The auditor assesses the overall risk of material error being present in the unaudited data and identifies the high-risk areas.

With the publication of reflection papers on risk management by FDA and EMA, inspectors will expect implementation of risk-based quality management systems in companies.

The quality management system should use a risk-based approach as described below.

- *Critical Process and Data Identification*: During protocol development, the sponsor should identify those processes and data that are critical to ensuring human subject protection and the reliability of trial results.
- *Risk Identification*: The sponsor should identify risks to critical trial processes and data. Risks should be considered at both the system level (e.g., standard operating procedures, computerized systems, personnel) and clinical trial level (e.g., trial design, data collection, informed consent process).
- *Risk Evaluation*: The sponsor should evaluate the identified risks against existing risk controls by considering:
  - a. The likelihood of errors occurring.
  - b. The extent to which such errors would be detectable.
  - c. The impact of such errors on human subject protection and reliability of trial results.
- *Risk Control*: The sponsor should decide which risks to reduce and/or which risks to accept. The approach used to reduce risk to an acceptable level should be proportionate to the significance of the risk.

- Risk reduction activities may be incorporated into protocol design and implementation, monitoring plans, agreements between parties defining roles and responsibilities, systematic safeguards to ensure adherence to standard operating procedures, and training in processes and procedures. Predefined quality tolerance limits should be established, taking into consideration the medical and statistical characteristics of the variables as well as the statistical design of the trial, to identify systematic issues that can impact subject safety or reliability of trial. Detection of deviations from the predefined quality tolerance limits should trigger an evaluation to determine if action is needed.
- *Risk Communication*: The sponsor should document quality management activities. The sponsor should communicate quality management activities to those who are involved in or affected by such activities, to facilitate risk review and continual improvement during clinical trial execution.
- *Risk Review*: The sponsor should periodically review risk control measures to ascertain whether the implemented quality management activities remain effective and relevant, taking into account emerging knowledge and experience.
- *Risk Reporting*: The sponsor should describe the quality management approach implemented in the trial and summarize important deviations from the predefined quality tolerance limits and remedial actions taken in the clinical study report.

8.10 NONCOMPLIANCE

Noncompliance with the protocol, SOPs, GCP, and/or applicable regulatory requirement(s) by an investigator/institution, or by member(s) of the sponsor's staff, should lead to prompt action by the sponsor to secure compliance. When significant noncompliance is discovered, the sponsor should perform a root cause analysis and implement appropriate corrective and preventive actions.

If required by applicable law or regulation the sponsor should inform the regulatory authority(ies) when the noncompliance is a serious breach of the trial protocol or GCP. If the monitoring and/or auditing identifies serious and/or persistent noncompliance on the part of an investigator/institution, the sponsor should terminate the investigator's/institution's participation in the trial. When an investigator's/institution's participation is terminated because of noncompliance, the sponsor should notify promptly the regulatory authority(ies).

FURTHER READING

CIOMS-WHO International ethical guidelines for biomedical research involving human subjects (Geneva 2002), <http://www.cioms.ch/frame_guidelines_nov_2002.htm>.

EMA Auditors' Working Group - Reflection paper on risk based quality management in clinical trials, <http://www.ema.europa.eu/docs/en_GB/document_library/Scientific_guideline/2013/11/WC500155491.pdf>.

FDA Regulations and guidelines. (a) 21 CRF Part 50 – Protection of Human Subjects. (b) 21 CFR Part 54 – Financial Disclosure. (c) 21 CFR Par 56 – Institutional Review Boards. (d) 21 CFR Part 312 – Investigational New Drug Application. (e) 21 CFR Part 11 – Electronic Records, Electronic Signatures. (f) 21 CFR Part 314.80 – Post reporting of adverse drug experiences.

ICH E6 Guideline on Good Clinical Practice, <http://www.ema.europa.eu/pdfs/human/ich/013595en.pdf>; 1995.

ICH ICH Good Clinical Practices (E6) – Consolidated guideline, May 1997.

ICH ICH Guideline (E8) – General considerations for clinical trials.

ICH Harmonised guideline integrated addendum to ICH E6(R1): guideline for good clinical practice E6(R2) Current Step 4 Version Dated 9 November 2016.

Reflection paper on risk based quality management in clinical trials draft EMA/269011/2013 dated 18 November 2013.

World Medical Association. Declaration of Helsinki: Ethical Principles for Medical Research Involving Human Subjects. Revision 2008 <http://www.wma.net/en/30publications/10policies/b3/index.html>.

SECTION

PHARMACEUTICAL LAW AND ETHICS III

CHAPTER

PHARMACEUTICAL MEDICINE AND LAW

9

Gursharan Singh

Life Sciences, SmartAnalyst India Private Limited, Gurgaon, Haryana, India

9.1 INTRODUCTION

Pharmaceutical products differ from ordinary items of commerce as a consumer cannot independently assess their safety, efficacy, and quality. To safeguard public health, countries have approved comprehensive laws and regulations as well as established national regulatory authorities to ensure that drug registration (including research & development), manufacturing, sale, distribution, and use of pharmaceutical products are regulated appropriately and that the public has access to accurate information on medicines. Apart from safeguarding public health, another critical consideration is to safeguard the investment made by pharmaceutical companies by protecting their intellectual property. The chapter also briefly discusses the criminal and civil law relevant to pharmaceutical medicine, including pharmaceutical product liability, data privacy, and protection and industry codes for ethical pharmaceutical promotion and communication [1].

9.2 SOURCES OF LAW

The sources of law vary across countries, with greater weight placed on some sources than others. The common sources of law are [2]:

- Constitution
- Legislative Enactment—Statute
- Judicial Decisions
- Treaties
- Other sources include academic writings of legal scholars, edicts from a reigning monarch, ruler, or religious group, etc.

9.3 THE LEGAL SYSTEMS

There are two main types of legal system in the world—Common Law and Civil law [2].

Pharmaceutical Medicine and Translational Clinical Research. DOI: http://dx.doi.org/10.1016/B978-0-12-802103-3.00010-9

The common law or case law system is followed by former British colonies as well as the United States. This system is primarily driven by judicial decisions; written constitution or codified laws are not mandatory. There is extensive freedom of contract and generally, anything that is not expressly prohibited by law is considered to be permitted [2].

The civil law system is followed by former French, Dutch, German, Spanish, and Portuguese colonies, including much of Central and South America, Central & Eastern Europe, and East Asia. The civil law system is a codified system of law. It takes its origins from Roman law. There is generally a written constitution based on specific codes (e.g., civil code, codes covering corporate law, administrative law, tax law, and constitutional law) enshrining basic rights and duties. Only legislative enactments are considered binding for all with limited role for judge-made case law. There is limited freedom of contract and writings of legal scholars get significant weightage [2].

9.4 STATUTES, LAWS, REGULATIONS, AND GUIDELINES

Statutes are laws (or acts) enacted by legislatures (Congress/Parliament/Assembly) that govern the country or state. To become law in some countries, statute must be agreed upon by the president of a republic or granted royal assent by a monarch. To enforce law, appropriate regulatory agencies are empowered. These agencies adopt regulations which serve to fill gaps in the legislation and their implementation. The regulations help agencies carry out their duties and mission, as defined in the law. As compared to law, the regulations are easier to pass and amend depending on the current and future requirements. The regulations, once passed, carry the force of law. Guidelines differ from regulations as they do not carry the force of law. However, they are even easier to modify and represent the agencies' current thinking on the subject. They help interpret and comply with laws and regulations [1,3].

9.5 AREAS COVERED BY A COMPREHENSIVE PHARMACEUTICAL LAW

The aim of pharmaceutical law is to provide a regulatory control on availability, marketing, and supply of medicines. Table 9.1 provides a list of areas covered by pharmaceutical laws [1].

9.6 PHARMACEUTICAL LAW IN US

In the US, the pharmaceutical law is covered under Title 21 of the Code of Federal Regulations. It is divided into nine volumes, three chapters, and parts 1–1499. Table 9.2 provides a brief overview of 21 CFR [4].

Table 9.3 provides a brief overview of notable sections of 21 CFR relevant to pharmaceutical medicine [4].

Table 9.1 Areas Covered by Pharmaceutical Law

S. No.	Areas
A	*Control of availability and marketing*
1	Drug Registration
2	National Essential Medicines List/Formulary
3	Prescription and Dispensing Authority
4	Labeling
5	Generic labeling, manufacturing, and substitution
6	Pharmacovigilance
7	Information and advertising
8	Public Education
9	Imposition of fees
10	Price control
11	Herbal medicines, medicines for clinical trials, and orphan drugs
B	*Control of supply mechanisms*
1	Importation of medicines
2	Exportation of medicines
3	Controls, incentives, and disincentives for local manufacture
4	Control of distribution, supply, storage, and sale

Source: Pharmaceutical legislation and regulation, http://apps.who.int/medicinedocs/documents/s19583en/s19583en.pdf; [accessed 15.01.17].

Table 9.2 CFR Composition

Title	Volume	Chapter	Parts	Regulatory Entity
Title 21 Food and Drugs	1	I	1–99	Food and Drug Administration, Department Of Health and Human Services
	2		100–169	
	3		170–199	
	4		200–299	
	5		300–499	
	6		500–599	
	7		600–799	
	8		800–1299	
	9	II	1300–1399	Drug Enforcement Administration, Department of Justice
		III	1400–1499	Office of National Drug Control Policy

Source: USGPO. Electronic code of federal regulations, http://www.ecfr.gov/cgi-bin/text-idx?SID=fac36cdb8bcb4e270695294577a7ff5c&mc=true&tpl=/ecfrbrowse/Title21/21tab_02.tpl; 2017 [accessed 15.01.17].

Table 9.3 Notable Sections of 21 CFR

S. No.	Part	Subject Matter
1	11	Electronic records; electronic signatures
2	21	Protection of privacy
3	26	Mutual recognition of pharmaceutical good manufacturing practice reports, medical device quality system audit reports, and certain medical device product evaluation reports: United States and the European Community
4	50	Protection of human subjects
5	54	Financial disclosure by clinical investigators
6	56	Institutional Review Boards
7	58	Good Laboratory Practice for Nonclinical Laboratory Studies
8	60	Patent term restoration
9	99	Dissemination of information on unapproved/new uses for marketed drugs, biologics, and devices
10	101	Food labeling
11	104	Nutritional quality guidelines for foods
12	105	Foods for special dietary use
13	106	Infant formula requirements pertaining to current good manufacturing practice, quality control procedures, quality factors, records and reports, and notifications
14	107	Infant formula
15	190	Dietary supplements
16	201	Labeling
17	202	Prescription drug advertising
18	203	Prescription drug marketing
19	208	Medication guides for prescription drug products
20	290	Controlled drugs
21	310	New drugs
22	312	Investigational new drug application
23	314	Applications for FDA approval to market a new drug
24	316	Orphan drugs
25	320	Bioavailability and bioequivalence requirements
26	328–369	Over-the-counter (OTC) products
27	500–589	Animal drugs, feeds, and related products
28	600–680	Biologics
29	700–740	Cosmetics
30	800–898	Medical devices

Source: USGPO. Electronic code of federal regulations, http://www.ecfr.gov/cgi-bin/text-idx?SID=fac36cdb8bcb4e270695294577a7ff5c&mc=true&tpl=/ecfrbrowse/Title21/21tab_02.tpl; 2017 [accessed 15.01.17].

9.7 PHARMACEUTICAL LAW IN EU

The body of European Union pharmaceutical law is compiled in Volume 1 and Volume 5 of the publication "The rules governing medicinal products in the European Union." The basic law is supported by a series of guidelines that are also published in the remaining volumes of the publication. Table 9.4 briefly describes the content of various EudraLex volumes [5].

Table 9.4 EudraLex Volumes and Contents

Volume	Topic	Notable Content
1	Pharmaceutical legislation for medicinal products for human use	Directive 2001/83/EC of the European Parliament and of the Council of 6 November 2001 on the Community code relating to medicinal products for human use, last amended in 2012 Directive 2001/20/EC OF the European Parliament and of the Council of 4 April 2001 on the approximation of the laws, regulations and administrative provisions of the Member States relating to the implementation of good clinical practice in the conduct of clinical trials on medicinal products for human use, last amended in 2009 Council Directive 89/105/EEC, of 21 December 1988, relating to the transparency of measures regulating the pricing of medicinal products for human use and their inclusion within the scope of national health insurance systems
2	Notice to applicants and regulatory guidelines for medicinal products for human use	
2A	Procedures for marketing authorization	Chapter 1—Marketing Authorization (latest version—December 2016) Chapter 2—Mutual Recognition (latest version—February 2007) Chapter 3—Union Referral Procedures (latest version—December 2016) Chapter 4—Centralized Procedure (deleted—July 2015). The European Medicines Agency is now responsible for the scientific evaluation of applications for European Union (EU) marketing authorizations for human and veterinary medicines in the centralized procedure. Therefore, this guidance is not a NTA document anymore. Chapter 5—Guidelines of 16 May 2013 on the details of the various categories of variations, on the operation of the procedures laid down in Chapters II, IIa, III, and IV of Commission Regulation (EC) No 1234/2008 of 24 November 2008 concerning the examination of variations to the terms of marketing authorizations for medicinal products and on the documentation to be submitted pursuant to those procedures Chapter 6—Community Marketing Authorization (latest version—November 2005)
2B	Presentation and content of the dossier	Includes information on eSubmission and eCTD (electronic Common Technical Document)

(Continued)

Table 9.4 EudraLex Volumes and Contents *Continued*

Volume	Topic	Notable Content
2C	Regulatory Guideline	Regulatory requirements such as renewal procedures, dossier requirements for Type IA/IB variation notifications, summary of product characteristics (SmPC), package information and classification for the supply, readability of the label, and package leaflet requirements
3	Scientific guidelines for medicinal products for human use	
4	Guidelines for good manufacturing practices for medicinal products for human and veterinary use	
5	EU pharmaceutical legislation for medicinal products for veterinary use	
6	Notice to applicants and regulatory guidelines for medicinal products for veterinary use	
7	Scientific guidelines for medicinal products for veterinary use	
8	Maximum residue limits	
9	Guidelines for pharmacovigilance for medicinal products for human and veterinary use	
9A	Pharmacovigilance for medicinal products for human use (latest version 2008) *With the application of the new pharmacovigilance legislation as from July 2012 Volume 9A is replaced by the Good Pharmacovigilance Practice (GVP) guidelines released by the European Medicines Agency*	
9B	Pharmacovigilance for medicinal products for veterinary use	
10	Guidelines for clinical trial	

Source: European Commission. EudraLex - EU Legislation, http://ec.europa.eu/health/documents/eudralex_en; 2017 [accessed 15.01.17].

9.8 PHARMACEUTICAL LAW IN JAPAN

In Japan, the revised Pharmaceutical Affairs Law was enacted on November 25, 2014 by the Pharmaceutical and Medical Devices Agency. The law covers regulations on drugs, medical devices, and regenerative medicine products. The Pharmaceutical Affairs Law was renamed as the "Act on Securing Quality, Efficacy and Safety of Pharmaceuticals, Medical Devices, Regenerative and Cellular Therapy Products, Gene Therapy Products, and Cosmetics" or "PMD Act." The revised Law consists of 17 chapters and 91 articles as outlined in Table 9.5 [6,7].

9.9 PHARMACEUTICAL LAW IN OTHER COUNTRIES

It is not possible to cover the pharmaceutical laws of all countries of the world in this chapter. The Table 9.6 provides the listing of regulatory agencies and respective websites of certain other countries of the world.

Table 9.5 Chapters and Articles of PMDA

Chapter	Articles	Content
1	1&2	General Provisions
2	3	Prefectural Pharmaceutical Affairs Councils
3	4–11	Pharmacies
4	12–23	Manufacturing/Marketing Businesses of Drugs, Quasi-drugs, and Cosmetics
5	23 up to 23–19	Manufacturing/Marketing Businesses, etc. of Medical Devices and in vitro Diagnostics
6	23–20 to 23–42	Manufacturing/Marketing Businesses of Cellular and Tissue-based Products
7	24–40	Retail sellers
8	41–43	Standards and Government Certification for Drugs
9	44–66	Handling of drugs
10	66–68	Advertising of drugs
11	68–2 to 68–15	Safety of drugs
12	68–16 to 68–25	Special handling of biological products
13	69 to 76–3	Supervision
14	76–4 to 77	Handling of designated substances
15	77–2 to 77–7	Designation of orphan drugs, orphan medical devices, and cellular and tissue-based orphan products
16	78 to 83–5	Miscellaneous provisions
17	83–6 to 91	Penal provisions

Sources: JPMA. Pharmaceutical Administration and Regulations in Japan, http://www.jpma.or.jp/english/parj/pdf/2015_ch02.pdf; 2015 [accessed 15.01.17]; MHLW. Outline of the Law for Partial Revision of the Pharmaceutical Affairs Law (Act No.84 of 2013), http://www.mhlw.go.jp/english/policy/health-medical/pharmaceuticals/dl/150407-01.pdf; 2014 [accessed 15.01.17].

9.10 INTELLECTUAL PROPERTY RIGHTS

Pharmaceutical research and development requires huge investment (up to US$2.6 billion per product reaching the market) and is time consuming (8–12 years). In return, the blockbuster pharmaceutical products like Humira (adalimumab) and Harvoni (ledipasvir/sofosbuvir) had peak annual sales of more than US$10 billion. With such high stakes, it is important that governments provide protection to the investments made by companies in pharmaceutical research and development. This is typically done through laws to protect intellectual property rights. Intellectual property is traditionally divided into two branches: "industrial property" and "copyright." While copyrights cover literary, artistic, and scientific works, industrial property is particularly important for pharmaceutical industry. The fields of industrial property protection include patents (for inventions), industrial designs, trademarks, service marks, commercial names & designations, and protection against unfair competition. The details of various intellectual property rights are covered in separate chapters in this book and the focus of this chapter is to briefly cover the

Table 9.6 Regulatory Agencies of Certain Other Countries of the World

S. No.	Country	Regulatory Agency	Website for Further Details on Pharmaceutical Law
1	UK	Medicines & Healthcare products Regulatory Agency	http://www.mhra.gov.uk/index.htm
2	Australia	Therapeutic Goods Administration	http://www.tga.gov.au/
3	Canada	Health Canada	http://www.hc-sc.gc.ca/dhp-mps/index-eng.php
4	Brazil	Anvisa	http://portal.anvisa.gov.br/contact-us
5	Russian Federation	Federal Service for Surveillance in Healthcare (Roszdravnadzor)	http://www.roszdravnadzor.ru/
6	India	Central Drugs Standard Control Organization	http://cdsco.nic.in/forms/Default.aspx
7	China	China Food and Drug Administration	http://eng.sfda.gov.cn/WS03/CL0755/
8	South Africa	Medicines Control Council	http://www.mccza.com/

Source: WHO. List of globally identified websites of medicines regulatory authorities, http://www.who.int/medicines/areas/quality_safety/regulation_legislation/list_mra_websites_nov2012.pdf; 2012 [accessed 15.01.17].

basic concepts of other areas of pharmaceutical law relevant to the practice of pharmaceutical medicine [9–12].

9.11 GLOBALIZATION AND HARMONIZATION IN THE FIELD OF PHARMACEUTICAL LAW

Laws and regulations evolve within countries over time and therefore, at a given point of time there occur differences in the laws and regulations across the countries. With the increasing cost of innovation, the pharmaceutical industry is continuously reorganizing on a worldwide scale. Over the last two decades, there has been an increase in globalization for innovative as well as generic drugs. Globalization has occurred with respect to both distribution of medicines in new markets as well as shifting of R&D and manufacturing to lower cost markets. There are many barriers to the process of globalization in pharmaceutical industry. These include variations with respect to intellectual property rights, reimbursement and pricing issues, provisions against counterfeit and substandard medicines, regulatory review standards and processes, and quality of locally available pharmaceutical products. In order to overcome these barriers, in recent years attempts have been made towards globalization and harmonization of pharmaceutical issues which affect national pharmaceutical laws. Some of these harmonization initiatives include the International Conference on Drug Regulatory Authorities (ICDRA), the International Council for Harmonization of Technical Requirements for Pharmaceuticals for Human Use (ICH), pharmacopoeias, the Trade Related aspects of Intellectual Property Rights (TRIPS) agreement etc. [1,13].

9.12 CRIMINAL AND CIVIL LAW RELEVANT TO PHARMACEUTICAL MEDICINE

Law is generally classified into civil and criminal. Civil law deals with the private rights of individuals, while criminal law deals with the rights and obligations of individuals in society. Civil litigation is generally brought by private individuals. Crime is an act committed in violation of a law prohibiting it, or omitted in violation of a law ordering it. The criminal act, omission to act, and criminal intent are essential elements or parts of a crime. A crime is punishable by the State. In some instances, an act may give rise to both a criminal offense and a civil liability. The standards of proof differ between criminal prosecution and civil litigation. To convict the accused of a crime, the prosecution must show there is proof beyond reasonable doubt that the person committed the crime. Civil cases are proven on a balance of probabilities—if it is more likely than not that the defendant caused harm or loss to the plaintiff, a court can uphold a civil claim [14–16].

Pharmaceutical laws of different countries create certain statutory offenses for certain acts of commission or omission. Criminal prosecution of pharmaceutical companies may occur in cases such as selling a misbranded drug or device or an adulterated product or fraud or forgery during conduct of clinical trials. There are many examples where pharmaceutical companies have faced both criminal and civil charges for violation of law. In 2012, GlaxoSmithKline LLC (GSK) pleaded guilty and agreed to pay US$3 billion to resolve its criminal and civil liability arising from illegal promotion of Paxil (paroxetine) and Wellbutrin (bupropion), its failure to report safety data about Avandia (rosiglitazone), and its civil liability for alleged pricing fraud allegations. Similarly, in 2009, Pfizer and its subsidiary Pharmacia & Upjohn Company Inc. paid a criminal fine of US $1.195 billion, the largest criminal fine ever imposed in the United States for promoting the sale of Bextra (valdecoxib) for uses and dosages that the FDA specifically declined to approve due to safety concerns. The company also forfeited US$105 million, for a total criminal resolution of US $1.3 billion. In addition, Pfizer paid US$1 billion to resolve allegations under the civil False Claims Act for illegally promoting Bextra (valdecoxib), Geodon (ziprasidone), Zyvox (linezolid), and Lyrica (pregabalin). In 2015, Dong-Pyou Han, a former scientist at Iowa State University in Ames, was sentenced to 57 months imprisonment and fined US$7.2 million for fabricating and falsifying data in HIV vaccine trials [17–19].

9.13 PHARMACEUTICAL PRODUCT LIABILITY

One of the most important areas of intersection of pharmaceutical medicine and civil law is the product liability law. Product liability law deals with the principles under which the consumers harmed by a product may seek compensation for their injuries. In the United States, there are various routes of product liability like strict liability, negligence, warranty, violation of specific statutory provisions of the country or state, and claims based on unfair and deceptive trade practices. Strict liability refers to liability without proof of fault while negligence requires the plaintiff to prove that the manufacturer failed to exercise "reasonable care" in making the product for its intended (normal) or foreseeable uses. Warranties are certain kinds of express or implied representations of fact that the law will enforce against the warrantor. There are three types of warranties

Table 9.7 Recent Pharmaceutical Product Liability Law Suits in the United States

S. No.	Drug/ Manufacturer	Injury	Compensation
1	Yasmin (drospirenone and ethinyl estradiol)/ Bayer	Blood clots, stroke etc.	According to Bayer, as of March 2014 it had settled 8250 cases for US$1.7 billion.
2	Avandia (rosiglitazone)/ GlaxoSmithKline	Heart attack or heart failure	2010: Company paid approximately US$520 million to settle 10,700 cases. 2011: GlaxoSmithKline paid US$250 million to settle 5500 claims where Avandia had resulted in death.
3	Vioxx (rofecoxib)/ Merck	Heart attacks, strokes and deaths	Company paid US$4.85 billion to settle a reported 50,000 claims.
4	Actos (pioglitazone)/ Takeda	Bladder cancer and congestive heart failure	In April 2015, company settled about 9,000 Actos-related lawsuits for US$2.37 billion.

Source: Drug watch. Drug and Medical Device Lawsuits, https://www.drugwatch.com/drug-lawsuits.php; 2016 [accessed 15.01.17].

involving the product's quality or fitness for use: express warranty, implied warranty of merchantability, and implied warranty of fitness for a particular purpose. Some recent pharmaceutical product liability law suits are listed in Table 9.7. Claims in these cases included negligence, strict product liability, breach of express and implied warranties, fraudulent and negligent misrepresentation, fraudulent concealment, medical monitoring, fraud and deceit, etc. [20–23].

The details of these cases are beyond the scope of this textbook. A brief description of the routes of product liabilities in various countries is presented below.

In the United Kingdom, product liability claims may be made under the Consumer Protection Act 1987, which implements the Product Liability Directive, 85/374/EEC and imposes strict liability on the producer of defective products for damage caused by the defect. Other routes of product liability in the UK include negligence, breach of contract, or breach of some statutory obligations, etc. [22,24,25].

In France, Defective Product Liability Law provided for by the Law n°98–389 of 19 May 1998 [art. 1386–1 to 1386–18 of French Civil Code (FCC)] implemented the European Directive 85/374/EEC. This regime is based on the strict liability of the producer for the damage caused by a defect of his product, whether he was bound to the victim by a contract or not. Other routes of product liability include breach of contract along with specific warranties against latent defects (Art. 1641 of the FCC) and warranties for the conformity of the product in matters between consumers and professionals—Article L 211–4 of the Consumer Code. Tortious Liability applies when damage is suffered by a party outside a contractual relationship on the ground of fault or negligence or on the ground of strict liability of the custodian [22,25].

In Germany, the routes of product liability include the Product Liability Act ("PLA"), a strict liability regime which implements the EU Product Liability Directive, 85/374/EC into national law, and a fault-based liability system under the law of tort. In addition, a special liability regime applies to medicinal products (pharmaceuticals) under the Federal Drug Act. The contractual liability only plays a role in the relationship between the end user/consumer and the final seller, and between the members of the supply chain [22,25].

In Japan, there are multiple routes of product liability. These include the Product Liability Law (Law No. 85 of 1994, the "JPLL") which provides for strict liability. Further, the Civil Code of Japan (the "CCJ") provides for breach of contract or tort claim. Where the claimant and the seller of the defective product have a direct contractual relationship, breach of contract claims or implied statutory warranties may be brought for "Liability for Incomplete Performance of Obligations" (CCJ Article 415) and "Warranty against Latent Defects" (CCJ Article 570). Where a claim has failed under the JPLL and no contractual relationship exists, a claimant may bring a tort claim [22–25].

In Australia, the various routes of product liability include the common law tort of negligence which is fault-based, breach of contract; and breach of the provisions of the Australian Consumer Law [22].

In Canada, product liability is grounded in the common law, except in the province of Québec, which is a civil law jurisdiction. The routes of product liability include law of tort and contract law. The common law does not provide for strict liability. However, in the Province of Québec product liability claims are grounded in strict liability. The province specific statutes provide statutory warranties of merchantability and fitness for purpose with respect to the consumer products marketed within the province. Additionally, in 2011, the federal government enacted a broad legislation called the Canada Consumer Product Safety Act (CCPSA), which affects all manufacturers, distributors, retailers, and importers of consumer products in Canada [22].

In Brazil, consumer relations are regulated by the Consumer Protection Code which provides for strict liability with regards to the product itself as well as for a flaw in the product. In cases where a consumer relationship does not exist, the Brazilian Civil Code shall provide for indemnity against illicit acts and also for contract liability [22,25].

In India, product liability is governed by the Consumer Protection Act, 1986, the Sales of Goods Act, 1930, the law of torts, and special statutes pertaining to specific goods [22,25].

In China, while strict liability applies for manufacturers, for other parties such as distributors, transporters, and storekeepers, the liability is fault based. Whenever the product fails to conform to safety regulations, the claimant may raise product liability disputes. While there is a concurrence of product liability and contractual liability, the claimant has to choose one of them [22,25].

In Russia, the tort is not recognized as a legal concept. Russia follows the doctrine of noncontractual obligations. The concept of product liability in Russia began with adoption of the Russian Civil Code (Part Two), and the Law on Consumer Rights Protection, in 1994 and 1992 respectively [25].

9.14 DATA PRIVACY AND PROTECTION

In the United States, the Health Insurance Portability and Accountability Act of 1996 (HIPAA) Privacy Rule sets national standards for the protection of individually identifiable health information by three types of covered entities: health plans, health care clearinghouses, and health care providers who conduct the standard health care transactions electronically. Compliance with the

Privacy Rule has been required since April 14, 2003. The Security Rule sets national standards for protecting the confidentiality, integrity, and availability of electronic protected health information. Compliance with the Security Rule was required since April 20, 2005 [26–28].

In the European Union, data protection is governed by Data Protection Directive (95/46/EC). The EU member states and members of the European Economic Area (EEA) have implemented the directive through domestic legislation. The Directive restricts processing (holding, collecting, using, or disclosing) of personal data (any information relating to an identified or identifiable natural person). In 2012, the European Commission proposed a new General Data Protection Regulation (GDPR). The GDPR is aimed to strengthen citizens' fundamental rights in the digital age, facilitate business by simplifying rules for companies in the Digital Single Market and reducing costly administrative burdens. The Regulation (EU) 2016/679 on the protection of natural persons with regard to the processing of personal data and on the free movement of such data was adopted by the European Parliament on 14 April 2016 repealing Directive 95/46/EC. It entered into force on 24 May 2016 and shall apply from 25 May 2018 establishing a modern and harmonized data protection framework across the EU [29].

From the perspective of pharmaceutical physician and industry, these laws are relevant for clinical trial and pharmacovigilance activities. Violations of data protection laws could result in criminal penalties including fines and imprisonment.

9.15 INDUSTRY CODES FOR ETHICAL PHARMACEUTICAL PROMOTION AND COMMUNICATION

It is the responsibility of pharmaceutical companies to provide scientifically accurate and fair information to healthcare professionals to support good patient care. Apart from national pharmaceutical laws covering these aspects, many multinational pharmaceutical companies adhere to international and national codes of practice on advertising medicines [30].

International Federation of Pharmaceutical Manufacturers and Associations (IFPMA), which represents research-based biopharmaceutical companies and regional and national associations across the world, enacted the IFPMA code in 1981 to promote ethical principles for the pharmaceutical industry. The same has been updated periodically with the latest revision done in September 2012. This revision expanded the code's scope beyond marketing practices to cover interactions with healthcare professionals, medical institutions, and patient organizations. All companies which are members of IFPMA or a member of an IFPMA member association are required to adhere to the code. IFPMA member associations have developed their own national codes, which as a baseline minimum reflect the standards of the IFPMA code. The alleged violations in geography are handled by respective IFPMA affiliated member associations. However, when violation occurs in countries where there is no national code or violation is done by a member company which is not a member of local/regional association, complaints fall within the scope of the IFPMA [30–32]. Table 9.8 lists regional and national codes of practice on advertising medicines.

The IFPMA code of practice covers the following aspects (Table 9.9) pertaining to promotion of medicines as well as all other interactions with the healthcare community [32].

Table 9.8 Regional and National Codes of Practice on Advertising Medicines

S. No.	Country(ies)	IFPMA-Member Association	National Codes
Regional			
1.	Europe	European Federation of Pharmaceutical Industries and Associations	EFPIA Code on the Promotion of Prescription Only Medicines to, and Interactions with, Healthcare Professionals EFPIA Code of Practice on Relationships between the Pharmaceutical Industry and Patient Organizations EFPIA Code on Disclosure of Transfers of Value from Pharmaceutical Companies to Healthcare Professionals and Healthcare Organisations
2.	Central America	Federación Centroamericana de Laboratorios Farmacéuticos	Code of Good Practices for the Promotion of Medicines
National			
1.	Australia	Medicines Australia	Medicines Australia Code of Conduct
2.	Brazil	Interfarma	Código de Conduta
3.	Canada	Rx&D	Code of Ethical Practices
4.	China	R&D-based Pharmaceutical Association in China	Code of Pharmaceutical Marketing Practices
5.	France	Les entreprises du médicament	Dispositions Déontologiques Professionnelles
6.	Germany	Verband Forschender Arzneimittelhersteller e.V. (German Association of Research-Based Pharmaceutical Companies)	FSA Code of Conduct on the Collaboration with Healthcare Professionals FSA Code of Conduct on the Collaboration with Patient Organizations
7.	India	Organisation of Pharmaceutical Producers of India	OPPI Code of Pharmaceutical Marketing Practices
8.	Italy	FARMINDUSTRIA Associazione delle Imprese del Farmaco	Codice deontologico Farmindustria (code of professional conduct)
9.	Japan	Japan Pharmaceutical Manufacturers Association	JPMA Promotion Code for Prescription Drugs
10.	Russia	Association of International Pharmaceuticals Manufacturers	Code of Marketing Practices of the Association of International Pharmaceutical Manufacturers
11.	South Africa	Marketing Code Authority	Code of Marketing Practice for the Marketing and promotion of medicines, medical devices, and in vitro diagnostics
12.	Spain	FARMAINDUSTRIA: The National Association of the Pharmaceutical Industry in Spain	Spanish Code of Good Practices for the Promotion of Medicines and Interaction with Healthcare Professionals Spanish Code of Practice on Relationships between the Pharmaceutical Industry and Patient Organizations

(*Continued*)

Table 9.8 Regional and National Codes of Practice on Advertising Medicines *Continued*

S. No.	Country(ies)	IFPMA-Member Association	National Codes
13.	United Kingdom	Association of the British Pharmaceutical Industry	Code of Practice for the Pharmaceutical Industry
14.	United States	Pharmaceutical Research and Manufacturers of America	Code on Interactions with Healthcare Professionals Principles on Conduct of Clinical Trials and Communication of Clinical Trial Results PhRMA Guiding Principles on Direct to Consumer Advertisements about Prescription Medicines PhRMA Principles on Interactions with Patient Organizations

Source: Francer J, Izquierdo JZ, Music T, Narsai K, Nikidis C, Simmonds H, et al. Ethical pharmaceutical promotion and communications worldwide: codes and regulations. Philosophy Ethics Humanities Med PEHM 2014; 9, 7.

Table 9.9 IFPMA Code of Practice

S. No.	Article	Content
1	3	Pre-Approval Communications and Off-Label Use
2	4	Standards of Promotional Information
	4.1	Consistency of Product Information
	4.2	Accurate and Not Misleading
	4.3	Substantiation
3	5	Printed Promotional Material
	5.1	All Printed Promotional Material including Advertisements
	5.2	Reminder Advertisements
4	6	Electronic Materials, including Audiovisuals
5	7	Interactions with Healthcare Professionals
	7.1	Events and Meetings
	7.2	Sponsorships
	7.3	Guests
	7.4	Fees for Service
	7.5	Gifts and other items
6	8	Samples
7	9	Clinical Research and Transparency
8	10	Support for Continuing Medical Education
9	11	Interactions with Patient Organizations
10	12	Company Procedures and Responsibilities
11	13	Infringements, Complaints and Enforcement

Source: IFPMA. IFPMA Code of Practice, http://www.ifpma.org/wp-content/uploads/2016/01/IFPMA_Code_of_Practice_2012_new_logo.pdf; 2012 [accessed 15.01.17].

REFERENCES

[1] Pharmaceutical legislation and regulation, <http://apps.who.int/medicinedocs/documents/s19583en/s19583en.pdf>; [accessed 15.01.17].

[2] World Bank Group PPPIRC. Legal systems overview, <https://ppp.worldbank.org/public-private-partnership/legislation-regulation/framework-assessment/legal-systems>; 2016 [accessed 15.01.17].

[3] Tobacco control legal consortium. Fact sheet laws, policies and regulations: key terms & concepts, <http://publichealthlawcenter.org/sites/default/files/resources/tclc-fs-laws-policies-regs-commonterms-2015.pdf>; 2015 [accessed 15.01.17].

[4] USGPO. Electronic Code of Federal Regulations, <http://www.ecfr.gov/cgi-bin/text-idx?SID=fac36cdb8bcb4e270695294577a7ff5c&mc=true&tpl=/ecfrbrowse/Title21/21tab_02.tpl>; 2017 [accessed 15.01.17].

[5] European Commission. EudraLex - EU Legislation, http://ec.europa.eu/health/documents/eudralex_en>; 2017 [accessed 15.01.17].

[6] JPMA. Pharmaceutical Administration and Regulations in Japan, <http://www.jpma.or.jp/english/parj/pdf/2015_ch02.pdf>; 2015 [accessed 15.01.17].

[7] MHLW. Outline of the Law for Partial Revision of the Pharmaceutical Affairs Law (Act No.84 of 2013), <http://www.mhlw.go.jp/english/policy/health-medical/pharmaceuticals/dl/150407-01.pdf>; 2014 [accessed 15.01.17].

[8] WHO. List of Globally identified Websites of Medicines Regulatory Authorities, <http://www.who.int/medicines/areas/quality_safety/regulation_legislation/list_mra_websites_nov2012.pdf>; 2012 [accessed 15.01.17].

[9] DiMasi JA, Grabowski HG, Hansen RA. Innovation in the pharmaceutical industry: new estimates of R&D costs. J Health Econ 2016;47:20–33.

[10] Evaluate. EP Vantage 2016 Preview, <http://info.evaluategroup.com/rs/607-YGS-364/images/epv-pb16.pdf>; 2016 [accessed 15.01.17].

[11] WIPO. Introduction (Chapter 1) accessed online at WIPO intellectual property handbook: policy, law and use, <http://www.wipo.int/export/sites/www/about-ip/en/iprm/pdf/ch1.pdf>; 2004 [accessed 15.01.17].

[12] WIPO. Fields of IP Protection (Chapter 2), <http://www.wipo.int/export/sites/www/about-ip/en/iprm/pdf/ch2.pdf>; 2004 [accessed 15.01.17].

[13] ITA. Challenges related to globalization for the J ff G Di t topic: challenges related to globalization for the pharmaceutical industry, <http://www.trade.gov/td/health/2010austin_gren.pdf>; 2010 [accessed 15.01.17].

[14] CSCJA. The Sources of our Law, <http://www.cscja-acjcs.ca/sources_of_law-en.asp?l=4>; 2006 [accessed 15.01.17].

[15] Saylor foundation. Criminal Law, <https://www.saylor.org/site/textbooks/Criminal%20Law.pdf>; n.d. [accessed 15.01.17].

[16] Storm LM. Criminal Law, version 10, Flat World Education, Inc, <http://catalog.flatworldknowledge.com/bookhub/reader/4373?e=storm_1.0-ch01_s03> [accessed 15.01.17].

[17] DOJ Justice News. GlaxoSmithKline to Plead Guilty and Pay $3 Billion to Resolve Fraud Allegations and Failure to Report Safety Data, <https://www.justice.gov/opa/pr/glaxosmithkline-plead-guilty-and-pay-3-billion-resolve-fraud-allegations-and-failure-report>; 2012 [accessed 15.01.17].

[18] DOJ. Justice Department announces Largest Health Care Fraud Settlement in Its History: Pfizer to Pay $2.3 Billion For Fraudulent Marketing, <https://www.justice.gov/sites/default/files/usao-ma/legacy/2012/10/09/Pfizer%20-%20PR%20%28Final%29.pdf>; 2009 [accessed 15.01.17].

[19] Reardon S. US vaccine researcher sentenced to prison for fraud. Nature 523, 138–139, <http://www.nature.com/news/us-vaccine-researcher-sentenced-to-prison-for-fraud-1.17660>; 2015 [accessed 15.01.17].

[20] NYSBA. Litigating the Products Liability Case: Law and Practice, <http://www.nysba.org/WorkArea/DownloadAsset.aspx?id=44097>; 2013 [accessed 15.01.17].

[21] Product Liability - Breach Of Warranty - Seller, Implied, Warranties, and Express - JRank Articles, <http://law.jrank.org/pages/9468/Product-Liability-Breach-Warranty.html#ixzz4VpXZHaM2> [accessed 15.01.17].

[22] ICLG. Product Liability 2016, <http://www.iclg.co.uk/practice-areas/product-liability/product-liability-2016>; 2016 [accessed 15.01.17].

[23] Drug watch. Drug and Medical Device Lawsuits, <https://www.drugwatch.com/drug-lawsuits.php>; 2016 [accessed 15.01.17].

[24] Consumer Protection Act. <http://www.legislation.gov.uk/ukpga/1987/43/pdfs/ukpga_19870043_en.pdf>; 1987 [accessed 15.01.17].

[25] Library of Congress. Tort Law Systems in Selected Countries, <http://www.loc.gov/law/help/reports/pdf/2013-008653%20FINAL.pdf>; 2012 [accessed 15.01.17].

[26] HIPAA. FAQs for Individuals, <https://www.hhs.gov/hipaa/for-individuals/index.html>; [accessed 15.01.17].

[27] Your Rights under HIPAA, <https://www.hhs.gov/hipaa/for-individuals/guidance-materials-for-consumers/index.html>; [accessed 15.01.17].

[28] HIPAA for Professionals, <https://www.hhs.gov/hipaa/for-professionals/index.html>; [accessed 15.01.17].

[29] European Commission. Reform of EU data protection rules, <http://ec.europa.eu/justice/data-protection/reform/index_en.htm>; 2016 [accessed 15.01.17].

[30] Francer J, Izquierdo JZ, Music T, Narsai K, Nikidis C, Simmonds H, et al. Ethical pharmaceutical promotion and communications worldwide: codes and regulations. Philosophy Ethics Humanities Med. PEHM 2014;9:7.

[31] IFPMA Code of Practice, <http://www.ifpma.org/subtopics/code-of-practice-2/?parentid=264>; [accessed 15.01.17].

[32] IFPMA. IFPMA Code of Practice, <http://www.ifpma.org/wp-content/uploads/2016/01/IFPMA_Code_of_Practice_2012_new_logo.pdf>; 2012 [accessed 15.01.17].

CHAPTER

PHARMACEUTICAL REGULATIONS IN THE UNITED STATES: AN OVERVIEW

10

Sunita Narang

Ranbaxy Laboratories Limited, Gurgaon, Haryana, India

The word pharmaceuticals is usually referred to as "drugs" in the United States. a drug is defined as an article whose intended use is for the diagnosis, cure, mitigation, treatment, or prevention of disease or as an article that is intended to affect the structure or function of the body.

10.1 HISTORY [1]

In the United States, the history of pharmaceutical regulation goes back as early as 1820. In 1820, a group of eleven doctors set up the US Pharmacopoeia and came up with the first list of standard drugs. In 1848, the Drug Importation Act was passed by Congress.

In 1906, the original Food and Drug Act was passed by Congress on June 30 and signed by President Theodore Roosevelt. The Act prohibits states from buying and selling food, drinks, and drugs that have been mislabeled and tainted. In 1930, the name of the Food, Drug and Insecticide Administration was shortened to the Food and Drug Administration (FDA).

In 1937, Elixir Sulfanilamide killed 107 persons, many of whom were children, dramatizing the need to establish drug safety before marketing and to pass the pending food and drug law.

In 1938, Congress passed the Federal Food, Drug, and Cosmetic (FDC) Act of 1938, which required that new drugs show safety before selling. This started a new system of drug regulation. The Act also required that safe limits be set for unavoidable poisonous matter and allowed for factory inspections. The Federal Trade Commission was given power to oversee advertising for all FDA regulated products except prescription drugs. The FDA stated that sulfanilamide and other dangerous drugs must be given under the direction of a medical expert. With this began the requirement for prescription-only drugs.

In 1941, nearly 300 deaths and injuries resulted from the use of sulfathiazole tablets, an antibiotic tainted with the sedative phenobarbital. In response, the FDA drastically changed manufacturing and quality controls. These changes led to the development of good manufacturing practices (GMPs).

Pharmaceutical Medicine and Translational Clinical Research. DOI: http://dx.doi.org/10.1016/B978-0-12-802103-3.00011-0

In 1951, Congress passed the Durham-Humphrey Amendment, which defined the kinds of drugs that cannot be used safely without medical supervision. This amendment limited sale of these drugs to prescription only by a medical professional. All other drugs were to be available without a prescription.

In 1962, Thalidomide, a new sleeping pill, caused severe birth defects of the arms and legs in thousands of babies born in Western Europe. The US media reported on how Dr. Frances Kelsey, a FDA medical officer, helped prevent approval and marketing of Thalidomide in the United States. These reports stirred up public support for stronger drug laws.

After this Congress passed the Kefauver-Harris Drug Amendments. For the first time, these laws required drug makers to prove their drug worked before FDA could approve them for sale.

The Advisory Committee on Investigational Drugs met for the first time. This was the first meeting of a committee to advise FDA on product approval and policy on an ongoing basis. In 1966, the FDA contracted the National Academy of Sciences/National Research Council to measure the effectiveness of 4,000 marketed drugs approved on the basis of safety alone between 1938 and 1962. In 1968, the FDA formed the Drug Efficacy Study Implementation (DESI) to carry out recommendations of the National Academy of Sciences investigation of the effectiveness of drugs first sold between 1938 and 1962.

In 1973, The US Supreme Court upheld the 1962 drug effectiveness law and approved the FDA's action to control entire classes of products.

In 1984, drug price competition and the Patent Term Restoration Act (Hatch-Waxman Act) increased the availability of less costly generic drugs by allowing the FDA to approve applications for generic versions of brand-name drugs without repeating the research that proved the safety and effectiveness of the brand-name drugs. The Act also allowed brand-name companies to apply for up to five years' additional patent protection for the new medicines they developed to make up for time lost while their products were going through the FDA's approval process.

In 1993, the FDA launched MedWatch, a system designed to collect reports from health professionals on problems with drugs and other medical products and in 1998 FDA introduced the Adverse Event Reporting System (AERS), a computerized database designed to store and study safety reports on already marketed drugs.

In 2004, the FDA advised medical professionals to limit the use of a pain reliever called Cox-2, a nonsteroidal anti-inflammatory drug (NSAID). Studies had shown that long-term use raised chances of heart attacks and strokes. The warning was also added to the over-the counter NSAIDs' Drug Facts label. Medicines used in hospitals were required to have a bar code to prevent patients from receiving the wrong medicine.

10.2 REGULATING BODIES

The US Department of Health and Human Services (HHS) is the principal agency responsible for protecting the health of all Americans and providing essential human services.

HHS has 11 operating divisions, including eight agencies in the US Public Health Service and three human services agencies. These divisions administer a wide variety of health and human services and conduct life-saving research for the nation.

The Food and Drug Administration (FDA), is one of the 11 Agencies working under the Department of Health and Human Services which ensures that food is safe, pure, and wholesome, that human and animal drugs, biological products, and medical devices are safe and effective, and that electronic products that emit radiation are safe.

The FDA consists of five different offices overseeing different operations [2]: Medical Products and Tobacco, Foods and Veterinary Medicine, Global Regulatory Operations and Policy, Operations and Policy, and Planning, Legislation and Analysis.

Please see below Fig. 10.1 for FDA Organization Chart.

Within the Division of Medical Products and Tobacco is a division called Center for Drug Evaluation and Research (CDER). The Center for Drug Evaluation and Research (CDER) performs an essential public health task by making sure that safe and effective drugs are available to improve the health of people in the United States. CDER regulates over-the-counter and prescription drugs, including biological therapeutics and generic drugs.

There are twelve different offices under the jurisdiction of Center for Drug Evaluation and Research which in turn have their subdivisions. Please refer to Fig. 10.2 for the CDER Organization chart.

CDER's job is to evaluate new drugs before they can be sold. The Center is involved in the review of new drug applications and also provides doctors and patients with the information they

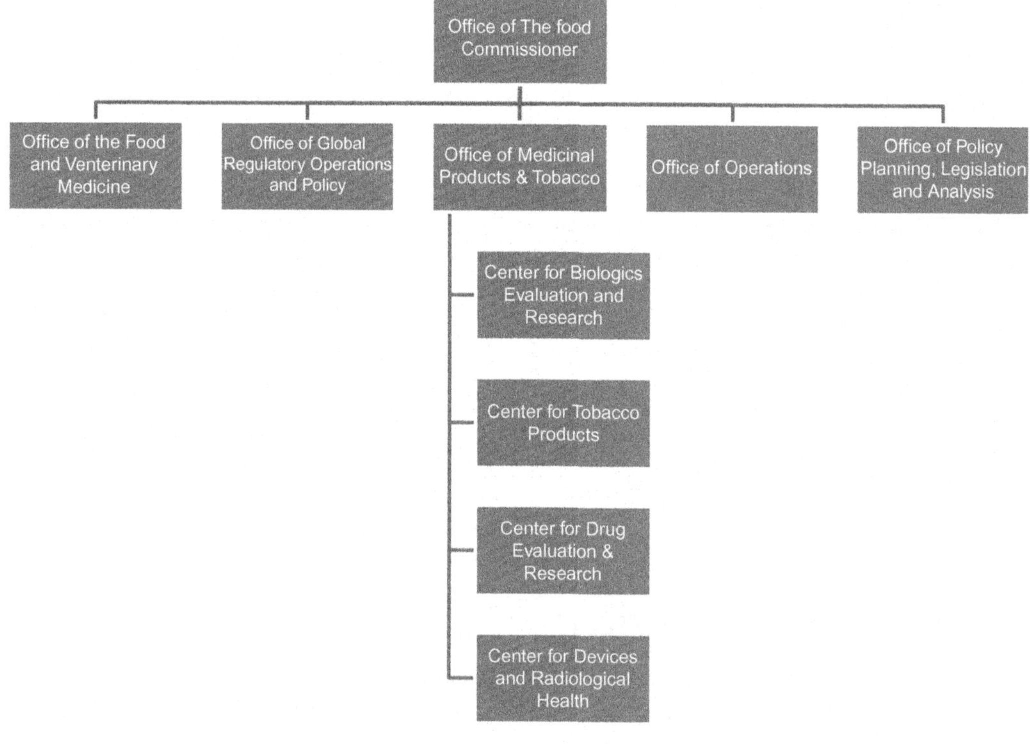

FIGURE 10.1

FDA organization chart.

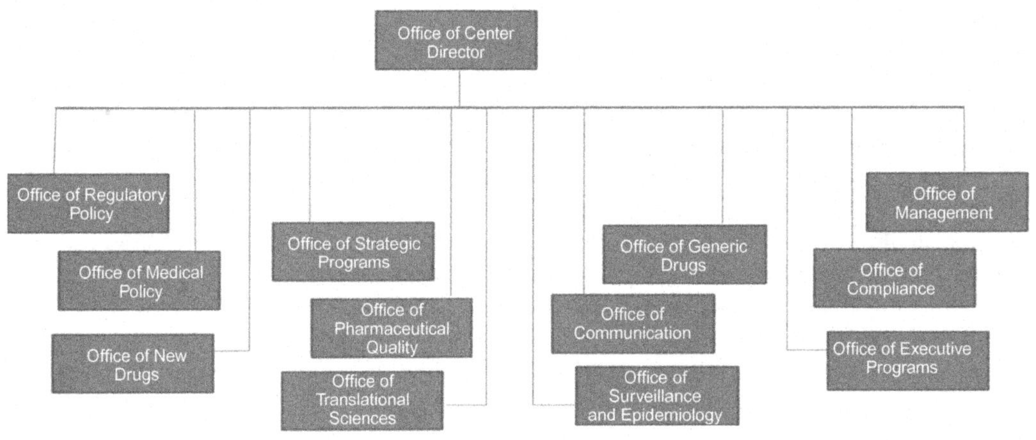

FIGURE 10.2

CDER organization chart.

need to use medicines wisely. CDER makes sure that safe and effective drugs are available to improve the health of consumers. CDER ensures that prescription and over-the-counter drugs, both brand-name and generic, work correctly and that the health benefits outweigh known risks.

The Center has oversight responsibilities for prescription, over-the-counter, and generic drugs. CDER ensures that consumers have access, as quickly as possible, to promising new treatments. The Center oversees the research, development, manufacture, and marketing of drugs. CDER ensures truth in advertising for prescription drugs and monitors the use of marketed drugs for unexpected health risks. If unexpected risks are detected after approval, CDER takes action to inform the public, change a drug's label, or if necessary remove a product from the market.

CDER regulates prescription drugs, generic drugs, and over-the-counter drugs.

All new drugs need proof that they are effective and safe before they can be approved for marketing. CDER decides through evaluation whether the studies submitted by the drug's sponsor (usually the manufacturer) show it to be safe and effective for its intended use. When a proposed drug's benefits outweigh the known risks, CDER considers it safe enough to approve. Once a drug gets CDER approval, the drug is on the market as soon as the firm gets its production and distribution systems going.

10.3 DRUG APPROVAL PROCESS [3]

For marketing a pharmaceutical product or a drug in United States, the applicants or the drug makers need to get the product approved by US FDA. For this approval, each product needs to go through a formal approval process set up by the Centre for Drug Evaluation and Research (CDER).

Two main divisions within CDER are responsible for review and approval of marketing authorization application for the new drugs as well as the generic drugs.

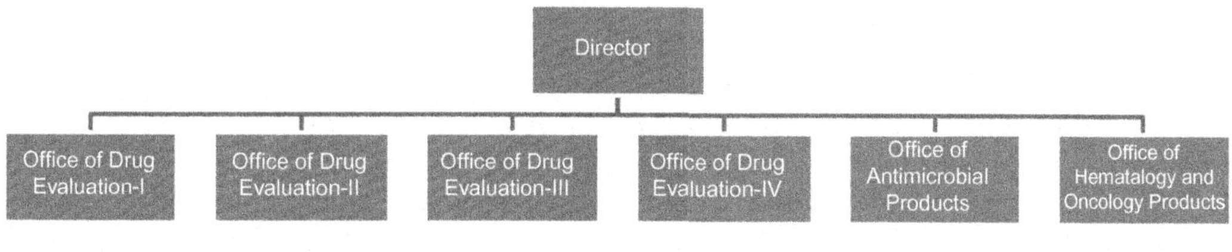

FIGURE 10.3

Office of New Drugs organization chart.

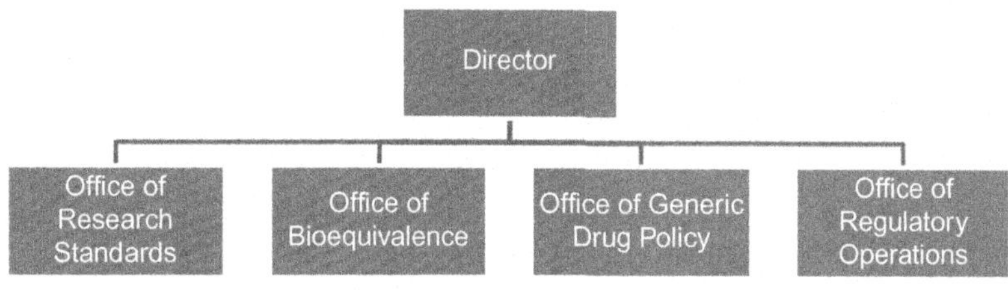

FIGURE 10.4

Office of Generic Drugs organization chart.

The Office of New Drugs (OND) is responsible for providing regulatory oversight for investigational studies during drug development and making decisions regarding marketing approval for new (innovator or nongeneric) drugs, including decisions related to changes to already marketed products. The Office of Generic Drugs (OGD) is responsible for providing regulatory oversight to expedite the availability of safe, effective, and high-quality generic drugs to patients.

Within these two divisions there are specific sub divisions responsible for different activities involved in reviewing various aspects of the marketing authorization application.

Apart from these two divisions, one more division which is largely involved in ensuring the approval of a product is the Office of Compliance. No drug product is given the go ahead for a formal approval by FDA unless the office of compliance certifies that the facilities associated with the pertinent drug product meet the standards as laid out by US law and 21 CFR part 210 and 211.

Please see Fig. 10.3 and Fig. 10.4 for the organization charts for the Office of New Drugs and the Office of Generic Drugs, respectively.

10.4 DRUG APPROVAL PROCESS FOR AN NEW DRUG APPLICATION (NDA)

Companies need submit a new drug application (NDA) to introduce a new drug product into the U.S. Market. It is the responsibility of the company seeking to market a drug to test it and submit evidence that it is safe and effective. A team of CDER physicians, statisticians, chemists, pharmacologists, and other scientists reviews the sponsor's NDA containing the data along with the proposed labeling.

The Federal Food, Drug, and Cosmetic Act is the basic food and drug law of the United States. The law is intended to assure consumers that foods are pure and wholesome, safe to eat, and produced under sanitary conditions; that drugs and devices are safe and effective for their intended uses; that cosmetics are safe and made from appropriate ingredients; and that all labeling and packaging is truthful, informative, and not deceptive.

The NDA review/approval process starts with the filing of an IND followed by the NDA. Before filing an IND, the FDA also provides a pre-IND consultation through mail or meetings to the sponsors regarding their initial development plan and future strategy. A pre-IND meeting or communication is the first step in the NDA process and is an excellent pathway for a sponsor to get some lead about if the FDA agrees with their basic study results and further plan of development and human testing.

10.5 INVESTIGATIONAL NEW DRUG (IND)

Current Federal law requires that a drug be the subject of an approved marketing application before it is transported or distributed across state lines. Because a sponsor will probably want to ship the investigational drug to clinical investigators in many states, it must seek an exemption from that legal requirement. The IND is the means through which the sponsor technically obtains this exemption from the FDA.

During a new drug's early preclinical development, the sponsor's primary goal is to determine if the product is reasonably safe for initial use in humans, and if the compound exhibits pharmacological activity that justifies commercial development. When a product is identified as a viable candidate for further development, the sponsor then focuses on collecting the data and information necessary to establish that the product will not expose humans to unreasonable risks when used in limited, early-stage clinical studies. The FDA's role in the development of a new drug begins when the drug's sponsor (usually the manufacturer or potential marketer), having screened the new molecule for pharmacological activity and acute toxicity potential in animals, wants to test its diagnostic or therapeutic potential in humans. At that point, an applicant needs to file an IND and seek legal permission to test this drug in humans.

The IND application must contain information in three broad areas:

1. Animal Pharmacology and Toxicology Studies
   Preclinical data to permit an assessment as to whether the product is reasonably safe for initial testing in humans. Also included are any previous experiences with the drug in humans (often foreign use).
2. Manufacturing Information
   Information pertaining to the composition, manufacturer, stability, and controls used for manufacturing the drug substance and the drug product. This information is assessed to ensure that the company can adequately produce and supply consistent batches of the drug.

3. Clinical Protocols and Investigator Information
Detailed protocols for proposed clinical studies to assess whether the initial-phase trials will expose subjects to unnecessary risks. Also, information on the qualifications of clinical investigators—professionals (generally physicians) who oversee the administration of the experimental compound—to assess whether they are qualified to fulfill their clinical trial duties. Finally, commitments to obtain informed consent from the research subjects, to obtain review of the study by an institutional review board (IRB), and to adhere to the investigational new drug regulations.

Once the IND is submitted, the sponsor must wait 30 calendar days before initiating any clinical trials. During this time, FDA has an opportunity to review the IND for safety to assure that research subjects will not be subjected to unreasonable risk.

A daily published record of proposed rules, final rules, meeting notices, and other important communication within FDA are collected in the Code Of Federal Regulations (CFR). The CFR is divided into 50 titles that represent broad areas subject to Federal regulations. The FDA's portion of the CFR interprets the The Federal Food, Drug, and Cosmetic Act and related statutes. Section 21 of the CFR contains most regulations pertaining to food and drugs. The regulations document all actions of all drug sponsors that are required under Federal law.

The CFR sections pertinent to an IND application are listed below:

21 CFR Part 312: Investigational New Drug Application
21 CFR Part 314: INDA and NDA Applications for FDA Approval to market a new drug
21 CFR Part 316: Orphan Drugs
21 CFR Part 58: Good Lab Practice for Nonclinical Laboratory Studies
21 CFR Part 50: Protection of Human Subjects
21 CFR Part 56: Institutional Review Boards
21 CFR Part 201: Drug Labeling
21 CFR Part 54: Financial Disclosure by Clinical Investigators

There are detailed guidelines available publically for the sponsors to make sure that the development process and the approval process is being taken up as per FDA acceptable rules and regulations.

A sponsor may need to file more than one IND during the clinical phases of a drug and these IND would need to be kept active through submission of updated information via annual reports to the INDs.

10.6 NEW DRUG APPLICATION (NDA)

Since 1938, every new drug has been the subject of an approved NDA before US commercialization. The NDA application is the vehicle through which drug sponsors formally propose that the FDA approve a new pharmaceutical for sale and marketing in the United States. The data gathered during the animal studies and human clinical trials of an Investigational New Drug (IND) become part of the NDA.

The goals of the NDA are to provide enough information to permit the FDA reviewer to reach the following key decisions:

- Whether the drug is safe and effective in its proposed use and whether the benefits outweigh the risks.
- Whether the proposed labeling is appropriate and supported with the submitted clinical data.
- Whether the methods used in manufacturing the drug and the controls used to maintain drug quality are adequate to preserve the drug's identity, strength, quality, and purity.

An NDA is supposed to tell the drug's whole story, including what happened during the clinical tests, what the ingredients of the drug are, the results of the animal studies, how the drug behaves in the body, and how it is manufactured, processed and packaged.

Guidance documents are available for FDA review staff and applicants/sponsors to provide assistance for the processing, content, and evaluation/approval of applications and also for the design, production, manufacturing, and testing of regulated products.

The Federal Food, Drug, and Cosmetic Act is the basic food and drug law of the United States. With numerous amendments which form the backbone of all regulations related to Foods and Drugs and thus for NDAs also, the basic governing law is The Federal Food, Drug, and Cosmetic Act. In addition, the CFR part which covers the approval process of NDA is

21 CFR Part 314: Applications for FDA Approval to Market a New Drug or an Antibiotic Drug.

An NDA essentially is submitted to the New Drug Division as per ICH CTD (Common Technical Document) structure. The CTD consists of five different modules which represent different sections of the NDA. These modules essentially contain all the information which the FDA may require to make a decision on approvability of the drug product being submitted.

The five modules are listed below:

Module 1: Administrative Information
Module 2: CTD Summaries
Module 3: Quality
Module 4: Nonclinical Study Reports
Module 5: Clinical Study Reports

Quite evident from the names themselves, each module contains information related to these different disciplines.

Module 1: Administrative Information

This module primarily and typically would contain all administrative information including but not limited to the following:

- Submission forms
- Fee receipts
- Proposed labeling
- Cross references
- Patent information
- Cover letter
- Access letters
- Debarment certifications

This module is a regional module and would contain all specific administrative documents required by the FDA as a part of an NDA.

Module 2: CTD Summaries

Module 2 essentially is the summary of whole NDA (Modules 3, 4, and 5). It contains three parts:

- Quality overall summary
- Nonclinical summary
- Clinical Summary

This module gives the reviewers an oversight of what is being provided in the detailed modules of the NDA.

Module 3: Quality

This module contains detailed information about the quality of the drug product. It includes all information related to chemistry, manufacturing, and controls of the drug product. All details of the facilities being used for manufacturing are also a part of this module. Essentially module 3 is divided into the three subsections:

- 32S—Drug Substance
- 32P—Drug Product
- 32R—Regional Information

Module 4: Nonclinical Study Reports

This module contains the detailed reports, evaluation, and conclusions of all the animal studies done during the drug development process to establish the safety of that particular drug product. All nonclinical pharmacological and toxicological data is a part of this module.

Module 5: Clinical Study Reports

This module contains the detailed reports, evaluation, and conclusions of all the human studies done during the drug development process to establish the efficacy of that particular drug product. All detailed clinical data from the studies done on humans is a part of this module.

Together all these five modules are submitted to the FDA as a compiled NDA either electronically or as a paper submission, and thus the review process begins.

NDAs are filed with USFDA under section 505 (b) of FD&C Act. There are two types of NDAs which can be filed under this section:

505 (b)(1) Applications: These applications are for products wherein the sponsor is the owner of all the data from nonclinical and clinical studies done on the drug product during the development process and it does not rely on any previous published literature.

505 (b)(2) Applications: These applications are for products wherein the sponsor's application is largely based on the published literature and can rely upon previous findings of safety and efficacy. Only limited studies are done by the sponsor themselves. The examples of 505 (b)(2) applications are listed below:

Changes from the previously approved drug in:

- Dosage form
- Strength
- Formulation
- Route of administration

- Dosage regimen
- Indication
- Salt of active ingredient

Substitution of an active ingredient in a combination product
Combination of two previously approved products
The Review Process of an NDA involves the following steps:

1. Submission of the application
2. Review for acceptance (preliminary screening)
3. Issuance of acceptance for review /refusal to review
4. Technical review by different divisions
5. Sponsor is provided with the deficiencies and requested to respond to FDA
6. Sponsor responds to the deficiencies
7. Office of New Drugs reviews responses and either accepts the responses or sends a second set of queries
8. Sponsor responds to the second set of queries
9. All technical review divisions provide their recommendation for approvable/nonapprovable status of the NDA
10. Final go ahead on facility status taken from the Office of Compliance
11. FDA issues either of the three: approval letter/approvable letter/nonapprovable letter

Please see Fig. 10.5 for details of the review process of an NDA.

10.7 DRUG APPROVAL PROCESS FOR AN ABBREVIATED NEW DRUG APPLICATION (ANDA)

An Abbreviated New Drug Application (ANDA) contains data which when submitted to FDA's Center for Drug Evaluation and Research, Office of Generic Drugs, provides for the review and ultimate approval of a generic drug product. Once approved, an applicant may manufacture and market the generic drug product to provide a safe, effective, low cost alternative to the American public.

A generic drug product is one that is comparable to an innovator drug product in dosage form, strength, route of administration, quality, performance characteristics, and intended use. Generic drug applications are termed "abbreviated" because they are generally not required to include preclinical (animal) and clinical (human) data to establish safety and effectiveness. Instead, generic applicants must scientifically demonstrate that their product is bioequivalent (i.e., performs in the same manner as the innovator drug).

Using bioequivalence as the basis for approving generic copies of drug products was established by the Drug Price Competition and Patent Term Restoration Act of 1984, also known as the Waxman-Hatch Act. This Act expedites the availability of less costly generic drugs by permitting FDA to approve applications to market generic versions of brand-name drugs without conducting costly and duplicative clinical trials.

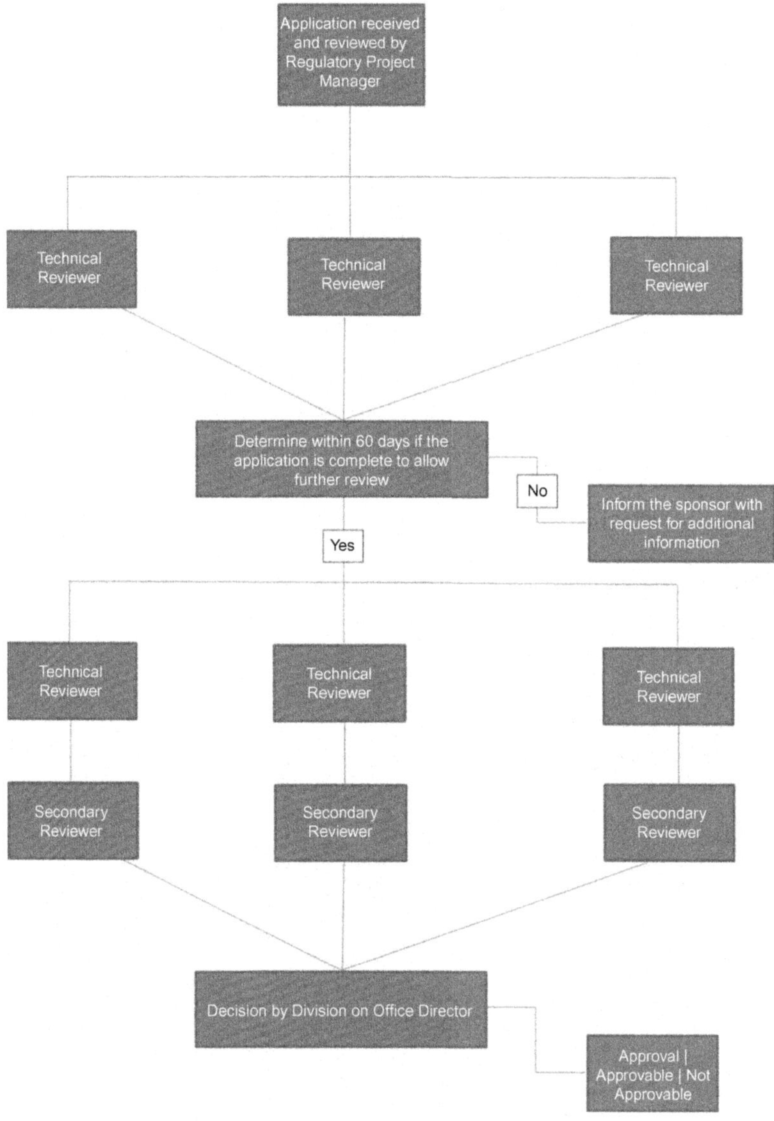

FIGURE 10.5

NDA review process.

All approved products, both innovator and generic, are listed in the FDA's Approved Drug Products with Therapeutic Equivalence Evaluations (Orange Book).

ANDAs are filed with USFDA under section 505 (j) of FD&C Act.

The following CFR sections apply to the ANDA process:

21 CFR Part 314: Applications for FDA Approval to Market a New Drug or/and Antibiotic Drug
21 CFR Part 320: Bioavailability and Bioequivalence Requirements

Numerous Guidance Documents are publically available to help sponsors with Generic Drug development and ANDA Filing and Approval Process.

An ANDA essentially is submitted to the Office of Generic Drugs as per ICH CTD (Common Technical Document) structure. The structure is the same as that of the NDA except that module 4 is usually not submitted and module 5 is much smaller and simpler than an NDA as it would contain only few studies done to prove the bioequivalence.

These modules essentially contain all the information which FDA may require to make a decision on approvability of the drug product being submitted.

The five modules are listed below:

Module 1: Administrative Information
Module 2: CTD Summaries
Module 3: Quality
Module 4: Nonclinical Study Reports (Usually not applicable for an ANDA and thus not submitted
Module 5: Clinical Study Reports (Human BE studies)

Again, as evident from the names themselves, each module contains information related to these different disciplines.

For detailed contents of an ANDA submission in each module, please refer to the details given under NDA.

The ANDA Review Process at Office of Generic Drugs (OGD) is comprised of the following steps:

1. Submission of ANDA
2. Initial screening or filing review
3. Issuance of refuse to receive or acceptance for review
4. Different OGD divisions review the ANDA
5. Deficiency letters issued to sponsor
6. Sponsor responds to deficiencies
7. OGD reviews the response and either accepts it or sends further deficiencies
8. Sponsor responds to second set of deficiencies
9. Office of compliance provides feedback to OGD about facilities
10. OGD issues the final approval/approvable/nonapprovable letter

Please see Fig. 10.6 for the details of review process of an ANDA.

10.8 ORANGE BOOK AND PATENT CERTIFICATIONS

10.8.1 ORANGE BOOK [4]

The publication Approved Drug Products with Therapeutic Equivalence Evaluations (commonly known as the Orange Book) identifies drug products approved on the basis of safety and effectiveness by the Food and Drug Administration (FDA) under the Federal Food, Drug, and Cosmetic Act (the Act) and related patent and exclusivity information.

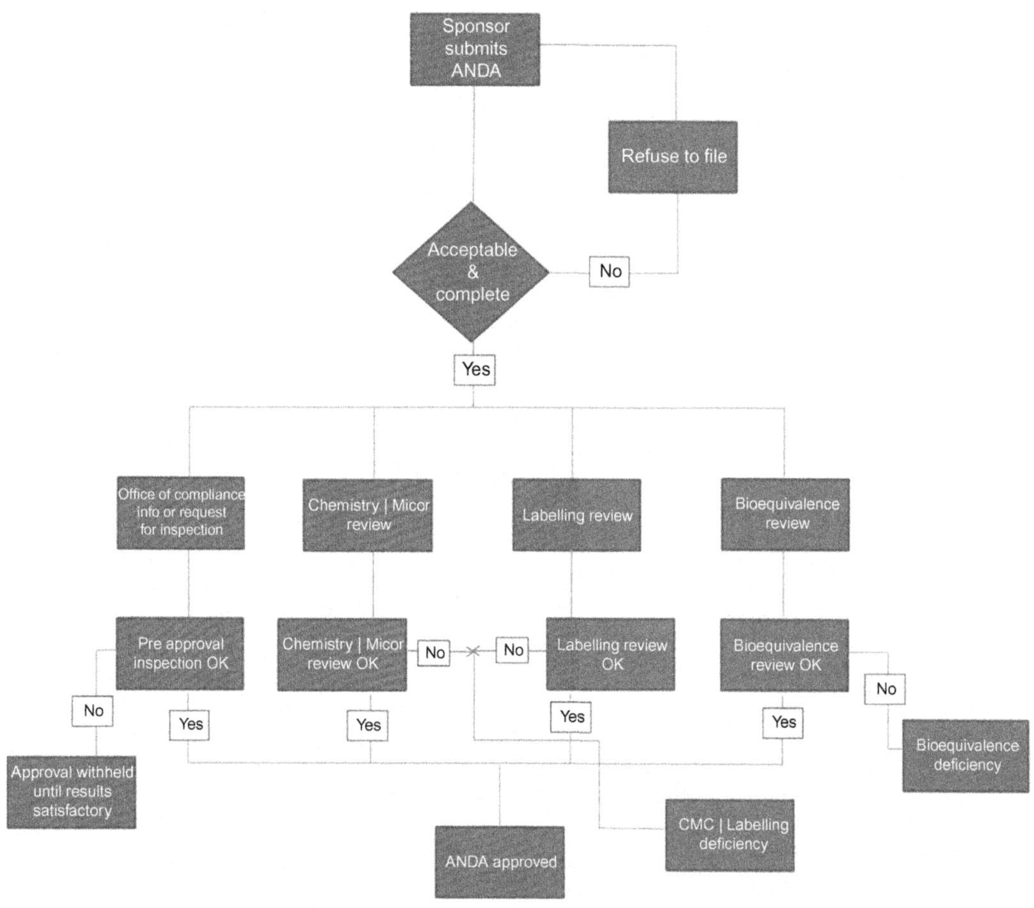

FIGURE 10.6

ANDA review process.

The main criterion for the inclusion of any product is that the product is the subject of an application with an effective approval that has not been withdrawn for safety or efficacy reasons. Inclusion of products on the list is independent of any current regulatory action through administrative or judicial means against a drug product. In addition, the list contains therapeutic equivalence evaluations for approved multisource prescription drug products. These evaluations are prepared to serve as public information and advice to state health agencies, prescribers, and pharmacists to promote public education in the area of drug product selection and to foster containment of healthcare costs.

The list comprises of four parts:

1. Approved prescription drug products with therapeutic equivalence evaluations
2. Approved over-the-counter (OTC) drug products for those drugs that may not be marketed without NDAs or ANDAs because they are not covered under existing OTC monographs

3. Drug products with approval under Section 505 of the FD&C Act administered by the Center for Biologics Evaluation and Research
4. A cumulative list of approved products that have never been marketed, are for exportation, are for military use, or have been discontinued from marketing

Under the FD&C Act, some drug products are given tentative approvals. Such products are not included in the list until the tentative approval becomes a full approval through a subsequent action letter to the applicant.

The Orange Book does not mandate the drug products that are purchased, prescribed, dispensed, or substituted for one another, nor does it, conversely, mandate the products that should be avoided. It does not constitute determinations that any product is in violation of the FD&C Act or that any product is preferable to any other. Exclusion of a drug product from the list does not necessarily mean that the drug product is either in violation of Section 505 of the FD&C Act, or that such a product is not safe or effective, or that such a product is not therapeutically equivalent to other drug products.

Products on the list are identified by the names of the holders of approved applications (applicants) who may or may not be the manufacturer of the product.

Drug products are considered to be therapeutic equivalents only if they are pharmaceutical equivalents and if they can be expected to have the same clinical effect and safety profile when administered to patients under the conditions specified in the labeling.

FDA classifies as therapeutically equivalent those products that meet the following general criteria:

1. They are approved as safe and effective.
2. They are pharmaceutical equivalents in that they:
   a. Contain identical amounts of the same active drug ingredient in the same dosage form and route of administration.
   b. Meet compendial or other applicable standards of strength, quality, purity, and identity.
3. They are bioequivalent in that:
   a. They do not present a known or potential bioequivalence problem, and they meet an acceptable in vitro standard; or
   b. If they do present such a known or potential problem, they are shown to meet an appropriate bioequivalence standard.
4. They are adequately labeled.
5. They are manufactured in compliance with Current Good Manufacturing Practice regulations.

The concept of therapeutic equivalence, as used to develop the list of products in the Orange Book, applies only to drug products containing the same active ingredient(s) and does not encompass a comparison of different therapeutic agents used for the same condition.

The FDA considers drug products to be therapeutically equivalent if they meet the criteria outlined above, even though they may differ in certain other characteristics such as shape, scoring configuration, release mechanisms, packaging, excipients (including colors, flavors, preservatives), expiration date/time, and minor aspects of labeling (e.g., the presence of specific pharmacokinetic information) and storage conditions.

The Orange book uses a coding system to describe the therapeutic equivalence evaluations of drug products being listed. It allows users to determine quickly whether the Agency has evaluated a particular approved product as therapeutically equivalent to other pharmaceutically equivalent products and provides additional information on the basis of the FDA's evaluations.

The generic applicants need to prove therapeutic equivalence against the drug product listed as Reference Listed Drug (RLD) in the Orange Book. A reference listed drug means the listed drug identified by FDA as the drug product upon which an applicant relies in seeking approval of its ANDA.

FDA has identified in the Prescription Drug Product and OTC Drug Product Lists those reference listed drugs to which the in vivo bioequivalence (reference standard) and, in some instances, the in vitro bioequivalence of the applicant's product is compared.

10.9 PATENT CERTIFICATIONS

Both NDA and ANDA applicants need to include the Patent related information or certification in their submissions.

NDA sponsor must identify in NDA those patents reasonably related to drug product, drug substance, or method of using drug for which approval is sought. Based on this NDA holder gets:

- 5-year New Chemical Entity (NCE) Exclusivity
- 3-year New Clinical Studies Exclusivity
- Patent Term Extension to account for time patented product is under review by FDA

Based on the approval information of the NDA:

- FDA "lists" patents identified by NDA sponsors in the "Orange Book" (OB)
- NDA product is referred to as "reference listed drug" or RLD

The ANDA Sponsor must certify with respect to each patent listed for that RLD in the Orange Book that:

- Patent Information has not been filed (Paragraph I Certification)
- The Patent has expired (Paragraph II Certification)
- The date the patent will expire with a certificate that the applicant seeks approval after the expiration of this date (Paragraph III Certification)
- The Patent is invalid or not infringed by the drug product proposed in the ANDA (Paragraph IV Certification)

In case of Paragraph I, II, and III certifications, the generic drug approval is based simply on the ANDA review process and the patent expiry date (for Para III).

In case of a Paragraph IV certification, after FDA notifies the applicant that ANDA is sufficiently complete to review, and the applicant must notify NDA/patent holder of Paragraph IV certification.

- The NDA sponsor can sue when it receives notice
- The Infringement lawsuit can start prior to ANDA approval and marketing
- If NDA sponsor sues within 45 days of notice, ANDA approval is stayed for 30 months
- If there is no lawsuit with 45 days, the FDA can approve ANDA when ready

The ANDA applicants that challenge patents get market entry before patent expiry and in case of First to File (FTF) ANDAs with Paragraph IV certification, the applicant gets 180 day marketing exclusivity for the product after approval.

10.10 OVER THE COUNTER (OTC) PRODUCT APPROVAL PROCESS

OTC drug products are those drugs that are available to consumers without a prescription. There are more than 80 therapeutic categories of OTC drugs, ranging from acne drug products to weight control drug products. The Center for Drug Evaluation and Research at the FDA oversees OTC drug approval process to ensure that they are properly labeled and that their benefits outweigh their risks.

A drug is considered to be an OTC if:

- Its benefits outweigh its risks
- The potential for misuse and abuse is low
- The consumer can use it for self-diagnosed conditions
- It can be adequately labeled
- Health practitioners are not needed for the safe and effective use of the product

Over-the-counter (OTC) drugs are developed under the OTC Monograph Process or through the New Drug Application (NDA) Process. The major difference between the two routes is that in the case of NDA process, formal preapproval from FDA is mandatory, however in case of the monograph process, no formal submission and approval process is necessary. The drug product has to comply completely to the monograph as well as the labeling requirements set forth in 21 CFR parts 330–358.

For OTC drug products developed through the NDA process, the applicant applies to the Division of Nonprescription Drug Products (DNDP) and submits the NDA for the drug product. DNDP oversees the drug development, including the review and regulatory action on Investigational New Drugs (INDs), and also obtains input from the specific subject matter review divisions during the development and review process. Based on the evaluation, the approval or non-approval of the drug product is recommended.

For OTC drug products developed under the OTC drug monograph process, DNDP is responsible for the development of the OTC drug monographs. Data supporting the safety and efficacy of OTC active ingredients in a particular drug monograph are reviewed by appropriate scientific personnel.

Although preapproval by the FDA for drugs marketed under a drug monograph is not required, many companies seek assurance that the product they intend to market under the drug monograph complies with the regulations. If a drug cannot comply with the drug monograph, an IND and approved NDA is necessary before the drug product may be marketed.

In case of over-the-counter drugs, clear and complete labeling information is a majorly important element. The FDA has published a final rule (Over-the-counter Human Drugs: Labeling requirement–Final Rule) to help the applicants finalize the labeling of OTC products which they intend to market. This final rule establishes a standardized format and standardized content requirements for the labeling of OTC drug products. All OTC drug products are required to follow the content and format for labeling as depicted in this rule.

10.11 POST APPROVAL ACTIVITIES

Once the drug product manufacturer obtains formal approval from US FDA to market the product in the United States, the applicant can launch it. From time to time, the manufacturer may require some changes to be made in the drug product that is originally approved.

The applicant must notify the FDA about each change in each condition established in an approved application. The holder of an approved application under section 505 of the Act must assess the effect of the change before distributing a drug product made with a manufacturing change. For each change, the supplement or annual report must contain information developed by the applicant that assesses the effects of the change. The FDA will determine whether this information is appropriate in support of the change.

Notification to the FDA is required for postapproval changes made to the Chemistry, Manufacturing and Controls (CMC) sections previously reviewed and approved, specifically for changes beyond the variations provided for in an approved NDA or ANDA. CDER guidances have been published, including the SUPAC (scale-up and postapproval changes) guidances that provide recommendations on reporting categories.

Section 506A of the FD&C Act and 21 CFR 314.70 provide for following reporting categories to notify changes made in the CMC section(s) of the application:

1. Changes that have a minimal potential to have an adverse effect on the identity, strength, quality, purity, or potency of a drug product are considered to be minor and must be documented and submitted by the applicant in the annual report to the ANDA (an Annual Report for each approved ANDA is required to be submitted every year within 60 days of the anniversary date of the approval).
2. Changes that have moderate potential to have an adverse effect on the identity, strength, quality, purity, or potency of a drug product are considered to be moderate and an applicant needs to submit a Changes Being Effected Supplement (CBE-0/ CBE-30) to the FDA.
   a. Certain moderate changes requiring submission at least 30 days prior to distribution of the drug product made using the hange, and for such changes a CBE 30 supplement is filed.
   b. Certain moderate changes for which distribution of drug product made with the change can occur when FDA receives the supplement, and for such changes a CBE 0 supplement is filed.

   If, after review, the FDA disapproves a CBE-30 or CBE-0, the FDA may order the manufacturer to cease distribution of the drug products made using the disapproved change.
3. Changes that have a substantial potential to have an adverse effect on the identity, strength, quality, purity, or potency of the drug product as these factors may relate to the safety or effectiveness of the drug product are considered to be major changes and a Prior Approval Supplement needs to be submitted to FDA for such changes. These changes cannot be implemented until the applicant receives formal approval from FDA.

In addition to the requirements in Section 506A of the FD&C Act and 21 CFR 314.70, applicants are required to comply with other applicable laws and regulations, including the Current Good Manufacturing Practice for Finished Pharmaceuticals (CGMP) regulations, stated in 21 CFR Parts 210 and 211.

Apart from this, the applicant also needs to keep a close watch on the postmarketing surveillance reports and adverse effects being reported for the drug product, and notify the FDA about these reports.

REFERENCES

[1] <www.fda.gov/AboutFDA/WhatWeDo/History/Milestones/ucm081229.htm>.

[2] <www.fda.gov/aboutfda/centersoffices/default.htm>.

[3] <www.fda.gov/drugs/developmentapprovalprocess/default.htm>.

[4] <www.fda.gov/drugs/informationondrugs/ucm129662.htm>.

CHAPTER

PHARMACEUTICAL REGULATIONS IN EUROPEAN UNION

11

Bharti Khanna
Pharmalex India Pvt. Ltd, New Delhi, India

11.1 EUROPEAN UNION (EU)

The European Union (EU) is a politico-economic union of 28 member states (Please refer to current Brexit process. UK currently is active member of the European Union.) that are located primarily in Europe. The EU covers over 4 million km^2 and has 508 million inhabitants—the world's third largest population after China and India. By surface area, France is the biggest EU country and Malta the smallest. [1,2]

The EU operates through a system of supranational institutions and intergovernmental-negotiated decisions by the member states. The main institutions of the EU are: the European Commission (EC), the Council of the EU, the Court of Justice of the EU, the European Central Bank, the European Court of Auditors, and the European Parliament. The European Parliament is elected by EU citizens every five years. [1,2]

European Council: The European Council brings together EU leaders to set the EU's political agenda. It represents the highest level of political cooperation between EU countries. [1,2]

European Commission: The European Commission (EC) is the EU's politically independent executive arm. It is alone responsible for drawing up proposals for new European legislation, and it implements the decisions of the European Parliament and the Council of the EU. [1,2]

Court of Justice of the European Union: The Court of Justice interprets EU law to make sure it is applied in the same way in all EU countries, and settles legal disputes between national governments and EU institutions. [1,2]

11.1.1 OTHER IMPORTANT ASSOCIATIONS

European Free Trade Association: Norway and Switzerland were among the founding Member States of European Free Trade Association (EFTA) in 1960. Iceland joined EFTA in 1970, followed by Liechtenstein in 1991. Norway, Iceland (from 1994), and Liechtenstein (from 1995) are also parties to the European Economic Area (EEA) Agreement with the EU, while Switzerland has signed a set of bilateral agreements with the EU. [3]

Pharmaceutical Medicine and Translational Clinical Research. DOI: http://dx.doi.org/10.1016/B978-0-12-802103-3.00012-2

European Economic Area (EEA): Norway, Iceland, and Liechtenstein form the EEA with the 28 Member States of the EU (Please refer to current Brexit process. UK currently is active member of the European Union.). These countries have, through the EEA agreement, adopted the complete Community acquis on medicinal products and are consequently parties to the Community procedures. [3]

The legally binding acts from the Community (e.g., Commission decisions) do not directly confer rights and obligations but have first to be transposed into legally binding acts in Norway, Iceland, and Liechtenstein. [3]

11.2 HISTORY OF DRUG REGULATIONS IN EU

The regulation of drugs in the EU started after the Thalidomide Catastrophe. It was the effect of this tragedy that enacted the first pharmaceutical Directive 65/65/EEC in the EU. The teratogenicity of this mild sedative shook public health authorities and the general public and made it very clear that no medicinal product must ever again be marketed without prior authorization. [4,5]

Directive 75/318/EEC and Directive 75/319/EEC were on the approximation of the laws of Member States relating to analytical, pharmacotoxicological and clinical standards and protocols in respect of the testing of proprietary medicinal products. These directives of the EU sought to bring the benefits of innovative pharmaceuticals to patients across Europe by introducing the mutual recognition, by Member States, of their respective national marketing authorizations. [4,5]

Directive 87/22/EEC introduced the concertation procedure which requires national authorities to have an opinion from EU level committees before authorizing innovative medicinal products. [4,5]

The Directive 91/356/EEC established principles of Good Manufacturing Practices for human products so as to improve the quality of medicines in the European Union.

Council Regulation (EEC) No 2309/93 of 22 July 1993 lays down Community procedures for the authorization and supervision of medicinal products for human and veterinary use and establishing a European Agency for the Evaluation of Medicinal Products (EMEA) (now known as EMA, European Medicines Agency) and by Council Directive 88/182/EEC amending Directive 83/189/EEC laying down a procedure for the provision of information in the field of technical standards and regulations. [4,5]

The EMA, based in London, UK, began its operations in 1995. The Agency brings together the scientific resources of the EU member states and works as a network to ensure the highest level of scientific evaluation and supervision of medicines developed by pharmaceutical companies for use in the European Union. [4,5]

The European Union introduced new legislation in 2000 which provides several incentives for the development of orphan and other medicinal products for the treatment of rare disorders. The orphan legislation aims at stimulating research and development of medicinal products for rare diseases by offering incentives to the sponsors. [4,5]

The Directive 2001/20/EC provides the requirements for the conduct of clinical trials on the approximation of the laws, regulations, and administrative provisions of the Member States in the EU. The Directive envisages good clinical practice in the conduct of clinical trials on medicinal products for human use. [4,5]

Directive 2001/83/EC is related to the Community code relating to medicinal products for human use. The Directive deals with the differences between certain national requirements, in particular between requirements relating to medicinal products, which directly affect the functioning of the internal market of the EU. [4,5]

In 2004, the EU became the first region in the world to set up a legal framework and a regulatory pathway for "biosimilars" or "similar biological medicinal products." The regulation on Advanced Therapy Medicinal Products was introduced in 2007. New EU pharmacovigilance rules strengthened the system for safety of medicines in 2010 for better prevention, detection, and assessment of adverse reactions to medicines, and direct patient reporting of adverse events. In 2014, the new Clinical Trial Regulation was introduced to simplify procedures across EU and facilitate cross-border cooperation in international clinical trials. [4,5]

11.3 EUROPEAN DRUG REGULATORY AUTHORITIES

11.3.1 EUROPEAN MEDICINES AGENCY (EMA)

The EMA is a decentralized agency of the European Union that is responsible for the scientific evaluation, supervision, and safety monitoring of medicines developed by the pharmaceutical companies for use in the EU. [6]

11.3.2 EMA STRUCTURE

EMA, headed by the Executive Director, is supported by the Deputy Executive Director, advisory functions, and seven divisions. [7]

11.3.3 EMA DIVISIONS [7]

- Human Medicines Research and Development Support Division
- Human Medicines Evaluation Division
- Inspections, Human Medicines Pharmacovigilance and Committees Division
- Veterinary Medicines Division
- Administration and Corporate Management Division
- Information Management Division
- Stakeholders and Communication Division

11.3.4 EMA SCIENTIFIC COMMITTEES

The EMA has seven scientific committees that carry out its scientific assessments. [6]

- Committee for Medicinal Products for Human Use (CHMP)
- Pharmacovigilance Risk Assessment Committee (PRAC)
- Committee for Medicinal Products for Veterinary Use (CVMP)
- Committee for Orphan Medicinal Products (COMP)

- Committee on Herbal Medicinal Products (HMPC)
- Committee for Advanced Therapies (CAT)
- Pediatric Committee (PDCO)

11.3.5 EMA FUNCTIONS [6]

EMA is responsible for:

- Scientific evaluation of applications for EU marketing authorizations for human and veterinary medicines in the centralized procedure
- Referral procedures
- Safety monitoring of medicines
- Inspections
- Telematics
- Stimulating innovation
- Scientific advice and other assistance
- Publishing guidelines

11.3.6 NATIONAL COMPETENT AUTHORITIES (HUMAN)

In Europe there is a drug regulatory authority at national level as given in Table 11.1. [6]

11.3.7 HEADS OF MEDICINES AGENCIES (HMA)

The Heads of Medicines Agencies (HMA) is a network of the heads of the National Competent Authorities (NCA) whose organizations are responsible for the regulation of medicinal products for human and veterinary use in the European Economic Area. The HMA cooperates with the EMA and the EC in the operation of the European medicines regulatory network and it is a unique model for cooperation and work-sharing on statutory as well as voluntary regulatory activities. [8]

11.3.8 MAIN ACTIVITIES [8]

The HMA:

- Addresses key strategic issues for the network, such as the exchange of information, IT developments, and sharing of best practices
- Focuses on the development, coordination and consistency of the European medicines regulatory system
- Ensures the most effective and efficient use of resources across the network. This includes developing and overseeing arrangements for work-sharing
- Coordinates the mutual recognition (MRP) and decentralized procedures (DCP).

Table 11.1 National Competent Authorities

Member State	Drug Regulatory Authority
Austria	Austrian Agency for Health and Food Safety (AGES)
Belgium	Federal Agency for Medicines and Health Products (Famhp)
Bulgaria	Bulgarian Drug Agency
Croatia	Agency for medicinal products and medical devices of Croatia (HALMED)
Cyprus	Ministry of Health—Pharmaceutical Services
Czech Republic	State Institute for Drug Control (SUKL)
Denmark	Danish Health and Medicines Authority (Sundhedsstyrelsen)
Estonia	State Agency of Medicines
Finland	Finnish Medicines Agency (Fimea)
France	National Agency for the Safety of Medicine and Health Products (ansm)
Germany	• Federal Institute for Drugs and Medical Devices (BfArM) • Paul Ehrlich Institute
Greece	National Organization for Medicines
Hungary	National Institute of Pharmacy
Iceland	Icelandic Medicines Agency
Ireland	Health Products Regulatory Authority (HPRA)
Italy	Italian Medicines Agency
Latvia	State Agency of Medicines
Liechtenstein	Office of Health / Department of Pharmaceuticals
Lithuania	State Medicines Control Agency
Luxembourg	Ministry of Health
Malta	Medicines Authority
Netherlands	Medicines Evaluation Board (MEB)
Norway	Norwegian Medicines Agency
Poland	• Office for Registration of Medicinal Products, Medical Devices and Biocidal Products • Main Pharmaceutical Inspectorate
Portugal	National Authority of Medicines and Health Products (Infarmed)
Romania	National Medicines Agency
Slovakia	State Institute for Drug Control (SUKL)
Slovenia	Agency for Medicinal Products and Medical Devices of the Republic of Slovenia (jazmp)
Spain	Spanish Agency for Medicines and Health Products
Sweden	Medical Products Agency (Lakemedelverket)
United Kingdom	Medicines and Healthcare Products Regulatory Agency (MHRA)

11.3.9 EUROPEAN DIRECTORATE FOR THE QUALITY OF MEDICINES (EDQM)

The EDQM is a leading organization that protects public health by enabling the development, supporting the implementation, and monitoring the application of quality standards for safe medicines and their safe use. Its standards are recognized as a scientific benchmark worldwide. The European Pharmacopoeia is legally binding in EU Member States.

The EDQM, established in 1994, is a Directorate of the Council of Europe and consists of four Departments and five Divisions. [9]

11.3.10 DEPARTMENTS OF THE EDQM

- European Pharmacopoeia Department (EPD)
- Publications and Multimedia Department (DPM)
- Laboratory Department (DLab)
- Biological Standardization, Network of Official Medicines Control Laboratories (OMCL) and HealthCare Department (DBO)

11.3.11 DIVISIONS OF THE EDQM

- Certification of Substances Division (DCEP)
- Reference Standards and Samples Division (DRS)
- Public Relations and Documentation Division (PRDD)
- Administration and Finance Division (DAF)
- Quality, Safety and Environment Division (QSED)

11.4 EU: LEGISLATIVE TOOLS

The EU legal framework for medicinal products for human use is intended to ensure a high level of public health protection and to promote the functioning of the internal market, with measures that encourage innovation. It is based on the principle that the placing on the market of medicinal products is made subject to the granting of a marketing authorization by the competent authorities. [1,10]

11.4.1 LEGALLY BINDING ACTS

Regulations: A regulation is an act of general application, binding in its entirety and directly applicable in all Member States. It does not require any transposition by the national authorities; e.g., Pediatric Regulation. [1,10]

- *Regulation (EC) No 726/2004*: Community procedures for the authorization and supervision of medicinal products for human and veterinary use and establishing EMA
- *Regulation (EC) No 141/2000*: Related to orphan medicinal products
- *Regulation (EU) No 536/2014*: Related to clinical trials on medicinal products for human use

Directives: A directive is a legal act binding upon the Member States to which it is addressed, as far as the results to be achieved are concerned; leaving the national authorities the choice of form and methods; e.g., Clinical Trials Directive. [1,10]

- *Directive 2001/83/EC* [11]: Community code relating to medicinal products for human use. In November 2001, the European Parliament and the Council adopted Directive 2001/83/EC on the Community Code relating to medicinal products for human use. The so-called "Community

Code Directive" combined in one legal act nearly all aspects of European law on medicinal products. This was an act of codification as the contents of all previous directives were fully preserved and amendments were only made as far as required by the merger. The Community Code Directive superseded former directives.
- *Directive 2003/94/EC*: Principles and guidelines of good manufacturing practice in respect of medicinal products for human use and investigational medicinal products for human use.

Decision: A decision is a legal act binding in its entirety upon those to whom it is addressed (Member State or natural or legal person). E.g., Council Decision 1999/468/EC laying down the procedures for the exercise of implementing powers conferred on the Commission. [1,10]

11.4.2 SOFT LAW

Soft law is the term applied to EU measures, such as guidelines, declarations, and opinions, which, in contrast to directives, regulations, and decisions, are not binding on those to whom they are addressed. However, soft law can produce some legal effects. [10]

11.4.3 GUIDELINES

- Community guidelines (EMA—Applicable to all member states)
- National guidelines (Applicable to single member state only)
- Notice to Applicants

11.4.4 NOTICE TO APPLICANTS

The body of EU legislation in the pharmaceutical sector is compiled in Volume 1 and Volume 5 of the publication "The rules governing medicinal products in the European Union." [1,10]

- Volume 1—EU pharmaceutical legislation for medicinal products for human use
- Volume 5—EU pharmaceutical legislation for medicinal products for veterinary use

The basic legislation is supported by a series of guidelines that are also published in the following volumes of "The rules governing medicinal products in the European Union":

- Volume 2—Notice to applicants and regulatory guidelines for medicinal products for human use:
  - Volume 2A—Procedures for marketing authorization
  - Volume 2B—Presentation and content of the dossier
  - Volume 2C—Regulatory Guidelines
- Volume 3 - Scientific guidelines for medicinal products for human use:
  It contains:
  - Introduction
  - Quality Guidelines
  - Biotechnology Guidelines
  - Nonclinical Guidelines
  - Clinical Efficacy and Safety Guidelines
  - Multidisciplinary Guidelines

- Volume 4—Guidelines for good manufacturing practices for medicinal products for human and veterinary use
- Volume 6—Notice to applicants and regulatory guidelines for medicinal products for veterinary use
- Volume 7—Scientific guidelines for medicinal products for veterinary use
- Volume 8—Maximum residue limits
- Volume 9—Guidelines for pharmacovigilance for medicinal products for human and veterinary use
- Volume 10—Guidelines for clinical trial

Note:

- Volume 9 is now replaced with Good Pharmacovigilance Practices guidelines
- Medicinal products for pediatric use, herbal, orphan medicinal products and advanced therapies are governed and controlled by specific rules

11.5 CLINICAL TRIALS IN EU

Clinical trials are research studies that test how well new medical approaches work in people. Each study answers scientific questions and tries to find better ways to prevent, screen for, diagnose, or treat a disease. Clinical trials may also compare a new treatment to a treatment that is already available.

11.5.1 EU CLINICAL TRIAL GUIDANCES

Volume10 of Eudralex contains guidance documents pertinent to clinical trials. [12]

11.5.2 CHAPTER I: APPLICATION AND APPLICATION FORM [12]

- Detailed guidance for the request for authorization of a clinical trial on a medicinal product for human use to the competent authorities, notification of substantial amendments, and declaration of the end of the trial
- Detailed guidance on the application format and documentation to be submitted in an application for an Ethics Committee opinion on the clinical trial on medicinal products for human use
- Detailed guidance on the European clinical trials database (EudraCT)

11.5.3 CHAPTER II: SAFETY REPORTING [12]

- Detailed guidance on the collection, verification, and presentation of adverse event/reaction reports arising from clinical trials on medicinal products for human use
- ICH guideline E2F—Note for guidance on development safety update reports

11.5.4 CHAPTER III: QUALITY OF THE INVESTIGATIONAL MEDICINAL PRODUCT [12]

- Good manufacturing practices for manufacture of investigational medicinal products
- Guideline on the requirements to the chemical and pharmaceutical quality documentation concerning investigational medicinal products in clinical trials

11.5.5 CHAPTER IV: INSPECTIONS [12]

- Guidance for the preparation of GCP inspections
- Guidance for the conduct of GCP inspections
- Annex VII to Guidance for the conduct of GCP inspections—Bioanalytical part, Pharmacokinetic, and Statistical Analyses of Bioequivalence Trials

11.5.6 CHAPTER V: ADDITIONAL INFORMATION [12]

- Guidelines on good clinical practice (ICH E6: Good Clinical Practice: Consolidated guideline, CPMP/ICH/135/95)
- Recommendation on the content of the trial master file and archiving

11.5.7 CHAPTER VI: LEGISLATION [12]

- Directive 2001/20/EC of the European Parliament and of the Council of 4 April 2001 on the approximation of the laws, regulations and administrative provisions of the Member States relating to the implementation of good clinical practice in the conduct of clinical trials on medicinal products for human use.
- Commission Directive 2005/28/EC of 8 April 2005 laying down principles and detailed guidelines for good clinical practice as regards investigational medicinal products for human use, as well as the requirements for authorization of the manufacturing or importation of such products.

11.5.8 EUDRACT [13]

EudraCT is a database of all clinical trials which commenced in the European Community from May 2004, and also includes clinical trials linked to European pediatric drug development. The EudraCT database has been established in accordance with Directive 2001/20/EC. The clinical trial Sponsors should see the EudraCT website to access the EudraCT application in order to:

- Obtain a EudraCT number
- Submit the clinical trial application form to the Ethics Committees and Competent Authorities

The information that needs to be provided about the clinical trial at the time of registration in EudraCT is similar, but less extensive than that provided in the CTA and includes the following:

- Trial identification

- Identification of the Sponsor responsible for the clinical trial
- Applicant identification
- Investigational medicinal product
- Description of the IMP
- Information on placebo
- Authorized site responsible in the community for the release of the investigational medicinal product in the community
- General information on the trial
- Population of trial subjects
- Proposed clinical trial sites in the member state concerned by this request
- Ethics committee in the member state concerned by this request
- Originator of the data provided electronically

11.5.9 CLINICAL TRIAL AUTHORIZATION (CTA) BY NATIONAL COMPETENT AUTHORITIES

Under current EU clinical trial framework, a sponsor wishing to conduct a clinical trial in several Member States must submit a clinical trial application to the Competent Authorities and Ethics Committees in each individual Member State and the authorization procedures are performed separately in each Member State. [14]

EU new clinical trials regulation was adopted on 16 April 2014 and was published on 27 May 2014 in the Official Journal. Under this new regulation, EU-portal will be the single entry point for the submission of all data and information relating to clinical trials. The EU-database will constitute the single repository of all submitted information through the EU-portal, related to a clinical trial and, unless confidentiality is justified, it will be publicly accessible.

11.5.10 DOCUMENTATION FOR CLINICAL TRIAL APPLICATION

MHRA submission package for a clinical trial application includes following: [14]

- Covering letter (when applicable, the subject line should state that the submission is for a Phase I trial and is eligible for a shortened assessment time, or if it is submitted as part of the notification scheme)
- A clinical trial application form in PDF and XML versions
- A clinical trial protocol document
- An investigator's brochure (IB) or document replacing the IB
- An investigational medical product dossier (IMPD) or a simplified IMPD
- A noninvestigational medicinal product dossier (if required)
- A summary of scientific advice from any Member State or the EMA, if available
- Manufacturer's authorization, including the importer's authorization and Qualified Person declaration on good manufacturing practice for each manufacturing site if the product is manufactured outside the EU
- A copy of the EMA's decision on the pediatric investigation plan (PIP) and the opinion of the pediatric committee, if applicable

- The content of the labeling of the investigational medicinal product (IMP) (or justification for its absence)
- Proof of payment (not required for applications made under the notification scheme)

11.5.11 EU CLINICAL TRIALS REGISTER [13]

The EU Clinical Trials Register contains information on interventional clinical trials on medicines conducted in the EU, or the EEA, which started after 1 May 2004.

Clinical trials conducted outside the EU/EEA are included if:

- They form part of a PIP, or
- They are sponsored by a marketing authorization holder (MAH), and involve the use of a medicine in the pediatric population as part of an EU marketing authorization.

The Register also provides information about older pediatric trials covered by an EU marketing authorization.

This database is used by national medicines regulators for data related to clinical trial protocols. The data on the results of these trials are entered into the database by the sponsors themselves and are published in this register once the sponsors have validated the data.

The EU clinical trials register has been a primary registry in the World Health Organization (WHO's) Registry Network since September 2011 and is a WHO Registry Network data provider. It is also available on the WHO International Clinical Trials Registry Platform.

11.6 MARKETING AUTHORIZATION IN EU

A medicinal product can only be marketed in the European Economic Area when a marketing authorization has been granted by the Competent Authority of a Member State for its own specific territory (national authorization) or when an authorization has been granted in accordance with Regulation (EC) No 726/2004 for the entire European Union (an Union authorization). The MAH must be established within the EEA. There are several alternative procedures which can be used to obtain the authorization depending on the type of the medicinal product and the countries where the product is intended to be marketed. [15,16]

11.6.1 CENTRALIZED AUTHORIZATION PROCEDURE

This procedure results in a *single marketing authorization* that is *valid in all EU countries*, as well as in the EEA countries Iceland, Liechtenstein, and Norway. The EMA is responsible for running the centralized authorization procedure for human and veterinary medicines. [15,16]

The centralized procedure is *mandatory for*:

- Human medicines for the treatment of human immunodeficiency virus (HIV) or acquired immune deficiency syndrome (AIDS), cancer, diabetes, neurodegenerative diseases, auto-immune and other immune dysfunctions, and viral diseases
- Veterinary medicines for use as growth or yield enhancers

- Medicines derived from biotechnology processes, such as genetic engineering
- Advanced-therapy medicines, such as gene-therapy, somatic cell-therapy, or tissue-engineered medicines
- Officially designated "orphan medicines" (medicines used for rare human diseases)

For medicines that do not fall within these categories, companies have the *option of submitting an application* for a centralized marketing authorization to the Agency. This is possible for medicines:

- That are a significant therapeutic, scientific, or technical innovation, or;
- Whose authorization would be in the interest of public or animal health

Applications filed through the centralized procedure are submitted directly to the EMA. The Agency's scientific committees take up to 210 active days plus "clock stops" for the evaluation, at the end of which the committee adopts an opinion on whether the medicine should be marketed or not. This opinion is then transmitted to the EC, which has the ultimate authority for granting the marketing authorizations in the EU.

11.6.2 NATIONAL AUTHORIZATION PROCEDURES [15,16]

Each EU Member State has its own national authorization procedures for the authorization, within their own territory, of medicinal products that fall outside the scope of the centralized procedure. Through National Procedure, an authorization is obtained in a single country.

11.6.3 DECENTRALIZED PROCEDURE [15,16]

Organizations can apply for the simultaneous authorization of a medicinal product in more than one EU country if it has not yet been authorized in any EU country and it does not fall within the compulsory scope of the centralized procedure.

11.6.4 MUTUAL-RECOGNITION PROCEDURE [15,16]

The organizations that have a medicinal product authorized in one EU Member State can apply for this authorization to be recognized in other EU countries. In addition, the marketing authorizations granted in Norway, Iceland, and Lichtenstein are eligible for mutual recognition procedure in the same way as marketing authorizations granted by EU Member States.

11.7 REGULATORY STRATEGY FOR FILING OF APPLICATION

In EU the strategy for filing of regulatory applications depends upon number of countries selected for marketing, type of medicinal product, innovation platforms, history of medicinal product, etc. The procedures and legal requirements for applying for a marketing authorization are set out in Directive 2001/83/EC and in Regulation (EC) No 726/2004. They are as follows: [15,16]

- According to Article 8(3) of Directive 2001/83/EC *(Complete Applications)*
- According to Article 10 of Directive 2001/83/EC, relating to *generic* medicinal products, "*hybrid*" medicinal products and *similar biological* medicinal products
- According to Article 10a of Directive 2001/83/EC, relating to applications relying on *well-established medicinal use* supported by bibliographic literature
- According to Article 10b of Directive 2001/83/EC, relating to applications for *new fixed combination of active substances* in a medicinal product
- According to Article 10c of Directive 2001/83/EC, relating to *informed consent* from a MAH for an authorized medicinal product

11.7.1 APPLICATIONS ACCORDING TO ARTICLE 8(3) OF DIRECTIVE 2001/83/EC [15,16]

11.7.1.1 Standalone application

Full applications (under Article 8(3)) must be accompanied by a dossier of information covering:

- Pharmaceutical (physico-chemical, biological, or microbiological) tests
- Preclinical (toxicological and pharmacological) tests
- Clinical trials

Any relevant published literature should also be included.

11.7.1.2 Mixed application

Where Module 4 and/or 5 of the application for marketing authorization (Nonclinical/Clinical Study Reports) consist of a combination of reports of limited nonclinical and/or clinical studies carried out by the applicant and of the bibliographical references.

11.7.2 APPLICATIONS ACCORDING TO ARTICLE 10 OF DIRECTIVE 2001/83/EC [15,16]

Article 10 constitutes a single legal base for the submission of applications. The content of such applications must comply with the requirements set out therein.

11.7.2.1 Application in accordance with paragraph 1 of article 10 (Generic medicinal product)

Directive 2001/83/EC defines a generic medicinal product in Article 10(2)(b) as a medicinal product which has:

- the same qualitative and quantitative composition in active substances as the reference medicinal product,
- the same pharmaceutical form as the reference medicinal product,
- and whose bioequivalence with the reference medicinal product has been demonstrated by appropriate bioavailability studies.

The different salts, esters, ethers, isomers, mixtures of isomers, complexes, or derivatives of an active substance must be considered to be the same active substance, unless they differ significantly in properties with regard to safety and/or efficacy.

11.7.2.2 Application in accordance with paragraph 3 of article 10 ("hybrid" medicinal product)

Article 10(3) of Directive 2001/83/EC requires that the results of the appropriate preclinical tests or clinical trials shall be provided in certain circumstances in the framework of an application under Article 10. These applications will thus depend in part on the results of preclinical tests and clinical trials for a reference product and in part on new data that has been generated.

11.7.2.3 Application in accordance with paragraph 4 of article 10 ("Similar biological medicinal product")

Article 10(4) of Directive 2001/83/EC requires that in the framework of an application under Article 10, where a biological medicinal product which is similar to a reference biological product, does not meet the conditions in the definition of generic medicinal products, owing to, in particular, differences relating to raw materials or differences in manufacturing processes of the similar biological medicinal product and the reference biological medicinal product, the results of appropriate preclinical tests or clinical trials relating to these conditions must be provided.

11.7.3 APPLICATIONS ACCORDING TO ARTICLE 10A OF DIRECTIVE 2001/83/EC (WELL-ESTABLISHED MEDICINAL USE) [15,16]

According to Article 10(a) of Directive 2001/83/EC it is possible to replace results of the preclinical and clinical trials by detailed references to published scientific literature (information available in the public domain) if it can be demonstrated that the active substances of a medicinal product have been in well-established medicinal use within the Union for at least ten years, with recognized efficacy and an acceptable level of safety.

11.7.4 APPLICATION ACCORDING TO ARTICLE 10(B) OF DIRECTIVE 2001/83/EEC (FIXED COMBINATION)

In accordance with Article 10(b) of Directive 2001/83/EC: "In the case of medicinal products containing active substances used in the composition of authorized medicinal products but not hitherto used in combination for therapeutic purposes, the results of new preclinical tests or new clinical trials relating to that combination must be provided in accordance with Article 8(3)(i), but it is not necessary to provide scientific references relating to each individual active substance."

The combination of active substances within a single pharmaceutical form of administration according to this provision is a so-called "fixed combination." In case of an application on the basis of Article 10b of Directive 2001/83/EC the applicant does not have to provide scientific references relating to each individual active substance.

11.7.5 APPLICATIONS ACCORDING TO ARTICLE 10(C) OF DIRECTIVE 2001/83 /EC (INFORMED CONSENT)

An exception from the requirements to submit all of the information required in Article 8(3)(i) is provided by Article 10c of Directive 2001/83/EC for so-called "informed consent" marketing authorization applications. According to Article 10(c): "Following the granting of a marketing authorization, the authorization holder may allow use to be made of the pharmaceutical, preclinical and clinical documentation contained in the dossier on the medicinal product, with a view to examine subsequent applications relating to other medicinal products possessing the same qualitative and quantitative composition in terms of active substances and the same pharmaceutical form."

11.8 CENTRALIZED PROCEDURE IN EU

A marketing authorization granted under the centralized procedure is valid for all the countries in the entire EU market, which means the medicinal product may be put on the market in all Member States subsequent to the grant of marketing authorization through the centralized procedure. The application is submitted to the EMA for the medicinal products which fall within the mandatory scope of the centralized procedure in accordance with the Annex to Regulation (EC) No 726/2004. An application may similarly be submitted to the EMA for the medicinal products that fall within the optional scope of the centralized procedure in accordance with Article 3(2) and 3(3) of Regulation (EC) No 726/2004 where the applicant wishes to obtain a Union marketing authorization. [15,16]

The centralized procedure is compulsory for certain types of human medicinal products such as designated orphans, advanced therapy medicinal products, those developed by certain biotechnological processes, and those containing new active substances for the treatment of cancer, acquired immune deficiency syndrome, diabetes, neurodegenerative disorders, auto-immune diseases, other immune dysfunctions, and viral diseases. [15,16]

The centralized procedure may also be used on a voluntary basis for medicinal products which constitute a significant scientific, therapeutic, or technical innovation, other medicinal products containing a new active substance, or that the granting of a Community marketing authorization would be in the interests of patients at EU level. The overview of centralized procedure is shown in Fig. 11.1. [15,16]

11.8.1 ELIGIBILITY REQUEST

Regardless of whether the product falls into the optional or mandatory scope, it is required to submit an "eligibility request" using the "presubmission request form" together with relevant additional justification in annex (e.g., draft SmPC, and for optional scope the justification for eligibility). [15,16]

EMA recommends that the eligibility request be provided preferably, at the earliest, 18 months before submission of the MAA and, at the latest, 7 months before the MAA is filed with the EMA, at which point it may be submitted along with the "letter of intent to submit." [15,16]

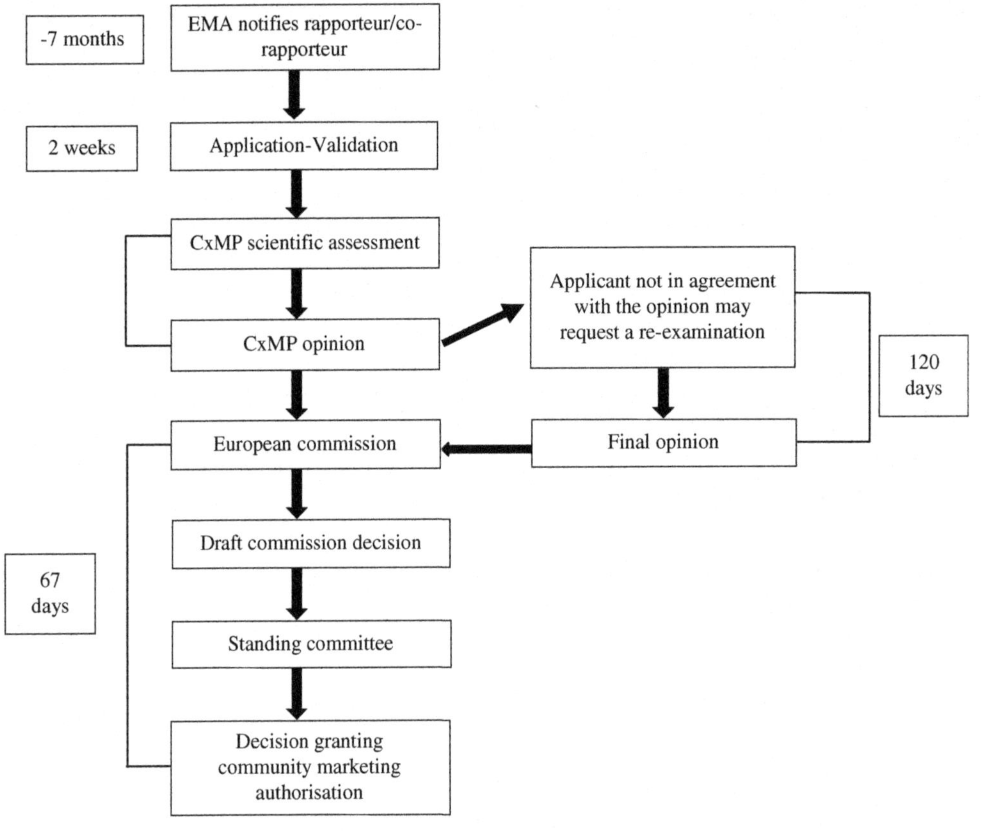

FIGURE 11.1

Overview of the centralized procedure.

Selection of rapporteur/corapporteur: For any scientific evaluation in the centralized procedure a "rapporteur," and if relevant a "corapporteur" will be appointed from the members of the CHMP or the alternates. The role of the (co-)rapporteur is to perform the scientific evaluation and to prepare an assessment report for the relevant committee according to an agreed timetable. [15,16]

The appointment process for rapporteur/corapporteur is usually initiated at the CHMP meeting following the receipt of the "presubmission request form" (intent to submit MA) and their request to assign rapporteurs, which should ideally be provided seven months before the MAA is intended to be submitted. [15,16]

11.8.1.1 Invented name of products evaluated via the centralised procedure

Medicinal products authorized via the centralized procedure will have the same name across the European Economic Area. The name of the medicinal product may be either an invented name, or a common name or scientific name accompanied by a trademark or the name of the Marketing

Authorization Holder. EMA assesses whether the invented name proposed for a medicinal product could create potential safety risks or a public-health concern. The details about the criteria used for checking the proposed invented names for human medicinal products are detailed in the guideline on the acceptability of names for human medicinal products processed through the centralized procedure (EMA/CHMP/287710/2014—Rev. 6) [15,16].

11.8.1.2 EMA contact point in the centralized procedure

A Procedure Manager (PM) is allocated for initial marketing authorization application (MAA) for human medicines, at the time of confirmation of eligibility to the centralized procedure and is the primary contact point for the applicant or Marketing Authorization Holder, prior to submission and throughout the procedure until the decision is granted by the EC. The project manager serves as the main interface between the Rapporteurs, EMA product team and the applicant/Marketing Authorization Holder. The project manager, in close cooperation with the EMA Product Lead (EPL) and the Rapporteurs, ensures that the applicant is kept informed of all aspects related to the MAA evaluation. [15,16]

11.8.1.3 EMA presubmission meeting

When preparing the submission of a marketing authorization application, applicants have the opportunity to meet the EMA to discuss procedural or regulatory issues in relation to the upcoming submission, and to establish contacts with the EMA staff that will be involved with the application. Presubmission meetings are free of charge and should take place approximately 6–7 months prior to the anticipated submission date of the marketing authorization application. [15,16]

11.8.1.4 Compilation of the application dossier

The application dossier for medicinal products for human use must be presented in accordance with the EU-CTD (common technical document) presentation outlined in volume 2B of the notice to applicants. [15,16]

The EU-CTD is organized in five modules: module 1 contains the specific EU administrative and prescribing information. The structure of modules 2, 3, 4, and 5 is common for all regions and will contain the high level summaries and quality, nonclinical and clinical documentation respectively. The EMA provides the applicant with a template of what must be included for the product information (SmPC, labeling and package leaflet texts), All applications must be submitted in English language. [15,16]

EMA has implemented electronic-only submission of applications for marketing authorization with electronic Common Technical Document (e-CTD) as the required format for medicinal products for human use. Since the 1st January 2010, eCTD is the only acceptable electronic format for all applications and all submission types. Non-eCTD electronic applications are no longer a valid format for submission of MAAs through centralized procedure.

The use of the e-Submission Gateway or Web client is mandatory for all eCTD submissions through the centralized procedure. EMA does not accept submissions on CD or DVD for any applications for human medicinal products.

11.8.1.5 Submission and validation of the application dossier

Target dates for submission for human and veterinary medicinal products are published on the EMA website. The applicant should immediately inform the EMA, rapporteur and corapporteur, if the original indicated submission date cannot be met.

The EMA will check if the application meets all relevant legal and procedural EU requirements ("validation"), before the start of the scientific evaluation. The EMA will issue an invoice on the date of the notification of the administrative validation to the applicant, and fees will normally be payable within 45 days of the date of the said notification. [15,16]

11.8.1.6 Evaluation of the application

The EMA starts the evaluation procedure of the MAA at the monthly starting date published on the EMA website once the application is validated. The EMA ensures that the evaluation is finalized within 210 days (excluding any clock-stops for the applicant to provide a response to questions raised during the procedure from the CHMP). The procedure is summarized in Table 11.2 as given below: [15,16]

Table 11.2 Flowchart of Centralized Procedure

Day	Action
1	Start of the procedure.
80	Receipt of the assessment report(s) from rapporteur and co-rapporteur(s) by CHMP members and EMA. Sent to applicant for information only.
100	Rapporteur, co-rapporteur, other CHMP members and EMA receive comments from members of the CHMP.
115	Receipt of draft list of questions (including the CHMP recommendation and scientific discussion) from rapporteur and co-rapporteur.
120	CHMP adopts the list of questions as well as the overall conclusions and review of the scientific data to be sent to the applicant by the EMA.
	Clock stop.
121	Submission of the applicant's responses, including revised SmPC, labeling and package leaflet texts in English.
	Restart of the clock.
150	Joint response assessment report from rapporteur and co-rapporteur received by CHMP members and the EMA. Sent to applicant for information only.
170	Deadline for comments from CHMP members to be sent to rapporteur and co-rapporteur, EMA, and other CHMP members.
180	CHMP discussion and decision on the need to adopt a list of "outstanding issues" and/or an oral explanation by the applicant. If an oral explanation is needed, the clock is stopped to allow the applicant to prepare the oral explanation.
	Clock stop.
181	Restart of the clock and oral explanation (if needed).
210	Adoption of CHMP opinion and CHMP assessment report (and timetable for the provision of product information translations).

If an applicant decides to withdraw their application before an opinion is adopted, the EMA makes this public on its website along with the relevant assessment report. [15,16]

Applicants are required to submit a risk management plan at the time of marketing application and keep it updated during the lifecycle of the product. The RMP will be subject to review by the pharmacovigilance risk assessment committee (PRAC) in parallel to the CHMP review.

11.8.1.7 Reexamination of the CHMP opinion

The applicant may notify in writing to the CHMP/EMA of their intent to request a re-examination of the CHMP opinion within 15 days of its receipt (after which the opinion becomes final in absence of such a request). Upon receipt of the notification of intent, the CHMP will appoint a new set of (co)rapporteurs to reexamine its opinion. The detailed grounds for the re-examination request must be forwarded to the EMA within 60 days after receipt of the opinion. The Marketing Authorization Holder or the applicant may also request a review by the relevant scientific advisory group. At the end of the re-examination procedure, the CHMP adopts a final opinion either changing that opinion or confirming its previous opinion on the marketing authorization application. [15,16]

11.8.1.8 Decision-Making process

After adoption of the CHMP opinion, the EMA has fifteen days to forward its final opinion to the Commission. This is the start of the "decision-making process," whereby the CHMP Opinion will be converted into a legally binding Commission decision for all Member States and the applicant. [15,16]

The Commission decision granting a marketing authorization to the medicinal product concerned includes the agreed SmPC, conditions for use, labeling and package leaflet texts (product information). The Commission decision is legally binding on all Member States, the product information must, therefore, be provided in all Community languages. The translations of the product information are typically provided by the applicant five days after adoption of the CHMP opinion. The Community marketing authorization for the medicinal product is granted in 67 days after adoption of the final CHMP opinion. [15,16]

The EMA will publish the CHMP assessment report on the medicinal product once the Community marketing authorization is granted. This assessment report includes the benefit risk assessment and the reasons for its opinion in favor of granting authorization. Any information of a commercially confidential nature is deleted before publishing of this document called the European public assessment report (EPAR). A Marketing Authorization for a medicinal product once granted is generally valid for five years. Subsequently it is renewed in accordance with the EU regulations. [15,16]

11.8.2 EXAMPLE OF CENTRALIZED APPROVED PRODUCTS

- *Biologics*: Epoetin zeta for the treatment of anemia in patients with chronic renal failure, cancer, etc., Everolimus for the treatment of breast cancer, pancreatic neuroendocrine tumors, advanced renal cell carcinoma, etc.
- *Cancer*: Docetaxel, Filgrastim

- *Orphan conditions*: Brentuximab vedotin for Hodgkin disease lymphoma and non-Hodgkin lymphoma, Aztreonam lysine for the treatment of cystic fibrosis .
- *HIV/AIDS*: Etravirine, Raltegravir
- *Neurodegenerative disease*: Pramipexole and Entacapone for Parkinson's disease
- *Viral diseases*: Daclatasvir and Peginterferon lambda-1a for the treatment of chronic hepatitis C
- *Immune dysfunction*: Human normal immunoglobulin for the treatment of Primary immunodeficiency (PID)
- *Diabetes*: Jentadueto (Linagliptin and Metformin hydrochloride) and Komboglyze (Saxagliptin and Metformin) for the treatment of diabetes mellitus, type 2
- *Centralized approved generic products*: Clopidogrel for the treatment of peripheral vascular diseases, Desloratadine for the treatment of allergic rhinitis

11.9 DECENTRALIZED PROCEDURE IN EU

The decentralized procedure can be used in cases where the product has never been authorized in any of the Member States, and the applicant wishes to obtain a license in a number of states simultaneously. The applicant must submit applications with the complete dossier to the Competent Authorities of each of the Member States where authorization is desired. A single Member State should be chosen as the reference member state to undertake the scientific assessment of the complete dossier, while the other states are designated as concerned states. The review process has many parallels with the centralized procedure, in that similar timelines exist, the Reference Member State (RMS) plays the role of the rapporteur, and the Concerned Member States (CMS) replace the CHMP. [15,16]

Once all States have validated that the dossiers are complete, the RMS is allowed 70 days to review the dossier and prepare a preliminary assessment report, which is circulated to the CMS and the applicant. Comments from the CMS and applicant responses are collected so that by Day 120 the RMS may issue a draft assessment report together with draft SmPC, label and leaflet texts. The clock may be stopped until requested responses from the applicant are received. The application then enters the second step in the assessment process, during which all the CMS consider the draft assessment report and the product information. During this phase, further fine tuning of the SmPC, leaflet and package insert may be required, depending on feedback from the Member States. At the end of 90 days the Member States must approve the assessment report, SmPC, leaflet and package insert, after which they are allowed a further 30 days to enact national legislation granting the authorizations. If there is a failure to reach agreement between the Member States the issues can be referred to the EMA/CHMP for arbitration. [15,16]

The Decentralized procedure is summarized in Table 11.3 given below: [17]

11.10 MUTUAL RECOGNITION PROCEDURE IN EU

An applicant must use the mutual recognition procedure to obtain marketing authorizations in other Member States, once an authorization has been granted by a Member State. The applicant must

Table 11.3 Flow chart of the Decentralized Procedure

Day	Action
Preprocedural Step	
Before Day-14	Applicant discussions with RMS.
	RMS allocates procedure number. Creation in CTS.
Day-14	Submission of the dossier to the RMS and CMSs
	Validation of the application. Positive validation should only be indicated in CTS, not via e-mail.
Assessment Step I	
Day 0	RMS starts the procedure. The CMS are informed via CTS.
Day 70	RMS forwards the Preliminary Assessment Report (PrAR) (including comments on SmPC, PL and labeling) on the dossier to the CMSs and the applicant.
Until Day 100	CMSs send their comments to the RMS, CMSs, and applicant. It may also be sufficient for the CMS to indicate in CTS only in case there are no additional comments.
Until Day 105	Consultation between RMS and CMSs and applicant.
	If consensus not reached RMS stops the clock to allow applicant to supplement the dossier and respond to the questions.
Clock-off period	Applicant may send draft responses to the RMS and agrees the date with the RMS for submission of the final response. Applicant sends the final response document to the RMS and CMSs within a period of 3 months, which can be extended by a further 3 months.
Day 106	RMS restarts the procedure following the receipt of a valid response or expiry of the agreed clock-stop period if a response has not been received. The CMS are informed via e-mail and CTS will be updated accordingly.
Assessment Step II	
Day 120 (Day 0)	RMS sends the DAR, draft SmPC, draft labeling and draft PL to CMSs and the applicant.
Day 145 (Day 25)	CMSs send comments to RMS, CMSs, and the applicant. It may also be sufficient for the CMS to indicate in CTS only in case there are no additional comments.
Day 150 (Day 30)	RMS may close procedure if consensus reached.
	Proceed to national 30 days step for granting MA.
Until 180 (Day 60)	If consensus is not reached by day 150, RMS to communicate outstanding issues with applicant, receive any additional clarification, prepare a short report, and forward it to the CMSs and the applicant.
Day 195 (at the latest)	A Break-Out Session (BOS) may be held at the European Medicines Agency with the involved MSs to reach consensus on the major outstanding issues.
Between Day 195 and Day 210	RMS consults with the CMSs and the applicant to discuss the remaining comments raised.
Day 210 (Day 90)	Closure of the procedure including CMSs approval of assessment report, SmPC, labeling and PL, or referral to Co-ordination group.
	Proceed to national 30 days step for granting MA.
Day 210 (at the latest)	If consensus on a positive RMS AR was not reached at day 210, points of disagreement will be referred to the Co-ordination group for resolution.

(*Continued*)

Table 11.3 Flow chart of the Decentralized Procedure *Continued*

Day	Action
Day 270 (at the latest)	Final position adopted by Co-ordination Group with referral to CHMP for arbitration in case of unsolved disagreement.
National Step	
5 days after close of procedure	Applicant sends high quality national translations of SmPC, labeling and PL to CMSs and RMS
30 days after close of the procedure	Granting of national marketing authorization in RMS and CMSs if outcome is positive and there is no referral to the Coordination group. (National Agencies will adopt the decision and will issue the marketing authorization subject to submission of acceptable translations).
30 days after close of CMD referral procedure	Granting of national marketing authorization in RMS and CMSs if positive conclusion by the Coordination group and no referral to the CHMP/CVMP. (National Agencies will adopt the decision and will issue the marketing authorization subject to submission of acceptable translations).

CTS, *Communication and Tracking System,* DAR, *Draft Assessment Report;* CMS, *Concerned Member State;* RMS, *Reference Member State*

submit the dossier to each of the CMS where authorization is requested, and identify the RMS that granted the original authorization. Before starting the process the applicant is advised to ensure that the existing member state dossier is fully up to date based on marketing experience. The applicant is also encouraged to engage in dialogue with the RMS on adjustments to the approved product information so as to ensure a smooth recognition procedure by the other member states. The RMS has 90 days to update the original assessment report based on dossier updates and forward it to the CMS together with the SmPC, label and leaflet text that it has approved. Within 90 days of the receipt of these documents, the CMS shall recognize the decision of the RMS and the approved SmPC, package leaflet and labeling by granting a marketing authorization with a harmonized SmPC, package leaflet and labeling. During this phase, adjustments to the product information may be agreed, with the RMS acting as the coordinating interface with the applicant. If there is disagreement between Member States regarding the granting of an authorization, the issue is referred to a coordinating group, consisting of Member States representatives. If it cannot be resolved there, then it is referred to the CHMP for review. [16,18]

It is possible to use the mutual recognition procedure more than once for subsequent applications to other Member States in relation to the same medicinal product (so called repeat use). It is recommended that, wherever feasible, the MAH considers involving all Member States where the product is intended to be marketed, in the first use of mutual recognition procedure or decentralized procedure. [16,18]

The mutual recognition and the repeat use procedure are summarized in Table 11.4 given below: [19]

The differences between different marketing authorization procedures are given in Table 11.5 [15–18]

Table 11.4 Flowchart for the Mutual Recognition Procedure (MRP) and Repeat Use Procedures (RUP)

Approx. 90 days before submission to CMS	Applicant requests RMS to update Assessment Report (AR) and allocate procedure number.
Day-14	Applicant submits the dossier to CMS. RMS circulates the AR including SmPC, PL, and labeling to CMSs. Validation of the application in the CMSs.
Day 0	RMS starts the procedure.
Day 30	CMSs send their comments to the RMS, CMSs, and applicant.
Day 40	Applicant sends the response document to CMSs and RMS.
Until Day 48	RMS evaluates and circulates a report on the applicant's response document to CMSs.
Day 55	CMSs send their remaining comments to RMS, CMSs, and applicant.
Day 55–59	The applicant and RMS are in close contact to clarify if the procedure can be closed at day 60 or if the applicant should submit a further response at day 60.
Day 60	MRP: If CMS have no remaining comments at Day 55, the RMS closes the procedure. RUP: If no potential serious risk to public health (PSRPH) has been outlined by CMS at Day 55, the RMS closes the procedure. In case a CMS has remaining comments (MRP) or PSRPH (RUP) at Day 55, the applicant sends the response document to CMSs and RMS.
Day 60-90	The period 60-90 will only be used if a CMS has remaining comments (MRP) or PSRPH (RUP) at Day 55.
Until Day 68	RMS evaluates and circulates a report on the applicant's response document to CMSs.
Day 75	CMSs send their remaining comments to RMS, CMSs, and applicant.
Until Day 80	A break-out session (BOS) can be organized around Day 75 (but may take place between days 73–80).
Day 85	CMSs send any remaining comments to RMS, CMS, and applicant.
Day 60–90	CMSs notify RMS and applicant of final position (and in case of negative position also the CMDh secretariat of the EMA). If consensus is reached, the RMS closes the procedure. If consensus is not reached, the points for disagreement submitted by CMSs are referred to CMDh by the RMS within 7 days after Day 90.
Day 150	Final position adopted by the CMDh: If consensus is reached at the level of CMDh, the RMS closes the procedure. If consensus is not reached at the level of CMDh, the RMS immediately refers the matter to EMA for CHMP arbitration
5 days after close of procedure	Applicant sends high quality national translations of SmPC, PL, and labeling to CMSs and RMS.
30 days after close of procedure	Granting of national marketing authorizations in the CMSs subject to submission of acceptable translations.

All days mentioned in this document should be regarded as calendar days. Whenever a response is received the RMS will send their position

Table 11.5 Differences Between Different Marketing Authorization Procedures

Process	Brief Description
National Authorization	Used for products that fall outside the scope of the EMA centralized procedure. Where MA is required in one country. Simple procedure involving one Member State.
Decentralized Procedure	Used for products that fall outside the scope of the EMA centralized procedure. Simultaneous authorization in numerous countries in the EU. A positive outcome results in numerous country approvals. Sponsor can select which countries to apply to. The number of countries is more than one and doesn't have to be all EU countries.
Mutual Recognition Procedure	Used for products that fall outside the scope of the EMA centralized procedure. Individual application to one country within the EU, followed by subsequent applications to other countries. Generally used for extending marketing approvals in other Member states. When the procedure is repeated to extend MAs in other Member States, it is called Repeat Use Procedure.
Centralized Procedure	Mandatory for certain types of Human Medicinal Products such as those developed by certain biotechnological processes, advanced therapy medicinal products, designated orphans, and those containing new active substances for the treatment of acquired immune deficiency syndrome, cancer, neurodegenerative disorders, diabetes, auto-immune diseases, other immune dysfunctions, and viral diseases. One application applies to all EU countries.

11.11 EMA REGULATORY TOOLS FOR EARLY ACCESS TO MEDICINES

11.11.1 ACCELERATED ASSESSMENT

Accelerated assessment reduces the timeframe for the EMA to review a marketing authorization application. Applications may be eligible for accelerated assessment if the CHMP decides the product is of major interest for public health and therapeutic innovation. Evaluation of a marketing authorization application can take up to 210 days under the centralized procedure, excluding clock stops when applicants have to provide additional information. On request, the CHMP can reduce the timeframe to 150 days if the applicant provides sufficient justification for an accelerated assessment. [16,20]

11.11.2 CONDITIONAL MARKETING AUTHORIZATION

If the medicinal product addresses an unmet medical need and targets a life-threatening or a seriously debilitating disease, a rare disease or is intended for use in emergency situations in response to a public health threat, a Conditional Marketing Authorization allows for the early approval of the medicinal product on the basis of less complete clinical data than generally required.

While less complete, the available data must still demonstrate that the medicine's benefits outweigh its risks and the applicant should be in a position to provide the comprehensive clinical data after authorization within a timeframe agreed with the CHMP. In addition, the benefit to public health must still outweigh the risk due to the limited availability of clinical data at the time of marketing authorization. [16,20]

Conditional marketing authorizations are valid for one year, on a renewable basis. The Marketing Authorization Holder will be required to complete the ongoing studies or to conduct new studies to confirm that the positive risk-benefit balance. In addition, specific obligations may be obligatory in relation to the collection and analysis of pharmacovigilance data. [16,20]

11.11.3 COMPASSIONATE USE

Compassionate use is a way of making available to patients with an unmet medical need a promising medicine which has not yet been authorized (licensed) for their condition. [20]

Compassionate use programs can only be put in place for medicines that are expected to help patients with life-threatening, long-lasting, or seriously disabling illnesses. These programs are expected to benefit seriously ill patients who currently cannot be treated satisfactorily with authorized medicines, or who have a disease for which no medicine has been authorized so far. The compassionate use route may be a way for patients who cannot enroll in an ongoing clinical trial to obtain treatment with a potentially life-saving medicine. [16,20]

The safety of the medicinal product may be limited at this stage in the development of the medicine. Generally, toxicology studies would have been completed and analyzed, and early studies observations at how the medicinal product is behaving in the body would have been conducted. However, there may still be some uncertainties about the best way to administer the medicinal product to patients, such as the exact dose to administer, dose frequency, and the medicinal product's safety profile is not yet fully established. [16,20]

11.12 GENERIC AND HYBRID MEDICINAL PRODUCTS

11.12.1 GENERIC MEDICINAL PRODUCT

According to Article 10 (1) of Directive 2001/83/EC, a generic medicinal product is defined as a medicinal product that has: [15,16]

- The same qualitative and quantitative composition in active substance(s) as the reference product
- The same pharmaceutical form as the reference medicinal product
- Demonstrated bioequivalence with the reference medicinal product in appropriate bioavailability studies

A company can only file a Marketing Authorization Application of a generic medicine for marketing once the period of exclusivity on the reference medicine has expired.

11.12.2 HYBRID MEDICINAL PRODUCT

Hybrid medicinal products are the products whose authorization depends partly on the results of tests on the reference medicine and partly on new data from clinical trials. This is possible when a manufacturer develops a generic medicine that is based on a reference medicine, but has a different strength or is given by a different route, such as by mouth or as an injection. It is also possible when a manufacturer develops a medicine with a different indication from the reference medicinal product. [15,16]

11.12.3 EUROPEAN REFERENCE MEDICINAL PRODUCT

The reference medicinal product is a medicinal product which has been granted a marketing authorization by a Member State or by the Commission on the basis of a complete dossier, i.e., with the submission of quality, preclinical, and clinical data in accordance with Articles 8(3), 10a, 10b or 10c of Directive 2001/83/EC and to which the application for marketing authorization for a generic/hybrid medicinal product refers, by demonstration of bioequivalence, usually through the submission of the appropriate bioavailability studies. Applicants will have to identify in the application form for the generic/hybrid medicinal product the reference medicinal product (product name, strength, pharmaceutical form, MAH, first authorization, Member State/Community). [15,16]

11.12.4 DOSSIER SUBMISSION REQUIREMENTS [15,16]

Generic Medicinal Product: According to Article 10 (1) of Directive 2001/83/EC, the Marketing Authorization Holder or the applicant is not required to provide the results of preclinical and clinical studies if it can be demonstrated that the medicinal product is a generic medicinal product of the reference medicinal product.

Hybrid Medicinal Product: According to Article 10 (3) of Directive 2001/83/EC, these applications will thus rely in part on the results of preclinical tests and clinical trials for a reference product and in part on new data.

Marketing authorization applications for a generic or a hybrid medicinal product should follow the format and structure of the ICH CTD format. CTD is an internationally agreed format for the preparation of applications to be submitted to regulatory authorities in the three ICH regions of Europe, USA and Japan. The CTD is organized into five modules.

11.12.5 MODULE 1 (ADMINISTRATIVE INFORMATION AND PRESCRIBING INFORMATION) [21]

The content of Module 1 is defined by the EC in consultation with the competent authorities of the Member States, the European Agency for the Evaluation of Medicinal Products and interested parties. Concerning the structure of Modules 2, 3, 4, and 5 they are common for all ICH regions.

1.0 Cover Letter
1.1 Comprehensive Table of Contents
1.2 Application Form

1.3 Product Information
  1.3.1 SmPC, Labeling and Package Leaflet
  1.3.2 Mock-up
  1.3.3 Specimen
  1.3.4 Consultation with Target Patient Groups
  1.3.5 Product Information already approved in the Member States
  1.3.6 Braille
1.4 Information about the Experts
  1.4.1 Quality
  1.4.2 Nonclinical
  1.4.3 Clinical
1.5 Specific Requirements for Different Types of Applications
  1.5.1 Information for Bibliographical Applications
  1.5.2 Information for Generic, "Hybrid" or Biosimilar Applications
  1.5.3 (Extended) Data/Market Exclusivity
  1.5.4 Exceptional Circumstances
  1.5.5 Conditional Marketing Authorization
1.6 Environmental Risk Assessment
  1.6.1 Non-GMO
  1.6.2 GMO
1.7 Information relating to Orphan Market Exclusivity
  1.7.1 Similarity
  1.7.2 Market Exclusivity
1.8 Information relating to Pharmacovigilance
  1.8.1 Pharmacovigilance System
  1.8.2 Risk Management System
1.9 Information relating to Clinical Trials
1.10 Information relating to Pediatrics
Responses to Questions
Additional Data

Note: A justification for the absence of a section or data should be provided in the respective section when certain elements are not included.

11.12.6 MODULE 2 (OVERVIEWS AND SUMMARIES)

Module 2 covers the Quality Overall Summary, Nonclinical Overview, and Clinical Overview. Nonclinical and Clinical Summaries are only mandatory if new additional studies have been provided within the Marketing Authorization Application. [22]

11.12.7 MODULE 3 (QUALITY)

11.12.7.1 3.2.S Drug substance

- 3.2.S.1 General Information

- 3.2.S.2 Manufacture
- 3.2.S.3 Characterization
- 3.2.S.4 Control of Drug Substance
- 3.2.S.5 Reference Standards or Materials
- 3.2.S.6 Container Closure System
- 3.2.S.7 Stability

11.12.7.2 3.2.P Drug product

- 3.2.P.1 Description and Composition of the Drug Product
- 3.2.P.2 Pharmaceutical Development
- 3.2.P.3 Manufacture
- 3.2.P.4 Control of Excipients
- 3.2.P.5 Control of Drug Product
- 3.2.P.6 Reference Standards or Materials
- 3.2.P.7 Container Closure System
- 3.2.P.8 Stability

11.12.7.3 3.2.A Appendices

- 3.2.A.1 Facilities and Equipment
- 3.2.A.2 Adventitious Agents Safety Evaluation
- 3.2.A.3 Excipients

11.12.7.4 Module 3.2.R Regional information

- Process Validation Scheme for the Drug Product
- Medical Device
- Certificate(s) of Suitability
- Medicinal products containing or using in the manufacturing process materials of animal and/or human origin

11.12.8 MODULE 4 (NONCLINICAL) AND MODULE 5 (CLINICAL)

- For a generic product, Module 4 is generally not required. The bioequivalence studies should be performed and should be included in Module 5, Section 5.3.1. When different salts, esters, ethers, isomers, mixtures of isomers, complexes or derivatives of the active substance of the reference medicinal product are used, additional information providing evidence that their safety and/or efficacy profile is not different from that of the reference medicinal product should be submitted in the CTD Dossier. [22]
- For a so-called "hybrid" of a reference medicinal product (Art 10.3), the results of appropriate preclinical and clinical studies should be provided in accordance to the requirements set out in the notice to applicants.

It should be noted that the responsibility for the quality of the submitted documentation lies with the applicant and is crucial to the overall process. [22]

11.13 EU ASMF AND EDQM CEP

11.13.1 EU ACTIVE SUBSTANCE MASTER FILE

The main objective of the Active Substance Master File (ASMF) procedure, formerly known as the European Drug Master File (EDMF) procedure, is to allow valuable confidential intellectual property or "know-how" of the manufacturer of the active substance (ASM) to be protected, while at the same time allowing the applicant or marketing authorization (MA) holder to take full responsibility for the medicinal product and the quality and quality control of the active substance. NCA/EMA thus have access to the complete information that is necessary for an evaluation of the suitability of the use of the active substance in the medicinal product. [23]

The ASMF procedure can be used for the following active substances, including herbal active substances/preparations, i.e., [23]

- New active substances
- Existing active substances not included in the European Pharmacopoeia (Ph. Eur.) or the pharmacopoeia of an EU Member State
- Pharmacopoeial active substances included in the Ph. Eur. or in the pharmacopoeia of an EU Member State

Note: The ASMF procedure cannot be used for biological active substances

11.13.2 SUBMISSION FORMAT

ASMFs linked to human medicinal products should be presented in the format of the Common Technical Document (CTD). The scientific information in the ASMF should be physically divided into two separate parts, namely the Applicant's Part (AP) and the Restricted Part (RP). The AP contains the information that the ASMF holder regards as nonconfidential to the Applicant/MA holder, whereas the RP contains the information that the ASMF holder regards as confidential. [23]

11.13.3 EDQM CERTIFICATE OF SUITABILITY

The Certificate of Suitability is used to demonstrate that the purity of a given substance produced by a given manufacturer is suitably controlled by the relevant monograph(s) of the European Pharmacopoeia. By demonstrating that they have been granted a CEP for their substance, suppliers of raw materials can prove this suitability to their pharmaceutical industry clients.

CEP procedure is applicable for:

- Organic or inorganic substances (active or excipients)—manufactured or extracted.
- Substances produced by fermentation as indirect gene products, which are metabolites of microorganisms, irrespective of whether or not the microorganisms have been modified by traditional procedures or r-DNA technology.
- Products with risk of transmitting agents of animal spongiform encephalopathies (TSE)

11.13.4 SUBMISSION OF A NEW CEP APPLICATION

The applicants must send the necessary documentation to the Certification of Substances Division (DCEP) of the EDQM to obtain a CEP. EDQM does not accept paper applications anymore. PDF, NeeS or eCTD formats are acceptable. [24]

Upon receipt, the application is validated and listed for assessment. After assessment, the EDQM may send queries to the applicant. When they are resolved, the EDQM sends the applicant a CEP, which is valid for 5 years from the date of first issue and valid indefinitely following the 5-year renewal.

EDQM is also responsible for inspection of Active Pharmaceutical Ingredients (API) manufacturers. [24]

11.14 EU GOOD MANUFACTURING PRACTICE AND GOOD DISTRIBUTION PRACTICE COMPLIANCE

Good manufacturing practice (GMP) and good distribution practice (GDP) are related aspects of the quality assurance for medicinal products in the EEA. The EMA plays an important role in coordinating these activities in collaboration with all Member States. [25]

GMP is a code of standards concerning the manufacture, processing, packing, release, and holding of a medicinal product or it active substance. GDP is a code of standards ensuring that the quality of the product or active substance is maintained throughout the distribution network, so that authorized medicinal products are distributed to the general public without any alteration of their properties. [25]

Good manufacturing practice (GMP) is "that part of quality assurance which ensures that products are consistently produced and controlled in accordance with the quality standards commensurate to their intended use." The principles and guidelines for GMP are specified in Directive 2003/94/EC for medicines and investigational medicines for human use.

GMP guidelines provide interpretation of these principles, outlined in the directive supplemented by a series of annexes that supplement or modify the detailed guidelines for certain types of products, or provide a detailed and more specific guidance on a particular topic.

Compliance with these GMP principles and guidelines is mandatory within the EEA.

Good distribution practice (GDP) ensures that the level of quality determined by GMP is ascertained throughout the distribution network, so that authorized medicines are distributed to retail pharmacists and others selling medicines to the general public without alteration of any of their properties. The principles of GDP are stated in Directive 92/25/EEC.

GDP should be applied through a quality system implemented by the distributor or wholesaler of medicinal products to ensure that:

- The medicinal products distributed by them are authorized in accordance with EU legislation
- Appropriate storage conditions are followed at all times, including during transportation
- Contamination from or of other products is avoided
- Products are stored in appropriately safe and secure areas
- An appropriate turnover of stored medicinal products takes place

The quality system should also ensure that the right products are delivered to the right addressee within a stipulated time period. A tracking and tracing system should enable any deficient products to be found and there should be an effective product recall procedure.

The guidance for the interpretation of the principles and guidelines of good manufacturing practices for medicinal products for human and veterinary use is presented in Eudralex: Volume 4.

The regulatory authorities in Member States must perform inspections of manufacturing-authorization holders located in EEA to ensure that they are adhering to the principles and guidelines of GMP. For products imported from countries outside the EU, the supervisory authority is responsible for verifying that the manufacturer conforms to standards of GMP equivalent to those in force in the EU, unless the country has negotiated an appropriate agreement with the EU establishing mutual recognition of GMP inspections.

11.14.1 MUTUAL RECOGNITION AGREEMENTS (MRA)

Mutual recognition agreements (MRAs) between the EC and MRA partner countries have been established to:

- Reduce technical barriers to trade by facilitating market access while ensuring consumer protection against poor-quality products
- Grant mutual acceptance of reports, certificates, authorizations, and conformity marks issued by the regulatory authorities
- Exchange information concerning procedures used by the conformity assessment bodies to ensure that they comply with agreed-upon requirements
- Encourage international harmonization

Mutual Recognition Agreements in relation to conformity assessment of regulated products include Sectoral Annexes on Good Manufacturing Practices (GMPs) for Human and Veterinary Medicinal Products. EMA status of Mutual Recognition Agreements (MRA) with different countries is given in Table 11.6. [26]

Table 11.6 Status of Mutual Recognition Agreements (MRA)

Country	Status
Australia	Fully operational
Canada	In operation, except for preapproval inspections and medicinal products derived from blood or blood plasma
Israel	Operational with some exclusions
Japan	Operational on 29 May 2004—with limited scope
New Zealand	Fully operational
Switzerland	Fully operational
USA	Not in operation

11.14.2 CERTIFICATION BY A QUALIFIED PERSON (Q.P.) AND BATCH RELEASE

The QP is typically a licensed pharmacist, biologist, or chemist (or a person with another permitted academic qualification) who has several years' experience working in pharmaceutical manufacturing operations, and has passed examinations attesting to her or his knowledge. Each batch of finished products must be certified by a Q.P. within the EC/EEA before being released for sale or supply in the EC/EEA or for export. This confirmation by other Q.P.s should be documented and should identify clearly the matters which have been confirmed. [27]

11.15 EU VARIATIONS, RENEWALS, AND SUNSET CLAUSE

11.15.1 EU VARIATIONS

A variation to the terms of a marketing authorization is an amendment to the contents of the documents referred to in Articles 8, 9, 10, 11, and Annex I of Directive 2001/83/EC. Variations are broadly classified into the following categories: [28,29]

- Minor variations of Type IA
- Minor variations of Type IB
- Major variations of Type II
- Extensions
- Urgent safety restriction

Type IA variations: Type IA variations are the minor variations which have only a small impact, or no impact at all, on the quality, safety, or efficacy of the medicinal product. These variations do not require prior approval before their implementation ("Do and Tell" procedure). [28,29]

These minor variations are further classified in two subcategories, which impact on their submission timelines:

- Type IA variations requiring immediate notification ("IA_{IN}").
- Type IA variations NOT requiring immediate notification ("IA"). Variations that do not require immediate notification can be submitted by the Marketing Authorization Holder (MAH) within 12 months of their implementation. [28,29]

Type IB Variations: Type IB includes the variations which are neither Type IA variations nor the Type II variations nor Extension to Marketing Authorizations. Such minor variations must be notified by the MAH to the National Competent Authority/EMA before their implementation, but do not require a formal approval. However, the MAH must wait a period of 30 days to ensure that the notification is deemed acceptable by the Agency before implementing the change ("Tell, Wait and Do" procedure).

Type II Variations: Commission Regulation (EC) No 1234/2008 defines major variation of Type II as a variation which is not an extension and which may have a significant impact on the Quality, Safety, or Efficacy of a medicinal product. [28,29]

Extension Applications: Changes to a marketing authorization listed in Annex I of Commission Regulation (EC) No 1234/2008 are regarded as "extensions" of the marketing authorization. Extension applications include the following:

- Changes to the active substance(s)
- Changes to strength, pharmaceutical form and route of administration

Urgent Safety Restrictions: Article 22 of the Variations Regulation foresees that in the event of a risk to public health in the case of medicinal products for human use or in the event of a risk to human or animal health or to the environment in the case of veterinary medicinal products, the holder may take provisional "urgent safety restrictions." Urgent safety restrictions concern interim change(s) in the terms of the marketing authorization due to new information having a bearing on the safe use of the medicinal product. These urgent changes must be subsequently introduced via a corresponding variation in the marketing authorization. [28,29]

11.15.2 GROUPING OF VARIATIONS

Variations Regulation sets out the possibility for a Marketing Authorization Holder (MAH) to group several Type-IA and Type-IA_{IN} variations under a single notification to the same relevant authority: [28,29]

11.15.3 WORKSHARING OF VARIATIONS

In order to avoid duplication of work in the evaluation of variations, a work-sharing procedure has been established under which one authority (the "reference authority"), chosen amongst the competent authorities of the Member States and the Agency, will examine the variation on behalf of the other concerned authorities. Under the Worksharing procedure, the MAH can submit the same Type IB or Type II variation, or the same group of variations affecting more than one marketing authorization from the same MAH in one application. Extensions are excluded from Worksharing. [28,29]

11.15.4 EU RENEWALS

A marketing authorization is valid for five years, except when a "conditional marketing authorization (for 1 year)" has been granted, and is renewable upon application by the MAH. After these five years, the marketing authorization may be renewed on the basis of a re-evaluation of the risk-benefit balance. Once renewed the marketing authorization will be valid for an unlimited period, unless the competent authority decides, on justified grounds relating to pharmacovigilance. The authorization may be renewed upon application by the MAH at least nine months before its expiry. [29]

11.15.5 SUNSET CLAUSE

"Sunset clause" is a provision leading to the cessation of the validity of the marketing authorization under either one of the following conditions:

- The medicinal product is not placed on the market within three years of grant of the marketing authorization
- Where a medicinal product that was previously placed on the market is no longer present on the market for three consecutive years

Exemptions may be granted on public health grounds and in exceptional circumstances if suitably justified. [29]

11.16 EU PHARMACEUTICAL PRICING AND REIMBURSEMENT

In the EU, national authorities are free to set the prices of medicinal products and to designate the treatments they wish to reimburse under their social security systems. At the same time, pricing and reimbursement systems are closely linked to the realization of European policy objectives such as the internal market, pharmaceutical competitiveness, sustainable research and development, and the protection of human health.

Upon the grant of a marketing authorization, the decisions about price and reimbursement take place at the level of each Member State judging the potential role and use of the medicinal product in the context of the national health system of that country.

The EMA has been working closely with Health Technology Assessment (HTA) bodies since 2008. Regional and National HTA bodies provide their recommendations on medicines and other health technologies that can be reimbursed or financed by the healthcare system in a particular Member State or that region.

In countries where health technology assessment is in place, pricing and reimbursement agencies, HTA bodies or payers rely upon these assessments to:

- Determine reimbursement status
- Provide information on benefits and risks of new treatments compared to currently available treatment options
- Support the process of price negotiation

The assessment criteria used by these HTA bodies differ between Member States, in accordance with national and regional legislation. [30]

11.17 EMA PHARMACOVIGILANCE AND RISK MANAGEMENT

Pharmacovigilance, or the surveillance of the safety of a medicinal product during its life on the market, is extensively regulated by EU directives and regulations. EU legislation requires Member States to establish national pharmacovigilance systems to collect and evaluate information on adverse reactions to medicinal products or their side effects and to take appropriate action where necessary. [16,31]

It also requires that the MAHs report suspected adverse reactions to the regulatory authorities in certain formats and within specified timeframes. Applicants and MAHs are also required to provide competent authorities with a description of their pharmacovigilance system and, where appropriate, of the risk management system applied to various products within the organization.

The pharmacovigilance system in the EU operates between the EC, EMA and regulatory authorities in Member States. In some Member States, regional centers are under the coordination of the national competent authority. The EMA pharmacovigilance system is described in the EMA pharmacovigilance system manual. [32]

Legal requirements related to the EMA (or the Agency) pharmacovigilance system for human medicines are laid down in Regulation (EC) No 726/2004, Directive 2001/83/EC, and Commission Implementing Regulation (EU) No 520/2012. [16,32]

Pursuant to the revised pharmacovigilance legislation, the EMA together with the Member States has drawn up Good Pharmacovigilance Practices (GVP), a new set of guidelines for the conduct of pharmacovigilance in the EU. The GVP modules replace Eudralex: rules governing medicinal products in the EU for medicinal products for human use Volume 9A.

11.17.1 GOOD PHARMACOVIGILANCE PRACTICES (GVP)

Good pharmacovigilance practices (GVP) apply to marketing-authorization holders (MAHs), the Agency and medicines regulatory authorities in EU Member States. They cover medicines authorized centrally via the Agency as well as medicines authorized at national level.

11.17.2 EUDRAVIGILANCE

EudraVigilance is a safety data processing network and management system, launched by the EMA in 2001, for reporting and evaluating suspected adverse reactions during development and following the marketing authorization of medicinal products in the EEA. [33]

11.17.3 EVWEB

The EudraVigilance system, in addition to automated message generation and processing, provides a web-based tool to allow for a manual safety and acknowledgment message creation as well as generation of medicinal product reports via a web interface, called EVWEB. [33]

11.17.4 RISK-MANAGEMENT PLAN (RMP)

RMP is a safety document which includes information on:

- A medicine's safety profile
- How its risks will be prevented or minimized in patients
- Plans for studies and other activities to gain more knowledge about the safety and efficacy of the medicine
- Risk factors for developing side effects
- Measuring the effectiveness of risk-minimization measures

In the EU, companies must submit an RMP to the Agency at the time of application for a marketing authorization. For medicines that do not have an RMP, it is likely that one will be required with any application involving a significant change to the marketing authorization. [34]

11.18 PATENT, DATA EXCLUSIVITY, AND DATA PROTECTION IN EUROPE

11.18.1 PATENTS

The EMA or any National Competent Authorities are not responsible for patents on medicines: issues regarding patent law are not within the Agency's remit. [35]

Member States are parties to the European Patent Convention 2000. The European Patent Office (EPO) grants patents only if the invention is patentable, i.e., the invention is novel, inventive, and susceptible to industrial application. Patent claims can be filed either with the national patent offices or with EPO, in which case, the patent will "confer on its proprietor from the date of publication of the mention of its grant, in each contracting state in respect of which it is granted, the same rights as would be conferred by a national patent granted in that State." The period of protection is 20 years from the date of the filing. [36]

11.18.2 DATA EXCLUSIVITY

The New EU Pharmaceutical Legislation adopted in 2004 has created a harmonized EU eight-year data exclusivity provision with an additional two-year market exclusivity provision. This effective 10-year market exclusivity can be extended by an additional one year maximum if, during the first eight years of those ten years, the MAH obtains an authorization for one or more new therapeutic indications which, during the scientific evaluation prior to their authorization, are held to bring a significant clinical benefit in comparison with existing therapies. This so-called 8 + 2 + 1 formula applies to new chemical entities NCEs in all procedures and to all Member States (unless certain new Member States are awarded derogations, which they can request following publication of the new law).

In practical terms, this means that a generic application for marketing authorization can be submitted after Year 8, but that the product cannot be marketed until after Year 10 or 11. [37]

11.18.3 SUPPLEMENTARY PROTECTION CERTIFICATES (SPC)

Supplementary protection certificates are an intellectual property right that serve as an extension to a patent right. They apply to specific pharmaceutical products that have been authorized by regulatory authorities. Supplementary protection certificates aim to offset the loss of patent protection for pharmaceutical that occurs due to the compulsory lengthy testing and clinical trials these products require prior to obtaining regulatory marketing approval.

An SPC can extend a patent right for a maximum of five years. A six-month additional extension is available in accordance with Regulation (EC) No 1901/2006 if the SPC relates to a medicinal product for children for which data has been submitted according to a PIP. PIPs are required to support the authorization of medicines for children. They ensure that enough data is collected on the effects of the medicine on children. The extension compensates for the additional clinical trials and testing that PIPs require. [38]

REFERENCES

[1] European Union. About the EU (cited December 22, 2016). Available from: <http://europa.eu/about-eu/facts-figures/living/indexen.htm>.

[2] European Commission. The European Union explained – How the European Union works; 2017 (updated June 2013; cited December 20, 2016). Available from <http://www.gr2014parliament.eu/Portals/6/PDFFILES/NA0113090ENC_002.pdf>.

[3] The European Free Trade Association. About EFTA; 2016 (cited May 20, 2016). Available from: <http://www.efta.int/>.

[4] Rägo L, Santoso B. Drug regulation: history, present and future. In: van Boxtel CJ, Santoso B, Edwards IR, editors. Drug Benefits and Risks: International Textbook of Clinical Pharmacology. 2nd ed. Amsterdam and Uppsala: IOS Press and Uppsala Monitoring Centre; 2008. p. 65–76. Available from: http://www.who.int/medicines/technical_briefing/tbs/Drug_Regulation_History_Present_Future.pdf>.

[5] European Commission. 50 years of EU pharmaceutical regulation milestones; 2014 (updated December 16, 2014; cited September 13, 2016). Available from: <http://ec.europa.eu/health/50_years_of_eu_milestones/timeline.htm>.

[6] European Medicines Agency. About us; 2016 (updated June 29, 2016; cited September 13, 2016). Available from: <http://www.ema.europa.eu/docs/en_GB/document_library/Other/2016/08/WC500211862.pdf>.

[7] European Medicines Agency. Organisation chart; 2017 (updated January 01, 2017; cited January 12, 2017). Available from: <http://www.ema.europa.eu/docs/en_GB/document_library/Other/2009/12/WC500017948.pdf>.

[8] Heads of Medicines Agencies. About HMA; 2016 (cited June 2, 2016). Available from: <http://www.hma.eu/abouthma.html>.

[9] European Directorate for the Quality of Medicines. EDQM Council of Europe; 2016 (cited June 5, 2016). Available from: <https://www.edqm.eu/>.

[10] European Commission. –Eudralex - EU Legislation; 2017 (updated January 18, 2017; cited January 19, 2017). Available from: <http://ec.europa.eu/health/documents/eudralex/index_en.htm>.

[11] European Commission. Directive 2001/83/EC of the European Parliament and of the Council of 6 November 2001 on the Community code relating to medicinal products for human use; November 6, 2001 (cited June 12, 2016). Available from: <https://ec.europa.eu/health/sites/health/files/files/eudralex/vol-1/dir_2001_83_consol_2012/dir_2001_83_cons_2012_en.pdf>.

[12] European Commission. EudraLex - Volume 10: Clinical trials guidelines; 2017 (updated January 18, 2017; cited January 19, 2017). Available from: <http://ec.europa.eu/health/documents/eudralex/vol-10_en>.

[13] European Union. EU Clinical Trials Register; 2016 (cited June 17, 2016). Available from: <https://www.clinicaltrialsregister.eu/>.

[14] Medicines and Healthcare products Regulatory Agency. Clinical trials for medicines: apply for authorisation in the UK; 2017 (updated January 12, 2017; cited January 20, 2017). Available from: <https://www.gov.uk/guidance/clinical-trials-for-medicines-apply-for-authorisation-in-the-uk>.

[15] European Commission. Notice to Applicants, Volume 2A (Revision 6): Procedures for marketing authorisation: Chapter 1: Marketing Authorisation; December 2016 (cited January 19, 2017). Available from: <https://ec.europa.eu/health/sites/health/files/files/eudralex/vol-2/vol2a_chap1_rev6_201612.pdf>.

[16] European Medicines Agency. User guide for micro, small and medium-sized enterprises (EMA/860940/2011); July 2016 (cited September 5, 2016). Available from: <http://www.ema.europa.eu/docs/en_GB/document_library/Regulatory_and_procedural_guideline/2009/10/WC500004134.pdf>.

[17] Co-ordination group for Mutual recognition and Decentralised procedures – human. Flow chart of the decentralised procedure (CMDh/080/2005/Rev2); March 2013 (cited July 15, 2016). Available from: <http://www.hma.eu/fileadmin/dateien/Human_Medicines/CMD_h_/procedural_guidance/Application_for_MA/DCP/CMDh_080_2005_Rev2_2013_03a_-_clean.pdf>.

[18] European Commission. Notice to Applicants, Volume 2A (Revision 5): Procedures for marketing authorisation: Chapter 2: Mutual Recognition; February 2007 (cited July 15, 2016). Available from: <http://ec.europa.eu/health//sites/health/files/files/eudralex/vol-2/a/vol2a_chap2_2007-02_en.pdf>.

[19] Co-ordination group for Mutual recognition and Decentralised procedures – human. Flow chart for the mutual recognition procedure (CMDh/081/2007, Rev.2); November 2016 (cited December 12, 2016). Available from: <http://www.hma.eu/fileadmin/dateien/Human_Medicines/CMD_h_/procedural_guidance/Application_for_MA/MRP_RUP/CMDh_081_2007_MRP_RUP_Flow_chart.pdf>.

[20] European Medicines Agency. Development support and regulatory tools for early access to medicines (EMA/531801/2015). March 1, 2016 (cited August 10, 2016). Available from: <http://www.ema.europa.eu/docs/en_GB/document_library/Other/2016/03/WC500202631.pdf>.

[21] European Commission. Notice to Applicants, Volume 2B. Medicinal products for human use: Presentation and format of the dossier; Common Technical Document (CTD); May 2008 (cited August 18, 2016). Available from: <https://ec.europa.eu/health/sites/health/files/files/eudralex/vol-2/b/update_200805/ctd_05-2008_en.pdf>.

[22] European Medicines Agency. Generic/hybrid applications: questions and answers; 2016 (cited December 19, 2016). Available from: <http://www.ema.europa.eu/ema/index.jsp?curl=pages/regulation/general/general_content_000179.jsp&mid=WC0b01ac0580022717>.

[23] European Medicines Agency. Guideline on Active Substance Master File Procedure (CHMP/QWP/227/02 Rev 3/Corr); May 31, 2013 (cited August 19, 2016). Available from: <http://www.ema.europa.eu/docs/en_GB/document_library/Scientific_guideline/2012/07/WC500129994.pdf>.

[24] European Directorate for the Quality of Medicines & HealthCare. New CEP Applications; 2016 (cited July 12, 2016). Available from: <https://www.edqm.eu/en/certification-new-applications-29.html>.

[25] European Medicines Agency. Good-manufacturing-practice and good-distribution-practice compliance; 2016 (cited December 13, 2016). Available from: <http://www.ema.europa.eu/ema/index.jsp?curl = pages/regulation/document_listing/document_listing_000154.jsp>.

[26] European Medicines Agency. Mutual Recognition Agreements; 2016 (cited December 10, 2016). Available from: <http://www.ema.europa.eu/ema/index.jsp?curl=pages/regulation/general/mutual_recognition_agreements.jsp&mid=WC0b01ac058006e013>.

[27] European Commission. Eudralex - Volume 4 - EU Guidelines for Good Manufacturing Practice for Medicinal Products for Human and Veterinary Use; Annex 16: Certification by a Qualified Person and Batch Release (Ref. Ares(2015)4234460); October 12, 2015 (cited December 13, 2016). Available from: <https://ec.europa.eu/health/sites/health/files/files/eudralex/vol-4/v4_an16_201510_en.pdf>.

[28] European Commission. Guidelines on the details of the various categories of variations, on the operation of the procedures laid down in Chapters II, IIa, III and IV of Commission Regulation (EC) No 1234/2008 of 24 November 2008 concerning the examination of variations to the terms of marketing authorisations for medicinal products for human use and veterinary medicinal products and on the documentation to be submitted pursuant to those procedures (2013/C 223/01); August 2, 2013 (cited August 29, 2016). Available from: <http://ec.europa.eu/health//sites/health/files/files/eudralex/vol-2/c_2013_2008/c_2013_2008_pdf/c_2013_2804_en.pdf>.

[29] European Medicines Agency. European Medicines Agency post-authorisation procedural advice for users of the centralised procedure (EMEA-H-19984/03 Rev. 63); March 7, 2016 (cited December 29, 2016). Available from: <http://www.ema.europa.eu/docs/en_GB/document_library/Regulatory_and_procedural_guideline/2009/10/WC500003981.pdf>.

[30] European Medicines Agency. Health technology assessment bodies; 2016 (cited December 15, 2016). Available from: <http://www.ema.europa.eu/ema/index.jsp?curl = pages/partners_and_networks/general/general_content_000476.jsp&mid=WC0b01ac0580236a57>.

[31] European Medicines Agency. Pharmacovigilance; 2015 (cited December 22, 2016). Available from: <http://www.ema.europa.eu/docs/en_GB/document_library/Leaflet/2011/03/WC500104236.pdf>.

[32] European Medicines Agency. EMA pharmacovigilance system manual (EMA/623550/2013); Version 1.2; October 13, 2016 (cited December 22, 2016). Available from: <http://www.ema.europa.eu/docs/en_GB/document_library/Other/2014/07/WC500170226.pdf>.

[33] European Medicines Agency. EudraVigilance; 2016 (cited August 22, 2016). Available from: <http://www.ema.europa.eu/ema/index.jsp?curl = pages/regulation/general/general_content_000679.jsp>.

[34] European Medicines Agency. Guideline on good pharmacovigilance practices (GVP); Module V – Risk management systems (Rev 1); April 15, 2014 (cited August 22, 2016). Available from: <http://www.ema.europa.eu/docs/en_GB/document_library/Scientific_guideline/2012/06/WC500129134.pdf>.

[35] European Medicines Agency. Frequently asked questions; 2016 (cited December 23, 2016). Available from: <http://www.ema.europa.eu/ema/index.jsp?curl=pages/about_us/q_and_a/q_and_a_detail_000114.jsp#section7>.

[36] European Patent Office; 2016 (cited September 9, 2016). Available from: <http://www.epo.org/service-support/faq/basics.html>.

[37] European Medicines Agency. EMA Procedural advice for users of the centralised procedure for generic/hybrid applications (EMEA/CHMP/225411/2006); August 4, 2016 (cited September 10, 2016). Available from: <http://www.ema.europa.eu/docs/en_GB/document_library/Regulatory_and_procedural_guideline/2009/10/WC500004018.pdf>.

[38] European Commission. Supplementary protection certificates for pharmaceutical and plant protection products. January 18, 2017 (cited January 19, 2017). Available from: <https://ec.europa.eu/growth/industry/intellectual-property/patents/supplementary-protection-certificates_en>.

CHAPTER

PHARMACEUTICAL REGULATIONS IN INDIA

12

Nidhi B. Agarwal[1] and Manoj Karwa[2]

[1]Jamia Hamdard University, New Delhi, India [2]Auriga Research Pvt. Ltd, New Delhi, India

12.1 HISTORY OF DRUG REGULATION IN INDIA

Until the 20th century, the drug manufacturing industries in India were at primary stages. The majority of drugs were imported from other countries. After the First World War, the demand for drugs increased enormously, giving rise to the manufacturing of spurious and substandard drugs in the market. The below table enlists some important acts that have paved the way towards the development of regulations in India (Table 12.1).

In 1947—the year of the independence of India—the pharmaceutical market was dominated by western MNCs. The supply of drugs into the market during this time was dependent upon 80%–90% import. Around 99% of all pharmaceutical products under patent in India were held by foreign pharmaceutical companies, and surprisingly the domestic Indian drug prices were among the highest around the globe. Later, in 1960, the government initiated policies emphasizing the production of drugs domestically. During that time many unscrupulous foreign manufacturers flooded the Indian market with adulterated and spurious drugs. Further, in response to the famous "Gigantic Quinine Fraud," the Government formed a drug inquiry committee under Sir Ram Nath Chopra also known as the "Chopra Committee," whose recommendations were later presented as "The Drug Bill" and amended to the Drugs and Cosmetic Act of 1940 (D and C Act) and Drugs and Cosmetic rules of 1945. To facilitate an independent supply of pharmaceutical products in the domestic market, the government of India founded five state-owned pharmaceutical companies. Government policy led to various actions, including the abolition of product patents on food, chemicals, and drugs; the limitation of multinational equity share in India pharmaceutical companies; the institution of process patents; and the imposition of price controls on certain formulations and bulk drugs. Due to lack of legal mechanisms to protect their patented products, many foreign pharmaceutical manufacturers left the Indian market. Accordingly, the share of the domestic Indian market held by foreign drug manufacturers declined to less than 20% in 2005. This policy resulted in local firms rushing in to fill the void, and by 1990, India was self-sufficient in the production of formulations and nearly self-sufficient in the production of bulk drugs. Today, India is the world's fifth largest producer of bulk drugs, ranks 3rd in the word in terms of production volume, and 13th in domestic consumption value, serving almost 95% of the country's pharmaceutical needs. This chapter will continue to shed light on the pharmaceutical regulations in India.

Pharmaceutical Medicine and Translational Clinical Research. DOI: http://dx.doi.org/10.1016/B978-0-12-802103-3.00013-4

Table 12.1 List of Acts

Act	Year	Regulation
Poisons Act	1919	Possession of substances or sale of substances as specified as poison.
The Dangerous Drugs Act	1930	Cultivation of the opium plant, manufacture and possession of opium, the import, export, transshipment, and sale of opium.
Drugs and Cosmetics Act	1940	The import, manufacture, distribution, and sale of drugs. This act covers allopathic, homeopathic, Unani, and Sidha drugs.
Drugs and Cosmetics Rules	1945	The manufacture of drugs for sale, and not for consumption, use, or possession.
Pharmacy Act	1948	The pharmacy profession of India.
Drugs and Magic Remedies (Objectionable Advertisements) Rules,	1955	The control of drug advertising in India.
Indian Patent Act	1970	Patent protection in India.
The Narcotic Drugs and Psychotropic Substances Act	1985	The operation of narcotic drugs and substances.
Drug Prices Control Order	1995	Controlling prices for consumers.
India joined Paris Cooperation Treaty	1999	The implementation of product patent.
Patent Amendment Act	2005	Provision for Black Box Application made.
Clinical Trial Registry-India	2007 (initiated) 2009 (voluntary measure)	Registration of CRO before the enrolment of the first patient for clinical trials.
Pharmacovigilance Programme of India (PvPI)	2010	Assurance of drugs safety for Indian patients.

12.2 INDIAN DRUG REGULATORY AUTHORITIES

The regulatory authorities of India and their functions are enlisted below:

- Central Drug Standard Control Organization (CDSCO)
- Indian Council of Medical Research (ICMR)
- Ministry of Health and Family Welfare

Central Drug Standard Control Organization (CDSCO)

The Central Drug Authority, Central Drugs Standard Control Organization (CDSCO), discharges functions appointed to the Central Government under the Drugs and Cosmetics Act. It works at both the central and state level and is responsible for ensuring safety, efficacy, and quality of drugs supplied to the public. The organization has six zonal offices, four sub-zonal offices, thirteen port offices, and seven laboratories under its control.

The structure of CDSCO is shown below.

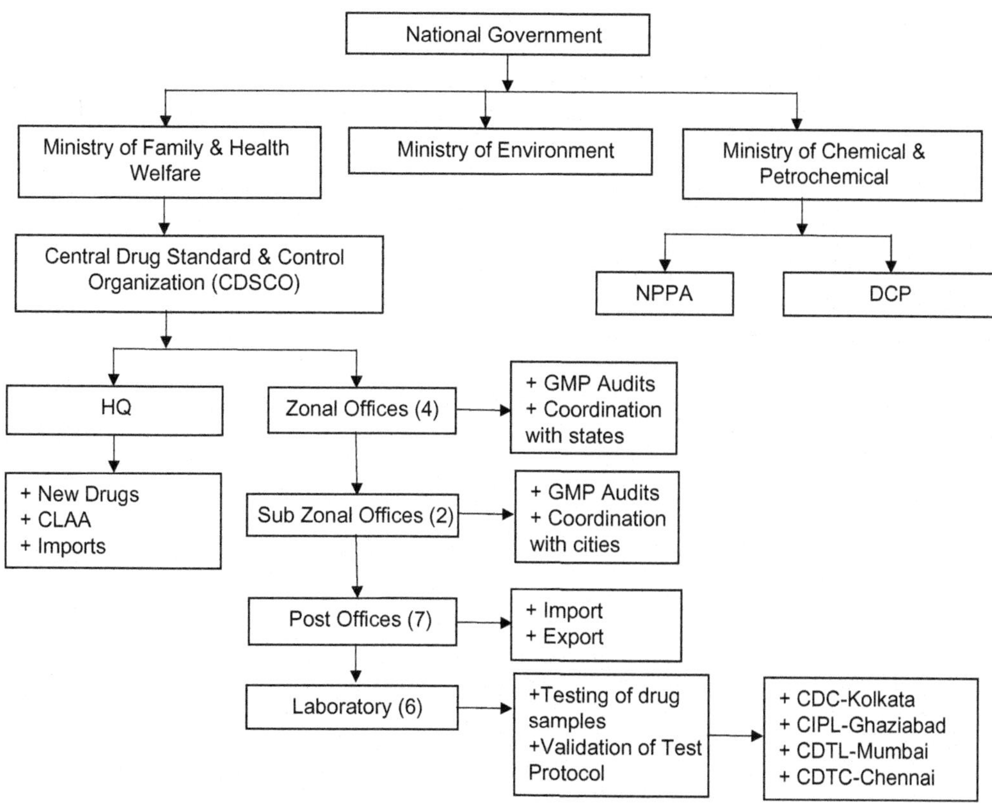

12.2.1 FUNCTION

CDSCO controls the import of drugs, the approval of new drugs and clinical trials, meetings of Drugs Technical Advisory Board (DTAB) and Drugs Consultative Committee (DCC). The approval of certain licenses through the Central License Approving Authority is exercised by the CDSCO headquarters. The state authorities are concerned with regulation of the manufacture, sale, and distribution of drugs, while the central authorities are responsible for approval of new drugs, clinical trials in the country, control over the quality of imported drugs, laying down the standards for drugs, coordination of the activities of state drug control organizations, and providing expert advice for the enforcement of the Drugs and Cosmetics Act. The Drug Controller General of India is responsible for approving licenses to particular categories of Drugs such as blood and blood products, Vaccine, I. V. Fluids, and Sera.

12.2.2 INDIAN COUNCIL OF MEDICAL RESEARCH (ICMR)

The Indian Council of Medical Research (ICMR), is amongst the oldest medical research bodies in the world and the apex body in India for the formulation, coordination, and promotion of biomedical research, and is funded by the government of India through the Ministry of Health and Family Welfare. The Indian Research Fund Association was built in 1911, and was responsible of sponsoring and coordinating medical research in the country. After independence, some important changes were made in the organization and the activities of the IRFA. It was re-named as ICMR in 1949 with a remarkable broadened scope of functions.

12.2.2.1 Structure

The governing body is overseen by the Union Health Minister. The Scientific Advisory Board aids in technical and scientific matters and comprises of distinguished experts in various biomedical disciplines. The board is further assisted by scientific advisory groups, scientific advisory committees, task forces, expert groups, steering committees, etc., which evaluate and monitor the various research activities of the Council. Biomedical research is promoted by the Council through intramural as well as extramural research. Over the decades, the base of extramural research and also its strategies have been broadened by the Council.

12.2.2.2 Function

The Council prefers research on management and control of communicable diseases, control of nutritional disorders fertility control, maternal and child health, forming alternative strategies for health care delivery, restraint amid safety limits of occupational and environmental health problems, research on major non-communicable diseases like cardiovascular disease, cancer, blindness, diabetes and other metabolic and hematological disorders, mental health research, as well as drug research (including traditional remedies). All these efforts are undertaken to reduce the total burden of disease and to promote the health and wellbeing of the population.

Intramural research is carried out by the council's 30 permanent research institutes. These institutes seek definitive research areas such as leprosy, tuberculosis, cholera, viral diseases including AIDS, kala-azar, malaria, vector control, nutrition, food and drug toxicology, immunohematology, reproduction, oncology, and medical statistics. Extramural research is promoted by the ICMR in the establishment of centers for advanced research in various research areas around current expertise and infrastructure in medical colleges, universities, and other non ICMR research institutes. It funds task force studies. Open-ended research is performed on the basis of applications for grants-in-aid. It inspires human resource development in biomedical research. The council offers the position of Emeritus Scientist for retired medical scientists and teachers, to allow them to continue or take up research on specific biomedical topics.

12.3 INDIAN CLINICAL TRIAL GUIDANCE

12.3.1 CLINICAL TRIALS REGISTRY-INDIA (CTRI)

The Clinical Trials Registry-India, an online, primary register of the WHO's International Clinical Trials Registry Platform, was launched on 20 July 2007. It was initiated as a voluntary measure. Nonetheless, the registry was made compulsory by the Drug Controller General of India (DCGI) on June 15, 2009 and is now accessible to the registration of prospective clinical trials on any intervention conducted in India that involves human participants. The registration is free of charge, and also the register is searchable free of cost. Disclosure of all the twenty items in the WHO Trial Registration Data Set is compulsory for allocation of a valid registration number. This registration number is required for publishing the results in the journals that endorse the International Committee of Medical Journal Editors' position on prospective trials registration. Trials registered in the CTRI-India are included in the central repository of the WHO's International Clinical Trials Registry Platform search portal. In addition to the 20 items, the CTRI-India also needs compulsory disclosure of details of regulatory clearances and ethics committee. Additional items regarding the methods to improve the internal validity of the trial are optional and aid in improving trial design and reliability of results.

All the trials conducted in India need to be registered. These include trials that involve human subjects, any intervention, surgical procedures, preventive measures, lifestyle modifications, devices, rehabilitation strategies, educational or behavioral treatment, and also the trials that are being conducted by the Department of AYUSH. The registration of clinical trials helps in improving the reliability of data generated, avoiding duplication of trials, helping clinicians to interpret and apply the research, and preventing the exposure of volunteers to potential risks related to trial.

12.3.2 CLINICAL TRIAL AUTHORISATION

A clinical trial in India can be initiated after obtaining written permission from Institutional Ethics Committee (IEC) and DCGI. Applications to conduct a trial are required to be submitted in Form 44 along with the requirements, as per schedule Y, which include documents of chemical, pharmaceutical product information, animal pharmacology, toxicology, and clinical pharmacology data. Additional, important documents include investigator's brochure, trial protocol, informed consent form, patient information sheet, case report form, and undertaking from investigator. To conduct studies in special population, e.g., pregnant women, nursing women, children, elderly patients, patients having renal or other organ system failure, and those on specific concomitant medication(s), specific requirements should be submitted. The protocol and other essential documents for clinical trial must be reviewed and approved by an IEC, which is constituted of seven members, including a medical scientist, a clinician, a statistician, a legal expert, a social scientist, and a layperson from the community. After obtaining approval from DCG(I) and IEC, an applicant can start clinical trial as per Schedule Y.

The import of active pharmaceutical ingredients from other countries requires a separate license called a 'T-license' (Trial License). The license is issued simultaneously with the approval of the clinical trial and it is valid for multiple shipments for a year. A No Objection Certificate (NOC) through separate application is required for the shipping of biological samples collected from the trial subjects out of India. Other important forms under the Drugs and Cosmetics Act are enlisted below:

Important Forms in the Drugs and Cosmetics Act

Important forms of the D and C Act	
Form 44	Application for grants of permission to import or manufacture a new drug or to undertake clinical trial
Form 12	Application for license to import drugs for the purposes of examination, test, or analysis
Form 11	License to import drugs for the purpose of examination, test, or analysis (Validity is one year)
Form 4	Issue of import certificate (Validity is six months)
Form 1	Application for the issue of import certificate for import of narcotic drugs and psychotropic substances

Important Rules of D and C Act	
Rule 34	Application for license for examination, test, or analysis
Rule 122-A	Application for permission to import new drug
Rule 122-B	Permission to manufacture new drug
Rule 122-D	Permission to import fixed dose combination
Rule 122-DA	Application for permission to conduct clinical trials for new drug/investigational new drug
Rule 122-DAA	Definition of clinical trial
Rule 122-E	Definition of new drug
Rule 55	Application for import certificate
Rule 56	Issue of import certificate
Rule 122-DAB	Compensation in case of injury or death during clinical trial
Rule 122 DAC	Permission to conduct clinical trial
Rule 122DB	Suspension or cancellation of Permission/Approval
Rule 122DD	Registration of Ethics committee
Rule 122E	Definition of new drugs

Important forms and rules of drug and cosmetic act for clinical research

12.3.2.1 Application for Permission to Conduct Clinical Trials for New Drug (122 DA)

No clinical trial involving a new drug, either for clinical experiment or any clinical investigation, should be started prior to the written permission of the Licensing Authority, and should be in accordance to clause (b) of rule 21. An application for grant of permission to conduct:

1. Human clinical trials (Phase-I) on a new drug shall be submitted to the Licensing Authority in Form 44 in addition to a fee of fifty thousand rupees and such information and data as required under Schedule Y.

2. Exploratory clinical trials (Phase-II) on a new drug shall be made on the basis of data emerging from Phase-I trial, accompanied by a fee of twenty-five thousand rupees.
3. Confirmatory clinical trials (Phase-III) on a new drug shall be made on the basis of the data emerging from Phase-II and, where necessary, data collected from Phase-I, and will be accompanied by a fee of twenty-five thousand rupees, provided that no additional fee shall be required to be paid along with the application for import/manufacture of a new drug based on successful completion of phases of clinical trials by the applicant. Moreover, no fee shall be required to be paid along with the application by Central Government or State Government institutes involved in clinical research for conducting trials for academic or research purposes.

(3) After being satisfied with the clinical trials the Licensing Authority shall grant permission in Form 45/45 A/46 or Form 46-A, subject to the conditions stated therein: if the data submitted on the clinical trials is inadequate, the Licensing Authority intimates the applicant within six months in writing, from the date of such intimation or such extended period, not exceeding six months, as the Licensing Authority may, for reasons to be recorded, in writing, permit, intimating the conditions which shall be satisfied before permission could be considered.

12.3.2.2 Suspension or Cancellation of Permission/Approval (122-DB)

If the importer or manufacturer under this fails to comply with any of the conditions of the permission or approval, the Licensing Authority may, therefore, after giving an opportunity to show the reason as to why such an order should not be passed by an order in writing stating the reasons, suspend or cancel it.

12.3.2.3 Appeal (122-DC)

Any person unsatisfied by an order passed by the Licensing Authority may within sixty days from the date of such order appeal to the Central Government, and the Central Government after an enquiry into the matter as considered necessary may pass such order in relation thereto as it thinks suitable.

12.4 DOCUMENTATION FOR CLINICAL TRIAL APPLICATION (CDSCO GUIDANCE)

For new drug substances that are discovered in India, clinical trials are needed to be carried out in India from Phase I. For new drug substances discovered in other countries, Phase I data is needed along with the application. After the submission of Phase I data that has been obtained from research outside India, to the licensing Authority, permission to either repeat Phase I trials and/or to conduct Phase II trials may be granted, followed by Phase III trials along with other global trials for that drug. Phase III trials are necessary to be conducted in India before obtaining permission to market the drug in India.

The data required with the application depends on the purpose of the new drug application.

For permission of clinical trials, the following documents are required to be submitted (www.cdsco.in):

1. Form 44
2. Treasury Challan of INR 50,000 (for Phase-I)/25,000/-(for Phase-II/III clinical trials).
3. Source of bulk drugs/raw materials.
4. Chemical and pharmaceutical information including:

   Information on active ingredients:

   Drug information (Generic Name, Chemical Name, or INN) and Physicochemical Data including:

   a. Chemical name and Structure—Empirical formula, Molecular weight.
   b. *Analytical Data*: Elemental analysis, Mass spectrum, NMR spectra, IR spectra, UV spectra, Polymorphic identification.
   c. *Stability Studies*: Data supporting stability in the intended container closure system for the duration of the clinical trial.

   Data on Formulation:

   a. Dosage form
   b. Composition
   c. Master manufacturing formula
   d. Details of the formulation (including inactive ingredients)
   e. In process quality control check
   f. Finished product specification and method of analysis
   g. Excipient compatibility study
   h. Validation of the analytical method
   i. Stability Studies: Data supporting stability in the intended container closure system for the duration of the clinical trial.

   Note: While adequate chemical and pharmaceutical information should be provided to ensure the proper identity, purity, quality, and strength of the investigational product, the amount of information needed may vary with the phase of clinical trials, proposed duration of trials, dosage forms, and amount of information otherwise available.

5. Animal Pharmacology
   a. Summary
   b. Specific pharmacological actions
   c. General pharmacological actions
   d. Follow-up and supplemental safety pharmacology studies
   e. Pharmacokinetics: absorption, distribution, metabolism, excretion
6. Animal Toxicology
   a. General aspects
   b. Systemic toxicity studies
   c. Male fertility study
   d. Female reproduction and developmental toxicity studies
   e. Local toxicity
   f. Allergenicity/hypersensitivity
   g. Genotoxicity
   h. Carcinogenicity

A. For Phase I Clinical Trials

   Systemic toxicity studies

   a. Single dose toxicity studies
   b. Dose ranging studies
   c. Repeat-dose systemic toxicity studies of appropriate duration to support the duration of proposed human exposure. (As per Clause1.8 of Appendix-III of Schedule Y to Drugs and Cosmetics Rules.

(Continued)

(CONTINUED)

Male fertility study

In vitro genotoxicity tests

Relevant local toxicity studies with proposed route of clinical application (duration depending on proposed length of clinical exposure)

Allergenicity/hypersensitivity tests (when there is a cause for concern or for parenteral drugs, including dermal application)

Photo-allergy or dermal photo-toxicity test (if the drug or a metabolite is related to an agent causing photosensitivity or the nature of action suggests such a potential)

B. For Phase II Clinical Trials

Provide a summary of all the non-clinical safety data (listed above) already submitted while obtaining the permissions for Phase I trial, with appropriate references.

In case of an application for directly starting a Phase II trial—complete details of the nonclinical safety data needed for obtaining the permission for Phase I trial, as per the list provided above, must be submitted.

Repeat-dose systemic toxicity studies of appropriate duration to support the duration of proposed human exposure.

In vivo genotoxicity tests.

Segment II reproductive/developmental toxicity study (if female patients of child bearing age are going to be involved).

C. For Phase III Clinical Trials

Provide a summary of all the non-clinical safety data (listed above) already submitted while obtaining the permissions for Phase I and II trials, with appropriate references.

In case of an application for directly initiating a Phase III trial—complete details of the non-clinical safety data needed for obtaining the permissions for Phase I and II trials, as per the list provided above, must be submitted.

Repeat-dose systemic toxicity studies of appropriate duration to support the duration of proposed human exposure.

Reproductive/developmental toxicity studies.

Segment I (if female patients of child-bearing age are going to be involved), and Segment III (for drugs to be given to pregnant or nursing mothers or where there are indications of possible adverse effects on fetal development).

Carcinogenicity studies (when there is a cause for concern or when the drug is to be used for more than 6 months).

D. For Phase IV Clinical Trials

Provide a summary of all the non-clinical safety data (listed above) already submitted while obtaining the permissions for Phase I, II, and III trials, with appropriate references.

In case an application is made for initiating the Phase IV trial, complete details of the non-clinical safety data needed for obtaining the permissions for Phase I, II, and III trials, as per the list provided above, must be submitted.

7. Human/Clinical pharmacology (Phase I)
  a. Summary
  b. Specific pharmacological effects
  c. General pharmacological effects
  d. Pharmacokinetics, absorption, distribution, metabolism, excretion
  e. Pharmacodynamic/early measurement of drug activity

8. Therapeutic exploratory trials (Phase II)
  a. Summary
  b. Study report(s) as given in Appendix II

9. Therapeutic confirmatory trials (Phase III)
  a. Summary
  b. Individual study reports with listing of sites and investigators as given in Appendix II.

10. Special studies
  a. Summary
  b. Bioavailability/Bioequivalence
  c. Other studies, e.g., geriatrics, pediatrics, pregnant or nursing women

(Continued)

(CONTINUED)

11. Regulatory status in other countries
 a. Countries where the drug is:
  i. Marketed
  ii. Approved
  iii. Approved as IND
  iv. Withdrawn, if any, with reasons
 b. Restrictions on use, if any, in countries where marketed/approved
12. Prescribing information (of the drug circulated in other countries, if any)
13. Application in Form 12 along with T-Challan of requisite fees (in case of import of investigational products)

NOTE:

For new drug substances discovered in India, for Phase I clinical trials data as per the items 1, 2, 3, 4, 5, 6, 7 (data, if any, from other countries) and 11 as mentioned above is required to be submitted.

For new drug substances discovered in countries other than India, for Phase I clinical trials data as per the items 1, 2, 3, 4, 5, 6, 7 (data from other countries) and 11 as mentioned above is required to be submitted.

A legal undertaking in the form of an affidavit should be submitted by the applicant (competent person from the Company) stating that the data submitted along with the application is scientifically valid and authentic.

12.5 COMMON TECHNICAL DOCUMENT

The Common Technical Document (CTD), describes the requirements for applications to be submitted for the registration of medicines, designed for use across Japan, Europe and the United States. It is a standardized format to prepare the applications of new drugs to be submitted to regional regulatory authorities in participating countries. The CTD is controlled by the International Conference on Harmonisation of Technical Requirements for Registration of Pharmaceuticals for Human Use (ICH). The CTD format is also required for submission to CDSCO.

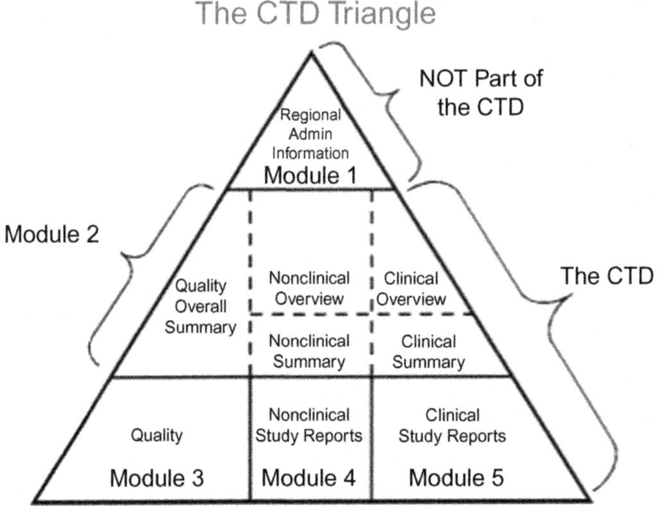

Module 1: General Information

This module should contain documents specific to India—for example, Form 44, Treasury challan fee, or the proposed label for use in India.

Module 2:

Introduction: This should include proprietary name, non-proprietary name or common name of the drug substance, company name, dosage form(s), strength(s), route of administration, and proposed indication(s).

Module 3: Quality

Information on Quality should be presented in the structured format according to all the subsections of quality guidelines. A. Drug Substance: Structure, General properties, Description of manufacturing process and process controls, Control of critical steps, Process validation, Impurities, Quality control of drug substance, Reference standards, Container closure system, Stability Data. B. Drug Product: Description and composition of drug product, Pharmaceutical development, Physicochemical and biological properties, Excipients, Microbiological attributes, Batch formula, Analytical Procedures, Control of excipients.

Module 4: Nonclinical Study Reports

The nonclinical study reports should be presented.

For Generic drugs complete non-clinical study is not required to be shown as the safety data are already provided during the approval of the branded drugs.

Module 5: Clinical Study Reports

The human study reports and related information should be presented.

MARKETING AUTHORIZATION (CDSCO guideline)

An application for approval of clinical trial or marketing authorization may comprise:

Entirely original data.

Data entirely from the literature.

Both original data and data from the literature ("hybrid").

For new drugs, it is likely that hybrid submissions will be the most common type.

12.6 DRUG APPROVAL PROCESS

An application for permission to conduct clinical trials in India is submitted to the DCGI along with the data of manufacturing, chemistry, control, and animal studies. The date concerning the trial protocol, informed consent documents, and investigator's brochures should also be submitted. A copy of the application is to be submitted to the ethical committee and the clinical trials can be conducted only after the approval of the DCGI and the ethical committee. Phase I clinical trials are conducted to determine the maximum tolerated dose in humans, adverse reactions, etc. on healthy human subjects. In Phase II trials involving 10–12 patients, the effective dose ranges and therapeutic uses are determined at each dose level. The confirmatory trials, called Phase III trials, are conducted to produce data regarding the efficacy and safety of the drug in ~100 patients, is metacentric (in 3–4 centers), to confirm efficacy and safety claims. The new drug registration (form # 44 besides full preclinical and clinical testing information) is to be applied after the completion of the clinical trials.

Stages of approval

1. Submission of Clinical Trial application for the evaluation of safety and efficacy.
2. Requirements for permission of the new drug approval.
3. Post approval changes in the biological products: quality, efficacy, and safety documents.
4. Preparation of quality information for the drug submission for new drug approval.

After the NDA is received by the agency, it undergoes the technical evaluation. This screening ensures that sufficient information and data have been submitted in each of the areas for the justification of "filing" the application, which is the FDA formal review. At the conclusion of an NDA review by FDA, 3 possible actions can be sent to the sponsor:

Not approvable—It states the list of deficiencies along with the explanation of the reasons.
Approvable—It means that the drug can be approved after correction of minor deficiencies that can be corrected, such as labeling changes and possibly requesting commitment for post-approval studies.
Approval—It states that the drug has been approved.

12.7 REGULATION OF GENERIC DRUGS

A generic drug can be defined as "a drug product that is comparable to brand/reference listed drug product in dosage form, strength, route of administration, quality and performance characteristics, and intended use." A generic drug should essentially contain the same active ingredients as in the original formulation. A generic drug is similar or bioequivalent to a brand drug in dosage form, route of administration, strength, quality, performance characteristics, safety, and intended use. In spite being chemically identical to generic drugs they are sold at considerable discount from the branded price.

12.7.1 GENERIC DRUG APPROVAL PROCESS IN INDIA

The following documents are required for the approval of generic drugs in India:

1. Form 44.
2. Treasury Challan of INR 15,000 is required, provided all the active ingredients have been approved in India for >1 year, or INR 50,000 if any of the active ingredients is approved for <1 year.
3. Source of bulk drugs/raw materials: For the ingredients which are approved and treated as new drugs. In case the applicant has a manufacturing license for bulk drugs, he needs to submit a copy of the same. Alternatively, the consent letter from the approved source is to be provided concerning the supply of material.

12.8 MEDICAL DEVICES

"A medical device is defined as an inert diagnostic of therapeutic article that does not achieve any of its principal intended purposes through chemical action, within or on the body unlike the medicated devices which contain pharmacologically active substances which are treated as drugs." Such devices include diagnostic test kits, electrodes, crutches, arterial grafts, pacemakers, intra-ocular lenses, orthopedic pins, and other orthopedic accessories.

Depending on the risks involved, the devices can be classified as follows:

1. *Noncritical devices*—Investigational devices which do not present substantial risk to the patients, e.g., thermometer, B.P. apparatus.
2. *Critical devices*—Includes investigational devices which present a significant risk to the health, safety, or welfare of the subject. For example, pacemakers, internal catheters, implants.

All the general principles of the clinical trials should also be implemented for trials of medical devices. Safety evaluation along with pre-market efficacy of devices for 1-3 years including data on adverse reactions has to be obtained before pre-market certification. The duration of the trial and the extent of use could be decided on a case-by-case basis by the appropriate authorities. Nevertheless, the following necessary factors that are unique to medical devices need to be taken into consideration while assessing the related research projects.

Guidelines

- Safety data of the medical device in animals should be obtained and likely risks posed by the device should be considered.
- A clinical trial of medical devices is different from drug trials, as the former cannot be done in healthy volunteers. Hence phase I of drug trial is not necessary for trial on devices.
- Medical devices used within the body may have a greater risk potential than those used on or outside the body, for example, orthopedic pins versus crutches.
- Medical devices not used regularly have less risk potential than those used regularly, for example, contact lenses versus intraocular lenses.
- Safety procedures to introduce a medical device in the patient should also be followed as the procedure itself may cause harm to the patient.
- Informed consent procedures should be followed as in drug trials. The patient information sheet should contain information on following procedures to be adopted if the patient decides to withdraw from the trial (IndianGCP).

12.9 PHARMACEUTICAL PRICING

The price controls for drugs and formulations in India have a long history. The very first price control order was issued under the Defense of India Act in 1963. However, from 1970 onwards price control orders have been issued under the Essential Commodities Act. The Drug Price Control Order (DPCO) in 1970 originated for the safeguard of the interests of the consumer, while providing for a restricted but reasonable return to the producers. Simultaneously, with the nullification of product patents in 1970, the measure also brought about an era of cheaper medicines in India, although at the expense of diluting intellectual property rights. Since then the DPCO has been amended thrice, the last time being in 1995. In 1995, the DPCO had introduced three parameters to ensure proper market conditions—turnover, market competition, and market monopoly. Under this, the prices of 74 bulk drugs and their formulations are being controlled presenting approximately 20% of the pharma market. Bulk drugs, with a turnover of over Rs.40 million, are under the attention of the DPCO, except those drugs with sufficient market competition. Sufficient market competition has been defined as "the presence of at least five bulk producers and 10 formulations, with no producer's market share exceeding 40 percent."

Industrial licensing has been abrogated for all drugs, formulations, and drug intermediates with the exception of five drugs which are reserved for public sector. Additionally, price controls have been waived for a period of five years for drugs which have been developed indigenously. The pricing methods for imported bulk drugs are based on landed cost, that is the maximum acceptable selling price. In case of a bulk drug manufactured locally, the price has been fixed based on the cost and capacity data of a particular production unit.

In the case of pricing of imported formulations, the price has been fixed according to the landed cost and the selling and distribution expenses, which will not exceed 50% of the landed price. In the case of locally produced formulations, the price has been determined using the retail price. This signifies that the DPCO fixes the price and monitors 74 bulk drugs and all the formulations which use these bulk drugs. Hence, the government controls, under DPCO, 50% of the pharmaceutical market in India (ICRA, 2000). The impact of prices of drugs during post-DPCO (1995) period evaluated by the ORG (1996) found that there had been a 4.6% increase in drug prices in the 12 months following the announcement of the 1995 DPCO, compared to an increase in the Consumer Price Index (CPI) of 9.8% for the same period. The study also revealed that the index of prices on products that had moved from the controlled to the decontrolled category under the DPCO had also registered around a 10.7% increase. Although there is a price control under DPCO, still most of drugs in the market are not regulated and the price rise during this period is considered to be minimal. In a nutshell, while the DPCO has evolved in a step-by-step ad hoc fashion, it has managed to strike a rough balance between regulating prices to ensure adequate access to essential medicines for the rural and urban poor while allowing the emergence of a globally competitive Indian domestic drug industry.

In 2002, a newer pharmaceutical policy was approved by the government in which the number of drugs under price control, and the span of price control, were sought to be decreased. Nonetheless, before this policy would have been implemented it was stayed by the Karnataka High Court in a PIL filed before the Court on the ground that a large number of essential drugs would go out of price control. An SLP was filed by Government in the Supreme Court against the order of the Karnataka High Court. The Supreme Court vide its interim order on 10th March, 2003, stayed the order of the Karnataka High Court. Nevertheless, it also ordered that "the petitioner shall consider and formulate appropriate criteria for ensuring essential and life saving drugs not to fall out of price control, and to review the drugs which are essential and life saving in nature until 2nd May, 2003."

After deliberation, in 2005, the government announced a new pharmaceutical policy, wherein it was proposed to bring an additional 354 drugs under a National List of Essential Medicines (NLEM) under price control. It was also proposed that patented drugs would be subjected to price negotiations, prior to grant of marketing approvals.

The important features of the National Pharmaceutical Pricing Policy, 2012 (NPPP-2012), are as follows:

- The regulation of the prices of drugs is on the basis of essentiality of drugs as specified under the National List of Essential Medicines (NLEM)-2011.
- The regulation of prices of drugs is on the basis of regulating the prices of formulations only.
- The regulation of prices of drugs is on the basis of fixing the ceiling price of formulations through Market Based Pricing (MBP).

NLEM-2011 contained 614 formulations of specified strengths and dosage forms, spread over 27 therapeutic categories, and satisfied the priority healthcare needs of the majority of the population of the country.

As per the provisions of NPPP-2012, all the manufacturers/importers manufacturing/importing the medicines as specified under NLEM-2011 shall be under the purview of price control. Such medicines shall have an MRP equal to or lower than the ceiling price (plus local taxes as applicable) as notified by the government for respective medicines.

The objective of National Pharmaceutical Pricing Policy-2012 is to put in place a regulatory framework for the pricing of drugs so as to ensure availability of required medicines—"essential medicines"—at reasonable prices even while providing sufficient opportunity for innovation and competition to support the growth of industry, thereby meeting the goals of employment and shared economic wellbeing for all.

During the 12th Five Year Plan, the Ministry of Health and Family Welfare proposes to instigate an initiative for free supply of essential medicines in public health facilities in the country for providing affordable healthcare to the people by decreasing out of pocket expenses on medicine. Additionally, to provide relief to the common man in the area of healthcare, a countrywide campaign, the 'Jan Aushadhi Campaign', was started by the Department of Pharmaceuticals, Government of India, in collaboration with the State Governments, by way of opening up Jan Aushadhi Generic Drug Stores to make available quality generic medicines at affordable prices to all. (Part of the National Pricing Authority.)

Functions of National Pharmaceutical Pricing Authority

1. To implement and enforce the provisions of the Drugs (Prices Control) Order in accordance with the powers delegated to it.
2. To deal with all legal matters arising out of the decisions of the Authority.
3. To monitor the availability of drugs, identify shortages if any, and to take remedial steps.
4. To collect/maintain data on production, exports and imports, market share of individual companies, profitability of companies, etc., for bulk drugs and formulations.
5. To undertake and/or sponsor relevant studies in respect of pricing of drugs/pharmaceuticals.
6. To recruit/appoint the officers and other staff members of the Authority, as per rules and procedures laid down by the Government.
7. To render advice to the Central Government on changes/revisions in the drug policy.
8. To render assistance to the Central Government in the parliamentary matters relating to the drug pricing.

12.10 PHARMACOVIGILANCE AND RISK MANAGEMENT IN INDIA

Pharmacovigilance is the best tool to establish the safety of a drug product. Every drug has some beneficial as well as undesirable or adverse effects. ADR are common clinical problems which need minimization to decrease the economic burden of the country. In this context the launch of the Pharmacovigilance Programme of India (PvPI) marks an essential milestone towards safeguarding public health. PvPI is one of the indispensable steps taken by the Ministry of Health and Family Welfare, Government of India for ensuring safe and rational use of medicines.

The CDSCO, under the Ministry of Health and Family Welfare, in collaboration with Indian Pharmacopeia commission, Ghaziabad, started a national Pharmacovigilance program for the protection of the health of patients by ensuring drug safety. The program is coordinated by the Indian Pharmacopeia Commission, Ghaziabad, as a National Coordinating Centre (NCC). The center operates under the supervision of a Steering Committee (CDSCO).

Steering Committee

Pharmacovigilance Programme of India

Chairman

Drugs Controller General (India), New Delhi, ex- officio

Members

1. Scientific Director, Indian Pharmacopoeia Commission, Ghaziabad, ex-officio
2. Head of Department, Pharmacology, AIIMS, ex-officio
3. Nominee of Director General, ICMR, ex-officio
4. Assistant Director General (Extended Programme of Immunization [ADG(EPI)] as representative of Directorate General Health Services
5. Under Secretary (Drugs Control) as representative of The Ministry of Health and family Welfare
6. Nominee of Vice Chancellor of Medical/Pharmacy University, ex-officio
7. Nominee of the Medical Council of India, ex-officio
8. Nominee of Pharmacy Council of India, ex-officio

Member Secretary

Officer-in-Charge (New Drugs), CDSCO, New Delhi, ex-officio

Goal and objective

Goal:

Ensure that the benefits of the use of medicine outweigh the risks and thus safeguard the health of the Indian population.

Objectives:

Monitor Adverse Drug Reactions (ADRs) in the Indian population.

Create awareness amongst healthcare professionals about the importance of ADR reporting in India.

Monitor benefit-risk profile of medicines.

Generate independent, evidence-based recommendations on the safety of medicines.

Support the CDSCO for formulating safety related regulatory decisions for medicines.

Communicate findings with all key stakeholders.

Create a national center of excellence on a par with global drug safety monitoring standards.

Program governance and reporting structures

The Pharmacovigilance Program of India is administered and monitored by the following two committees:

1. Steering Committee
2. Strategic Advisory Committee

Technical support will be provided by the following committees:

1. Signal Review Panel
2. Core Training Panel
3. Quality Review Panel

Governance Structure
ADR MONITORING CENTRES MCI
Approved Medical Colleges and Hospitals
Private Hospitals
Public Health Programs
Autonomous Institutes (ICMR etc.)

Collaboration with World Health Organization-Uppsala Monitoring Centre (UMC).

WHO and UMC work with and/or provide technical support to more than 94 countries worldwide. The PvPI National Coordinating Centre has collaborated with the WHO Collaborating Centre, Uppsala Monitoring Centre, Sweden, to achieve its long-term objective of establishing a "Centre of Excellence" for Pharmacovigilance in India.

FURTHER READING

Chittaranjan A, Nilesh S, Sarvesh C. The new patent regime: implications for patients in India. Indian J Psychiatry 2007 January–March;49(1):56–9.

Draft guidance on approval of clinical trials & new drugs. Available from <www.cdsco.nic.in/writereaddata/Guidance_for_New_Drug_Approval-23.07.2011.pdf>; 2011 [accessed 25.02.17].

Guidance for industry on preparation of common technical document for import/manufacture and marketing approval of new drugs for human use (new drug application – NDA). Available on <www.cdsco.nic.in/writereaddata/CTD%20Guidance%20Final.pdf>; [accessed 25.02.17].

<http://cdsco.nic.in>.

<http://www.nppaindia.nic.in/>.

Imran M, Najmi AK, Mohammad F. Clinical research regulation in India-history, development, initiatives, challenges and controversies: still long way to go. J Pharm Bioallied Sci. 2013 January–March;5(1):2–9.

Kumar D, Muthu M, Pandey BL. Pharmacovigilance programme in India: current status and its per-spectives. RRJPTS 2015;3:20–2.

Naishi K, Dilip M. Documentation requirements for generic drug application to be marketed in India- a review. JPSBR 2014;4(4):237–42.

Narayan S. Price controls on pharmaceutical products in India. ISAS Working Paper. Available online on <https://www.isas.nus.edu.sg/ISAS%20Reports/20.pdf>; [accessed 24.02.17].

Satyanarayana K, Anju S, Purvish P. Statement on publishing clinical trials in Indian biomedical journals. Indian J Ophthalmol 2008 May–June;56(3):177–8.

Sil A, Nilay K. How to register a clinical trial in India? Indian J Dermatol 2013 May–June;58(3):235–6.

Shekar V, Reddy G, Ramu B, Rajkamal B. Regulatory aspects of generic market in india and developed countries. World J Pharmacy Pharmaceut Sci 2016;5(5):1312–21.

Tharyan P, Ghersi D. Registering clinical trials in India: a scientific and ethical imperative. Natl Med J India 2008 January–February;21(1):31–4.

<www.icmr.nic.in/>.

William G. The Emergence of India's Pharmaceutical Industry and Implications for the U.S. Generic Drug Market. Office of economics working paperUS international trade commission. 2007. Available from <https://www.usitc.gov/publications/332/EC200705A.pdf>; [accessed 26.02.17].

CHAPTER

PHARMACEUTICAL REGULATIONS FOR COMPLEMENTARY MEDICINE

13

Preeti Vyas and Divya Vohora
Jamia Hamdard University, New Delhi, India

13.1 INTRODUCTION: COMPLEMENTARY MEDICINE AROUND THE WORLD

According to the World Health Organisation (WHO), "Traditional medicine is the sum total of the knowledge, skills and practices based on the theories, beliefs and experiences indigenous to different cultures, whether explicable or not, used in the maintenance of health, as well as in the prevention, diagnosis, improvement or treatment of physical and mental illnesses" [1]. In many countries, the terms "complementary" or "alternative" medicine (CAM) are used instead of "traditional medicine"; however, many states distinctively define them. For example, if used in conjunction with conventional medicine, it is referred as "complementary medicine" and if used in place of conventional therapy, it is considered as "alternative medicine" [2]. Such traditional systems of medicine are practiced in different countries based on their own traditional healing practices entrenched in their civilization, like traditional Chinese medicines, Islamic Unani medicine, and Indian Ayurvedic medicine. Other complementary systems of medicine include homeopathy, acupuncture, chiropractic and osteopathic manipulation, naturopathy, etc. [1].

Many years ago, these medicines were only used in low-income countries, where costly allopathic treatment was unaffordable, though this trend has completely changed over the years. Now, these traditional medicines and healing practices have gained worldwide recognition, and are sometimes preferred over the allopathic drugs that produce different side effects. In contrast, these natural products and practices are usually believed to be free of adverse reactions and, therefore, are generally tagged "safe," and this is responsible for their elevated use. However, these medicines are also liable to show certain side effects depending upon their individual compositions and qualities, which can be reported as adverse events. Besides, such traditional products can be inappropriately substituted or adulterated with substandard herbal drugs. They may also contain undeclared conventional drugs with a known history of adverse drug reactions based on their pharmacological properties. Other reasons may include the presence of contaminants, viz heavy metals, aflatoxins, microbial contaminants, or agricultural residues including pesticides. Moreover, the incorporation of these traditional medicines in conventional healthcare systems has elevated the risk of drug-herb interactions. These medicines, being the amalgamation of various chemical constituents influencing their therapeutic efficacy, may show the constitutive variability in their chemical composition and

Pharmaceutical Medicine and Translational Clinical Research. DOI: http://dx.doi.org/10.1016/B978-0-12-802103-3.00014-6

may not be bioequivalent. The active ingredients obtained from the same species of herb procured from various geographical sources may also show dissimilarities in the percentage of specific chemical constituents. This fluctuation may also lead to serious adverse events and, consequently, proper identification and standardization are requisite prior to the formulation of any dosage form (Box 13.1).

Pertaining to the reasons mentioned above, the stringent regulatory control, facilitating the safe and effective use of these medicines, is required to provide better healthcare to patients [2,3]. Additionally, the current era of globalization has also mandated the presence of common international standards to shape the synchronized rules and regulations for diverse countries to assure the highest quality standards for these medicines under the same roof (Table 13.1). To work in this direction, the WHO has designed global guidelines for herbal and other traditional medicinal products that serve as the reference standards for the member states to design their respective rules and policies, and determines the legal statuses of these medicines all over the world. Besides this, the WHO has also published four volumes of monographs of the selected medicinal plants (containing 28, 30, 31, and 28 monographs respectively) [4–7], which provide technical assistance to these states by illustrating the proper use of the extensively exploited herbal drugs. In 2010, "WHO monographs on medicinal plants commonly used in the Newly Independent States (NIS)" was published to assist the Newly Independent States (NIS) and the Countries of Central and Eastern Europe (CCEE) in promoting research and development in this field. This document was finalized after receiving the comments and opinions of around 256 experts and national regulatory authorities of 99 countries in addition to nongovernment organizations.

Furthermore, the "National policy on traditional medicine and regulation of herbal medicines: Report of a WHO global survey" came out, which summed all the CAM-related national policies and highlighted their regulatory statuses in all the responsive member states [8]. However, 2005 was not the first time that WHO made efforts to publish the worldwide review of the regulatory status of these medicines. Prior to that, WHO also contacted its member states in 1998 to collect the relevant data, but only 52 members showed their interest, based upon which a document named "Regulatory situation of herbal medicines: a worldwide review" was released [9]. Later on, another publication entitled "Legal status of traditional medicine and complementary/alternative medicine: a worldwide review" also came out in 2001, giving the summary of 123 countries [10]. This data shows the remarkable progress of several member states displaying their remarkable support towards the WHO campaigns. All these states came forward and made efforts to develop the regulations and policies related to traditional and complementary medicines. Many national and regional regulations were also developed in order to check the legal status of these medicines and to improve the safety, efficacy, practice, and education systems of all these systems of medicines. Prior to 1988, there were only 14 WHO member states regulating these traditional medicines, but the scenario has improved drastically over the past few years. The WHO traditional medicine strategy (2014–23) showed the increased participation and the development of CAM regulations by various WHO member states. In addition, 80% of the total members now recognize acupuncture as a traditional system of medicine with around 29 states having regulations for the therapists and 18 member states covering it in their health insurance policies. On the other hand, 30% of the surveyed states were also reported to have provisions for the high-level education for these systems of medicines. More importantly, "National policy and regulation of traditional medicine: report of the second WHO global survey," which is anticipated to unveil the current regulatory statuses of these

BOX 13.1 THE SCHEMATIC VIEW OF QUALITY CONTROL OF RAW MATERIAL AND FINISHED HERBAL PREPARATIONS ALONG WITH THE REQUIRED GMP COMPLIANCE: [130]

Part I: Quality control of raw materials and finished products:

1. Macroscopic examination:
   a. Biological name of the plant.
   b. Shape, size, color, surface characteristics, texture, fracture characteristics and appearance of the cut surface, other organoleptic characters, etc.
   c. QC of good agricultural and cultivation practices.
2. Microscopic examination:
   Types of stomata, determination of stomatal index, palisade ratio, vein islet number, stomatal number, microsublimation study, etc.
3. General tests for identity markers: (by HPLC, TLC, etc.).
4. Physico-chemical examination:
   a. Determination of specific gravity, pH value, melting and boiling range, optical rotation.
   b. Determination of ash value: total ash value, water soluble ash, acid insoluble and sulfated ash).
   c. Determination of extractable matter: alcohol-soluble extract, water-soluble extract, ether-soluble extract.
   d. Determination of moisture content: azeotropic distillation, gravimetric determination (loss on drying).
   e. Determination of volatile oil content.
   f. Determination of bitterness value.
   g. Determination of haemolytic activity (saponins).
   h. Qualitative and quantitative determination of tannins.
   i. Determination of swelling and foaming index.
   j. Qualitative and quantitative determination of other constituents, like alkaloids, starch, carbohydrates, proteins etc.
   k. Tests for sterility (as required).
   l. Determination of contaminants:
      i. Heavy metals: mercury, arsenic, cadmium, lead, etc.
      ii. Microbial contaminants: total viable aerobic count and specific microorganisms (ex:*Escherichia coli, Salmonella* spp.).
      iii. Mycotoxins : aflatoxins (B1, B2, G1 and G2), ochratoxin A.
      iv. Endotoxins.
      v. Radioactive contamination.
   m. Determination of residues:
      i. Pesticidal residues: phosphates, chlorates, etc.
      ii. Residual solvents.
   n. Determination of any unprofessed substance (if present).
   o. Justified use of appropriate substituent (if used).
5. Additional specifications for finished herbal product:
   a. Chromatographic profile.
   b. Specific tests for various dosage forms.
   c. Safety and efficacy considerations.
   d. Stability of the preparation/product.

(*Continued*)

BOX 13.1 (CONTINUED)

Part II: Compliance with GMP guidelines:

1. Premises (adequate space, quality control sections, goods store, etc.).
2. Personnel (qualification, training, number of employees, etc.).
3. Qualification of the equipments used.
4. Proper sanitation and hygiene.
5. Water supply (specifications for the quality of water used for manufacturing, cleaning, etc.).
6. Waste disposal.
7. Cleaning of the containers and equipments.
8. Good practices for the in-process and product quality control (sampling, testing methods, etc.) to be used for Part I.
9. Adequate storage of raw material and finished goods.
10. Validation.
11. Batch manufacturing records.
12. Distribution records.
13. Complaints and product recall.
14. Outsourced activities.
15. Self-inspection.
16. Documentation.

Note*: Wherever applicable, the analytical methods, the respective reference standards, and the limits for the acceptance used for these tests have been specified in different pharmacopoeias. These official methods are followed for the same. In case of any deviation, the justified reason should be approved prior to the conduct of the tests.*

Ref. 1. WHO guidelines on good manufacturing practices (GMP) for herbal medicines. World Health Organization; 2007. Available at: www.who.int [accessed 25.08.15].

2. WHO guidelines for accessing the quality of herbal medicines with reference to contaminants and residues. World Health Organization. Geneva; 2007. Available at: www.who.int [accessed 30.08.15].

3. Yau WP, Goh CH, Koh HL. Quality control and quality assurance of phytomedicines: key considerations, methods, and analytical challenges. In: Ramzan I, editor, Phytotherapies: Efficacy, Safety, and Regulation. Willey Publishers; 2015, vol 2, p. 2.2.

medicines in various countries, is currently under preparation and is expected to be released soon [11]. For the present chapter, the legal statuses and the current regulatory scenario of complementary medicines in the five main member states of WHO is detailed in the subsequent sections.

13.2 REGULATORY STATUS OF COMPLEMENTARY MEDICINE IN USA

The National Health Interview Surveys (NHIS) conducted in 2002, 2007, and 2012 substantiated the increased use and popularity of traditional medicinal approaches in the United States. These reports provided a comparative sketch determining their utilization by the adult US-population, which were found to be 32.3% [12], 35.5% [13], and 33.2% [14]. Not much difference in the percentage use was observed in children in 2007 (12%) and 2012 (11.6%) [15]. In addition, natural products (dietary supplements excluding vitamins and minerals) were estimated as the most popular complementary and alternative medicine (CAM) approach among the American adult population (17.7%). The same report has also highlighted yoga, tai chi, chiropractic or osteopathic manipulation, meditation, massage, homeopathy, etc. as other popular methods [16].

Table 13.1 WHO Limits for the Microbial Contaminants (Per Gram) in Herbal Raw Material, Herbal Medicines, and Other Herbal Preparations

	Raw material for processing	Pretreated herbal material	Other herbal materials (for internal use)	Herbal medicines to which boiling water is added before use	Other herbal medicines
Echerichia coli	10^4	10^2	NA	10	Absent
mold propagules	10^5	NA	NA	NA	NA
Yeasts and molds	NA	10^4	NA	10^3	10^3
Aerobic bacteria	NA	10^7	10^5	10^7	10^5
Other enterobacteria	NA	10^4	10^3	10^3	10^3
Clostridia	NA	Absent	Absent	Absent	Absent
Salmonellae	NA	Absent	Absent	Absent	Absent
Shigella	Absent	Absent	Absent	Absent	Absent

Ref: WHO guidelines for accessing the quality of herbal medicines with reference to contaminants and residues. World Health Organization. Geneva; 2007. Available at www.who.int [accessed 30.08.15].

Principally, these traditional herbal medicines are regulated in the USA by the United States Food and Drug Administration (US-FDA) as the dietary supplements under the "Dietary Supplement Health and Education Act of 1994" (DHSEA of 1994) [17], however, their regulatory statuses may vary depending upon their intended use. It took more than 100 years to develop this present regulatory framework, which commenced with the enactment of the "Pure Food and Drugs Act of 1906" [18]. Later, the "Federal Food, Drug and cosmetics act of 1938" ("FDCA of 1938" or "the Act") was passed under which herbal products were regulated as food or drug products [19]. Afterwards, the DHSEA of 1994 was signed in the 103rd US congress, and came out as a historic amendment of the latter, under which a special category for the "Food for dietary use," including the botanicals or herbal products used in complementary therapies, was added.

The National Centre for Complementary and Alternative Medicine (NCAAM) investigates the safety and efficacy of complementary and integrative medicines (including mind and body interventions, natural products, etc.) in USA. It conducts as well as sponsors advanced research in various scientific institutions and circulates the allied information to the public in order to upgrade the treatment standards. To date, it has released three strategic plans in the years 2000 [20], 2005 [21], and 2011 [22], where its goals and objectives have been published. The "NCCAM Third Strategic Plan: 2011–2015," the most recent one, comprise of five major strategic goals which collectively aim to boost present comprehension by developing and sharing the evidence-based data enhancing the current understanding of these approaches and their possible outcomes [22]. According to NCCAM, CAM is defined as "a group of diverse medical and healthcare systems, practices, and products that are not presently considered to be part of conventional medicine" [23] and categorizes all these "complementary health approaches" into two classes: (1) Natural products including herbs, vitamins, minerals, probiotics (sold as the dietary supplements under the DHSEA of 1994), and (2) Mind and Body Practices including Acupuncture, Movement therapies (like Alexander technique and Pilates), Relaxation techniques (like breathing exercises), Yoga, Chiropractic and osteopathic manipulations, Tai chi and Qui gong practices (from traditional Chinese medicines), Massage therapies, and other traditional procedures administered or trained by the practitioners. Other approaches viz Indian ayurvedic medicines, Traditional Chinese medicines, homeopathy, naturopathy, not falling under any of the above categories, are separately listed as "Other Complementary Health Approaches" by NCAAM [2]. In December 2014, it has been renamed as the "National Centre for Complementary and Integrative Health" as per the provision of the "omnibus appropriations bill, the Consolidated and Further Continuing Appropriations Act, 2015" signed by the former president Obama [24].

Furthermore, the FDA's perspective towards the complementary and alternative medicines became even clearer in 2006 with the publication of the most awaited draft guidance outlining the regulations on CAM products. These consolidated regulations, prepared by the U.S. Department of Health and Human Services, FDA, were released under the name of "Complementary and Alternative Medicine Products and their Regulation by the Food and Drug Administration" [23]. It specifically stated that CAM practices along with the products or medical devices used in such therapies are subjected to the regulations of the FDCA of 1938 or the PHS Act, depending upon the anticipated use of the product. For example, if a particular botanical is used to treat some diseased condition, it will be regulated as a drug under the FDCA of 1938. Likewise, if the same herb is used as a cosmetic, food, food additive, or a dietary supplement, it will be regulated accordingly. It was also shown that for any product used in these therapies meeting the minimum legislative definition of drug, food, or medical device, the biological substance will be mandatorily regulated under any of these acts and is not allowed to be manufactured, distributed, or marketed without the following the stated laws [25]. If a new botanical preparation is anticipated to treat, alleviate, or diagnose any disease or its symptoms, then it will also be regulated as a new drug and the investigational new drug (IND) application has to be filed [26]. Even the medical devices used for these therapies, like acupuncture needles, are also regulated as Class II medical devices under "the Act" and all specifications regarding the labeling (i.e., for single use only), biocompatibility, and sterility, were added in the Federal Register in December 1996 [25].

In contrast to conventional medicinal products, US-FDA is not legally involved in the premarketing of dietary supplements and it is the sole duty of the manufacturer to ensure their safety and efficacy prior to the sale of the product. Under the DHSEA of 1994, the legitimate role of FDA comes into picture only if any concern over the safety of these products is observed after the marketing [17]. In addition, there are provisions for the submission of the safety evidence by the manufacturer for the marketing of any new dietary ingredient not recognized as an article used for food without chemical alteration or not marketed in the United States before 1994 [27]. Moreover, the manufacturers are allowed to make certain claims on their merchandise, though these claims are likewise governed by law, and any fictitious or misleading information related to the same is not countenanced. It may include health, nutritional content, and structural/functional claims, but any claims related to any disease are not legitimate [28,29]. These formulations can be claimed as health benefits for certain deficiency diseases if the prevalence of the disease in the US is also informed in the label. Moreover, the nutritional claims may include both descriptive (such as cholesterol free or high fiber content, low level of fat, etc.) and the percentage claims (such as 30% of "X," 100 mg of "substance Y" per unit, etc.) [30,31]. The DHSEA regulations for structural and functional claims are associated with the effect of a particular nutrient or the dietary supplement on the normal structure or function of the human body (such as the drug "purifies blood" or "improves cognition"). Petitions notifying these claims are submitted to FDA before 30 days of product marketing. Such claims, being informative and consumer-friendly, are also advantageous to consumers suffering from deficiency diseases. Moreover, their label specifications should be in accordance with the FDA guidelines directly stating that the claim has not been evaluated by FDA [32].

In 2004, the "ephedra controversy" [33,34] profoundly highlighted the serious safety issues associated with the use of certain herbal products and led to the amendment of "the Act" in 109th U.S. congress. Also, the "Dietary Supplement and Nonprescription Drug Consumer Protection Act" or "DSNDCPA (Public Law 109–462)" was signed [35], which mandated the reporting of all adverse events within 15 business days of their receipt by the responsible person (manufacturer, packer, or retailer/distributor). This act also commands the responsible person to surrender any new medical information received within 1 year of the preliminary reporting. Such information is submitted to the FDA within 15 business days of its receipt in a med-watch form 3500A and the related records should be maintained for a period of 6 years. In this form, the information regarding the identifiable injured person, the initial reporter, contact information of the responsible person, name of the suspected product, and the adverse event/fatal outcomes, etc., have to be reported [36].

The FDA has also released a final rule in the federal register for the establishment of the 21 CFR part 111 regulations for the current good manufacturing practices (cGMP) of the dietary supplements (72 FR 34752) [37]. It consists of the introduction, the discussion (specifying the compliance dates and organization of rules), written procedures and records required by the rule followed by various subparts (subpart A to subpart P) specifying the cGMP guidelines. The subparts enumerate the rules for the personnel (particularly for their qualifications and hygienic practices), physical plant and grounds (specifying the sanitary requirements for the grounds as well as the whole the physical plant to protect against the microbial contamination of the manufactured products), equipment and utensils (used in the manufacturing, packaging or labeling of such products), production and process control systems (requirements for the batch production, laboratory operations, manufacturing, labeling, distributing, etc.), Product complaints, record-keeping, etc. Notably, trained practitioners and healers like naturopaths, herbalists, and acupuncturists formulating the traditional preparations in limited quantities (small scale) for their individual patients are exempted from these rules if they are not preparing their formulations in batches and selling them to individual patients and these batches are free from any ingredient with reported safety concern. Moreover, many Asian practitioners using traditional Asian medicines (not considered as dietary supplements) are not essentially subjected to these rules.

Subsequent to the release of cGMP guidelines, the USP Dietary Supplements Compendium (DSC) was also launched by the United States Pharmacopoeial Convention in the year 2009. It helps in ensuring the safety and efficacy of new dietary products by providing the quality standards for their development, manufacturing, quality control, packaging, labeling, and storing, etc.

Table 13.2 Acceptable Limits for Heavy Metal and Aflatoxins Contamination in Traditional Products

	APHA	USP [1]		BP	Ph. Eur.	China	HSA [2]
	PDE	PDE	Individual limits				
Lead	6 mcg day^{-1}	10 mcg day^{-1}	1.0 ppm	5 ppm	5 ppm	10 ppm	20 ppm
Cadmium	4.1 mcg day^{-1}	5 mcg day^{-1}	0.5 ppm	1 ppm	1 ppm	1 ppm	–
Mercury	2.0 mcg day^{-1} (as Methyl mercury)	15 mcg day^{-1} (total), 2 mcg day^{-1} (as methyl mercury)	1.5 ppm (total), 0.2 ppm (as methyl mercury)	0.1 ppm	0.1 ppm	0.5 ppm	0.5 ppm
Arsenic	10 mcg day^{-1}	15 mcg day^{-1}	1.5 ppm	–	–	2 ppm	5 ppm
Copper							150 ppm
Total Heavy metal content	–	–	–	–	–	20 ppm	–
Aflatoxin B1	5 ppb	5 ppb		2 ppb	2 ppb	–	–
Aflatoxin B1 + G1 + B2 + G2 (sum)	20 ppb	20 ppb		4 ppb	4 ppb	–	–

APHA, *American Herbal Products association;* BP, *British Pharmacopoeia;* HAS, *Health Sciences authority (Singapore);* Ph Eur., *European Pharmacopoeia;* PDE, *Permitted daily exposure;* ppm, *parts-per-million,* 10^{-6}; ppb, *parts-per-billion,* 10^{-9}; USP, *United States Pharmacopoeia.*
Ref: (1, 561) Articles Of Botanical Origin. The United States Pharmacopeial Convention; 2014. Available on: http://www.usp.org/sites/default/files/usp_pdf/EN/USPNF/gc_561.pdf [accessed 5.11.15].
(2) Requirements of Test Reports. Chinese Proprietary Medicines. Complementary health products. Health Products Regulations. Health Sciences Authority. Singapore Government. Available on: http://www.hsa.gov.sg/content/hsa/en/Health_Products_Regulation/Complementary_Health_Products/Chinese_Proprietary_Medicines/Application_and_Registration/requirements-of-test-reports-.html [accessed 2.11.15].

These quality standards were formerly available only by the "United States Pharmacopeia and the National Formulary" (USP–NF) or the Food Chemicals Codex (FCC) compendia. Alongside, USP convention also circulates the referral documents/industry guidelines [38] and grants the product certification to the manufacturers under its Dietary Supplement Verification Program (USP-DSVP), covering all the dietary supplements governed by the DHSEA and 21 CFR Part 111 [39]. This mark guarantees the product quality authenticating the manufacturer's declaration on the label [40]. It includes the validation of the specified ingredients (their concentrations and label claims) and the quality standards of the product, like acceptable limits of the contaminants, dissolution, or the disintegration standards. The official acceptable limits are specified in Tables 13.2 and 13.3.

Table 13.3 Permissible Limits of Microbial Contamination in USA

	United States Pharmacopoeia (USP)			American Herbal Products Association (AHPA) [1]	
	Raw material (botanical)	Powdered botanical extracts to be used as raw materials	Finished products	Dried and unprocessed herbs to be used as ingredients (and containing finished products)	Powdered and soft extracts to be used as ingredients (and containing finished products)
Staphylococcus aureus	NA	NA	NA	Absent in 25 g	Absent in 25 g
Escherichia coli	Absent in 10 g	Absent in 10 g	Absent in 10 g	Absent in 10 g	Absent in 10 g
Salmonella spcs.	Absent in 10 g	Absent in 10 g	Absent in 10 g	Absent in 25 g	Absent in 25 g
Total aerobic plate count (TPC)	10^5 CFU g^{-1}	10^4 CFU g^{-1}	10^4 CFU g^{-1}	10^7 CFU g^{-1}	10^4 CFU g^{-1}
Total Yeast & Mold	10^3 CFU g^{-1}	10^3 CFU g^{-1}	10^3 CFU g^{-1}	10^5 CFU g^{-1}	10^3 CFU g^{-1}
Enterobacterial count (bile tolerant gram negative)	10^4 CFU g^{-1}	–	–	10^4 CFU g^{-1} (coloniforms)	10^2 CFU g^{-1} (coloniforms)

(1) AHPA Guidance Policies. American Herbal Products Association; 2015. Available on: http://www.ahpa.org/default.aspx?tabid = 223#section_heavy_metals [accessed 5.11.15].

13.3 REGULATORY STATUS OF COMPLEMENTARY MEDICINE IN THE EUROPEAN UNION (EU)

Traditional medicines are regulated in the European Union by the well-known European Medicines Agency (EMA) as per the general European legislations assembled in various volumes of "The rules governing medicinal products in the European Union" [41]. The basic rules are established in Volume 1 (EU pharmaceutical legislation for medicinal products for human use) [42] supported by the sequential guidelines specified in volume 2 (Notice to applicants and regulatory guidelines for medicinal products for human use) [43], volume 3 (Scientific guidelines for medicinal products for human use) [44], volume 4 (Guidelines for good manufacturing practices for medicinal products for human and veterinary use) [45], volume 8 (Maximum residue limits) [46], volume 9 (Guidelines for pharmacovigilance for medicinal products for human and veterinary use) [47] and volume 10 (Guidelines for clinical trial) [48]. Moreover, in accordance with the EU treaties signed for their functioning [49], many member states are also equipped with their independent laws and policies (potentially influenced by the common EU directives) developed by their respective national regulatory bodies [50].

Considering the complicacy of this legal framework, EMA simplified these common EU legislations by launching a perspicuous procedure for the "traditional-use registration" and marketing authorization of Traditional Herbal Medicinal Products (THMP) [51] with recognized safety and efficacy under the directives 2004/24/EC [52] and 2001/83/EC [53]. The former came into full effect after a transitional period of 7 years (on 30 April 2011) restricting the sale and manufacture of unlicensed and noncertified herbal products in the European Union. It also led to the establishment of the Committee on Herbal Medicinal Products (HMPC) under EMA [52,54,55], which introduced a simplified registration procedure for herbal products in order to implement synchronized legislations in the EU member states. Prior to this aforementioned amendment, the conduct of pharmacological and clinical studies was obligatory for the applying the marketing authorization of any such product. In this scenario, this procedure came out as a benchmark allowing the applicants to apply for the same in a simplified manner (before the end of the transitional period) by merely detailing the published scientific literature corroborating the established medicinal use of the applied or the equivalent product for the last 30 years with no less than 15 years within the EU. These applications, marking the product safety within the acceptable range as per Directive 2001/83/EC, are not required to be accompanied by the preclinical and clinical data [52], however, the submission of quality reports, including the physicochemical, biological, and microbiological testing, are mandatory. Such products should meet the quality standards mentioned either in European pharmacopoeia or in pharmacopoeial standards of the member state (Section 4, Part III of Annex I to Directive 2001/83/EC). Moreover, the HMPC, in association with the "Committee For Medicinal Products For Human Use" (CHMP) and the "Committee For Medicinal Products For Veterinary Use" (CVMP), has also provided separate guidelines for the herbal products and preparations including the guidelines for quality assessment, labeling requirements, etc. [56–58]. These specifications include the limits or the acceptable criteria for the analytical methods used to substantiate the product quality. Furthermore, if any such application not accompanied by the bibliographical evidence of product safety marking its public use for the last 15 years is received, the member state can refer the case to HMPC (as per Article 16c [4] of Directive 2001/83/EC), however, such product should qualify other parameters of being eligible for the simplified registration [54,59]. In addition to this, HMPC is also associated with the organization of the herbal monographs and the "European Union list of herbal substances, preparations, and combinations thereof for use in traditional herbal medicinal products." It is first reviewed and concluded by HMPC before final submission to the European commission [60], after which the applicants can refer the same before submitting their applications [54]. This entire network has been designed to expedite the registration procedure for herbal preparations.

Apart from this simplified and registration, herbal products can also be registered as products with "well-established use," for which the applicant is required to submit additional bibliographic evidence substantiating the acceptable safety and efficacy of the "active ingredients" (Article 10a of directive 2001/83/EC). The applicant is also required to show that these active ingredients are in medicinal use for no less than ten years preceding the submission of application. Lastly, a "stand-alone" application accompanied by adequate clinical and/or clinical and safety studies and data by the manufacturing company can also be submitted to register the product (as per Article 8 [3] of directive 2001/83/EC). Additionally, supporting bibliographical evidence consolidated in the form of a mixed application can also be submitted for product registration. Unlike the **simplified** procedure, these categories of products can be registered by either the member state (via a decentralized procedure) or the EMA (a centralized procedure) [61]. Moreover, certain herbal products, which are exempted from the basic definition of THMPs, like herbal products containing new active ingredients intended to be used for specific disease conditions or the herbal drugs used for the orphan diseases, are required to be registered via a centralized procedure in the EU [55,59]. It includes herbal drugs for AIDS, HIV, and other viral infections, malignancies (as classified under the current International Classification of Diseases for Oncology), neurological disorders like Alzheimer's and Parkinson's disease (medicines for symptomatic; curative or neuroprotective treatments), diabetes, and autoimmune diseases (like Systemic lupus erythematosus, Autoimmune thrombocytopenia, Myasthenia gravis, etc.). The detailed lists of diseases coming under these categories are available in annexure I–IV [62]. Furthermore, all types of applications for the registration of these products should mandatorily follow the official format of the Common Technical Document (CTD) for the medicinal products for human use (given in Volume 2B in the Note to applicants) [63,64]. Akin to HPMC, the Homeopathic Medicinal Products Working Group (HMPWG) working under the Heads of Medicines agency (HMA) functions for the regulatory control over the quality, safety, and use of homeopathic medicinal products for human and veterinary use. These medicines are also registered via the similar procedure under Directive 2001/83/EC [53,65].

The foundation of the good manufacturing practices for the products of medicinal human use in the European Union was decreed in the Commission Directive 91/356/EEC of 1991. However, the Directive 2003/94/EC came out as a benchmark, subsequent to which, the GMP guidelines were published in Volume 4 of the rules governing medicinal products in the European Union. According to the latter, "All medicinal products for human use manufactured or imported into the community, including medicinal products intended for export, are to be manufactured in accordance with the principles and guidelines of good manufacturing practice." It particularly affected the manufacturers requiring marketing authorization for the medicinal and investigational products for human use under the directive 2001/83/EC [66]. The general GMP guidelines are contained in three parts followed by 19 annexures and other documents [45]. The GMP part I contains nine chapters including the regulations for the basic requirements and regulations for the medicinal products [67–75]. The "basic requirements for the substances as starting materials" are listed in part II [76] and the GMP part III contains information regarding the GMP-related documents [45]. Moreover, the specific GMP guidelines for herbal products, contained in Annex 7 of Volume 4 of the EU rules governing medicinal products, are centralized around the dogma of ensuring the quality and safety standards for these products (Table 13.4). The implementation of these guidelines is based on the purposive use of the herbal material by the manufacturer holding the authorization. For the cultivation, collection, and harvesting of these herbs, only the guidelines for "Good Agricultural and Collection Practice for starting materials of herbal origin" (GACP) are pertinent [77], however, parts I and II of the GMP guide also come into picture from the subsequent steps viz cutting and drying (GCAP + Part I + Part II GMP), distillation (Part I + Part II GMP), comminution, processing of exudates, extraction, fractionation, purification, concentration, or fermentation of herbal substances (Part I + Part II GMP), and additional processing steps like packaging (part I GMP) [78]. The required good manufacturing steps, ranging from the selection of the seeds to the packaging of the finished product, are covered under the above-mentioned guidelines.

Despite all this, there is still a need for better harmonization and recognition of CAM policies, practices, research, and patient care between the borders of all the EU member states, considering

Table 13.4 Acceptable Limits for the Microbiological Quality of the Finished Herbal Medicinal Products (Oral Use) in EU[a]

	Category A		Category B		Category C	
	Acceptance criterion.	Maximum acceptable count.	Acceptance criterion.	Maximum acceptable count.	Acceptance criterion.	Maximum acceptable count.
TAMC	10^7 CFU g^{-1}	50,00,0000 CFU g^{-1}	10^4 CFU g^{-1} or CFU mL^{-1}	50,000 CFU g^{-1} or CFU mL^{-1}	10^5 CFU g^{-1} or CFU mL^{-1}	500000 CFU g^{-1} or CFU mL^{-1}
TYMC	10^5 CFU g^{-1}	500,000 CFU g^{-1}	10^2 CFU g^{-1} or CFU mL^{-1}	500 CFU g^{-1} or CFU mL^{-1}	10^4 CFU g^{-1} or CFU/mL	50 000 CFU g^{-1} or CFU mL^{-1}
E. coli	10^3 CFU g^{-1}		Absent in 1 g or 1 mL		Absent in 1 g or 1 mL	
Salmonella	Absent in 25 g		Absent in 25 g or 25 mL		Absent in 25 g or 25 mL	
Bile-tolerant gram-negative bacteria.	–		–		10^4 CFU g^{-1} or CFU mL^{-1}	

A: Herbal medicinal products containing herbal drugs, with or without excipients, intended for the preparation of infusions and decoctions using boiling water.
B: Herbal medicinal products containing, for example, extracts and/or herbal drugs, with or without excipients, where the method of processing (for example, extraction) or, where appropriate, in the case of herbal drugs, of pretreatment reduces the levels of organisms to below those stated for this category.
C: Herbal medicinal products containing, for example, extracts and/or herbal drugs, with or without excipients, where it can be demonstrated that the method of processing (for example, extraction with low-strength ethanol or water that is not boiling, or low-temperature concentration) or, in the case of herbal drugs, of pretreatment, would not reduce the level of organisms sufficiently to reach the criteria required under B.
TAMC: Total aerobic microbial count, Total combined yeasts/molds count, CFU: Colony forming Units.
Ref: Microbiological quality of herbal medicinal products for the oral use and extracts used in their preparation. (01/2014:50108) European Pharmacopoeia. 2014; 8: 5.1.8.
[a]*Other microorganisms can be tested based on the general nature of the herbal preparation, manufacturing process, and the anticipated use of the product. The testing methods for these limits have also been specified in the general chapters 2.6.12, 2.6.13, and 2.6.31 of the Eur. Ph 8.0. In addition, separate limits have been provided in monographs of different drugs.*

which, the Directive 2011/24/EU on the application of patients' rights in cross-border healthcare was passed. It focused upon the rights of the patients and allowed the conditional reimbursement of the patient's healthcare costs permitted by law of the allied state [79]. More recently, a report on the operation of the latter illustrating its contribution in addition to certain voids in the functioning has also been published [80]. Along with the patient's rights, the European directive 2005/36/EC was published to accomplish some degree of synchronized professional conduct of healthcare therapies (including CAM). This directive, along with the subsequent amendments, aimed to provide the mutual recognition of professional qualifications in the EU member states [81].

Furthermore, CAMbrella, a pan-European research network for complementary and alternative medicine (CAM), was also launched in 2010 under the Seventh Framework Programme (FP7). The main purpose of this project was reviewing various aspects (for example, the regulatory control, government supervision, reimbursement status, citizen's attitude, patient's perspective, etc.) of common CAM therapies in the 27 EU member and 12 associated states. The main objective of this project was to establish harmonized regulatory control in all the EU member states via development of a common EU network (linking various research centers) and terminologies describing CAM. It also aimed to scrutinize the patient's perspective as well as the need for these therapies in the EU boosting present comprehension [82]. Under the CAMbrella project, the information was collected from the health ministries, CAM associations, Education and Law ministries of these 39 countries, and searches were performed on their national databases and web portals. Additionally, the data were also collected via emails, telephone conversations, conferences and personal visits, and published in different work packages [50,83–90]. According to these reports, 17 EU member states were reported to have general CAM legislation, out of which 11 states follow the specific laws for these medicines with other 6 countries abiding the CAM legislations specified in their general law. In addition to that, several EU member states have specific regulations for different CAM therapies (like discrete laws for naturopathy, Osteopractic physical therapy, Ayurveda, etc.), which are very different from each other [50], which further reflects the need for better harmonization. Furthermore, the seventh framework program of the European Commission also led to the launch of the "Good Practice in Traditional Chinese Medicine (GP-TCM) Research Association" in 2012 to promote good practices in the field of TCM research, development, and practice [91]. Apart from that, the European Federation of Complementary and Alternative Medicine (EFCAM), an alliance of all the European federations of specific CAM modalities and national CAM organizations, is also working in the same direction by promoting the use of holistic medicinal approaches. The prime objective of this organization is to safeguard the rights of all the lawfully trained and qualified practitioners [92].

Furthermore, the provisions of Article 26 of the Regulation (EC) No 726/2004 [55] and article 6 of directive 2001/83/EC, revised by the directives 2004/24/EC [52] and 2004/27/EC [93], highlighted a need for regulatory guidelines for the collection, verification, and assessment of all the adverse events reported in the European union. As a result, general pharmacovigilance guidelines in volume 9A of Eudralex came into existence as a collaborative effort of the European commission, EMEA, and the member states [47,94]. It includes the guidelines for marketing authorization holders, the competent authorities, and the agency—guidelines for the last three with respect to the electronic exchange of the information in the European Union and the pharmacovigilance communication. There is no separate system for the pharmacovigilance of the traditional drugs and the common procedures of ADR reporting are followed in their case. The details of the risk management plan (EU-RMP), including the safety specifications, the details of pharmacovigilance system, evidences for the services of the "Qualified Person Responsible for Pharmacovigilance" (QPPV), etc. are required to be submitted to the CTD for the marketing of a medicinal product (Article 8, Directive 2001/83/EC) [94,95]. All "the serious adverse reactions, including any suspected transmission by a medicinal product of an infectious agent" are mandatorily submitted to the member state within the 15 days of receipt by the marketing authorization holder registered via a centralized, decentralized, or referral procedure in the European Union [94]. The legislation also defines an electronic data processing and management network for the pharmacovigilance of the medicinal products, called EudraVigilance. This mandatory e-reporting system was first launched in 2001 under Article 24 of Regulation (EC) No 726/2004, and currently is one of the finest pharmacovigilance systems of the world [96].

In short, the European legislations allow the regulation of herbal medicines via three channels. As discussed in this section, these categories include the products for "traditional use," "well-established use," and "new products." However, many of the EU member states have made their individual efforts in the promotion of the CAM practices by launching the related legislations and policies, a void indicating that the need for improved harmonization still exists.

13.4 REGULATORY STATUS OF COMPLEMENTARY MEDICINES IN THE UNITED KINGDOM

The traditional medicines in the United Kingdom are subjected to both the EU directives regulated by the EMA and the exclusive UK legislations regulated by the member state authority titled "Medicines and Healthcare Products Regulatory Agency" (MHRA). Earlier, traditional medicines in the United Kingdom were regulated as per section 12 (S12(1) and S12(2) and section 56 of the "Medicines Act 1968," in relation to which the herbal drugs were exempted from the general rules enforced for the issue of certificates and license for the sale, supply, or manufacture and dealing with medicinal products. Moreover, these exemptions were subjected to certain conditions specified within the same. As per section 12(1), the herbal products manufactured in the personal business area that are sold only upon personal request are excused from the general legislations [97]. Presently, this section has been replaced by Regulation 3 of the Human Medicines Regulations 2012 [98]. On the other hand, as indicated by section 12(2) of the Medicines Act 1968, only the herbal products containing plants or parts of plants (not prohibited by MHRA [99]) with the processing method confined to drying, crushing, or comminuting, are allowed to be sold with the name of the constituting herbs or process, however no brand name or the name of the associated ailment is allowed to be specified on such products [97].

Furthermore, the enforcement of the Directive 2004/24/EC led to the mandatory registration of such herbal products either by the product licensing via the stand alone/mixed application or by the simplified registration procedure for Traditional Herbal Registration (THR) [100]. It is generally granted to the well-established applicant upon the submission of the application in the official format of electronic CTD (specified in Schedule 12 of Human Medicines Regulations 2012) accompanied by the proof of established safety profile consistent with the EU legislations [98]. These evidence of product safety can be submitted in the form of a review report prepared by the registered doctor, registered pharmacist, herbal practitioner (associated with the regulatory control of herbal medicines), or other competent professional, such as a toxicologist. An outline of the product characteristics (as per the EU guidelines), label design, and payment proof are required [100]. A herbal preparation containing additional mineral or a vitamin with secondary action and well-documented safety profile is also eligible for the THR (regulation 126 of Human Medicines Regulations 2012). The THR license is valid for a period of 5 years from the date of the grant, subsequent to which the marketing authorization holder is required to submit an application for renewal prior to the last nine months before expiry. Even after the grant of the THR license based on the quality and safety assessment of the product, the MHRA holds the right to cancel, postpone, or alter the same if any safety issue is observed. It can also be done if the product does not meet the qualitative or quantitative standards mentioned in the application. In addition, the marketing authorization holder should essentially update the product specification as per the revised official monographs, and notify the same to the authority [98]. Notably, the registration fee for the THR is subject to variation depending upon the number of active ingredients in the herbal product [100].

In contrast to the common EU legislations where the translational period of seven years was allotted for "traditional-use" registration, the government of the United Kingdom decided to prolong the period up to April 30, 2014, allowing retailers to "sell-through" their stocks. Alongside, it was also suggested that the extension of this period would allow the firms to upgrade their quality standards to meet the obligations of the directive 2004/24/EC [101]. Moreover, the traditional practitioners selling the self-prepared medications in their own premises subsequent to the personal patient consultation are not required to obtain a license or apply the THR before the sale (S12(2) of Medicines Act 1968) [100]. For these traditional herbal registrations, 460 applications were received by MHRA from 2006 to 2014, out of which 326 applications have been approved (30 applications were referred to HMPC) [102]. Furthermore, the section 4 of the Medicines Act 1968 (now regulated by Part 1, schedule 11 of the Human Medicines Regulations 2012) led to the organization of an "Advisory Non-Departmental Public Body" (ANDPB) named as the "Herbal Medicines Advisory Committee" or HMAC ("the Committee"). The formal establishment of this committee took place in 2005 when "The Herbal Medicines Advisory Committee Order 2005" was passed [103,104]. It counsels on the issues related to safety, quality, and efficacy of herbal drugs for human use including the herbal medicinal products registered under "traditional-use" category (under Directive 2004/24/EC) in addition to the unauthorized ones. Technically, for the grant of THR, the work of HMAC is only restricted to the quality assessment of the product [104]. More importantly, the conclusion of a cross-government review (which was announced in 2010) changed the status of HMAC [105] from ANDPB to a committee of experts under MHRA [102,106].

Alike HMAC, the "Advisory Board on the Registration of Homeopathic Products" (ABRHP), also referred as "the Board," also gained the status of an expert committee under MHRA on Nov 1, 2012 [106]. It works analogously to HMAC in the arena of homeopathic medicines [107]. Alongside, the British Herbal Medicine Association (BHMA), established even before the enactment of the Medicines Act, is supporting the use of herbal medicines, for 51 years fostering research and development in this field. For the same purpose, it has produced a wide variety of scientific literature in the form of classic herbal pharmacopoeias and compendia [108]. In addition, it also represents the United Kingdom in the European Scientific Cooperative on Phytotherapy (ESCOP), which is a federation of the organizations representing national herbal medicine or phytotherapy in EU [109].

Except for the common EU legislations [69] for GMP and GDP, which are still followed in the United Kingdom, MHRA has separately published general rules and GMP guidelines in the form of eBooks, commonly known as the Orange [110] and the Green [111] guides, which are specially designed for the pharmaceutical manufacturers and distributors. These eBooks contain updated versions of all the legislations, directives, and related guidelines that can be followed for the manufacture and distribution of medicines in the United Kingdom. The private laboratories conducting the contract quality control testing of authorized or investigational products on behalf of the manufacturers holding the license or the authorization (for new products) are also required to follow these guidelines. The MHRA perform the routine inspections of these "standalone" laboratories and provide separate guidelines for their usage [112], prepared in reference to EU legislations. In a similar manner, influenced by EU legislations [95], the MHRA has also proposed general guidelines for good pharmacovigilance practices (GPvP) which also applies to traditional medicines. The pharmacovigilance systems of the manufacturing firms are also inspected on a priority basis [113]. The suspected adverse drug reactions caused by all drugs available in the United Kingdom, including the medical devices used in traditional Chinese therapies, herbal, and homeopathic medicines, are reported by the yellow card system to the MHRA [114]. The collected information is accessed and evaluated by the respective committees like the HMAC and the Homeopathic and the Advisory Board of the Registration of herbal and homeopathic products, respectively. According to a recent committee report, 529 and 10 suspected adverse reactions, with respect to herbal (126 ADRs) and homeopathic (131 ADRs) medicines respectively, have been reported in the United Kingdom in the past 10 years. Out of these 529 reports received for herbal medicines, 400 represented serious suspected adverse reactions that further highlighted the significance of regulatory control over traditional products. The data stated for the herbal products do not include the adverse reactions caused by food products, homeopathic medicines, unlicensed medicines containing nonherbal components, medical devices, and other authorized medicines not marketed as traditional products [102]. The detailed listings of all such reports received from healthcare professionals, patients, manufacturing firms, etc. via a yellow card system are contained in Drug Analysis Prints (DAPs), by the name of the active constituent, in alphabetical order. Trade names of the medicines are not used in DAPs. Notably, the massive scientific evaluation is carried out by MHRA upon the receipt of these reports, following which they are uploaded on the web portal [115].

To encapsulate, the regulatory scenario of traditional medicinal products in the United Kingdom is very similar to the European Union, however, MHRA is also working exclusively to implement new changes to improve the present picture.

13.5 REGULATORY STATUS OF COMPLEMENTARY MEDICINE IN INDIA

It has been estimated that around 70% of the total Indian population is dependent upon the traditional system of medicine for their basic healthcare needs [11] and therefore, their proper regulatory control in the Indian subcontinent is indispensable. The utilization of these medicines, including ayurvedic, siddha, unani, and homeopathy (ASU&H drugs) is regulated in India as per chapter IVA (inserted by act 13 of 1964) of the "Drugs and Cosmetics Act, 1940 and rules, 1945." It includes regulations related to the manufacturing of these drugs for sale and the distribution (part XVI—Rule 151−160), institutional approval meant for testing the purity and quality of raw material and the finished products (part XVIA—Rule 160A-J), labeling, packing, and permissible alcohol limits (part XVII—Rule 161-161A), regulations for the government analysts and inspectors for this category of drugs (part XVIII—Rule 162−167), quality standards (part XIX—Rule 162−163F), etc. [116]. The Ayurvedic, Siddha and Unani Drugs Consultative Committee (ASUDCC) constituted under this act also plays a key role by counseling the state government, the central government, and the Ayurvedic, Siddha and Unani Drugs Technical Advisory Board (ASUDTAB) on related matters, and thereby indirectly substantiates the homogeneous execution of this act all over the country. This act is amended from time-to-time and new rules are inserted, substituted, or omitted as per requirement, and these amendments are notified as the general statutory rules (G.S.R.) in the official gazette of India published by the Department of Publication and printed from the Government of India Printing Presses recurrently [117]. This act has undergone a number of amendments and many revisions have been made to improve drug standards. To ensure the safety of these medicines, poisonous ASU substances are listed in Schedule E [1] of this act (added by notification number 1-23/67-D, published on 2-2-1970) [116]. As per this act, the labels of medicines should clearly specify them as the ayurvedic, unani, or siddha medicines. Alongside this, all the ingredients—with the quantity used—should be listed on the label, in addition to other label requirements (including the batch number, manufacturing date, etc.). Importantly, the use of drugs covered under schedule E1 should be indicated separately. Moreover, the amendments regarding the inclusion of the provisions related to shelf-life, expiry [under rule 161(B)] [118], and the issuance of the certificate for Good Manufacturing Practices (GMP) of ASU drugs to the license holders were also made [119]. The grant of this license is subject to the fulfillment of the GMP guidelines for ASU drugs, as specified in schedule T. The minimum requirements for the internal quality control have also been specified, however the quality testing of such drugs can also be carried out in government-approved laboratories on the behalf of the licensee. Earlier, the validity of this GMP certificate was 3 years, which was later extended to 5 years in consideration of the manufacturer's demand [120].

The Indian Medicines Central Council Act of 1970 (IMCC Act, 1970) led to the establishment of the Central Council of Indian Medicine (CCIM) to regulate the education and practice of the ASU system of medicine in India [121]. In 2012, the "Sowa Rigpa System of Medicine," which is one of the oldest traditional systems of medicine practiced in the Himalayan regions and other neighboring countries like China, Nepal, and Bhutan, was also appended to this council via gazette notification [122]. The CCIM maintains and updates the registers of the approved medical practitioners of Indian systems of medicine (ISM) [48] registered under the IMCC Act, 1970 [116]. The CCIM also firms the minimum educational guidelines and standards for the colleges and institutes offering courses in these subjects [123]. Akin to CCIM, central council of homeopathy (CCH) also works as a legislative council for education and practice in homeopathy [124].

Later, the Pharmacopoeial Laboratory for Indian Medicine (PLIM) and the Homeopathic Pharmacopoeial Laboratory (HPL) were also established as central laboratories. They lay down the quality standards for the single ASU&H drugs as well as for their compound formulations and are involved in the drug testing routines of official samples provided by the drug control authorities [125−128]. The general protocol for testing specifications includes the pharmacognostic identification of the drug or powder characteristics (if in powder form), the loss on drying (at 105 degrees), calculation of total ash contents and the acid soluble ash, chromatographic profiling of the drug,

Table 13.5 Permissible Limits of Contamination (Heavy Metals, Microbial Contaminants, and Aflatoxins) for the Grants of AYUSH Standard and Premium Marks in India [126]

Parameter	AYUSH Standard Mark	AYUSH Premium Mark
Lead	10 ppm	10 ppm
Cadmium	0.3 ppm	0.3 ppm
Mercury	1 ppm	1 ppm
Arsenic	3 ppm	3 ppm
Staphylococcus aureus/g	Absent	As per the API/UPI/SP/HPI/ Importing country regulations
Salmonella sp./g	Absent	
Pseudomonas aeruginosa/g	Absent	
Escherichia coli	Absent	
Total microbial plate count (TPC)	10^5/g* (10^7 g^{-1} for topical use)	
Total Yeast & Mold	10^3 g^{-1}	
Aflatoxin B1	0.5 ppm	5.0 ppb
Aflatoxin G1	0.5 ppm	–
Aflatoxin B2	0.1 ppm	–
Aflatoxin G2	0.1 ppm	
Aflatoxin B1 + G1 + B2 + G2 (sum)	–	10 ppb

AYUSH, *Ayurveda, Yoga and Naturopathy, Unani, Siddha and Homoeopathy;* API, *Ayurvedic pharmacopoeia of India;* UPI, *Unani Pharmacopoeia of India;* SP, *Siddha Pharmacopoeia;* HPI, *Homeopathic Pharmacopoeia of India;* ppm, *parts-per-million,* 10^{-6}; ppb, *parts-per-billion,* 10^{-9}.

tests for heavy metals, microbial contamination, pathogens, pesticides, and aflatoxins [126] (Table 13.5). The standard data obtained by these laboratories are generally released in their respective pharmacopoeias are periodically updated. Both PLIM and HPL are approved under the D & C Act, 1940, and are situated in Ghaziabad, Uttar Pradesh (India) [125,127]. In 2003, the "Department of the Indian System of Medicine & Homeopathy (ISM & H)" was renamed as the "Department of AYUSH," which functions as the main regulatory authority for the ASU & H drug. In Nov 2014, a separate Ministry of AYUSH was also formed in India with an aim to strengthen the existing regulatory scenario of ISM & H in India [128]. The product certifications for these medicines are provided by this department as the "AYUSH standard mark" and the "premium mark" [Table 13.5]. The standard mark is awarded only if the drug product complies with the regulations prescribed under the Drugs and Cosmetics Act, 1940 for the AYUSH products. On the other hand, the award of the premium mark is subject to the fulfillment of (1) the WHO guidelines for good manufacturing practices (GMP) and permissible limits of heavy metals aflatoxins, pesticides, and microbial contaminants and/or (2) the regulations of the importing nation (only if they are more stringent). However, it should also be noted that the metallic raw materials anticipated for internal marketplaces are exempt from these limits for heavy metals. This certification procedure is carried out by autonomous certification bodies accredited in line with the ISO/IEC Guide 65, by National Accreditation Board for Certification Bodies (NABCB) and/or recommended by the Quality Council of India (QCI) [129,130].

The guidelines for Good Clinical Practices (GCP) for ASU drugs (GCP-ASU guidelines) have also been issued by the Department of AYUSH [131]. These guidelines were prepared in line with the "GCP Guidelines for Clinical Trials on Pharmaceutical Products" published by the Central Drugs Standard Control Organization (CDSCO) in 2001. The aims and objectives of the study, along with the details of both the investigational drug with proper justification of the drug product being used as control, the preclinical data favoring the scope of the study, information regarding the risks and benefits of drug, the whole study design, the inclusion and exclusion criteria, have to be provided separately in addition to the ethical considerations and the statement of the informed consent of the subjects [132]. These guidelines also necessitate the documentation of the physicochemical and the pharmaceutical properties of such drug formulations in order to acquire an enhanced knowledge to undertake the safety measures in the clinical study. Similar to GCP-ASU, it also introduced the guidelines for "Good Agricultural Practices Standard for Medicinal Plants—Requirements" [133] based on the WHO guidelines on "Good Agricultural and Collection Practices (GACP) for Medicinal Plants." In 2013, the Essential Drug Lists (EDLs) for ayurveda, siddha, and unani medicines were released by the Department of AYUSH to ease the manufacturing process [128]. More recently, the National AYUSH Mission was also launched with an aim to provide better AYUSH services (with improved access) and to upgrade the educational standards. Additionally, it supported the GAP guidelines to attain an unrestricted supply of genuine raw material for the manufacture of the certified finished products [134]. Also in 2013, New Delhi (India) witnessed the organization of an "International Conference on Traditional Medicine" by the Government of India in alliance with WHO-SEARO where the Delhi declaration was signed by the 11 member states [135]. These countries agreed to mutually collaborate and develop a synchronized approach to the education, practice, and regulation of traditional medicines. It also included the adaptation and the execution of the WHO Traditional Medicine Strategy 2014–2023 in view of in-house regulations, and emphasized the strengthening of the pharmacovigilance systems to ensure the safe use of such drugs in future.

To work in this direction, the "Protocol for National Pharmacovigilance Programme for Ayurveda, Siddha and Unani (ASU) Drugs" was released by the Department of AYUSH, Ministry of Health & Family Welfare, Government of India in association with the WHO Country Office for India, New Delhi in the year 2008, and the "Institute for P. G. Teaching & Research in Ayurveda" (Gujarat Ayurveda University, Jamnagar, Gujarat) was announced as the National Pharmacovigilance Resource Centre for the ASU Drugs (NPRC-ASU) [136]. This protocol also aimed to set 8 Regional Pharmacovigilance Centres (RPC), 30 Peripheral Pharmacovigilance Centres (PPC) for ASU drugs all over India [137]. A study conducted by the Pharmacovigilance Programme of India (PvPI) during July 2011 to December 2013 also highlighted the need for a robust pharmacovigilance system and concluded uninformed poly-pharmacy as the major reason for increasing adverse drug reactions (ADRs) from herbal products [138]. Another cross-sectional, observational, and questionnaire-based study conducted in a Tertiary Care Teaching Hospital in Rajasthan, India showed that only 24% of such patients informed their physicians about the use of traditional medicines. In addition, 31.5% of patients disclosed that the physician has not enquired regarding the same [139]. Overall, the regulatory scenario for complementary medicine in India is currently in the nascent stage; however, it is expected to escalate in the forthcoming years. The manufacturing criteria, as well as the quality control guidelines of such medicines, have been described, yet improved and stringent control along with better pharmacovigilance can improve the existing scenario.

13.6 REGULATORY STATUS OF COMPLEMENTARY MEDICINES IN CHINA

There is no denying the fact that Chinese herbal medicines, also known as traditional Chinese medicines (TCM), have gained worldwide attention in past few years. Beyond China, these medicines have also been adopted by many other countries, which also led to improvement in their regulatory statuses and expansion of their market all over the world [11]. In china, these medicines are currently regulated by the "China Food and Drug Administration" or CFDA [140] (sometimes also referred as "State Food and Drug Administration," or SFDA) as per section 102 of the "Drug Administration Law of the People's Republic of China," 2001 [8,141]. Subsequent to the enactment of this law, the SFDA also released separate regulations for its implementation, which came into full effect in 2002 [142]. These herbal drugs can be registered under the Department of Drug Registration (Dept. of TCMs & Ethno-Medicines Supervision), however, they can also be registered as functional foods (health foods) depending upon the intended usage. The regulatory approval for the manufacture of these food products is granted by the "Department of Food License" of the SFDA [140].

The requirements for the registration of these "drugs" (in accordance with the "Drug Administration Law") have been specified in the "Provisions for Drug Registration" (order no 28) of the SFDA [143]. These rules are pertinent to all applications for the manufacture, import, quality control (for drug registration), clinical trials, and other related dealings in mainland China (Article 2). Later on, SFDA also came out with "Supplementary Requirements on the Registration of TCM" and specifically highlighted the distinctive features of TCM registration to promote the registration as well as the development of the traditional and the minority medicines [144]. For the grant of this registration, a dossier, including the proof of safety, efficacy, and quality of the product, is required to be submitted (Article 13) in the format prescribed in Annex 1 and 2 for traditional Chinese and natural medicines respectively (Article 170). It can be applied for new Chinese and natural medicines (that have not been marketed in the state), generic medicines (for new indications, new routes or changes in the dose), and for the import or export of these drugs (also in case of Hong-Kong and Macau, which are not the part of the mainland). Furthermore, evidences for the preclinical efficacy of the unknown active ingredients extracted from the plants, new crude Chinese drugs, or their preparations, should be submitted with the study of the source of the drug as well as the pretreatment and processing details. (Article 21). Upon submission, these applications are reviewed and if approved, a unique drug approval number is allotted in a SFDA specified format: Name Z + Four-digit year number + four-digit sequence number, where, the alphabet following the name represents the category of the product. As examples, the traditional Chinese medicines are represented by "Z," whereas the other drugs, like pharmaceutical products, biological products, and the repacked imported drugs are represented by the alphabets H, S, and J respectively (Article 171) [143].

According to the Drug Administration Law, the use of any drug substance without the drug approval number or the pharmaceutical or import license is not permitted for the manufacture of drug products. However, no such approval number is granted for Chinese crude drugs and their prepared slices (Article 6), which are required to comply with the national drug standards. In the case that the respective crude drug or the slices are not covered by the latter, the manufacturer should essentially follow the processing procedures devised by the drug regulatory department of the province, autonomous regions, or municipality directly under the Central Government (Article 10).

Moreover, all the manufactured products should mandatorily comply with the quality standards specified in the national pharmacopoeia, failing which they cannot be released into the market (Article 12) [141]. To ensure the quality and safety of these drugs, the SFDA also issued separate guidelines (2002) for the agricultural practices for Chinese medicines that specify the species and nutritional and other requirements [145]. Alongside, the revised good manufacturing practices guidelines constituting 313 articles (listed in 14 chapters) have also been released, which also applies to the traditional Chinese medicines [146]. In addition, the distribution and labeling requirements for Chinese crude drugs and their slices have also been specified in drug administrative law. For the distribution of crude Chinese drugs, the source of the drug should be disclosed (Article 9). Moreover, the label should also indicate the name, name of the dispatcher, and the certification mark as proof of the quality (Article 53). These crude drugs can be sold in community fairs if their sale is restricted by SFDA. The sale of other drug products are not permitted in such country fairs, however they can be sold by possessing the "Drug Supply Certificate" in a limited business scope, which is again controlled by the state council (Article 21) [141].

Notably, the People's Republic of China has also adopted the "Regulations on Protection of Traditional Chinese Medicines" [147] in 1993, which became a highspot in the regulatory history of traditional Chinese medicines. These unique regulations were also designed to raise the quality standards of the TCMs, to safeguard the legal rights of the firms involved in their production, and to encourage research related activities (Article 1). Upon prior approval of health administrative departments of the State Council, the applications for the issue of the license can be submitted with respect to all the varieties of the standardized TCMs listed in the official standards (Article 5). Based on these regulations, two types of protections are obtainable viz. Class I and Class II. The Class I license can be applied for TCMs (1) having a special healing effect in a diseased condition, (2) formulated using a variety of the medicinal plants registered with Class I protection, or (3) effective in the prevention and restoration of a specific disease (Article 6). On the other hand, a Class II license can be applied if (1) the TCM complies with the grade 1 requirements but is now annulled from the category, (2) the TCM shows the conspicuous curative effect in the specified disease, or (3) it is extracted or purposely orchestrated using the active ingredients of the medicinal plants (Article 6). The Grade 1 protection license shields for a time period of 30, 20, or 10 years, whereas the Grade 2 is granted for 7 years, subsequent to which the renewal can also be applied in the last 6 months of the protection period approved by the health administrative department of the state council (Article 8) [147,148]. Furthermore, the dossier to obtain a "Certificate for TCM protected species" is required to be submitted in the layout approved by the Department of drug registration, SFDA. It should be accompanied by the supplementary application form for protected TCM, approval documents, Drug production license, GMP license, justified reason for protection, statement of patent ownership, medical information and allied data (pre- and postmarketing), related pharmacological and toxicological reports and the anticipated improvement planning and implementation programs. The application is reviewed by the working staff of the Administrative Service Centre of SFDA and can be immediately rejected if it lies beyond scope. In case of errors, deficiencies, or noncompliance with the legal requirements, the applicant can be allowed to correct the same on the spot or within the stipulated period of 5 days. In the case of acceptance, the application is forwarded to the Office of the National Committee on the Assessment of the Protected Traditional Chinese Medicinal Products (NCAPTCMP) for the specialized evaluation by the committee, which is usually completed within 120 days. Supplementary information can also be requested for re-evaluation to complete the whole process, which takes an extra 40 days. After the completion of this process, the SFDA takes its final decision in the context of the application, which takes 20 more days. This final decision is delivered to the applicant within 10 days of approval, after which, a certificate of the Protected TCM Species is granted [148]. Moreover, for the TCMs holding Class I protection, the constituents, the secret formulae, and the applied preparatory knowledge are disclosed or publicized by the concerned departments and the individuals involved in the registration process (Article 13), violation of which is a punishable offense (Article 22) [147].

The proper control over the cultivation, processing, and packaging (indicating the contraindications, label warnings, etc.), by employing standardized procedures in compliance with the SFDA rules and regulations, may aid in reduction of associated adverse effects. All these guidelines may promote the safe use of the TCM ensuring the quality and efficacy of such medicines. Moreover, the state has also developed an ADR surveillance system with the aim of detecting the adverse events at early stages, thereby promoting the safe use of such medicines. Since 1989, the ADR monitoring centers are working in this direction, but the associated regulations came into existence in 1999, only after it joined the WHO International Drug Monitoring Programme. Furthermore, the online ADR surveillance information system was also organized, which allowed the easy and early reporting of ADRs [149,150]. A major highlight in this regard came in 2011 when the SFDA revised the "Provisions for Adverse Drug Reaction Reporting and Monitoring" [151]. Despite of all these efforts, there is a major gap in the ADR reporting of complementary drugs in the practical scenario. As per the statistical evidences, these medicines were found responsible for 13.8% of the total ADRs reported in 2010. In addition, most of these ADRs were attributed to the drug preparations (99.7 %) and crude drugs only contributed to 0.4% of cases, moreover, they also accounted for 12.2% unexpected and life-threatening ADR reports [149]. In 2012, a minor increase in reporting was observed by the National ADR Monitoring Network (17.1%) [152], however, further improvement is still required.

To work in this direction, the nation also extends support to the international organizations for the promotion of the CAM services all over the world. In 2008, the WHO's inaugural congress on traditional medicine was also held in the People's Republic of China, where the "Beijing Declaration" was signed. The latter recommend the governments of all the member states to respect, preserve, promote, and exchange the knowledge of traditional medicines. Alongside, the creation of health policies, legislations, and standards, in addition to the integration of these medicines in the conventional systems, was suggested [153].

13.7 CONCLUSION

In the present era of drug discovery, when a number of synthetically-engineered drug molecules are incessantly flooding the market, the significance of ancient complementary medicinal approaches has not tapered. Conversely, these traditional approaches have become more popular and are used by a large number of people worldwide. For this reason, many countries have formulated their own legislations and regulatory guidelines for ensuring the safety and quality of these established medicines; however, many challenges are still hindering current progress, the overcoming of which may increase the statuses of these medicines in the future.

REFERENCES

[1] General Guidelines for Methodologies on Research and Evaluation of Traditional Medicine. World Health Organisation. Available at: <http://apps.who.int/iris/bitstream/10665/66783/1/WHO_EDM_TRM_2000.1.pdf>; 2000 [accessed 25.10.15].

[2] Complementary, Alternative, or Integrative Health: What's In a Name? National Center for Complementary and Integrative Health. National Institutes of Health. U.S. Department of Health and Human Services Food and Drug Administration. Available on: <https://nccih.nih.gov/sites/nccam.nih.gov/files/Whats_In_A_Name_08-11-2015.pdf> [accessed 16.10.15].

[3] WHO guidelines on safety monitoring of herbal medicines in pharmacovigilance systems. World Health Organization, Geneva. Available at: <www.who.int>; 2004 [accessed 20.08.15].

[4] WHO Monographs on Selected Medicinal Plants - Volume 1. World Health Organization, Geneva. Available on: <www.who.int>; 1999 [accessed 25.10.15].

[5] WHO Monographs on Selected Medicinal Plants - Volume 2. World Health Organization, Geneva. Available on: <www.who.int>; 2004 [accessed 25.10.15].

[6] WHO Monographs on Selected Medicinal Plants - Volume 3. World Health Organization, Geneva. Available on: <www.who.int>; 2007 [accessed 25.10.15].

[7] WHO Monographs on Selected Medicinal Plants - Volume 4. World Health Organization, Geneva. Available on: <www.who.int>; 2009 [accessed 25.10.15].

[8] National policy on traditional medicine and regulation of herbal medicines: report of a WHO global survey. World Health Organization, Geneva. Available at: <www.who.int>; 2005 [accessed 27.08.15].

[9] Regulatory Situation of Herbal Medicines - A Worldwide Review. World Health Organization, Geneva. Available at <www.who.int>; 1998 [accessed 5.10.15].

[10] Legal Status of Traditional Medicine and Complementary/Alternative Medicine: A Worldwide Review. World Health Organization, Geneva. Available at: <www.who.int>; 2001 [accessed 5.10.15].

[11] WHO traditional medicine strategy 2014–2023. World health organisation. Available on: <www.who.int>; 2013 [accessed 20.08.15].

[12] National Health Interview Survey (NHIS). Public use data release. National Center for Health Statistics. Centers for Disease Control and Prevention U.S. Department of Health and Human Services; 2003. Available at: <ftp://ftp.cdc.gov/pub/Health_Statistics/NCHS/Dataset_Documentation/NHIS/2002/srvydesc.pdf>; 2002 [accessed 5.10.15].

[13] National Health Interview Survey (NHIS). Public use data release. National Center for Health Statistics. Centers for Disease Control and Prevention U.S. Department of Health and Human Services; 2008. Available at: <ftp://ftp.cdc.gov/pub/Health_Statistics/NCHS/Dataset_Documentation/NHIS/2007/srvydesc.pdf>; 2007 [accessed 5.10.15].

[14] National Health Interview Survey (NHIS). Public use data release. National Center for Health Statistics. Centers for Disease Control and Prevention U.S. Department of Health and Human Services; 2013. Available at: <ftp://ftp.cdc.gov/pub/Health_Statistics/NCHS/Dataset_Documentation/NHIS/2012/srvydesc.pdf>; 2012 [accessed 5.10.15].

[15] Black LI, Clarke TC, Barnes PM, Stussman BJ, Nahin RL. Use of complementary health approaches among children aged 4–17 years in the United States: National Health Interview Survey, 2007–2012. National Health Statistics Report; no 78. Hyattsville, MD: National Center For Health Statistics; 2015.

[16] Clarke TC, Black LI, Stussman BJ, Barnes PM, Nahin RL. Trends In the use of complementary health approaches among adults: United States, 2002–2012. National Health Statistics Report; no 79. Hyattsville, MD: National Center For Health Statistics; 2015.

[17] Dietary Supplement Health and Education Act of 1994. Public Law 103-417. 103rd Congress. Available at: <https://ods.od.nih.gov/About/DSHEA_Wording.aspx> [accessed 14.10.15].

[18] U.S. Food and Drug Administration. Available at: <http://www.fda.gov/AboutFDA/WhatWeDo/History/> [accessed 5.10.15].

[19] Federal Food, Drug, and Cosmetic Act (FD&C Act). U.S. Food and Drug Administration.U.S. Department of Health and Human Services. Available at: <http://www.fda.gov/regulatoryinformation/legislation/federalfooddrugandcosmeticactfdcact/> [accessed 17.10.15].

[20] Expanding Horizons of Health Care. Five year Strategic Plan 2001–2005. National center for Complementary and Alternative Medicine. National institute of Health. U.S. Department of Health and Human Services. Available at: <https://nccih.nih.gov/sites/nccam.nih.gov/files/about/plans/fiveyear/fiveyear.pdf>; 2000 [accessed 17.10.15].

[21] Expanding Horizons of Health Care: Strategic Plan 2005–2009. National center for Complementary and Alternative Medicine. National institute of Health. U.S. Department of Health and Human Services. Available at : <https://nccih.nih.gov/sites/nccam.nih.gov/files/about/plans/2005/strategicplan.pdf>; 2005 [accessed 17.10.15].

[22] Exploring the Science of Complementary and Alternative Medicine. NCCAM Third Strategic Plan: 2011–2015. National center for Complementary and Alternative Medicine. National institute of Health. U.S. Department of Health and Human Services. Available at: <https://nccih.nih.gov/sites/nccam.nih.gov/files/about/plans/2011/NCCAM_SP_508.pdf>; 2011 [accessed 17.10.15].

[23] Guidance for Industry on Complementary and Alternative Medicine Products and their Regulation by the Food and Drug Administration. U.S. Department of Health and Human Services. Food and Drug Administration. Available on: <http://www.fda.gov/OHRMS/DOCKETS/98FR/06D-0480-GLD0001.PDF>; 2006 [accessed 10.10.15].

[24] Mission. National center for complementary and Integrative Health. National institute of Health. Available at: <http://www.nih.gov/about/almanac/organization/NCCIH.htm> [accessed 13.10.15].

[25] 21 CFR § 880.5580. Code of Federal Regulations.U.S. Government Publishing Office (GPO). Available on: <http://www.ecfr.gov> [accessed 16.10.15].

[26] Guidance for Industry Botanical Drug Products. Food and Drug Administration. U.S. Department of Health and Human Services. Available on: <http://www.fda.gov/downloads/drugs/guidancecompliancer-egulatoryinformation/guidances/ucm070491.pdf>; 2004 [accessed 14.10.15].

[27] Levitt J.A. FDA's Progress with Dietary Supplements., Center for Food Safety and Applied Nutrition. Food and Drug Administration. Presented before the House Committee on Government Reform. Available on: <http://www.fda.gov/NewsEvents/Testimony/ucm115229.htm>; 2001 [accessed 16.10.15].

[28] 21 eCFR §101.14(a)(1). Electronic Code of Federal Regulations. U.S. Government Publishing Office (GPO). Available on: <http://www.ecfr.gov> [accessed 16.10.15].

[29] 21 eCFR §101.70(f). Electronic Code of Federal Regulations. U.S. Government Publishing Office (GPO). Available on: <http://www.ecfr.gov> [accessed 16.10.15].

[30] 21 eCFR§101.13.Electronic Code of Federal Regulations. U.S. Government Publishing Office (GPO). Available on: <http://www.ecfr.gov> [accessed 16.10.15].

[31] 21 eCFR §101.69(o). Electronic Code of Federal Regulations. U.S. Government Publishing Office (GPO). Available on: <http://www.ecfr.gov> [accessed 16.10.15].

[32] Sharfstein J.M. Oversight of Dietary Supplements. Food and Drug Administration Department of Health and Human Services. Presented Before the Special Committee on Aging U.S. Senate. Available on: <http://www.fda.gov/newsevents/testimony/ucm213531.htm#_ftn2>; 2010 [accessed 16.10.15].

[33] Guidance for Industry: Final Rule Declaring Dietary Supplements Containing Ephedrine Alkaloids Adulterated Because They Present an Unreasonable Risk; Small Entity Compliance Guide. Food and Drug Administration. U.S. Department of Health and Human Services. Available on: <http://www.fda.gov/Food/GuidanceRegulation/GuidanceDocumentsRegulatoryInformation/DietarySupplements/ucm072997.htm>; 2008 [accessed 15.10.15].

[34] Soller RW, Bayne HJ, Shaheen C. The regulated dietary supplement industry: myths of an unregulated industry dispelled. HerbalGram: J Am. Bot. Council 2012;93:42–57. Available on: http://www.herbalgram.org [accessed 22.09.15].

[35] Dietary Supplement and Nonprescription Drug Consumer Protection Act. Public Law 109–462. In: 109th Congress. Office of Dietary supplements. National institute of Health. Available on: <https://www.congress.gov/109/plaws/publ462/PLAW-109publ462.pdf>; 2006 [accessed 4.10.15].

[36] Guidance for Industry: Questions and Answers Regarding Adverse Event Reporting and Recordkeeping for Dietary supplements as Required by the Dietary Supplement and Nonprescription Drug Consumer Protection Act. Draft Guidance. U.S. Food and Drug Administration. U.S. Department of Health and Human Services. Center for Food Safety and Applied Nutrition. Available on: <http://www.fda.gov/Food/GuidanceRegulation/GuidanceDocumentsRegulatoryInformation/DietarySupplements/ucm179018.htm>; 2009 [accessed 21.10.15].

[37] Final Rule: current Good Manufacturing Practice in manufacturing, packaging, labeling, or holding operations for dietary supplements. *FedRegist*; 2007; 72: 34752–34958.

[38] USP Dietary Supplements Compendium. Dietary Supplements. U.S. Pharmacopeial Convention. Available on: <http://www.usp.org/dietary-supplements/dietary-supplements-compendium> [accessed 31.10.15].

[39] USP Dietary Supplement Verification Program. Manual for Participants. U.S. Pharmacopeial Convention. Available on: <http://www.usp.org/sites/default/files/usp_pdf/EN/dsvp_manual_2013-10.pdf>; 2009 [accessed 31.10.15].

[40] USP Verified Mark. U.S. Pharmacopeial Convention. Available on: <http://www.usp.org/usp-verification-services/usp-verified-dietary-supplements/usp-verified-mark>; [accessed 31.10.15].

[41] EU Legislation – Eudralex. Directorate General for Health and Food Safety. European Commission. Available on: <http://ec.europa.eu/health/documents/eudralex/index_en.htm>; [accessed 29.10.15].

[42] EudraLex Volume 1 - Pharmaceutical Legislation Medicinal Products for Human Use. Eudralex. Directorate General for Health and Food Safety. European Commission. Available on: <http://ec.europa.eu/health/documents/eudralex/vol-1/index_en.htm#dir>; [accessed 30.10.15].

[43] EudraLex - Volume 2 - Pharmaceutical Legislation Notice to applicants and regulatory guidelines medicinal products for human use. Eudralex. Directorate General for Health and Food Safety. European Commission. Available on: <http://ec.europa.eu/health/documents/eudralex/vol-2/index_en.htm>; [accessed 30.10.15].

[44] EudraLex - Volume 3 Scientific guidelines for medicinal products for human use. Eudralex. Directorate General for Health and Food Safety. European Commission. Available on: <http://ec.europa.eu/health/documents/eudralex/vol-3/index_en.htm>; [accessed 30.10.15].

[45] EudraLex - Volume 4 Good manufacturing practice (GMP) Guidelines. Eudralex. Directorate General for Health and Food Safety. European Commission. Available on: <http://ec.europa.eu/health/documents/eudralex/vol-4/index_en.htm>; [accessed 30.10.15].

[46] EudraLex - Volume 8 Maximum residue limits guidelines (MRL). Eudralex. Directorate General for Health and Food Safety. European Commission. Available on: <http://ec.europa.eu/health/documents/eudralex/vol-8/index_en.htm>; [accessed 31.10.15].

[47] EudraLex - Volume 9 Pharmacovigilance guidelines. Eudralex. Directorate General for Health and Food Safety. European Commission. Available on: <http://ec.europa.eu/health/documents/eudralex/vol-9/index_en.htm>; [accessed 31.10.15].

[48] EudraLex-Volume 10 Clinical trials guidelines. Eudralex. Directorate General for Health and Food Safety. European Commission. Available on: <http://ec.europa.eu/health/documents/eudralex/vol-10/index_en.htm>; [accessed 31.10.15].

[49] The Lisbon Treaty. Treaty on the Functioning of the European Union & comments. Part 3 - Union policies and internal actions. Title XIV - Public health (Article 168). Available on: <http://www.lisbon-treaty.org/wcm/the-lisbon-treaty/treaty-on-the-functioning-of-the-european-union-and-comments/part-3-union-policies-and-internal-actions/title-xiv-public-health/456-article-168.html>; [accessed 3.10.15].

[50] Wiesener S, Falkenberg T, Hegyi G, Hok J, Roberti di Sarsina P, Fønnebø V. Legal status and regulation of CAM in Europe. Part I - CAM regulations in the European Countries. Final report of CAMbrella Work Package 2. Available on: <https://fedora.phaidra.univie.ac.at/fedora/get/o:291583/bdef:Content/get>; [accessed 5.10.15].

[51] Herbal Medicinal Products. Medicinal products for human use. Directorate General for Health and Food Safety. European Commission. Available on: <http://ec.europa.eu/health/human-use/herbal-medicines/index_en.htm>; [accessed 29.10.15].

[52] Directive 2004/24/EC of the european parliament and of the council. Official Journal of the European Union. L 136, 30.4.2004, pp. 34–56. Available on: <http://eur-lex.europa.eu/LexUriServ/LexUriServ.do?uri = OJ:L:2004:136:0085:0090:en:PDF>; [accessed 30.10.15].

[53] Directive 2001/83/EC of the european parliament and of the council. Official Journal of the European Union. L 311, 28/11/2004, pp. 67–128. Available on: <http://eur-lex.europa.eu/LexUriServ/LexUriServ.do?uri = OJ:L:2001:311:0067:0128:en:PDF>; [accessed 30.10.15].

[54] HMPC: Overview. European Medicines Agency. Available on: <http://www.ema.europa.eu/ema/index.jsp?curl = pages/about_us/general/general_content_000122.jsp&mid = WC0b01ac0580028e7d>; [accessed 16.10.15].

[55] Regulation (EC) No 726/2004 of the European Parliament and of the council. Official Journal of the European Union. L 136, 30.4.2004, pp. 1–33. Available on: <http://eur-lex.europa.eu/LexUriServ/LexUriServ.do?uri = OJ:L:2004:136:0001:0033:en:PDF>; [accessed 30.10.15].

[56] Guideline On Quality Of Combination Herbal Medicinal Products/ Traditional Herbal Medicinal Products. (EMEA/HMPC/CHMP/CVMP/214869/2006). European Medicines Agency. Available on: <http://www.ema.europa.eu/docs/en_GB/document_library/Scientific_guideline/2009/09/WC500003286.pdf>; [accessed 26.10.15].

[57] Guideline on specifications: test procedures and acceptance criteria for herbal substances, herbal Preparations and herbal medicinal products/traditional herbal medicinal products. (EMA/CPMP/QWP/2819/00 Rev. 2) European Medicines Agency. Available on: <http://www.ema.europa.eu/docs/en_GB/document_library/Scientific_guideline/2011/09/WC500113209.pdf>; [accessed 26.10.15].

[58] Guideline on declaration of herbal substances and herbal preparations in herbal medicinal products/traditional herbal medicinal products. (EMA/HMPC/CHMP/CVMP/287539/2005 Rev.1). Committee On Herbal Medicinal Products (HMPC), Committee For Medicinal Products For Human Use (CHMP), Committee For Medicinal Products For Veterinary Use (CVMP). European Medicines Agency. Available on: <http://www.ema.europa.eu/docs/en_GB/document_library/Scientific_guideline/2009/09/WC500003272.pdf>; [accessed 26.10.15].

[59] Questions & Answers on the EU framework for (traditional)herbal medicinal products, including those from a 'non-European' tradition. (EMA/HMPC/402684/2013). European Medicines Agency. Available on: <http://www.ema.europa.eu/docs/en_GB/document_library/Regulatory_and_procedural_guideline/2014/05/WC500166358.pdf>; 2014 [accessed 29.10.15].

[60] Procedure for the preparation of an entry to the 'Community list of herbal substances, preparations and combinations thereof for use in traditional herbal medicinal products. European Medicines Agency. Available on: <http://www.ema.europa.eu/docs/en_GB/document_library/Regulatory_and_procedural_guideline/2009/12/WC500017053.pdf>; 2007 [accessed 26.10.15].

[61] Marketing Authorisation. Procedures for marketing authorisation. Chapter 1, Volume 2A. Available on: <http://ec.europa.eu/health/files/eudralex/vol-2/a/vol2a_chap1_2013-06_en.pdf>; 2013 [accessed 16.10.15].

[62] Scientific Aspects and Working Definitions for the Mandatory Scope of the Centralised Procedure. [Regulation (EC) No 726/2004 of the European Parliament and of the council of 31 march 2004]. (EMEA/CHMP/121944/2007) Committee For Medicinal Products For Human Use (CHMP). European Medicines Agency. Available on: <http://www.ema.europa.eu/docs/en_GB/document_library/Regulatory_and_procedural_guideline/2009/10/WC500004085.pdf>; 2007 [accessed 31.10.15].

[63] Notice to Applicants. Volume 2B. Medicinal products for human use. Presentation and format of the dossier. Common Technical Document (CTD). Available on: <http://ec.europa.eu/health/files/eudralex/vol-2/b/update_200805/ctd_05-2008_en.pdf>; 2003 [accessed 23.10.15].

[64] Guideline on the use of the CTD format in the preparation of a registration application for traditional herbal medicinal products. (EMA/HMPC/71049/2007 Rev. 1) Committee on Herbal Medicinal Products. Available on: <http://www.ema.europa.eu/docs/en_GB/document_library/Regulatory_and_procedural_guideline/2012/08/WC500130862.pdf>; 2012 [accessed 20.10.15].

[65] Homeopathic Medicinal Products Working Group (HMPWG). Heads of medicines Agency. Available on: <http://www.hma.eu/>; [accessed 21.09.15].

[66] Commission Directive 2003/94/EC. Official Journal of the European Union. L 262, 14.10.2003, pp. 22–26. Available on: <http://ec.europa.eu/health/files/eudralex/vol-1/dir_2003_94/dir_2003_94_en.pdf>; 2012 [accessed 27.10.15].

[67] Eudralex. Volume 4. EU Guidelines to Good Manufacturing Practice Medicinal Products for Human and Veterinary Use. Part I - basic requirements for medicinal products. Chapter 1. Pharmaceutical quality system. Available on: <http://ec.europa.eu/health/files/eudralex/vol-4/vol4-chap1_2013-01_en.pdf>; 2012 [accessed 30.10.15].

[68] Eudralex. Volume 4. EU Guidelines to Good Manufacturing Practice Medicinal Products for Human and Veterinary Use. Part I - basic requirements for medicinal products. Chapter 2. Personnel. European Commission Available on: <http://ec.europa.eu/health/files/eudralex/vol-4/2014-03_chapter_2.pdf>; 2014 [accessed 30.10.15].

[69] Eudralex. Volume 4. EU Guidelines to Good Manufacturing Practice Medicinal Products for Human and Veterinary Use. Part I - basic requirements for medicinal products. Chapter 4. Documentation. European Commission. Available on: <http://ec.europa.eu/health/files/eudralex/vol-4/chapter4_01-2011_en.pdf>; 2010 [accessed 30.10.15].

[70] Eudralex. Volume 4. EU Guidelines to Good Manufacturing Practice Medicinal Products for Human and Veterinary Use. Part I - basic requirements for medicinal products. Chapter 3. Premise and Equipment. European Commission. Available on: <http://ec.europa.eu/health/files/eudralex/vol-4/chapter_3.pdf>; 2015 [accessed 30.10.15].

[71] Eudralex. Volume 4. EU Guidelines to Good Manufacturing Practice Medicinal Products for Human and Veterinary Use. Part I - basic requirements for medicinal products. Chapter 5. Production. European Commission. Available on: <http://ec.europa.eu/health/files/eudralex/vol-4/chapter_5.pdf>; 2015 [accessed 31.10.15].

[72] Eudralex. Volume 4. EU Guidelines to Good Manufacturing Practice Medicinal Products for Human and Veterinary Use. Part I - basic requirements for medicinal products. Chapter 6. Quality Control. European Commission. Available on: <http://ec.europa.eu/health/files/eudralex/vol-4/2014-11_vol4_chapter_6.pdf>; 2014 [accessed 2.11.15].

[73] Eudralex. Volume 4. EU Guidelines to Good Manufacturing Practice Medicinal Products for Human and Veterinary Use. Part I - basic requirements for medicinal products. Chapter 7. Outsourced activities. European Commission. Available on: <http://ec.europa.eu/health/files/eudralex/vol-4/vol4-chap7_2012-06_en.pdf>; 2012 [accessed 30.10.15].

[74] Eudralex. Volume 4. EU Guidelines to Good Manufacturing Practice Medicinal Products for Human and Veterinary Use. Part I - Basic Requirements for Medicinal Products. Chapter 8. Complaints, quality

defects and product recall. European Commission. Available on: <http://ec.europa.eu/health/files/eudralex/vol-4/2014-08_gmp_chap8.pdf>; 2014 [accessed 30.10.15].

[75] Eudralex. Volume 4. EU Guidelines to Good Manufacturing Practice Medicinal Products for Human and Veterinary Use. Part I - basic requirements for medicinal products. Chapter 9. self inspection. Available on: <http://ec.europa.eu/health/files/eudralex/vol-4/pdfs-en/cap9_en.pdf>; [accessed 30.10.15].

[76] Eudralex. Volume 4. EU Guidelines to Good Manufacturing Practice Medicinal Products for Human and Veterinary Use. Part II. Basic requirements for active substances used as starting materials. European Commission. Available on: <http://ec.europa.eu/health/files/eudralex/vol-4/2014-08_gmp_part1.pdf>; 2014 [accessed 01.11.15].

[77] Guideline on Good Agricultural and Collection Practice (GACP) for Starting Materials of Herbal Origin. (EMEA/HMPC/246816/2005) Committee on Herbal Medicinal Products (HMPC). Available on: <http://www.pitdc.org.tw/member/%E5%90%84%E5%9C%8B%E6%B3%95%E8%A6%8F/EU/guideline%20on%20GACP.pdf>; 2006 [accessed 30.10.15].

[78] Eudralex. Volume 4. EU Guidelines to Good Manufacturing Practice Medicinal Products for Human and Veterinary Use. Annex 7. Manufacture of herbal medicinal products. Directorate General for Health and Food Safety. European Commission. Available on: <http://ec.europa.eu/health/files/eudralex/vol-4/vol4_an7_2008_09_en.pdf>; [accessed 30.10.15].

[79] Directive 2011/24/EU of the European Parliament and of the Council. Official Journal of the European Union. L 88, 4.4.2011, pp. 45–65. Available on: <http://eur-lex.europa.eu/LexUriServ/LexUriServ.do?uri = OJ:L:2011:088:0045:0065:EN:PDF>; [accessed 16.10.15].

[80] Commission report on the operation of Directive 2011/24/EU on the application of patients' rights in cross-border healthcare. Report From the Commission to the European Parliament And The Council, European Commission.. Available on: <http://ec.europa.eu/health/cross_border_care/docs/2015_operation_report_dir201124eu_en.pdf>; 2015 [accessed 20.10.15].

[81] Directive 2005/36/EC of the European Parliament and of the Council. Official Journal of the European Union. L 255, 30.9.2005, Pg 22–41. Available on: <http://eur-lex.europa.eu/LexUriServ/LexUriServ.do?uri = OJ:L:2005:255:0022:0142:en:PDF>; [accessed 16.10.15].

[82] CAMbrella. Available on: <http://www.cambrella.eu/home.php?>; [accessed 16.10.15].

[83] Fonneb V., Kristiansen T., Falkenberg T., Hegyi G., Hök J., Roberti di Sarsina P., et al. Legal status and regulation of CAM in Europe. Part II –Herbal and Homeopathic Medicinal Products. Final report of CAMbrella Work Package 2. Available on: <https://fedora.phaidra.univie.ac.at/fedora/get/o:291682/bdef:Content/get>; 2012 [accessed 5.10.15].

[84] Wiesener S, Falkenberg T, Hegyi G, Hok J, Roberti di Sarsina P, Fonnebo V. Legal status and regulation of CAM in Europe. Part III - CAM regulations in the CAM regulations in EU/EFTA/EEA. Final report of CAMbrella Work Package 2. Available on: <https://fedora.phaidra.univie.ac.at/fedora/get/o:291585/bdef:Content/get>; 2012 [accessed 5.10.15].

[85] Nissen N, Johannessen H, Schunder-Tatzber S, Lazarus A, Weidenhammer W. Citizens' needs and attitudes towards CAM. Final report of CAMbrella Work Package 3. Available on: <https://fedora.phaidra.univie.ac.at/fedora/get/o:264407/bdef:Content/get>; 2012 [accessed 5.10.15].

[86] Eardley S, Bishop F, Prescott P, Cardini F, Brinkhaus B, Santos-Rey K, et al. CAM use in Europe – The patients' perspective. Part I: A systematic literature review of CAM prevalence in the EU. Final report of CAMbrella Work Package 4. Available on: <https://fedora.phaidra.univie.ac.at/fedora/get/o:292161/bdef:Content/get>; [accessed 5.10.15].

[87] Eardley S, Bishop F, Cardini F, Santos-Rey K, Jong M, Ursoniu S, et al. CAM use in Europe – The patients' perspective. Part II: A pilot feasibility study of a questionnaire to determine EU wide CAM use. Final report of CAMbrella Work Package 4. Available on: <https://fedora.phaidra.univie.ac.at/fedora/get/o:292164/bdef:Content/get>; 2012 [accessed 15.10.15].

[88] von Ammon K, Cardini F, Daig, U, Dragan S, Frei-Erb M, Hegyi G, et al. Health Technology Assessment (HTA) and a map of CAM provision in the EU. Final report of CAMbrella Work Package 5. Available on: <https://fedora.phaidra.univie.ac.at/fedora/get/o:300096/bdef:Content/get>; [accessed 20.10.15].

[89] Hök J, Fønnebø V, Lewith G, Santos Rey K, Vas J, Wiesener S, Weidenhammer W, Falkenberg T. Global stakeholders view on CAM research and development: Implications for the EU roadmap. Final report of CAMbrella Work Package 6. Available on: <https://fedora.phaidra.univie.ac.at/fedora/get/o:290227/bdef:Content/get>; [accessed 20.10.15].

[90] Reiter B, Baumhöfener F, Dlaboha M, Madsen J, Regenfelder R, Weidenhammer W. CAMbrella strategy for dissemination of project findings and future networking. Final Report of CAMbrella Work Package 8. Available on: <https://fedora.phaidra.univie.ac.at/fedora/get/o:291795/bdef:Content/get>; [accessed 16.10.15].

[91] The GP-TCM Research Association. Availabel on: <http://www.gp-tcm.org>; [accessed 22.09.15].

[92] European Federation for Complementary and Alternative Medicines. Available on: <http://www.efcam.eu/>; [accessed 19.09.15].

[93] Directive 2004/27/EC of the European Parliament And Of The Council. Official Journal of the European Union. L 136, 30.4.2004, p. 34. Available on: <http://www.biosafety.be/PDF/2004_27.pdf>; [accessed 29.10.15].

[94] Volume 9A of The Rules Governing Medicinal Products in the European Union. Guidelines on Pharmacovigilance for Medicinal Products for Human Use. Available on: <http://ec.europa.eu/health/files/eudralex/vol-9/pdf/vol9a_09-2008_en.pdf>; [accessed 3.10.15].

[95] Questions and Answers to support the implementation of the Pharmacovigilance legislation-Update, November 2012. (EMA/228816/2012−v.3) European medicines agency. Available on: <http://www.ema.europa.eu/docs/en_GB/document_library/Regulatory_and_procedural_guideline/2012/05/WC500127658.pdf>; [accessed 19.10.15].

[96] Eudravigilance. Mandatory e-reporting essentials. Available on: <https://eudravigilance.ema.europa.eu/human/>; [accessed 14.10.15].

[97] Medicines Act 1968. Chapter 67. Available on: <http://www.legislation.gov.uk/ukpga/1968/67/section/12> [accessed 20.10.15].

[98] The Human Medicines Regulations 2012. S.I. 2012/1916. Available on: <http://www.legislation.gov.uk/uksi/2012/1916/pdfs/uksi_20121916_en.pdf>; 2012 [accessed 29.09.15].

[99] Banned and restricted herbal ingredients. Guideline. Medicines and Healthcare products Regulatory Agency. Available on: <https://www.gov.uk/government/publications/list-of-banned-or-restricted-herbal-ingredients-for-medicinal-use/banned-and-restricted-herbal-ingredients>; 2014 [accessed 27.10.15].

[100] Traditional herbal medicines: registration form and guidance. Medicines and Healthcare products Regulatory Agency. Available on: <https://www.gov.uk/guidance/apply-for-a-traditional-herbal-registration-thr>; [accessed 21.10.15].

[101] Medicines watchdog takes further action to protect public from unlicensed herbal medicines. Press Release. Medicines and Healthcare products Regulatory Agency. Available on: <http://www.atcm.co.uk/PDF%20files/Press%20Releases/MHRA%20Press%20Release%20con341199.pdf>; 2013 [accessed 23.10.15].

[102] Herbal Medicines Advisory Committee Annual Report 2014. Medicines and Healthcare products Regulatory Agency. Available on: <https://www.gov.uk/government/uploads/system/uploads/attachment_data/file/443430/ABRHP_HMAC_Annual_Report_2014.pdf>; 2014 [accessed 17.10.15].

[103] The Herbal Medicines Advisory Committee Order 2005. S.I. 2005/2791. U.K. Laws Legal Portal. Available on: <http://www.uklaws.org/statutory/instruments_34/doc34203.htm>; 2005 [accessed 15.10.15].

[104] Herbal Medicines Advisory Committee. The National Archives. Medicines and Healthcare products Regulatory Agency. Available on: <http://webarchive.nationalarchives.gov.uk/20141008013752/http://www.mhra.gov.uk/Committees/Medicinesadvisorybodies/HerbalMedicinesAdvisoryCommittee/index.htm>; [accessed 21.10.15].

[105] Public bodies reform. Government Guidance. Available on: <https://www.gov.uk/guidance/public-bodies-reform>; [accessed 15.10.15].

[106] Public Bodies Reform-Proposals for Change. Available on: <https://www.gov.uk/government/uploads/system/uploads/attachment_data/file/62125/Public_Bodies_Reform_proposals_for_change.pdf>; 2011 [accessed 16.10.15].

[107] Advisory Board on the Registration Of Homeopathic Products Annual Report 2014. Medicines and Healthcare products Regulatory Agency. Available on: <https://www.gov.uk/government/uploads/system/uploads/attachment_data/file/443430/ABRHP_HMAC_Annual_Report_2014.pdf>; 2014 [accessed 17.10.15].

[108] British Herbal Medicine Association. Available on: <http://bhma.info/index.php/about-the-bhma/>; [accessed 25.09.15].

[109] European Scientific Cooperative on Phytotherapy. <http://escop.com/members/>; [accessed 25.09.15].

[110] Rules and Guidance for Pharmaceutical Manufacturers and Distributors 2015. The orange Guide. Pharmaceutical press. Medicines and Healthcare products Regulatory Agency. Available on: <http://www.pharmpress.com/product/9780857111715/orangeguide> [accessed 20.09.15].

[111] Rules and Guidance for Pharmaceutical Distributors. The green Guide. Available on: <http://www.pharmpress.com/product/MC_GREEN/rules-and-guidance-for-pharmaceutical-distributors>; [accessed 20.09.15].

[112] Guidance for UK Manufacturer's Licence and Manufacturer's Authorisation (for Investigational Medicinal Products) Holders on the use of UK Stand Alone Contract Laboratories. Medicines and Healthcare products Regulatory Agency. Available on: <https://www.gov.uk/government/uploads/system/uploads/attachment_data/file/450561/Guidance_for_UK_Manufacturers_licenceand_Manufacturers_authorisation.pdf>; 2014 [accessed 25.09.15].

[113] Good pharmacovigilance practice (GPvP). Medicines and Healthcare products Regulatory Agency. Available on: <https://www.gov.uk/guidance/good-pharmacovigilance-practice-gpvp>; 2014 [accessed 20.10.15].

[114] Yellow card. Medicines and Healthcare products Regulatory Agency. Available at: <https://yellowcard.mhra.gov.uk/>; [accessed 20.09.15].

[115] Drug Analysis Prints. Medicines and Healthcare products Regulatory Agency. Available on: <http://www.mhra.gov.uk/drug-analysis-prints/drug-analysis-prints-a-z/index.htm?indexChar = S#retainDisplay>; [accessed 24.09.15].

[116] The Drugs and Cosmetics Act, 1940. Central drugs standard control organization. Ministry of health and family welfare. Government of India. Available on: <http://www.cdsco.nic.in>; [accessed 29.09.15].

[117] The Gazette of India. Department of Publication. Ministry of Urban development. Government of India.Available on: <http://egazette.nic.in/default.aspx?AcceptsCookies = yes>; [accessed 10.10.15].

[118] G.S.R. 764(E). Drug and Cosmetics (6th amendment) Rules, 2009. Ministry Of Health And Family Welfare. Department of AYUSH. Published in: The Gazette of India. Department of Publication, Government of India, 2009.

[119] G.S.R. 561(E). Drugs and Cosmetics (Amendment) Rules, 2000. Ministry of Health and Family Welfare. Department of Indian Systems of Medicine & Homoeopathy. Published in: The Gazette of India. Department of Publication, Government of India, 2000.

[120] Gazette Notification issued under the Drugs & Cosmetics Rule 1945. Available on: <http://www.indianmedicine.nic.in/writereaddata/mainlinkFile/File337.pdf> [Accessed 3.10.15].

[121] The Indian Medicine Central Council Act, 1970.(Act 48 of 1970). Ministry of AYUSH. Government of India. Available on: <http://www.ccimindia.org/actandammendment.php>; [accessed 29.09.15].

[122] Central Council For Research in Ayurvedic Sciences. Ministry of AYUSH, Government of India. Available on: <http://www.ccras.nic.in/Sowa_Rigpa/sowarigpa.html>; [accessed 28.09.15].

[123] Central Council of Indian Medicine. Ministry of AYUSH, Government of India. Available on: <http://www.ccimindia.org/actandammendment.php>; [accessed 28.09.15].

[124] The central council of homeopathy. Available on: <http://www.cchindia.com/history.htm>; [accessed 20.08.15].

[125] Pharmacopoeial Laboratory for Indian Medicine. Ministry of Health and Family Welfare. Government of India. Available on: <http://www.plimism.nic.in>; [accessed 29.09.15].

[126] Protocol for the testing of Ayurvedic, Siddha and Unani Medicines. Pharmacopoeial Laboratory For Indian Medicines, Department of AYUSH, Ministry of Health & Family Welfare, Government of India, Ghaziabad. Available on: <http://www.plimism.nic.in>; [accessed 29.09.15].

[127] Homeopathic Pharmacopoeia Laboratory. Ministry of Health and Family Welfare. Government of India. Available on: <http://www.hplism.nic.in>; [accessed 29.09.15].

[128] Ministry of AYUSH, Government of India. Available on: <http://indianmedicine.nic.in/>; [accessed 5.10.15].

[129] G.S.R.157(E). Drug and Cosmetics (3rd Amendment) Rules, 2009. Ministry Of Health and Family Welfare. Department of AYUSH. Published in: The Gazette of India. Department of Publication, Government of India, 2009.

[130] Guidelines for Inspection of GMP Compliance by ASU Drug Industry. Dept. of Ayush, Ministry of Health & Family Welfare, Government of India.. Available on: <http://indianmedicine.nic.in/writereaddata/mainlinkFile/File779.pdf>; 2014 [accessed 29.09.15].

[131] Good Clinical Practice Guidelines for clinical trials in ayurveda, siddha and unani medicine. Department of AYUSH, Ministry of Health & Family Welfare, Government of India. Available on: <http://indianmedicine.nic.in/>; 2013 [accessed 5.09.15].

[132] G.S.R.702 (E). Drugs and Cosmetics (5th Amendment) Rules, 2013. Ministry of Health and Family Welfare. Department of Health and Family Welfare. Published in: The Gazette of India. Department of Publication, Government of India, 2013.

[133] Good Agricultural Practices Standard For Medicinal Plants-Requirements. National Medicinal Plants Board, Department of AYUSH, Ministry of Health and Family Welfare, Government of India. Available on: <http://www.nmpb.nic.in>; [accessed 9.09.15].

[134] National AYUSH Mission. Department of AYUSH. Ministry of Health & Family Welfare. Government of India. Available on: <http://www.indianmedicine.nic.in>; [accessed 20.09.15].

[135] Traditional Medicine: Delhi Declaration. World Health Organization, South-East Asia Region. Available on: <http://www.searo.who.int>; 2013 [accessed 15.09.15].

[136] National Pharmacovigilance Programme for Ayurveda, Siddha and Unani (ASU) Drugs. Institute for Post Graduate Teaching & Research in Ayurveda. Gujarat Ayurved University. Available on: <http://www.ayushsuraksha.com/>; [accessed 7.10.15].

[137] Baghel M. The national pharmacovigilance program for ayurveda, siddha and unani drugs: current status. Int J Ayurveda Res 2010;1:197–8.

[138] Kalaiselvan V, Saurabh A, Kumar R, Singh GN. Adverse reactions to herbal products: an analysis of spontaneous reports in the database of the pharmacovigilance programme of India. J Herb Med 2015;5:48–54.

[139] Sharma A, Agrawal A. Complementary and alternative medicine (CAM) use among patients presenting in out-patient Department at Tertiary Care Teaching Hospital in Southern Rajasthan, India-a questionnaire based study. Altern Integr Med 2015;4:1.

[140] China Food and Drug Administration. P.R. China. Available on: <http://eng.cfda.gov.cn/WS03/CL0755/>; 2001 [available 28.10.15].

[141] Drug Administration Law of the People's Republic of China. Order no. 45. State Food and Drug Administration. P.R. China. Available on: <http://eng.sfda.gov.cn/WS03/CL0766/61638.html>; 2001 [accessed 28.10.15].

[142] Regulations for Implementation of the Drug Administration Law of the People's Republic of China. Decree of the State Council of the People's Republic of China. State Food and Drug Administration. P. R. China. Available on: <http://eng.sfda.gov.cn/WS03/CL0767/61640.html>; 2002 [accessed 29.10.15].

[143] Provisions for Drug Registration. Order no. 28. State Food and Drug Administration. P.R. China. Available on: <http://eng.sfda.gov.cn/WS03/CL0768/61645.html>; 2007 [accessed 30.10.15].

[144] SFDA issues Supplementary Requirements on the Registration of TCM. State Food and Drug Administration. P.R. China. Available on: <http://eng.sfda.gov.cn/WS03/CL0757/62097.html>; 2008 [accessed 2.11.15].

[145] Good Agricultural Practice for Chinese Crude Drugs (Interim). Order No. 32. State Food and Drug Administration. P.R. China. Available on: <http://eng.sfda.gov.cn/WS03/CL0768/61642.html>; 2002 [accessed 6.11.15].

[146] Good Manufacturing Practice for Drugs (2010 Revision). MOH Decree No. 79. Ministry of Health. P. R. China. Available on: <http://eng.sfda.gov.cn/WS03/CL0768/65113.html>; 2011 [accessed 11.11.15].

[147] Regulations on Protection of Traditional Chinese Medicines. Ministry Of Commerce People's Republic of China. Available on: <http://english.mofcom.gov.cn/article/lawsdata/chineselaw/200211/20021100050739.shtml>; 1993 [accessed 4.11.15].

[148] Review and (initial) issuance of Certificate for protected TCM species. State Food and Drug Administration. P.R. China. Available on: <http://eng.sfda.gov.cn/WS03/CL0769/98074.html>; 2013 [accessed 5.11.15].

[149] Zhanga L, Yanb J, Liuc X, Yed Z, Yangb X, Meyboome R, et al. Pharmacovigilance practice and risk control of Traditional Chinese Medicine drugs in China: Current status and future perspective. J Ethnopharmacol 2012;140:519–25.

[150] Yan-min, Development of ADR Reporting and Monitoring in China. Department of Drug Safety & Inspection. State Food and Drug Administration. P.R. China. Available on: <http://www.who.int/medicines/areas/quality_safety/regulation_legislation/icdra/WG-3_2Dec.pdf>; [accessed 11.11.15].

[151] The revised Provisions for Adverse Drug Reaction Reporting and Monitoring issued. State Food and Drug Administration. P.R. China. Available on: <http://eng.cfda.gov.cn/WS03/CL0757/62683.html>; 2011 [accessed 9.11.15].

[152] China Food and Drug Administration issued the "Notice on Good Food and Drug Administration during Institutional Reform." China Pharmaceutical Newsletter. China Center for Pharmaceutical International Exchange & Servier (Tianjin) Pharmaceutical Co., Ltd.; 4:2. Available on: <http://www.ccpie.org/news/download/2013pharm-4.pdf>; 2013 [accessed 12.11.15].

[153] The Bejing Declaration. Adopted by the WHO Congress on Traditional Medicine, Beijing, China. Available on: <http://www.who.int/medicines/areas/traditional/congress/beijing_declaration/en/>; 2008 [accessed 26.09.15].

CHAPTER

ETHICAL CONSIDERATIONS IN CLINICAL RESEARCH

14

Divya Vohora
Jamia Hamdard University, New Delhi, India

14.1 HISTORY OF HUMAN RESEARCH ETHICS

The word "ethics" is derived from the Greek word, *ethos*, which means custom or character. Ethics always says, "*Not I, but thou.*" Its motto is, "*Not self, but non-self,*" by Swami Vivekananda. The history of human research dates back to the 1700s when Edward Jenner tested a smallpox vaccine on his own son and children from the neighborhood. In the 1900s, Walter Reed's experiments to develop an inoculation for yellow fever included evidence of voluntary consent. In 1966, Henry K. Beecher reported in the New England Journal of Medicine twenty-two examples of unethical practices in clinical research, which raised public outrage and interest amongst researchers to improve the ethical practices in human research [1]. Beecher also pointed out that journal editors reject such papers where unethical practices have been followed. It is only inspired by this that the Vancouver group (International Committee of Medical Journal Editors) met in 1978 and published the "*Uniform Requirements for Manuscripts Submitted to Biomedical Journals*" (revised many times thereafter) and including a section on "*Ethical Considerations in the Conduct and Reporting of Research*" [2]. The unethical practices quoted by Beecher included either lack of informed consent or increased risk to research subjects, among others [1,3]. Another study that is of importance with respect to violation of research ethics was the Tuskegee Syphilis study conducted in the US between 1932 and 1972 (longest study) on a large group of black men. It was aimed at discovering how blacks react to syphilis and how long a human being can live with untreated syphilis. The subjects in this study were denied treatment, even after penicillin was available as a standard cure for syphilis in the mid-1940s. Moreover, they were ill informed of having "bad blood" and were promised free care. The study was finally stopped in 1972 after a public outrage. The study raised several ethical issues in that era among which were informed consent, racism, unfair subject selection, maleficence, truth-telling and justice, etc. [3]. All this outrage led to the development of the National Research Act (1974), which further established Institutional Review Boards (IRBs) and realization among people regarding the complexities of establishing ethical standards for human research [3].

14.1.1 THE NUREMBERG CODE 1947

The historical event that majorly influenced human research ethics due to violation of fundamental human rights was the experiment on war prisoners at the end of World War II by Nazi physicians

Pharmaceutical Medicine and Translational Clinical Research. DOI: http://dx.doi.org/10.1016/B978-0-12-802103-3.00015-8

at Nuremberg, Germany. The trials were conducted by the US on physicians convicted on the basis of various torturous experiments, the most horrifying experiment being placing the subjects in a pressure chamber to simulate conditions that German pilots might encounter under high-altitude, low-pressure conditions with no oxygen or pressure suits [4]. The judgment of these trials formed the Nuremberg code in 1947 [5]. It is considered to be the foundation document for ethics in clinical research and the first document to set out ethical principles/regulations on human experimentation based on informed consent. Later, however, it was shown that informed consent in human experimentation guidelines existed well before the Nuremberg code [6] and that six of the ten principles of Nuremberg code were based on the Guidelines for Human Experimentation of 1931 [7]. The ten ethical principles laid down in Nuremberg code are shown in Box 14.1. Though most of the principles are the cornerstone of human research ethics, principle number 5, involving self-experimentation, was debated. It wasn't considered justifiable to put the lives of others at risk just because the investigator is willing to risk his own life [8]. The Nuremberg code is not legally binding, and unlike the Declaration of Helsinki, it is not regularly reviewed and updated.

BOX 14.1 THE NUREMBERG CODE

1. The voluntary consent of the human subject is absolutely essential.
2. The experiment should be such as to yield fruitful results for the good of society.
3. The experiment should be so designed and based on the results of animal experimentation and a knowledge of the natural history of the disease or other problem under study that the anticipated results justify the performance of the experiment.
4. The experiment should be so conducted as to avoid all unnecessary physical and mental suffering and injury.
5. No experiment should be conducted where there is an a priori reason to believe that death or disabling injury will occur; except, perhaps, in those experiments where the experimental physicians also serve as subjects.
6. The degree of risk to be taken should never exceed that determined by the humanitarian importance of the problem to be solved by the experiment.
7. Proper preparations should be made and adequate facilities provided to protect the experimental subject against even remote possibilities of injury, disability or death.
8. The experiment should be conducted only by scientifically qualified persons. The highest degree of skill and care should be required through all stages of the experiment of those who conduct or engage in the experiment.
9. During the course of the experiment the human subject should be at liberty to bring the experiment to an end if he has reached the physical or mental state where continuation of the experiment seems to him to be impossible.
10. During the course of the experiment the scientist in charge must be prepared to terminate the experiment at any stage, if he has probable cause to believe, in the exercise of the good faith, superior skill and careful judgment required of him, that a continuation of the experiment is likely to result in injury, disability, or death to the experimental subject.

Source: https://history.nih.gov/research/downloads/nuremberg.pdf, accessed on March 28, 2017, "Trials of War Criminals before the Nuremberg Military Tribunals under Control Council Law No. 10", Vol. 2, pp. 181-182. Washington, D.C.: U.S. Government Printing Office, 1949.

14.1.2 DECLARATION OF HELSINKI 1964

The Declaration of Helsinki (DoH), developed by the World Medical Association (WMA), is a statement of ethical principles for medical research involving human subjects. It was adopted by the 18th WMA General Assembly held in Helsinki, Finland in 1964, and subsequently amended in the WMA General Assembly meetings held in Japan (1975), Italy (1983), Hong Kong (1989), South Africa

(1996), Scotland (2000), Seoul (2008), and more recently Brazil (2013). A note on clarification (on two paragraphs—P29 and P30) was added in the WMA meetings held in Washington (2002) and Tokyo (2004). According to DoH, it is the duty of the physician to promote and safeguard the health of patients, including those who are involved in medical research, and that the wellbeing of the individual research subject must take precedence over all other interests. The declaration also included protection for the vulnerable population, the need for proper designing of research protocol, its approval and monitoring by research ethics committees, principles of maintaining confidentiality and privacy of research subjects, and described a written informed consent process [9]. However, the FDA in 2008 abandoned the DoH fourth revision (1996) by restricting the use of placebos where proven interventions had become established. This had major implications for research in resource-poor nations, where placebos were being used in such situations. The International Conference on Harmonization Good Clinical Practice (GCP) was designated as the new regulatory standard by the FDA [10]. The recent 7th revision of DoH in 2013 includes 37 principles categorized under various subsections for clarity [11,12]. Some of the notable changes in the new version of the DoH are [13]: (1) need to include under-represented groups in research (P13); (2) ensuring compensation to research participants (P15); (3) increasing protection for the vulnerable population (P20); (4) posttrial access to care (P34); (5) post-study requirements for reporting results to participants (P36); and (6) use of unproven interventions (P37). The new version, thus, aimed to prevent exploitation of research subjects and calls for greater protection of vulnerable groups, and for the first time declares compensation and treatment for those injured during the study [9]. The DoH is again not a legal document. Nevertheless, many organizations follow the same while framing ethical guidelines and the document is revised and updated.

14.1.3 BELMONT REPORT 1979

The four basic ethical principles were described in biomedical ethics in 1979 [14] and the Belmont Report was published by the National Commission for the Protection of Human Subjects of Biomedical and Behavioral Research, US [15]: (1) *Beneficence*: meaning "maximize possible benefits and minimize possible harms"; (2) *Non-maleficence*: meaning "avoidance of harm" (*included under Beneficence in Belmont Report*); (3) *Autonomy*: meaning free and informed participation, respect for persons, persons with diminished autonomy (vulnerable) are entitled to protection; and (4) *Justice*: meaning benefits a burden of disease to be fairly distributed among entire population (for selection of subjects, etc.). The above principles of the conduct of research require:

1. *Informed consent*: This has a further three important components:
   a. *Information*: concerning procedure, purposes, risks, anticipated benefits, alternative procedures (where therapy is involved), and a statement offering the subject the opportunity to ask questions and to withdraw at any time from the research, etc.
   b. *Comprehension*: the manner in which information is conveyed is equally important and is the responsibility of the Investigators.
   c. *Voluntariness*: An agreement to participate in research means informed consent given free of coercion and undue influence.
2. *Assessment of risks and benefits*: Brutal or inhumane treatment of human subjects is never justified. Risks should be reduced by careful attention to alternative procedures. Review committees should evaluate thoroughly the justification given in case of significant risks of

serious impairment. Relevant risks and benefits must be thoroughly provided in documents and procedures used in the informed consent process.

3. *Selection of subjects*: Certain groups, such as racial minorities, the economically disadvantaged, the very sick, and the institutionalized may continually be sought as research subjects, due to their easy availability. The vulnerable population, owing to their dependent status and their compromised capacity for free consent, should be protected against the danger of being involved in research.

14.1.4 ICH-GCP, CIOMS, AND OTHERS

Many other guidelines were subsequently developed including the *International Ethical Guidelines for Biomedical Research involving Human Subjects* (1982; revised in 1993, 2002), prepared by the Council for International Organizations of Medical Sciences (CIOMS) in collaboration with the World Health Organization (WHO) to help developing countries apply the principles of the Declaration of Helsinki and the Nuremberg Code. The CIOMS guidelines (2002) (http://www.cioms.ch/publications/layout_guide2002.pdf) [16] include 21 guidelines (15 in the original report) and are under the process of revision since, despite several changes in the field of research ethics, the guidelines were not revised after 2002. In 2011, a working group was constituted to revise the guidelines, and the major revisions recommended were the guideline on risks and benefits (Guideline 8), choice of control (Guideline 11), and women (Guideline 16). Guidelines 4 and 6, which are both on informed consent, were proposed to be merged. Further, the Green Book (the CIOMS Guidelines for Biomedical Research) with the Blue Book (the CIOMS Guidelines for Epidemiological Research) were also proposed to be merged together [16].

The Guidelines for Good Clinical Practice (ICH-GCP or ICH-E6, 1996) prepared by the International Conference on Harmonization (ICH) (now the International Council on Harmonization) of Technical Requirements for Registration of Pharmaceuticals for Human Use brought together the US, the EU, and Japan to follow a unified standard. The objective of the harmonization was to increase the speed of development and availability of new medicines, while maintaining safeguards on quality, safety, and efficacy, and ethical obligations to protect public health. The ICH-GCP guidelines for international ethical and scientific quality standards for designing, conducting, recording, and reporting trials involving human subjects, aimed to provide public assurance that the rights, safety, and wellbeing of trial subjects are protected, consistent with the principles of the Declaration of Helsinki, in addition to the credibility of clinical trial data [17]. Currently, the E6 (R2) integrated addendum updates the 1996 guidelines to enable implementation of innovative approaches to clinical trial design and conduct, as well as better human subject protection. A finalized integrated addendum is available [18].

In the USA, the federal codes for the protection of human subjects in clinical studies are contained in 21 CFR Part 50 (Title 21, Part 50 of the electronic codes of federal regulations) [19] and the 45 CFR Part 46 (Title 21, Part 45 of the electronic codes of federal regulations). The 21 CFR part 50 contains subpart A (the general provisions), Subpart B (Informed Consent of Human Subjects), Subpart C (reserved for future use) and the Subpart D (Additional Safeguards for Children in Clinical Investigations) [19,20]. Similarly, the ethical requirements for the conduct of the clinical trials in the European union are contained in Directive 2001/20/EC [21] of the European parliament and of the council, which was amended by Regulation (EC) No 1901/2006 [22] of the European Parliament and of the Council of 12 December 2006 and the Regulation (EC)

No 596/2009 of the European Parliament and of the council of 18 June 2009 [23]. Moreover, the GCP directive 2005/28/EC details the comprehensive guidelines for good clinical practice with respect to investigational medicinal products for human use and the requirements for their authorisation of the manufacturing or importation of such investigational products [24]. Notably, the "Regulation (EU) No 536/2014 of the European Parliament and of the Council of 16 April 2014 on clinical trials on medicinal products for human use" came out as the repealing directive enlisting the specific considerations for the vulnerable population and Chapter V containing the regulations for the protection of subjects and their informed consent (Article 10) [25].

Based on these directives, the European member states have established their own laws and regulations. The Medicines for Human Use (Clinical Trials) Regulations of 2004, SI 2004/1031 was enforced in the United Kingdom as a national legislation wherein the regulations have been arranged in nine parts and twelve schedules [26]. At the time of enforcement, it was based upon the Directive 2001/20/EC and the GMP directive 2003/94/EC and later amended by the SI 2006/1928 with the inclusion of the regulations from the GCP directive 2005/28/EC [27]. Soon after that, the SI 2006/2984 came out as an amendment in the 2004/1031 for the conduct of the clinical trials in the incapacitated adults without prior consent of the legal representative under certain circumstances [28].

In India, the "Ethical Guidelines for Biomedical Research on Human Subjects" were released by the Indian Council of Medical Research in 2000 (revised in 2006) [29], while Indian GCP guidelines were released in 2001. Schedule Y of Drugs and Cosmetics Act and Rules there under (amended in 2005) regulates clinical trials in India. Several amendments in rules have been notified since then. The ICMR guidelines "*National Ethical Guidelines for Biomedical and Health Research Involving Human Participants, 2017*" were released in October 2017 (http://www.icmr.nic.in/guidelines/ICMR_Ethical_Guidelines_2017.pdf) [30].

14.2 THE INFORMED CONSENT

The informed consent process is considered to be vital to ethical conduct of clinical research. The basic principle of an informed consent is that individuals, after receiving and understanding the necessary information (without any undue inducement or intimidation) are entitled to choose freely whether to participate in research or not, thus following the principles of autonomy. Prior to participation, the subject or the subject's legal representative should receive a copy of the signed and dated informed consent form [18]. There are several key elements of an effective informed consent process summarized below which have been defined in various guidelines including the CIOMS, ICH-GCP and Schedule Y guidelines. As per the Code of Federal Regulations, CFR (title 21, volume 1, part 50B) [31], there are eight basic elements of an informed consent document (Table 14.1). Briefly, it should include: (1) A statement that the study involves research, explaining its purpose, expected duration, and description of procedures to be followed; (2) Defining any foreseeable risks and discomfort to the subjects; (3) Any benefit to the subject or to society; (4) Any alternative procedure or treatment that might be advantageous to the subject; (5) A statement regarding maintaining confidentiality of records and that they may be subjected to inspection; (6) Details regarding compensation and treatment provided in case of injury to the subjects; and (7) A

Table 14.1 A Comparative View of Elements of Informed Consent Document as Per 21 CFR50, ICH-GCP and Indian-GCP (mostly similar with only minor differences)

	Code of Federal Regulations (21 CFR 50)	ICH-GCP (ICH-GCP E6, R2 Guidelines)	Indian GCP and/or Schedule Y of Drugs and Cosmetic Act and Rules
Basic/Essential Elements	**Basic Elements**	**Not Categorized Into Basic or Essential Element in ICH-GCP E6, R2**	**Essential Elements**
1. "Statement regarding research, purpose, procedures, duration, etc."	1. "A statement that the study involves research, an explanation of the purposes of the research"	1. "That the trial involves research and the purpose of the trial"	1. "Statement that the study involves research"
	"Description of the procedures to be followed"	"The trial procedures to be followed, including all invasive procedures"	"Description of the procedures to be followed, including all invasive procedures"
	"The expected duration of the subject's participation"	"The expected duration of the subject's participation in the trial"	"Expected duration of the subject's participation"
	"Identification of any procedures which are experimental"	"Those aspects of the trial that are experimental"	Not mentioned in Schedule Y
	Not mentioned	"The trial treatment(s) and the probability for random assignment to each treatment"	"Trial treatment schedule(s) and the probability for random assignment to each treatment (for randomized trials)"
	Not mentioned	"The subject's responsibilities"	"Subject's responsibilities on participation in the trial"
2. "Description of risks or discomforts to subject"	2. "A description of any reasonably foreseeable risks or discomforts to the subject"	2. "The reasonably foreseeable risks or inconveniences to the subject and, when applicable, to an embryo, fetus, or nursing infant"	2. "Description of any reasonably foreseeable risks or discomforts to the Subject"
3. "Any benefits"	3. "A description of any benefits to the subject or to others which may reasonably be expected from the research"	3. "The reasonably expected benefits. When there is not intended clinical benefit to the Subject, the subject should be made aware of this"	3. "Description of any benefits to the Subject or others reasonably expected from research. If no benefit is expected Subject should be made aware of this"
4. "Alternative procedures"	4. "A disclosure of appropriate alternative procedures or courses of treatment, if any, that might be advantageous to the subject"	4. "The alternative procedure(s) or course(s) of treatment that may be available to the subject, and their important potential benefits and risks"	4. "Disclosure of specific appropriate alternative procedures or therapies available to the Subject"

Table 14.1 A Comparative View of Elements of Informed Consent Document as Per 21 CFR50, ICH-GCP and Indian-GCP (mostly similar with only minor differences) ***Continued***

	Code of Federal Regulations (21 CFR 50)	ICH-GCP (ICH-GCP E6, R2 Guidelines)	Indian GCP and/or Schedule Y of Drugs and Cosmetic Act and Rules
Basic/Essential Elements	**Basic Elements**	**Not Categorized Into Basic or Essential Element in ICH-GCP E6, R2**	**Essential Elements**
5. "Confidentiality"	5. "A statement describing the extent, if any, to which confidentiality of records identifying the subject will be maintained and that notes the possibility that the Food and Drug Administration may inspect the records"	5. "That the monitor(s), the auditor(s), the IRB/IEC, and the regulatory authority (ies) will be granted direct access to the subject's original medical records for verification of clinical trial procedures and/or data, without violating the confidentiality of the subject, to the extent permitted by the applicable laws and regulations and that, by signing a written informed consent form, the subject or the subject's legally acceptable representative is authorizing such access"	5. "Statement describing the extent to which confidentiality of records identifying the subject will be maintained and who will have access to Subject's medical records"
	Not mentioned	"That records identifying the subject will not be made publicly available even when the results of the trial are published, the subject's identity will remain confidential"	"No such clause on confidentiality of subject in publication mentioned in the guideline"
6. "Compensation"	6. "For research involving more than minimal risk, an explanation as to whether any compensation and an explanation as to whether any medical treatments are available if injury occurs and, if so, what they consist of, or where further information may be obtained"	6. "The compensation and/or treatment available to the subject in the event of trial related injury"	6. "Compensation and/or treatment(s) available to the subject in the event of a trial related injury"
	Not mentioned	"The anticipated prorated payment, if any, to the subject for participating in the trial"	"The anticipated prorated payment, if any, to the subject for participating in the trial"
7. "Whom to contact"	7. "An explanation of whom to contact for answers to pertinent questions about the research and research subjects' rights, and whom to contact in the event of a research-related injury to the subject"	7. "The person(s) to contact for further information regarding the trial and the rights of trial subjects, and whom to contact in the event of trial-related injury"	7. "An explanation about whom to contact for trial related queries, rights of Subjects and in the event of any injury"

(*Continued*)

Table 14.1 A Comparative View of Elements of Informed Consent Document as Per 21 CFR50, ICH-GCP and Indian-GCP (mostly similar with only minor differences) ***Continued***

	Code of Federal Regulations (21 CFR 50)	ICH-GCP (ICH-GCP E6, R2 Guidelines)	Indian GCP and/or Schedule Y of Drugs and Cosmetic Act and Rules
Basic/Essential Elements	**Basic Elements**	**Not Categorized Into Basic or Essential Element in ICH-GCP E6, R2**	**Essential Elements**
8. "Voluntary participation"	8. "A statement that participation is voluntary, that refusal to participate will involve no penalty or loss of benefits to which the subject is otherwise entitled, and that the subject may discontinue"	8. "That the subject's participation in the trial is voluntary and that the subject may refuse to participate or withdraw from the trial, at any time, without penalty or loss of benefits to which the subject is otherwise entitled"	8. "Statement that participation is voluntary, that the subject can withdraw from the study at any time an tat refusal to participate will not involve any penalty or loss of benefits to which the Subject is otherwise entitled"
	"Participation at any time without penalty or loss of benefits to which the subject is otherwise entitled"		
"Other essential elements"	"When seeking informed consent for applicable clinical trials, the following statement shall be provided to each clinical trial subject in informed consent documents." The statement is: *"A description of this clinical trial will be available on* http://www/ClinicalTrials.gov, *as required by U.S. Law. This Web site will not include information that can identify you. At most, the Web site will include a summary of the results. You can search this Web site at any time."*		"Audio-visual recording of informed consent process is required for vulnerable subjects in NCE/NME trials; and for anti-HIV & anti-leprosy drug trials, only audio recording is required" *"[As per recent amendments in schedule Y] (2015 GSR 611 dt. 31-7-2015) for essential Elements in informed consent]"* "Any other pertinent information may be included"
	Not mentioned	Not mentioned	"A statement that there is possibility of failure of investigational product to provide intended therapeutic benefit"
	Not mentioned	Not mentioned	"Statement that incase of placebo controlled trials, the placebo administered to the subject shall not have any therapeutic effect"

Table 14.1 A Comparative View of Elements of Informed Consent Document as Per 21 CFR50, ICH-GCP and Indian-GCP (mostly similar with only minor differences) ***Continued***

Additional Elements	Additional Elements	Not Categorized as Additional Elements in ICH-GCP E6, r2	Additional Elements
1. "Unforeseeable risks to fetus"	1. "A statement that the particular treatment or procedure may involve risks To the subject (or to the embryo or fetus, if the subject is or may become pregnant) which are currently unforeseeable."	1. "The reasonably foreseeable risks or Inconveniences to the subject and, when applicable, to an embryo, fetus, or nursing infant (please see point no.2 under Basic/ essential Elements)"	1. "A statement that the particular treatment or procedure may involve risks to the Subject (or to the embryo or fetus, if the Subject is or may become pregnant), which are currently unforeseeable"
2. "Termination of participation"	2. "Anticipated circumstances under which the subject's participation may be terminated by the investigator without regard to the subject's consent."	2. "The foreseeable circumstances and/or reasons under which the subject's participation in the trial may be terminated"	2. "Statement of foreseeable circumstances under which the Subject's participation may be terminated by the Investigator without the Subject's consent"
3. "Additional costs"	3. "Any additional costs to the subject that may result from participation in the research."	3. "The anticipated expenses, if any, to the subject for participating in the trial"	3. "Additional costs to the Subject that may result from participation in the study"
4. "Consequences of withdrawing participation"	4. "The consequences of a subject's decision to withdraw from the research and procedures for orderly termination of participation by the subject."	4. Not mentioned in ICH-GCP	4. "The consequences of a Subject's decision to Withdraw from the research and procedures for Orderly termination of participation by Subject"
5. "Sharing of new findings developed during study"	5. "A statement that significant new findings developed during the course of the research which may relate to the subject's willingness to continue participation will be provided to the subject."	5. "That the subject or the subject's legally acceptable representative will be informed in a timely manner if information becomes available that may be relevant to the subject's willingness to continue participation in the trial"	5. "Statement that the Subject or Subject's representative will be notified in a timely manner if significant new findings develop during the course of the research which may affect the Subject's willingness to continue participation will be provided"
6. "Number of subjects"	6. "The approximate number of subjects involved in the study."	6. "The approximate number of subjects"	6. "Approximate number of Subjects enrolled in the study"

Table 14.1 is derived from following guidelines:

1. 21CFR §50.25. Elements of informed consent. Electronic Code of Federal Regulations. U.S. Government Publishing Office (GPO). Available on https://www.accessdata.fda.gov/scripts/cdrh/cfdocs/ccfr/CFRSearch.cfm?CFRPart = 50&showFR = 1&subpartNode = 21:1.0.1.1.20.2; [accessed 17.03.17].
2. Integrated addendum to ICH E6 (R1): guideline for good clinical practice. E6(R2), ICH harmonised guideline. International conference on harmonisation of technical requirements for registration of pharmaceuticals for human use. (2015) Available on: http://www.ich.org/fileadmin/Public_Web_Site/ICH_Products/Guidelines/Efficacy/E6/E6_R2__Step_4.pdf; [accessed 17.03.17].
3. http://cdsco.nic.in/html/D&C_Rules_Schedule_Y.pdf; [accessed 17.03.17].

statement regarding voluntary participation and that the subject may discontinue at any time without penalty or loss of benefit. Some additional elements include a statement regarding unforeseeable risks, circumstances under which subject's participation may be terminated, consequences of a subject's decision to withdraw from the research, providing information to the subject regarding any new findings developed during the course of study, approximate number of subjects involved in the study, etc. [29,30]. Though the basic elements of the informed consent document as specified in 21CFR50 are similar in other guidelines, a comparison of these with ICH-GCP and Indian-GCP guidelines is depicted in Table 14.1 with only minor differences.

One of the significant issues concerning effective informed consent is the participant's understanding of the project ("understood consent") especially among the illiterate, and hence it is particularly important to prevent the 'therapeutic misconception' where participants believe that they are receiving treatment. This is particularly true in developing countries [32]. In a recent study conducted by Koonrungsesomboon et al. in the Asia-pacific and African region, it was found that the participants' understanding of the required ICF elements was limited and there is a need to improve training among researchers and IEC members [33].The patient often participates in trial without fully understanding its implications. To address this issue, the government of India released a notification in 2014 regarding mandatory audio-visual recording of Informed Consent process while adhering to the principle of confidentiality [34]. Such audio-visual recording and related documentation was required to be preserved. Though the advantages of AV recording like reliability, transparency, etc. are certain, there were many issues related to infrastructure, the subject's refusal to be videographed, confidentiality, cost implications, etc., affecting clinical and academic research. Later, on the 31st July 2015 (the Fifth Amendment rules 2015 to DCR), an AV recording of the informed consent was required only in case of vulnerable subjects in clinical trials of a new chemical entity or a new molecular entity [35]. Also, in anti-HIV or anti-leprosy trials, only audio recording of the informed consent of individual subjects is required.

14.3 THE INSTITUTIONAL REVIEW BOARDS/ETHICS COMMITTEES

The National Research Act of 1974 led to the development of IRBs to specifically address research involving the vulnerable population. A "*common rule*" was adopted for 16 federal agencies (Subpart A, Part 46: Protection of Human Subjects, of Title 45: Public Welfare, in the Code of Federal Regulations, 46 CFR 45) with some common requirements, including a minimum of 5 members, with varying backgrounds representing different racial and cultural backgrounds, with one member completely independent of the Institution [36]. The ICH-GCP described IRBs as Independent Ethics Committees (IECs) which could be institutional, regional, national, or international, whereas IRBs were primarily institutional. As per ICH-GCP E6 (R1) [17], an IRB/IEC should safeguard the rights, safety, and wellbeing of all trial subjects. Special attention should be paid to trials that include vulnerable subjects. The committee should consist of at least 5 members, with at least one member from a non-scientific area and one member independent of the institution/trial site (Table 14.2). The committee may invite nonmembers as subject experts for a particular study. The IRB/IEC should perform its functions according to written standard operating procedures and should maintain proper records of its activities, all relevant records, and minutes of its

Table 14.2 A Comparative View of Composition of Ethics Committees (IRB/IEC) As Per 21CFR50, ICH-GCP and Indian-GCP

	Code of Federal Regulations (21 CFR 50)	ICH-GCP (ICH-GCP E6, R2 Guidelines)	Indian-GCP/ Schedule Y (2005) with subsequent amendments
Minimum number of members"	"EC must have at least 5 members."	"EC must have at least 5 members."	"EC must have at least 7 members."
"Quorum"	Not mentioned.	"An IRB/IEC should make its decisions at announced meetings at which at least a quorum, as stipulated in its written operating procedures, is present."	"The quorum should be at least 5 members with following representation: 1. basic medical scientists (preferably one pharmacologist). 2. clinicians 3. legal expert 4. social scientist / representative of non-governmental voluntary agency / philosopher / ethicist / theologian or a similar person 5. lay person from the community."
"Qualification/ Expertise of members"	"Each IRB Shall include at least one member whose primary concerns are in the scientific are a and at least one member in non-scientific areas." "In addition, the IRB shall include persons knowledgeable to comply with regulations, applicable laws and professional conduct and practice."	"At least one member whose primary area of interest is in a non-scientific area."	"Members should be a mix of medical/non-medical, scientific and non-scientific persons, including lay public. to reflect the different view points" (as mentioned above). "The members representing medical scientists and clinicians should have postgraduate qualifications and adequate experience in their respective fields. They should be conversant with the provisions of clinical trials under Schedule Y, GCP guidelines and other regulatory requirements."
"Training of ethics committee members"	Not mandatory	Not mandatory	"Mandatory (as per GSR 72E, Feb 8, 2013) Proof of GCP training required for re-registration of ethics committee (as per order dated Feb 01, 2016)"

(*Continued*)

Table 14.2 A Comparative View of Composition of Ethics Committees (IRB/IEC) As Per 21CFR50, ICH-GCP and Indian-GCP *Continued*

	Code of Federal Regulations (21 CFR 50)	ICH-GCP (ICH-GCP E6, R2 Guidelines)	Indian-GCP/ Schedule Y (2005) with subsequent amendments
"Member independent of Institution"	"Each IRB shall include at least one member who is not affiliated with the institution and who is not part of the immediate family of a person affiliated with the institution"	"At least one member who is independent of the institution/trial site."	"Ethics Committee should appoint, from among its members, a Chairperson (who is from outside the institution) and a Member Secretary."
"Diversity of members and Gender representation"	"Adequate consideration of race, gender, cultural backgrounds and diversity. No IRB may consist entirely of members of one profession."	–	"There should be appropriate age and gender representation."
"Vulnerable category"	"In case of a vulnerable category of subjects, such as children, prisoners, pregnant women, or handicapped or mentally disabled persons, consideration shall be given to the inclusion of one or more individuals who are knowledgeable about and experienced in working with those subjects."	–*	"For clinical trials conducted in the pediatric population, the reviewing ethics committee should include members who are knowledgeable about pediatric, ethical, clinical and psychosocial issues." "Based on the requirement of research area, e.g. HIV AIDS, genetic disorders etc. specific patient groups may also be represented in the Ethics Committee as far as possible."
"Members on invitation"	"An IRB may, in its discretion, invite individuals with competence in special areas to assist in the review of complex issues. These individuals may not vote with the IRB."	"An IRB/IEC may invite non-members with expertise in special areas for assistance."	"If required, Subject experts may be invited to offer their views."
Source:	"eCFR, Title 21, Chapter 1, Sub-chapter A, Part 56, http://www/ecfr.gov/cgi-bin/text-idx?SID = 2f0abfe3ada70e76b76b7dc682ca85c9a97 &mc = true&node = pt21.1.56&rgn = div5, updated as on March 24, 2017, accessed on March 28, 2017"	"Integrated addendum to ICH E6 (R1): guideline for good clinical practice. E6(R2). ICH harmonised guideline. International conference on harmonisation of technical requirements for registration of pharmaceuticals for human use. (2015). Available on: http://www.ich.org/fleadmin/ Public_Web_Site/ICH_Products?Guidelines/ EfficaCy/E6/E6_R2__Step_4.pd [Accessed on: March 17, 2017]."	"Schedule Y (amended version)—CDSCO file:///Gl/RGCB%20IHEC/Schedule%20Y (amended%20version)%20-%20CDSCO.htm [80-05-2014 10:53:51] and ICMR Ethical Guidelines for Biomedical Research (2006)."

**However, separate guidelines are available for vulnerable subjects.*

meetings, for a period of at least 3 years after completion of the trial. The committee is responsible for approving/disapproving the initiation of a proposed clinical study, its review at intervals appropriate to the degree of risk to human beings, is responsible for ensuring informed consent, reviewing payment to subjects in order to rule out coercion, approving any deviations from the protocol initially approved, reviewing all serious and unexpected adverse drug reactions (ADRs), and reviewing the reasons for premature termination or suspension of a trial. It should also ensure the compensation and/or treatment available to the subject in the event of trial-related injury, as stated in the informed consent. In India, as per rule 122 DD [37], no Ethics Committee shall review and accord its approval to a clinical trial protocol without prior registration with the Drugs Controller General of India [38].

14.4 COMPENSATION TO TRIAL SUBJECTS

As per ICH-GCP E6 (4.8.10), the compensation to the subject in the case of trial-related injury is recommended [18]. As per US regulations too (45 CFR 46.116), an explanation of any compensation and treatment if injury occurs during study participation needs to be incorporated in the informed consent [39]. Similarly, in EU Directive 2001/20/EC, it is the duty of the ethics committee to determine compensation in case of any injury or death of subject due to the clinical trial. However, there are no explicit guidelines regarding the quantum of compensation [40].

In India, this subject has been dealt in much detail after the negative coverage in the media about flouting of ethical rules in some of the trials. As per the rule 122 DAB, the Drugs and Cosmetics (first amendment) rules released on 30th Jan 2013 [41], compensation was included in case of injury/ death during clinical trial and free medical management of the participant for as long as required. In addition to this, some of the debatable issues were compensation for "any injury" or death during clinical trial due to failure of the investigational product to provide the intended therapeutic effect, and use of placebo in a placebo-controlled trials. All these regulations were acting as a deterrent for many pharmaceutical multinational companies and academic Institutions to undertake clinical trials. All above amendments contributed to a decline in the number of clinical trials in India with sponsors starting diverting their trials to other countries. The following statements were added as "*Essential elements*" in the informed consent document as per the Fifth Amendment Rules of Drugs and Cosmetics Act in July 2015 [35]: (1) there is a possibility of failure of investigational product to provide intended therapeutic effect; and (2) in case of placebo controlled trials, the placebo administered to the subjects shall not have any therapeutic effect (please also see Table 14.1). The amendment to rule 122 DAB (dated 12th December 2014 and called Drugs and Cosmetics Sixth Amendment Rules) [42] was released on 30th December 2014 with modifications shown in italics: (1) In case of an injury occurring to the subject during the clinical trial, free medical management shall be given *as long as required or until such time it is established that the injury is not related to clinical trial, whichever is earlier*; (2) *In case there is no permanent injury, the quantum of compensation shall be commensurate with the nature of non-permanent injury and loss of wages of the subject*; (3) Compensation for any injury or death during clinical trial due to failure of investigational product to provide intended therapeutic effect *where the standard care, though available, was not provided to subject as per clinical trial protocol*; and (4) Compensation for any injury or death during clinical trial due to use of placebo in placebo-controlled trials *where the standard care, though available, was not provided to subject as per clinical trial protocol.* In addition to above, the formulas for the determination of the quantum of compensation in the case of the clinical trials related to the serious adverse events of deaths or other injuries related to the same were derived for the first time in India in 2013 [43] and 2014 [44]. These formulae are mentioned below:

1. In case of the clinical trials related to the serious adverse events of deaths:

$$\text{Compensation} = (B \times F \times R)/99.37$$

   Where, B = base amount (Rs 8 lacs), F = factor depending on the age of the subject (based on workmen compensation act), here it is shown for a 65 year old person. R = risk factor factor depending on severity of disease, comorbid conditions, and duration of disease at the time of enrollment in CT between the scale of 0.5–4 as under: 0.50: terminally ill patient (expected survival not more than (NMT) 6 months), 1.0: Patient with high risk (expected survival between 6 and 24 months), 2.0: Patient with moderate risk, 3.0: Patient with mild risk, 4.0: Healthy Volunteers or subject of no risk.
2. In case of clinical trials related to the serious adverse events of injury other than deaths (clinical trial-related SAE) in the clinical trials:
   a. *Serious adverse event causing permanent disability to the subject:*
      Compensation = $(C \times D \times 90)/(100 \times 100)$ Where, C = quantum of compensation which would have been due for the payment to the subject nominee(s) in case of the death of the subject, D = percentage of disability the subject has suffered.
   b. *Serious adverse event causing permanent anomaly or the birth defects:*
      It may be due to the participation of parent (or parents) in the clinical trial. It may include stillbirths, early deaths, correctable deformity (via appropriate therapy), and permanent disability (mental or physical). In such cases, the total compensation should be a sum of money, the fixed deposit of which should yield half the monthly wages of an unskilled laborer in Delhi. For the correctable deformity of the permanent disability, the medical expenses are required to be borne by the sponsor or their legal representative, which can also be considered as the financial compensation for the same.
   c. *In case of serious adverse event causing life-threatening disease or reversible serious adverse event (if resolved):*

   In both cases, the quantum of compensation is related to the total time of the hospitalization, which is compensated for by calculating the quantum of compensation as per the total working hours of the subject, and is associated with the loss of daily wages. The daily working wages are calculated according to the minimum wages of an unskilled worker in the capital city.

   Compensation = $2 \times W \times N$ Where, W = Minimum wage per day of unskilled worker (in Delhi), N = Number of days of hospitalization.

14.5 ETHICS IN THE VULNERABLE POPULATION

"Vulnerable population" refers to (but is not limited to) those who are incapable of protecting their own interests and hence pregnant women, neonates, children, fetuses, prisoners, physically handicapped, mentally challenged, economically disadvantaged, institutionalized and very sick patients, etc., can all come under this group. The ethical issues in involving the vulnerable population for research are still debatable. The justification for involving the vulnerable population in research is mainly when a particular problem affects that group. However, it is important to understand that participation in biomedical research not only poses risks, but also may provide benefit to participants and society. Moreover, in today's world of evidence-based medicines, we must have evidence in the vulnerable population as well. One of the important points of concern in the vulnerable population is that some lack the ability to consent or understand. In that case, a legally acceptable representative (LAR) should be involved in the decision. Further, vulnerable persons may require significant and repeated education/information about the research, benefits and risks, and alternatives, if any. As per the ICMR National Ethical Guidelines for Biomedical and Health Research involving Human Participants, 2017 [30], research must be planned in vulnerable populations only if that population will benefit from research. Care should be taken that participants are not exploited. LAR shouldn't be given any reward for encouraging the participation of their dependents. It is the duty of the IRBs/IECs and clinical investigators to give special attention to protecting the welfare of vulnerable subjects, and to ensure voluntariness and freedom from coercion, to review justification for inclusion of such subjects, and to suggest additional safeguards for them.

Children are one of the several classes under vulnerable population due to their lack of ability to understand, and the fact that they are under the authority of others. Though involving children in research raises serious ethical concerns, restricting children is also not appropriate as their participation is necessary to develop new treatments or prevention methods [45]. Other advantages include detection of pediatric-specific adverse events, variable pharmacokinetics, etc. It has been seen that sponsors are generally reluctant to carry out studies in pediatrics due to the small market and the high cost of the trial. There are other obvious problems like difficult recruitment, noncompliance, excessive liability, etc. To work in this direction, the "*NIH policy and guidelines on the inclusion of children as participants in research involving human subjects*" was published in 1998 (http://grants.nih.gov/grants/guide/notice-files/not98-024.html) [46], last updated on Oct 13, 2015 (https://grants.nih.gov/grants/guide/notice-files/NOT-OD-16-010.html) [47]. As per the national ethics guidelines for biomedical research involving children, 2017 [48], research in children can be carried out on diseases exclusively seen in childhood or in other conditions (of both adults and children) where the results are expected to be significantly different in children or where the risk-benefit ratio is low and there is minimal risk to the child. The US code of federal regulations on research involving children (sub-part D of 45CFR46) classify the trial into one of four risk categories to determine the rules: (1) minimal risk and no direct benefit to subjects; (2) greater than minimal risk, but with direct benefit to subjects; (3) greater than minimal risk and no direct benefit to subjects, but likely to yield generalizable knowledge about condition; and (4) research not otherwise approvable which represents an opportunity to understand, prevent, or alleviate a serious problem affecting the health or welfare of children [49]. Decisions about the participation of a child in research are expected to be taken by the parents/LAR, in the best interests of their children. The ethics committee must carefully assess if there are any other extraneous factors involved in giving such consent. In addition to consent from parents/LAR, an oral or written assent, as approved by the ethics committee, should be obtained from children of 7–18 years of age (except under certain circumstances). Parental permission and a child's assent remain critically important protections for a child's participation in clinical research, and it is the responsibility of the investigator and ethics committee to ensure the same. As per eCFR §50.55, in determining whether children are capable of providing assent, the IRB must take into account the ages, maturity, and psychological state of the children involved. However, the assent of the children is not a necessary condition for proceeding with the clinical investigation if the IRB determines that the capability of the child is so limited that they cannot reasonably be consulted, or the intervention or procedure involved in the clinical investigation holds out a prospect of direct benefit that is important to the health or wellbeing of the children. Even where the IRB determines that the subjects are capable of assenting, the IRB may still waive the assent requirement if it finds that the clinical investigation involves no more than minimal risk to the subjects and will not adversely affect their rights and welfare, or when the clinical investigation could not practicably be carried out without the waiver.

Similar to children, there should be proper justification for inclusion of pregnant and nursing women in clinical trials designed to address the health needs of such women or their fetuses or nursing infants. For instance, a trial may be designed to test the safety and efficacy of a drug for reducing perinatal transmission of HIV infection from mother to child, or a trial of a device for detecting fetal abnormalities, etc. [45]. Other than research directed towards the health of a pregnant women and/or her fetus, it is recommended that pregnant women be actively excluded from the clinical research that involves greater than minimal risk [46]. In the USA, research involving pregnant women, fetuses, and neonates is covered in sub-part B of 45CFR46 according to which such a research may be conducted if previous research has been done on non-pregnant women to assess potential risks, and there should be evidence of a direct benefit to women or fetuses, or there should be minimal risk. If there is benefit only to the fetus, then consent of both parents may be required.

14.6 CONCLUSION

Clinical research helps in advancing our understanding of disease or science for promoting human health. However, it is important to protect the rights and welfare of those subjects who volunteer to participate in research. Ethical guidelines are therefore established, ethics committees are formulated, and it is the obligation of all stakeholders to follow these regulations and responsibilities. The scientific justification and rationale to include human participants in the study is of paramount importance. Informed consent, confidentiality, privacy, and safety of subjects are key considerations in ethical research. Finally, it is the moral duty of the investigator, the members of the ethics committee, sponsors, clinicians, or any other person directly or indirectly involved in clinical study, to follow the principles of clinical research ethics.

REFERENCES

[1] Beecher HK. Ethics and clinical research. N Engl J Med 1966;274:1354–60.

[2] Fischer BA. A summary of important documents in the field of research ethics. Schizophr Bull 2006;32:69–80.

[3] Rice TW. The historical, ethical, and legal background of human-subjects research (symposium paper). Respir Care 2008;53:1325–9.
[4] Emaneul EJ, Grady CC, Crouch RA, Lie RK, Miller FG, Wendler DD. The Oxford textbook of clinical research ethics. New York, NY: Oxford University Press; 2008.
[5] The Nuremberg Code 1947. BMJ 1996; 313: 1448.
[6] Informed consent in human experimentation before the Nuremberg code. BMJ1996; 313:1445.
[7] Ghooi RB. The Nuremberg code - a critique. Perspect. Clin Res 2011;2:72–6.
[8] Editorial. Self-experimentation and the Nuremberg Code. BMJ 2010; 341: c7103.
[9] Arie S. Revision of Helsinki declaration aims to prevent exploitation of study participants. BMJ 2013;347:f6401.
[10] Goodyear MD, Lemmens T, Sprumont D, Tangwa G. Does the FDA have the authority to trump the declaration of Helsinki? BMJ 2009;338:b1559.
[11] World Medical Association Declaration of Helsinki Ethical Principles for Medical Research involving human subjects (Adopted by the 18th WMA General Assembly, Helsinki, Finland, June 1964, and amended by the: 29th WMA General Assembly, Tokyo, Japan, October 1975, 35th WMA General Assembly, Venice, Italy, October 1983, 41st WMA General Assembly, Hong Kong, September 1989, 48th WMA General Assembly, Somerset West, Republic of South Africa, October 1996, 52nd WMA General Assembly, Edinburgh, Scotland, October 2000, 53rd WMA General Assembly, Washington 2002 (Note of Clarification on paragraph 29 added), 55th WMA General Assembly, Tokyo 2004 (Note of Clarification on Paragraph 30 added), 59th WMA General Assembly, Seoul, October 2008, 64th WMA General Assembly, Fortaleza, Brazil, October 2013).
[12] World Medical Association. Declaration of Helsinki: ethical principles for medical research involving human subjects. JAMA 2013;310:2191–4.
[13] Ndebele P. The declaration of Helsinki, 50 years later. JAMA 2013;310:2145–6.
[14] Beauchamp TL, Childress JF. Principles of biomedical ethics. 6th ed. Oxford: Oxford University Press; 2008.
[15] Belmont Report: Ethical Principles and Guidelines for the Protection of Human Subjects of Research, Report of the National commission for protection of human subjects of biomedical and behavioral research, US, 1979.
[16] <http://www.cioms.ch/publications/layout_guide2002.pdf>, and <http://www.cioms.ch/index.php/2012-06-10-08-47-53/ethics/cioms-guidelines-working-group> [accessed 23.10.15].
[17] <http://www.ich.org/fileadmin/Public_Web_Site/ICH_Products/Guidelines/Efficacy/E6/E6_R1_Guideline.pdf> [accessed 23.10.15].
[18] Integrated addendum to ICH E6 (R1): guideline for good clinical practice. E6(R2). ICH harmonised guideline. International conference on harmonisation of technical requirements for registration of pharmaceuticals for human use. (2015). Available on: <http://www.ich.org/fileadmin/Public_Web_Site/ICH_Products/Guidelines/Efficacy/E6/E6_R2__Step_4.pdf> [accessed 17.03.17].
[19] 21 eCFR §50. Electronic Code of Federal Regulations. U.S. Government Publishing Office (GPO). Available on: <http://www.ecfr.gov/cgi-bin/text-idx?SID = d9b85b838a768e56e100ed8d43c0b7bd&mc = true&tpl = /ecfrbrowse/Title21/21cfr50_main_02.tpl> [accessed 25.02.16].
[20] 45 CFR§46. Electronic Code of Federal Regulations. U.S. Government Publishing Office (GPO). Available on: <http://www.ecfr.gov/cgi-bin/text-idx?SID=7c9dde68b1e9b364d2a2f7c30fea9470&mc=true&node = pt45.1.46&rgn = div5> [accessed 13.10.15].
[21] Directive 2001/20/EC of the European parliament and of the council of 4 April 2001. OJ L 121, 1.5.2001, p. 34. Available on: <http://ec.europa.eu/health/files/eudralex/vol-1/dir_2001_20/dir_2001_20_en.pdf> [accessed 22.01.16].

[22] Regulation (EC) No 1901/2006 of the European Parliament and of the Council of 12 December 2006. OJ L 378, 27.12.2006, p 1. Available on: <http://ec.europa.eu/health/files/eudralex/vol-1/reg_2006_1901/reg_2006_1901_en.pdf> [accessed 22.01.16].

[23] Regulation (EC) No 596/2009 of the European Parliament and of the Council of 18 June 2009. OJ L 188, 18.7.2009, p 14. Available on: <http://eur-lex.europa.eu/LexUriServ/LexUriServ.do?uri = OJ:L:2009:188:0014:0092:EN:PDF> [accessed 25.01.16].

[24] Commission Directive 2005/28/EC of 8 April 2005. OJ L 91, 9.4.2005, P 13. Available on: <http://eur-lex.europa.eu/LexUriServ/LexUriServ.do?uri = OJ:L:2005:091:0013:0019:en:PDF> [accessed 1.02.16].

[25] Regulation (EU) No 536/2014 of the European Parliament and of the Council of 16 April 2014 on clinical trials on medicinal products for human use, and repealing Directive 2001/20/EC. OJ L 158, 27.05.2014, p 1. Available on: <http://www.therqa.com/assets/js/tiny_mce/plugins/filemanager/files/Committees/Good_Clinical_Practice_GCP/Clinical_Trials_Regulation_EU_No_5362014.pdf> [accessed 7.02.16].

[26] The Medicines for Human Use (Clinical Trials) Regulations 2004. SI 2004/1031. Available on: <http://www.legislation.gov.uk/uksi/2004/1031/pdfs/uksi_20041031_en.pdf> [accessed 20.02.16].

[27] The Medicines for Human Use (Clinical Trials) Amendment Regulations 2006. SI 2006/1928. Available on: <http://www.legislation.gov.uk/uksi/2006/1928/pdfs/uksi_20061928_en.pdf> [accessed on 20.02.16].

[28] The Medicines for Human Use (Clinical Trials) Amendment (No.2) Regulations 2006 Available on: <http://www.legislation.gov.uk/uksi/2006/2984/pdfs/uksi_20062984_en.pdf> [accessed 20.02.16].

[29] Ethical guidelines for biomedical research on human participants. Available on: <http://www.icmr.nic.in/ethical_guidelines.pdf>; 2006 [accessed 25.12.16].

[30] <http://www.icmr.nic.in/guidelines/ICMR_Ethical_Guidelines_2017.pdf> [accessed 22.10.17].

[31] 21CFR §50.25. Elements of informed consent. Electronic Code of Federal Regulations. U.S. Government Publishing Office (GPO). Available on: <https://www.gpo.gov/fdsys/granule/CFR-2012-title21-vol1/CFR-2012-title21-vol1-sec50-25> [accessed 13.10.16].

[32] Bhutta ZA. Beyond informed consent. Bull World Health Organ 2004;82:771−7.

[33] Koonrungsesomboon N, Laothavorn J, Karbwang J. Understanding of essential elements required in informed consent form among researchers and institutional review board members. Trop Med Health 2015;43:117−22.

[34] Draft Guidelines On Audio-Visual Recording Of Informed Consent Process In Clinical Trial. Central Drugs Standard Control Organisation, Directorate general of health services, Government of India. Available on: <http://www.cdsco.nic.in/writereaddata/Guidance_for_AV%20Recording_09.January.14.pdf>; 2014 [accessed 13.03.16].

[35] GSR 611 (E) dtd 31.7.2015. The Drugs and Cosmetics (Fifth Amendment) Rules, 2015. Available on: <http://www.cdsco.nic.in/writereaddata/Gazette%20Notification%2031%20July%202015.pdf>; [accessed 14.10.16].

[36] Fischer BA. A summary of important documents in the field of research ethics. Schizophrenia Bull 2006;32:69−80.

[37] G.S.R 72(E) dtd 08.02 2013. The Gazette of India. Department of Publication, Government of India. Available on: <http://cdsco.nic.in/writereaddata/G.S.R%2072(E)%20dated%2008.02.2013.pdf>; [accessed 15.10.16].

[38] Minutes of the 3rd meeting of technical committee held on 29-04-2013. Central drugs standard control organization. Director general of health services, ministry of health and family welfare, Government of India. Available on: <http://www.cdsco.nic.in/writereaddata/3rd%20Minutes_Technical_Committee_29_04_13.pdf> [accessed 15.10.16].

[39] 45 CFR§46.116 General requirements for informed consent. Available on: <http://www.hhs.gov/ohrp/regulations-and-policy/regulations/45-cfr-46/#46.116> [accessed 15.10.16].

[40] Directive 2001/20/EC of the European parliament and of the council of 4 April 2001. OJ L 121, 1.5.2001, p. 34. Available on: <http://ec.europa.eu/health/files/eudralex/vol-1/dir_2001_20/dir_2001_20_en.pdf> [accessed 14.10.16].

[41] GSR 53(E) dtd 30.01.2013. The Gazette of India. Department of Publication, Government of India. Available on: <http://www.cdsco.nic.in/writereaddata/GSR%2053(E).pdf> [accessed 17.10.16].

[42] GSR 889(E) dtd 12.12.2014. Drugs and Cosmetics Sixth Amendment Rules. Available on: <http://www.cdsco.nic.in/writereaddata/Notificatiohn%20on%20Compensation%20on%20clincial%20trial%20(1).pdf>; 2014 [accessed 15.10.16].

[43] Formula to determine the quantum of compensation in the cases of clinical trial related serious adverse events (SAEs) of deaths occurring during clinical trials. Central Drugs Standard Control Organisation, Directorate general of health services, Government of India. Available on: <http://www.cdsco.nic.in/writereaddata/formula2013SAE.pdf> [accessed 16.10.15].

[44] Formula to determine the quantum of compensation in the cases of clinical trial related serious adverse events (SAEs) of injuries other than death occurring during the clinical trials. Central Drugs Standard Control Organisation, Directorate general of health services, Government of India. Available on: <http://www.cdsco.nic.in/writereaddata/ORDER%20and%20Formula%20to%20Determine%20the%20quantum%20of%20compensation%20in%20the%20cases%20of%20Clinical%20Trial%20related%20serious%20Adverse%20Events%28SAEs%29%20of%20Injury%20other%20than%20Death.pdf> [accessed 12.03.16].

[45] Schwenzer KJ. Protecting vulnerable subjects in clinical research: children, pregnant women, prisoners, and employees. Respir Care 2008;53:1342–9.

[46] NIH Policy and Guidelines on the Inclusion of Children Participants in Research Involving Human Subjects. National institutes of Health. Available on: <http://grants.nih.gov/grants/guide/notice-files/not98-024.html>; 1998 [accessed 7.11.16].

[47] Inclusion of Children in Clinical Research: Change in NIH Definition. (NOT-OD-16-010). National institutes of Health. Available on: <https://grants.nih.gov/grants/guide/notice-files/NOT-OD-16-010.html>; 2015 [accessed 5.11.16].

[48] <http://www.icmr.nic.in/guidelines/National_Ethical_Guidelines_for_BioMedical_Research_Involving_Children.pdf> 2017 [accessed 22.10.17].

[49] 45 CFR Part 46, Subpart D - Additional Protections for Children Involved as Subjects in Research. Office of human research protection. U.S. Department of Health & Human Services. Available on <http://www.hhs.gov/ohrp/regulations-and-policy/regulations/45-cfr-46/#subpartd> [accessed 12.11.16].

SECTION

PHARMACEUTICAL INDUSTRY AND INTELLECTUAL PROPERTY RIGHTS IV

CHAPTER

PATENT

15

Bindu Sharma
Origiin IP Solutions LLP, Bengaluru, Karnataka, India

Patent is basically a set of exclusive rights granted by a state (national government) to an inventor for a limited period of time (generally 20 years from the date of filing) in exchange for a public disclosure of an invention. Patents are granted to the inventions that are novel, inventive, and have industrial application. The Office of the Controller General of Patents, Designs, and Trade Marks is a subordinate office under the Department of Industrial Policy and Promotion, Ministry of Commerce and Industry, Government of India. This office is responsible for administration of the Patents Act, including all matters relating to the filing, processing, and granting of patents in India. There are four patent offices in India, the with head office at Kolkata and the branch office at Delhi, Chennai, and Mumbai.

Patent rights are territorial, i.e., the patent granted in India is valid only in India. To have protection in other countries, the inventor is required to file a patent application separately in those other countries.

15.1 CRITERIA OF PATENTABILITY

Patents are granted to inventions, and the "invention" has been defined in Section 2 of The Patent Act, 1970 as:

> A new product or process involving an inventive step and capable of industrial application.

This indicates that in order to get a patent, invention shall fulfill following three conditions:

- Novelty
- Inventive step/nonobviousness
- Industrial application

The first requirement to get a patent is that the invention should be new or novel, because a patent cannot claim something that already exists. Before filing for a patent, prior art in the form of publication or patent application are often used to assess the novelty of the invention. To be more precise, a novel invention or technology is one that has not been anticipated by publication or patent on the date of filing. The date of filing a patent application, also called the priority date, plays a

Pharmaceutical Medicine and Translational Clinical Research. DOI: http://dx.doi.org/10.1016/B978-0-12-802103-3.00016-X

very important role in determining the term of a patent as well as calculating various other timelines, such as international filing, examination, etc.

The prior art search may be done by searching databases of patents, patent applications, and other documents such as articles, publications, etc. No search can possibly cover every single publication on earth, and therefore one cannot actually say that an invention is "new." A prior art search may, for instance, be performed using a keyword search of large patent databases, scientific papers, and publications. However, it is impossible to guarantee the novelty of an invention, even if a patent has been granted, since some obscure, little-known publication may have disclosed the claimed invention.

The second requirement to getting a patent is that the invention shall have an inventive step—or we can also say that the invention should be nonobvious to the person skilled in the art. One of the most complex aspects of patent law is the inventive step or obviousness of an invention. Novelty is determined before the inventive step because the creative contribution of the inventor can be assessed only by knowing the novel elements of the invention.

Inventive step (nonobviousness) means a feature of an invention that involves a technical advance as compared to the existing knowledge, or having economic significance, or both, and that makes the invention nonobvious to a person skilled in the art. Invention should take technology one step ahead and should not be a replica or repetition of previous inventions. Or we can say that invention must be nonobvious to a person skilled in the art.

The inventive step is defined in Section 2(1)(ja) of India Patents Act, 1970 as:

> "Inventive step" means a feature of an invention that involves technical advance as compared to the existing knowledge or having economic significance or both and that makes the invention not obvious to a person skilled in the art.

For example, a drug formulation which is a mere combination of two well-known drugs may be a new formulation for the reason that there is no prior art in the form of publication or patent that disclosed this formulation but this may be very obvious to a person who is an expert in the subject. In such a case, even though the combination of formulation is new, it fails to be nonobvious to a person skilled in the art. Hence, getting a patent for such inventions is tough. However, if there was synergy between these two drugs and the combination leads to better efficacy, the chances of getting a patent become better.

The third criterion of patentability is that the invention should be capable of industrial application. It is defined in Section 2 (1) (ac) of the Patents Act, 1970:

> "Capable of Industrial application", in relation to an invention, means that the invention is capable of being made or used in an industry.

If the subject matter of the invention is devoid of industrial application, it does not satisfy the definition of "invention" for the purpose of the Act. The purpose of granting a patent is not to reserve an unexplored field of research for an applicant. Methods of testing are generally regarded as capable of industrial application if the test is applicable to the improvement or control of a product, apparatus, or process, which itself is capable of industrial application. It is therefore advisable to indicate the purpose of the test if this is not otherwise apparent.

An invention for a method of treatment of the human or animal body by surgery or therapy or of diagnosis practiced on the human or animal body is not taken to be capable of industrial application. Parts/pieces of the human or animal body to be used in transplants are objected to as not being capable of industrial application.

15.2 INVENTIONS NOT PATENTABLE

Other than the definition of invention under Section 2 (j), the patents Act 1970 doesn't define the category of the inventions that can be patented, but provides a list of inventions that cannot be patented and are exempted from being patented. Such inventions, even though they fulfill the basic criteria of patentability, i.e., novelty, industrial utility, and nonobviousness, are not granted patents. Such inventions are listed in Sections 3 and 4 of the Act and are termed as "Inventions not patentable," and the purpose is:

1. To discourage and prevent monopoly over inventions which are injurious to health, environment, morality, national defense, and security.
2. Grant patents only for inventions which are useful for society and the progress of science and technology.

Section 3: What are not inventions?
The following are not inventions within the meaning of this Act:

1. *An invention which is frivolous or which claims anything obviously contrary to well established natural laws.*
2. *An invention the primary or intended use or commercial exploitation of which could be contrary to public order or morality or which causes serious prejudice to human, animal, or plant life or health, or to the environment.*
3. *The mere discovery of a scientific principle or the formulation of an abstract theory or discovery of any living thing or nonliving substance occurring in nature;*
4. *The mere discovery of a new form of a substance which does not result in the enhancement of a known efficacy of that substance or the mere discovery of a new property or new use of a known process, machine, or apparatus, unless such known process results in a new product or employs at least one new reactant.*
   a. *Explanation: For the purpose of this clause, salts, esters, ethers, polymorphs, metabolites, pure form, particle size, isomers, mixtures of isomers, complexes, combinations, and other derivatives of known substance shall be considered to be the same substance, unless they differ significantly in properties with regard to efficacy.*
5. *A substance obtained by a mere admixture resulting only in the aggregation of the properties of the components thereo,f or a process for producing such substance.*
6. *The mere arrangement or re-arrangement or duplication of known devices each functioning independently of one another in a known way.*
7. *Omitted by Act 38 of 2002, Sec 4 (wef 20-05-2003).*
8. *A method of agriculture or horticulture.*

9. *Any process for the medicinal, surgical, curative, prophylactic, diagnostic, therapeutic, or other treatment of human beings or any process for a similar treatment of animals to render them free of disease or to increase their economic value, or that of their products.*
10. *Plants and animals in whole or any part thereof other than microorganisms but including seeds, varieties, and species and essentially biological processes for production or propagation of plants and animals.*
11. *A mathematical or business method or a computer program per se, or algorithms.*
12. *A literary, dramatic, musical, or artistic work or any other aesthetic creation whatsoever, including cinematographic works and television productions.*
13. *A mere scheme or rule or method of performing a mental act or a method of playing a game.*
    a. *A presentation of information.*
14. *A topography of integrated circuits.*
15. *An invention which in effect is traditional knowledge or which is an aggregation or duplication of known properties or a traditionally known component or components.*

Section 3 (d) of the Indian Patent Act is very important and has been a very controversial section and was interpreted by the Court in Novartis case.

Novartis International AG is a multinational pharmaceutical company based in Basel, Switzerland that filed a patent application in the Indian Patent Office (Patent number: 1602/MAS/1998) related to a beta crystal form of salt, *imatinib mesylate*. This form is the most stable version which Novartis formulated into a pharmaceutically useful drug, Glivec. The drug was proven to be effective for innumerable patients' drugs and was approved by the FDA in 2001. Novartis applied for an Exclusive Marketing Right (EMR) pending grant of a product patent, and was granted one in November 2003.

In 2005, an amendment to India's patent regime introduced product patents for pharmaceuticals and the mailbox application by Novartis was opened and examined. The grant of a patent was opposed [pre-grant opposition, U/S 25 (1)] by several generic drug companies (and an NGO, the Cancer Patients Aid Association (CPAA)), on several grounds, including:

- Lack of novelty/anticipation
- Lack of significantly enhanced "efficacy" under section 3(d)
- Obviousness
- Wrongful priority, as before 2005 Switzerland was not convention country

Assistant controller of patents in pre-grant proceedings issued 5 distinct orders in January 2006 refusing grant of patent. In June 2009, application was rejected by IPAB [Intellectual Property Appellate Board] primarily because of violation of Section 3(d) that aims to prevent "ever-greening" by prohibiting the patenting of new forms of existing pharmaceutical substances that do not demonstrate significantly enhanced efficacy.

After a complete study, IPAB found the claims to be inventive and novel, but failing to qualify the requirements of section 3 (d), and Assistant Controller of Patents rejected the patent application. Aggrieved by this rejection, Novartis AG, along with its Indian subsidiary, Novartis India, filed two writ petitions in the Madras High Court. These petitions not only sought a reversal of the Assistant Controller's order, but also a declaration that Section 3(d) was unconstitutional and in violation of India's obligations under TRIPS.

The High Court transferred the first petition to the IPAB, a specialist tribunal set up to deal with appeals from the various intellectual property offices across the country, and the biggest issue was whether Novartis' beta crystalline form is patentable or not under section 3 (d).

Article 27 of TRIPs says that the patents shall be available for any inventions, whether products or processes, in all fields of technology, provided that they are new, involve an inventive step, and are capable of industrial application. Novartis claimed that the active ingredient in Glivec (a beta crystalline form of imatinib mesylate) is more effective than the imatinib free base, since it displays better bioavailability properties, i.e., it is absorbed more easily into the blood. To this effect, it submitted evidence before the Assistant Controller demonstrating an increase in bioavailability of up to 30%. However, the Assistant Controller held that this was not sufficient to constitute "increased efficacy."

As per the affidavit, the technical expert conducted studies to compare the relative bioavailability of the free base with that of beta crystalline form of imatinib mesylate, and said that the difference in bioavailability is only 30% and also the difference in bioavailability may be due to the difference in their solubility in water. The present patent specification does not bring out any improvement in the efficacy of the beta crystal form over the known substances, rather it states that the base can be used equally in the treatment of diseases of in the preparation of pharmacological agents wherever the beta crystal is used. Even the affidavit submitted on behalf of the Applicant did not prove any significant enhancement of known efficacy.

Novartis finally moved to the Supreme Court in September 2009. In April 2013, the Supreme Court rejected the Novartis cancer drug Glivec patent plea.

15.3 COMPULSORY LICENSE AND ITS RELEVANCE TO PHARMACEUTICAL INDUSTRY

Patents are granted to encourage the inventors to disclose their inventions and also to grant them monopolistic right to exploit the invention. The objective of the Patent Grant in India is to ensure that the inventions are worked in India on a commercial scale and to the fullest extent without any undue delay. If the patentee is not commercializing the invention and, as a result, the reasonable requirements of the public are not met, or the patented product is not available to public at a reasonable price, the compulsory license is available as a remedy against abuse of patent right.

The provisions for compulsory licenses are made to prevent the abuse of patent as a monopoly and to make the way for commercial exploitation of the patented invention by an interested person. If the invention is not commercialized for 3 years, or in case of national emergency or urgency, the compulsory license may be granted to the person who requests the controller for it. After thorough assessment of the entire situation, such as necessity and demand of commercializing the invention, and capacity of the applicant to manufacture the patented product or process, the controller may grant compulsory license.

Many patent law systems provide for the granting of compulsory licenses in various situations. The Paris Convention of 1883 provides that each contracting State may take legislative measures for the grant of compulsory licenses. Article 5A(2) of the Paris Convention reads:

> Each country of the Union shall have the right to take legislative measures providing for the grant of compulsory licenses to prevent the abuses which might result from the exercise of the exclusive rights conferred by the patent, for example, failure to work.

The Agreement on Trade-Related Aspects of Intellectual Property Rights (TRIPs) also sets out specific provisions that shall be followed if a compulsory license is issued. All significant patent systems comply with the requirements of TRIPs.

When a patent is granted to the patentee, he must commercialize the invention so that invention is available to the public at reasonable cost and the reasonable requirements of the public are met. However, if he doesn't commercialize the invention for 3 years, any person interested can request the controller to grant him a compulsory license in order to manufacture and commercialize the patented invention so that the reasonable requirements of the public are met and the product is available to the public at reasonable cost. The application for compulsory license under Section-84 can be filed only after expiry of 3 years from the date of grant of the patent.

As per Section-84 [Compulsory license], a compulsory license shall be granted only if 3-year period has expired after the grant of the patent, and the patentee himself is not working the invention. However, there are certain circumstances, such as national emergencies or extreme urgency, wherein the compulsory license is granted without waiting for the expiry of 3 years.

For the pharmaceutical industry in particular, though patents have been playing an important role in protecting the innovations, the law ensures the balance between patent rights and public health.

CHAPTER

COPYRIGHT

16

Bindu Sharma
Origiin IP Solutions LLP, Bengaluru, Karnataka, India

The history of copyright law started with early privileges and monopolies granted to printers of books. The British Statute of Anne 1709, full title "*An Act for the Encouragement of Learning, by Vesting the Copies of Printed Books in the Authors or Purchasers of such Copies, during the Times therein mentioned*," was the first copyright statute. Initially, copyright law only applied to the copying of books, but over time other uses such as translations and derivative works were made subject to copyright, and copyright now covers a wide range of works, including maps, performances, paintings, photographs, sound recordings, motion pictures, and computer programs.

16.1 REQUIREMENTS FOR COPYRIGHT PROTECTION

The copyright law protects the expression of the idea and not the idea. Apart from this, the work shall be original and fixed in a tangle form.

16.1.1 ORIGINALITY

In order to get protection as copyright, the work has to be original and should not be copied from any source. Originality here doesn't refer to novelty, but the work shall not be a copy of another work. It must be a result of an author's independent skill and efforts. If the work is a compilation, the compilation must involve some originality beyond mere alphabetic sorting of all available works.

16.1.2 FIXATION

The work shall be fixed in a tangible form or reduced to material form, recorded, written, painted, or typed. A work is considered fixed when it is stored on some medium in which it can be perceived, reproduced, or otherwise communicated. For example, a story is considered fixed when it is written down on paper. The paper is the medium on which the story can be perceived, reproduced, and communicated. Similarly, a song is fixed when it is recorded, and a computer program code is fixed when it is typed or written.

Pharmaceutical Medicine and Translational Clinical Research. DOI: http://dx.doi.org/10.1016/B978-0-12-802103-3.00017-1

16.1.3 EXPRESSION OF IDEA AND NOT IDEA ITSELF

Copyright law does not protect the idea but only expression of the idea.

16.1.3.1 Idea-expression dichotomy in copyright

It is interesting to note that copyright law does not protect the ideas, but only expressions of the idea. An idea is considered to be a thought, as a mental image, as a conception of a theory, and cannot be a formulation of thought on a particular subject, whereas an expression would constitute implementing the said idea. An idea can have numerous expressions and all expressions individually are protected as copyright, provided they fulfill the requirements of copyright law. It is sometimes very challenging to draw a line between idea and expression.

For example, there are hundreds of movies based on love stories. Here the idea is a love story, which is not protected by copyright law. Expression of an idea when expressed by different directors in different ways is considered as different expression and hence receives protection under copyright law. Each expression and each movie will have individual copyrights. Additionally, the work shall have a substantial amount of skill, judgment, and labor, and the work shall be fixed in a tangible form, e.g., if it is a song or drama it should be recorded, if it is a computer program it should be written.

16.2 WHICH WORKS ARE PROTECTED BY COPYRIGHT?

The Indian Copyright Act, 1957, provides protection to the following categories of works:

16.2.1 LITERARY WORK

"Literary work" includes books, periodicals, journals, magazines, computer programs, tables, and compilations, including computers and databases.

16.2.2 DRAMATIC

"Dramatic work" includes recitation, scenic arrangement, and work capable of being performed by action.

16.2.3 MUSICAL WORK

A "Musical work" means a work consisting of music and includes any graphical notation of such work, but does not include any words or any action intended to be sung, spoken, or performed with the music. A musical work need not be written down to enjoy copyright protection.

16.2.4 ARTISTIC WORKS

An "Artistic work" includes:

- A painting, a sculpture, a drawing (including a diagram, map, chart, or plan), an engraving or a photograph, whether or not any such work possesses artistic quality

- A work of architecture
- Any other work of artistic craftsmanship

16.2.5 CINEMATOGRAPH FILMS

"Cinematograph film" means any work of visual recording on any medium produced through a process from which a moving image may be produced by any means and includes a sound recording accompanying such visual recording and "cinematograph" shall be construed as including any work produced by any process analogous to cinematography including video films.

16.2.6 SOUND RECORDINGS

"Sound recording" means a recording of sounds from which sounds may be produced regardless of the medium on which such recording is made or the method by which the sounds are produced. A phonogram and a CD-ROM are sound recordings.

16.2.7 TERM OF PROTECTION

The general rule is that copyright lasts for 60 years. In the case of original literary, dramatic, musical, and artistic works, the 60-year period is counted from the year following the death of the author. In the case of cinematograph films, sound recordings, photographs, posthumous publications, anonymous and pseudonymous publications, works of government, and works of international organizations, the 60-year period is counted from the date of publication.

16.2.8 NOTICE OF COPYRIGHT

Notice of copyright in a proper format is important to be included in the work. It is an identifier placed on copies of the work to inform the world of copyright ownership. While use of a copyright notice was once required as a condition of copyright protection, it is now optional, though it is highly recommended for various reasons. Use of the copyright notice is the responsibility of the copyright owner and does not require advance permission from, or registration with, the Copyright Office. This means that even if the work is not registered with copyright registry, the copyright notice can be put on the work.

Copyright notice informs the public that the work is protected by copyright, identifies the copyright owner, and shows the year of first publication of the work. Furthermore, in the event that a work is infringed, if a proper notice of copyright appears on the published copy or copies to which a defendant in a copyright infringement suit had access, then no weight shall be given to such a defendant's defense based on innocent infringement. Innocent infringement occurs when the infringer did not realize that the work was protected.

The commonly used format to apply copyright notice comprises of the following elements:

- The copyright symbol © (the letter C in a circle), or the word "Copyright" can be used.
- The year of first publication of the work shall be added.
- Further to this, a line such as "All rights reserved" can also be used.

Examples:

Copyright © 2012 XYZ Pvt Ltd, All rights reserved.

In case of DVD or CD, usually you may see following format:

Copyright © 2012 XYZ entertainment, all rights of the producer and owner of the CD content reserved. Unauthorized copying, public performance, and broadcasting of this content/recording is prohibited and punishable under the Copyright Act, 1957.

16.3 WHO OWNS THE COPYRIGHT IN A WORK?

Copyright protection subsists from the time the work is created in fixed, tangible form. The copyright in the work of authorship immediately becomes the property of the author who created the work. Only the author or those deriving their rights through the author can rightfully claim copyright. In the case of works "made for hire," where an author has created the work while in their capacity of an employee, the employer and not the employee is considered to be the author and copyright holder.

- In the case of a literary or dramatic work—the author, i.e., the person who creates the work.
- In the case of a musical work—the composer.
- In the case of a cinematograph film—the producer.
- In the case of a sound recording—the producer.
- In the case of a photograph—the photographer.
- In the case of a computer-generated work—the person who causes the work to be created.

The owner of the copyrighted work can transfer their rights to anyone by means of:

- Assignment
- License

Assignment is a permanent transfer of rights where the license agreement can be executed for a specified duration of time. The assignment shall be in writing signed by the assignor (copyright owner) or by their duly authorized agent. It shall mention the specific works and specify the rights assigned and the duration and territorial extent of such assignment. It shall also specify the amount of royalty payable to the author or their legal heirs during the currency of the assignment and the assignment shall be subject to revision, extension, or termination on terms mutually agreed upon by the parties. If the period of assignment is not stated, it shall be deemed to be five years from the date of assignment. If the territorial extent of the assignment of the rights is not specified, it shall be presumed to extend throughout the whole of India.

16.4 RIGHTS OF AN AUTHOR

Rights of a copyright owner usually include economic rights, such as right to reproduce the work, to issue copies of the work to the public, to perform the work in public, to communicate the work to the public, to make a cinematograph film or sound recording in respect to the work, to make any translation of the work, to make any adaptation of the work.

Other than economic rights, the author also has moral rights. The author of a work has the right to claim authorship of the work and to restrain or claim damages in respect of any distortion, mutilation, modification, or other act in relation to the said work if such distortion, mutilation, modification, or other act would be prejudicial to their honor or reputation. Moral rights are available to the authors even after the economic rights are assigned. The moral rights are independent of the author's copyright and remains with them even after assignment of the copyright.

Amar Nath Sehgal Versus Union of India

The Union Government of India commissioned Amar Nath Sehgal to design a mural (a mammoth 40 feet high and 140 feet long) for Vigyan Bhawan. The design was given the green flag by Pandit Jawahar Lal Nehru, and the mural was completed in 1962. The mural won widespread acclaim, and gave the world a glimpse of the "real" India—its farmers, artisans, women, and children, their daily chores and celebrations, frozen in time, and molded from tons of solid bronze. For nearly 20 years the mural attracted dignitaries and art connoisseurs from all over the world and became a landmark of the cultural life of the capital.

Later Vigyan Bhawan buildings were renovated and in the process, the mural was ripped off the walls and the remnants put into store. Distressed by the destruction of his artistic work, and after petitioning the authorities for years without a response, Mr. Sehgal brought a lawsuit against the government. Sehgal had assigned his copyright to the government in an agreement dated 31st October 1960 and the government had purchased all the rights from Sehgal, and was consequently free to do as it pleased with the mural. The mural was already damaged in a fire in the Vigyan Bhawan. As per the agreement, any grievance should be referred to an arbitrator appointed by the governmental decision of Delhi High Court—all rights of the mural shall henceforth vest with Mr. Sehgal. The court ordered the return of the remains of the mural to the sculptor, and also slapped on damages of Rs.500,000.

16.5 REGISTRATION OF COPYRIGHT

The most frequently asked question about copyright is whether it is necessary to register copyright or not. The answer is no, because copyright is statutory as well as an inherent right, and comes into being automatically after competition of the work. The legal copyright notice can be put even without registration of copyright but registration of copyright with copyright registry is important under certain circumstances, such as:

1. The certification of copyright registration is an authentic proof of valid title and ownership that can be produced in the court as *prima facie* evidence of ownership in case of any dispute or litigation. In some jurisdictions, copyright registration is a prerequisite for bringing a copyright infringement lawsuit. Therefore, when a work is to be used commercially; it is advisable to get copyright registration done. However, it is highly recommended to put a copyright notice on the work even if it is not registered.
2. Copyright registration is even more important in the case of computer software programs because the copied work looks identical to the original work and it is extremely difficult to differentiate between the two.
3. Copyright registration establishes a public record of the copyright claim.

4. A registration certificate is very important to have if it becomes necessary for the copyright owner to obtain a preliminary injunction against a copyright infringer, such as the immediate cessation of the distribution of the infringer's work. The presumption of validity will only apply if the work has been registered.

Therefore, even though the copyright registration is not mandatory, it is a good idea to get it done, especially when the work has to be used commercially. It is proof of ownership and valid title. The copyright can be registered at Copyright Registry, New Delhi.

16.6 WHAT IS COPYRIGHT INFRINGEMENT?

Anyone who exploits any of the exclusive rights of copyright without the copyright owner's permission commits copyright infringement. Infringement could be either literal or nonliteral. Literal infringement is the exact copying of original work and it can be verified by comparing original and copied work, while "non-literal" copying is paraphrased, or loosely paraphrased.

In order for a court to determine that a copyright in a work has been infringed, it must find that:

1. The infringing work is "substantially similar" to the copyrighted or the original work, and
2. The alleged infringer had access to the copyrighted work, meaning they actually saw it or heard it. There are no clear rules for deciding when "substantial similarity" exists between the two works. Courts look for similarities in appearance, sound, words, format, layout, sequence, and other elements of the works.

In case of infringement, copyright owner can enforce their rights in the form of:

- *Civil remedies*: Injunction, damages, or account of profit.
- *Criminal remedies*: Imprisonment, fine, or both.

16.6.1 SUBSTANTIAL SIMILARITY IN COPYRIGHT

Copyright infringement is violation of the exclusive rights of the copyright holder, the unauthorized or prohibited use of works under copyright such as the right to reproduce or perform the copyrighted work, or to make derivative works. To establish copyright infringement in court of law, a copyright owner must establish proof of copyright ownership and proof of copying by direct evidence of copying or by indirect evidence showing access to the original work, and "substantial similarity" between the original and allegedly infringing work.

In case of any copyright infringement, the plaintiff (the party who initiates the lawsuit) must prove that the defendant's (a person or party against whom an action or claim is brought in a court) work is "substantially similar" to the plaintiff's work. Hence, the infringement test involves two important components. First, did the defendant actually copy the plaintiff's work? And secondly, whether the copied elements would protect the expression and is sufficiently important to be actionable. In simple words, the aim of the test is to determine if the copying constitutes any infringement.

16.6.2 ACCESS: NECESSARY TO PROVE IN INFRINGEMENT MATTERS

To prove copyright infringement, the infringer shall have access to the plaintiff's work. In this context, access means whether the infringer had a suitable opportunity to witness the original work or not. Therefore, while determining infringement, the courts often compare all possible elements of both the works created by the plaintiff and the defendant & exclude all public domain elements from work and look only to the key elements that are protectable. Hence, in establishing the protectable elements, the court would distinguish between the idea underlying in the work created and its expression. For that reason the term "substantial similarity" causes confusion in the copyright infringement analysis because the same term has different meanings at two different points in the infringement analysis.

An interesting copyright infringement battle was fought in an Indian court (*Barbara Taylor Bradford & Anr vs Sahara Media Entertainment Ltd*), which delayed the telecast of the famous teleserial "Karishma: The miracle of destiny" by two months. Barbara Taylor Bradford, the well-known New York based novelist, sued Sahara TV, alleging that its television series "Karishma: The miracle of destiny" infringed copyright on her novel. Mrs. Taylor argued that she had never authorized the Sahara to make or produce any serial or film based on the novel and that the series in question amounted to a reproduction of her copyright work. Sahara TV argued that the story was the original work of Hindi film writer Sachin Bhowmick. After lengthy argument, the Calcutta High Court dismissed Mrs. Taylor's appeal of interim injunction restraining Sahara TV from broadcasting, and allowed the telecast of the serial.

The court in this case tried to draw a thin line between the idea and the expression. The Indian copyright act gave protection to the "expression" of the idea and concept, and not to the "idea" and "concept" itself. Mrs. Taylor appeared to be seeking protection for the idea or concept rather than the expression because she does not have a property over the story of a woman who achieves fame against all odds. The court opined that the exclusive right granted under the copyright act is infringed only if a substantial part of the work is reproduced. As previously mentioned, the court finally decided in favor of Sahara TV.

16.7 FAIR USE

Reproduction of copyrighted work for news reporting, criticism or comment, teaching, scholarship or research, in connection with judicial proceeding, and performance by an amateur club or society if the performance is given to a non-paying audience, is not considered infringement in law. The Fair Use Doctrine provides for limited use of copyrighted materials for educational and research purposes without permission from the owners. Reproduction of the work is not considered as infringement if used:

1. For the purpose of research or private study
2. For criticism or review
3. For reporting current events
4. In connection with judicial proceeding
5. Performance by an amateur club or society if the performance is given to a non-paying audience
6. The making of sound recordings of literary, dramatic, or musical works under certain conditions

CHAPTER

TRADEMARK

17

Bindu Sharma
Origiin IP Solutions LLP, Bengaluru, Karnataka, India

A trademark is extremely important for businesses as it identifies and distinguishes the goods/services of one entity from the other. It is a symbol of goodwill and indicates the source of goods or services. For a customer, a trademark makes purchasing decision easy and guaranties consistency in quality. Logos and trademarks have a significant role to play in the advertisement of goods and services and help companies to create their image in the market. Some famous Trademarks include Nike, Reebok, Coca-cola, Pepsi, etc.

Trademarks may be a single word or a combination of words, letters, numerals, etc. Trademarks are especially important when consumers and producers are far away from one another. Children ask for Barbie dolls, Lego building blocks, and Hot Wheels toy cars. Some adult's dream of Ferrari automobiles, but more can afford to buy Toyota or Honda brands. These consumers need trademarks to seek or avoid the goods and services of particular firms.

Selection of the right trademark for a business and formulating logo or tag line is a foundation and starting point. It is always advisable to avoid generic, suggestive, or descriptive words. Fancy names as trademarks are considered to be the strongest type of trademark. The fancy trademark could be combination of two words that do not result in a word that is already been used by other business, and which have a dictionary meaning. For example, the trademark, Kodak, Inalsa etc., are fancy and distinct marks.

Arbitrary trademarks are names that use a word which does not relate to the product or service for which it is being used. For example, usage of the word "Apple" with respect to the Computer, Blackberry for cell phone, Mango for garments, Ivory for soap. Arbitrary marks, like fanciful marks, are considered to be distinct and strong marks.

A suggestive trademark is a kind of trademark where the name gives indication or suggestion about the product or services in relation to which the trademark is being used. The trademark describes traits and characterizes the product. For example, the mark All-Clear and Junior are suggestive in nature. All-clear indicates that it is used for cleaning and Junior indicates that it is a product that is meant for children.

Another category of trademark is the descriptive mark, which is very similar to the suggestive mark. The difference between the two is that where the suggestive mark makes a person think and imagine the product or services in relation to whom the trademark is used, the descriptive mark doesn't leave any room for imagination but clearly describes the product. For example, Uncle Chips.

The trademark is not registered by the trademark office if such a trademark is likely to deceive the public or cause any kind of confusion. Nor if the mark is likely to hurt religious susceptibility

Pharmaceutical Medicine and Translational Clinical Research. DOI: http://dx.doi.org/10.1016/B978-0-12-802103-3.00018-3

of any section of citizens of India, or if it comprises of scandalous or obscene matter, or if its use is prohibited under the Emblems and Names Act 1950.

Before finalizing the trademark to be used for a business, it is critical to perform a trademark search to find out if the trademark selected by you is already in use or it is confusingly similar to an existing mark. It is also important to know which class your trademark falls into. A class specific and general search is important to be performed to assess the strength of your mark, and increases the chances of getting it granted by trademark office without much objection. The basic concept is to have a unique or fancy name and avoid descriptive and suggestive names.

17.1 INFRINGEMENT OF TRADEMARK

Infringement of the trademark takes place when the same or a similar trademark is used by any third party(ies), and such infringement usually involves likelihood-of-confusion or dilution of trademark. Likelihood-of-confusion is the term used when the infringed trademark is capable of confusing the consumer or both marks are confusingly similar to each other. The trademark office would often object to the grant of a trademark that is confusingly similar to any existing trademark.

In *Horlicks Limited & Ors.* vs. *Kartick Sadhukan*, Delhi High Court 2002 (25) PTC 126 Del, the trademark "HORLICKS," was registered in India. This mark has been used with respect to foods for infants, children, and biscuits. Another company called Kartick Confectionery started manufacturing similar products under the brand "HORLIKS" infringing the trademark rights enjoyed by "HORLICKS." Both marks are very similar to each other and are likely to confuse the consumer, specifically because the products are similar in nature. A Single Judge Bench of the Delhi High Court comprising of Justice B Chaturvedi found that HORLICKS is an old and registered brand with lots of goodwill and reputation in the market. The Court barred Kartick Confectionery from using the mark.

Dilution of trademark is a very interesting form of infringement where a well-known mark is infringed. This usage of trademark may not affect the business of a well-known brand but it may dilute the trademark. Dilution of trademark typically occurs as the result of blurring or tarnishing of the famous mark and dilution of trademark results in weakening of distinctiveness and reputation of the well-known trademark, whereas tarnishing means weakening of the trademark by using it with respect to the sexual or other offensive things. Blurring generally indicates weakening of the well-known mark. In such cases, interest of the well-known trademark can be protected by taking appropriate legal action and preventing dilution.

In the *Daimler Benz Aktiegesellschaft vs. Hybo Hindustan* case, the defendant used the word BENZ for an underwear product. BENZ is a famous mark, and usage of this mark with respect to any product or services is likely to dilute this mark, and such dilution shall not be permitted. The judge stated in the judgment as follows:

> I think it will be a great perversion of the law relating to Trade Marks and Designs, if a mark of the order of the "Mercedes Benz", its symbol, a three pointed star, is humbled by indiscriminate colorable imitation by all or anyone; whether they are persons, who make undergarments like the defendant, or anyone else. Such a mark is not up for grabs—not available to any person to apply upon anything or goods.

17.2 REGISTRATION OF TRADEMARK IS NECESSARY

Typically, the process of trademark registration starts with preparing and filing an application for the trademark in the national office. Before filing the application, it is highly recommended to perform a trademark search to assess the strength of the trademark. It should not be confusingly or deceptively similar to any existing mark.

Upon filing application for the trademark, the application number is allocated by the trademark office. Following the filing of the application, publication takes place, followed by examination of the trademark to ensure that the application has been filed in accordance with the statute, with assessment of the distinctiveness of the mark with respect to the existing marks. During examination, there may be a few objections raised by the trademark office which need to be replied to within specified period of time. If the trademark office is satisfied with the reply submitted, the trademark is granted and a certificate of registration is received by the applicant. After a trademark is accepted for registration, it is published in the trademark journal and is open for opposition. If a trademark is opposed, the trademark is not registered until opposition proceedings are completed.

Objection to the grant of a trademark by the trademark office could be relative or absolute. The relative ground of refusal is when the trademark is similar or identical to an earlier trademark for the same or similar goods/products or services or it is similar or identical to an earlier trademark in respect to different goods/products or services. Absolute grounds for refusal of a trademark grant are that the trademark lacks any distinctive character, or it consists exclusively of the marks or indications which may serve in trade to designate the quantity, intended purpose, values, geographical origins, or the time of productions of rendering of the service or other characteristics of the goods or services.

Trademark protection, unlike other forms of intellectual property, is perpetual provided the trademark is renewed after every ten years. It is interesting to note that after the trademark is registered, it shall be used for the trade. Inability to use the trademark for 5 years from the date of registration may result in losing rights over the trademark.

17.3 TRADEMARK MARKING

When the name is to be used as trademark, it is necessary to write TM as superscript over the trademark. This could be an indication to the public that this name is being used by you as a trademark. TM can be used before filing application for registration of a trademark or before getting a trademark registration certificate. However, once the mark is registered, the applicant shall write ® over the trademark as superscript. This marking helps to distinguish the trademark from other text that might be there in a document.

17.4 TRADEMARK AND DOMAIN NAME

In today's digital world, no business is complete without its presence on Internet by having a website or an e-commerce portal. It is obvious for companies to use trademarks (or word marks) as a domain name. This makes the domain name as important as a trademark.

In *Dr. Reddy's Laboratories Limited* vs. *Manu Kosuri & Another*, Delhi High Court 2001 (58) DRJ 241, it was established that having registered the trademark can restrict third parties from using the same name as a domain name. Dr. Reddy's Lab is a well-known pharmaceutical company with lots of goodwill and reputation and its trademark is registered. A person, Manu Kosuri, registered the domain name "drreddyslab.com," which comprised of trademark of Dr. Reddy's Lab. He registered this domain name primarily for Internet-related business. It is evident that this domain name is confusingly similar to Dr. Reddy's Lab and likely is to confuse the consumer. Dr. Reddy's Lab filed for a suit seeking a permanent injunction restraining Manu Kosuri from using the domain name 'drreddyslab.com' or any other domain name which is similar or identical to the Dr. Reddy's Lab trademark.

Based on reputation and goodwill acquired by Dr. Reddy's Lab, single Judge Bench of the Delhi High Court stated that the function of a domain name is very similar to a trademark and often the consumer can guess the domain name. With time, the usage of the Internet has increased and the usage of Dr. Reddy's Lab trademark as a domain name might confuse the consumer and damage the reputation of Dr. Reddy's Lab.

Accordingly, Manu Kosuri was restrained by a permanent injunction from registering a domain name or operating any business on the Internet and elsewhere under the domain name "drreddyslab.com" or any other domain name which is identical or similar to Dr. Reddy's Lab trademark Dr. Reddy's Lab. Manu Kosuri was also ordered to pay the cost of suit to Dr. Reddy's Lab and other profits that they made as a result of using the challenged domain name.

CHAPTER

TRADE SECRET

18

Bindu Sharma
Origiin IP Solutions LLP, Bengaluru, Karnataka, India

Trade secrets generally give the business a competitive edge over its rivals. Almost any type of data, process, or information can be referred to as a trade secret so long as it is intended to be and be kept a secret, and involves an economic interest for the owner. For example, a business may have certain internal business processes that it follows for its day-to-day operations that give it an edge over its competitors. This could be regarded as a trade secret.

The trade secret is probably the oldest form of IP protection. In Roman times, the law afforded relief against a person who induced another's employee (slave) to disclose secrets relating to the master's commercial affairs. The modern trade secret law evolved in England in the early 19th century in response to the growing accumulation of technology and know-how and the increased mobility of employees.

In Williams vs. Williams, a son sold medicines for his own account, although he had prepared them from the formulas given to him by his father on the understanding that the two would use the formulas for their joint benefit. The Chancery trial court issued an injunction restraining the son from using or divulging the trade secret and from selling the medicines. Even on appeal, the court said that the son cannot breach the contract set up by his father, but the court did not go into the injunction much because the formula had already been given out.

The Agreement on Trade-related Aspects of Intellectual Property Rights (TRIPS) under the auspices of the World Trade Organization lays down the following criteria for regarding any information as undisclosed information (or trade secrets):

- It must not be generally known or readily accessible by people who normally deal with such types of information.
- It must have commercial value as a secret. For example, the Coca-Cola formulation.
- The lawful owner must take reasonable steps to keep it secret.

Intellectual property being intangible in nature requires registration in order to obtain ownership. One of the steps involved in the registration of IP is that at the time of registration, the owner has to disclose IP which they wish to protect and register. For example, patent specification insists inventor to disclose the best mode of working the invention, copyright registration of the work requires submission of a copy of the work. Trade secrets and confidential

Pharmaceutical Medicine and Translational Clinical Research. DOI: http://dx.doi.org/10.1016/B978-0-12-802103-3.00019-5

information are very unique forms of IP that may lose their value once disclosed and hence are not registered. Therefore, the only way to protect them is to keep them confidential and not to disclose them to anyone. Hence it becomes extremely important to take precautionary measures to protect all such IP.

18.1 LEGAL FRAMEWORK IN INDIA

There is no specific law in India that protects trade secrets and confidential information. Nevertheless, Indian courts have upheld trade secret protection on the basis of principles of equity, and at times, upon a common law action of breach of confidence, which in effect amounts to a breach of contractual obligation.

The law of trade secrets is derived from the basic principles of the law of torts, restitution, agency, quasi-contract, property, and contracts. There are sound economic reasons for trade secret protection. Failure to protect obligations of confidentiality could inhibit both the quantity of information exchanged and its quality. The "owner" of the secret will in most cases have expended substantial resources to discover the secret and hence has a clear economic interest in its remaining secret. In addition, the economic rationale behind such protection is to offer an incentive to invest in the creation of information. In India it is possible to contractually bind a person not to disclose any information that is revealed to them in confidence.

18.2 CONFIDENTIAL INFORMATION VERSUS TRADE SECRET

The terms "confidential information" and "trade secret" are used synonymously and are usually confused with each other. Confidential information is the information which company would like to keep secret. It may or may not be essential for business. In other words, all trade secrets are confidential in nature, but all confidential information cannot be termed as a trade secret. Confidential Information is generally limited to a single event in the conduct of a business, whereas a trade secret is a process or device for continuous use in the operation of a business.

Merging or acquisition of the company is confidential information and may be related to a single event. The day merging or acquisition has been announced formally, this information is no more confidential information. In contrast, the use of a combination of yeast strains to make softer bread may be a trade secret as it is a process to be used continuously in the operation of a business.

18.2.1 EXAMPLES OF TRADE SECRET

- The fish medicine for chronic asthma patients, a traditional treatment known only to Bathini Gowd brothers of Hyderabad. The medicine is given out only once a year on an auspicious day determined by astrological calculations.

- The Coca-Cola formula is The Coca-Cola Company's secret recipe for Coca-Cola. As a publicity marketing strategy started by David W. Woodruff, the company presents the formula as a closely held trade secret known only to a few employees.
- Recipes for other soft drinks of Coca-Cola and other products—such as KFC chicken and McDonald's special sauce—are also closely-guarded trade secrets, but the Coke formula is the oldest.

Its important to note that confidential information should always be clearly communicated to be "confidential." Such information cannot be proved to be so, if not communicated to be confidential, and if appropriate measures have not been taken to protect it.

But on the other hand do remember that even if the document is not labeled or communicated as "confidential," and from its contents it looks confidential to an educated person like you, it is your implied duty to maintain its confidentiality and safeguard it.

18.3 PECULIAR FEATURES

All trade secrets are confidential in nature, but all confidential information may or may not be a trade secret. Peculiar features of trade secret are:

- Usually there may or may not be any defined legal framework to protect it, but misappropriation (by unauthorized passion, disclosure, or communication) may be an offence. In India there is no legal framework to protect it, but it is protected under contract law or law of torts, etc.
- It is not disclosed like other forms of IP and it loses its value in the market once the secret is disclosed and there is no way to retrieve it back. Since it cannot be disclosed, it cannot be registered.
- It is very crucial for business (trade secret). There are numerous examples of valuable trade secrets, such as the Google AdWords algorithm, Green Chartreuse liqueur protected by confidential information of the ingredients, etc.
- The duration not fixed as in the case of patent or copyright. The life of a trade secret lasts as long as it is kept a secret. Once the secret is out, the value is lost forever.
- It is easily misappropriated because it represents nothing more than information, which can be memorized, scribbled down, e-mailed, or copied onto some tangible medium and then quietly removed from company premises.
- Independent creation of a trade secret is not infringement. If company X has a trade secret and company Y creates an identical trade secret independently without stealing it from company X, there is no infringement.
- Once a trade secret is disclosed in public, its value is lost forever. One cannot "unring the bell."
- It's often overlooked as an intellectual property asset and most of the time companies fail to identify and assess its value properly. A Periodic IP audit may be extremely helpful to identify new innovations and the decision to keep them a secret or file for a patent may be taken. This process may help to identify and document a trade secret. This also makes the process and steps to maintain secrecy better and more stringent.

- The creation of trade secret is never announced in public like other forms of IP. In fact the creation of a trade secret is often not realized by the organization. It may be created accidentally too. Henry Ford insisted that parts for his cars be delivered to his factory in a wooden crate of a particular size. Later, a piece of that crate was the perfect size for the floorboard in the Model A. Henry figured out how to get a car part for free, and that is a trade secret.
- Independent creation or reverse engineering of a trade secret is not an offence.
- The validity of a trade secret is only proved in case of litigation. One can create a process, and consider it a trade secret. Since it is not disclosed, there is no way to validate it by third parties.

18.4 REVERSE ENGINEERING OF A TRADE SECRET

It is interesting to note that unlike other forms of IP, a trade secret can be reverse engineered or back engineered and such reverse engineering is not an offence. Therefore, companies have to opt to protect innovation in the form of a trade secret very carefully. In other words, we can say that if the process can be easily reverse engineered, it may not be a strong trade secret and the competitors can easily replicate the same. Often companies find it tough to decide between patent and trade secret. The advantage of protecting innovations in the form of a trade secret is that it doesn't need to be filed for registration formally and hence unlike patents, there is no attorney or government expense. The term of a patent is usually 20 years from the date of filing but a trade secret remains a trade secret as long as secrecy of the same is maintained. However, protecting the secrecy of the trade secret and preventing it from reverse engineering may be a big challenge.

18.5 HOW DO ORGANIZATIONS PROTECT TRADE SECRETS?

Protection of a trade secret is a very tricky business. Organizations have to be very smart and should have unique and effective ways to protect the same. Execution of a Non-Disclosure Agreement (NDA) with a strict and leak-proof clause on confidentiality is one of the approaches the organization uses to avoid misappropriation of trade secrets by employees, external vendors, and consultants.

Apart from execution of agreement, physical methods are also effective in protecting the trade secret. These methods include classifying the trade secret into low, moderate, and high confidential categories and accordingly limiting/restricting access to them. If too many people know about the trade secret, it could prove to be a very risky affair. Whenever such information is accessed, a log book shall be updated with time and date of access. List of employees that can access such information has to be clearly maintained along with their photographs and employee ID, etc. Visitors and external parties shall be restricted from entering the unauthorized areas.

Physical security also plays a critical role wherein important documents can be saved in lockers with biometric authentication of the user at the time of accessing the documents and CCTV camera provisions. Additionally, a good IP policy in writing to provide guidelines on protection of the trade secrets and company policies should be implemented. Having an IP policy is not useful until the time contents of the policy are not simplified and communicated to the employees clearly.

Educating employees and sensitizing them about the protection of the trade secret plays a very important role and could prove to be very effective in the protection of the trade secret. Periodic external and internal sessions from the experts and knowledge of the latest case studies shall be made a regular practice. Labeling the confidential document as "Confidential" to serve as a legal notice describing its nature.

At the time of resignation or termination of the employees, it is a good idea to have a discussion about the confidentiality agreement that they signed at the time of joining the company. This discussion can become part of the exit interview and help them recall obligations and duties. Other than this, employees shall also be asked to return all information that they possessed in tangible form, like documents, hard disk, laptop, etc.

Though the best way to protect a trade secret is to prevent disclosure, it is often essential to share it to third parties in order to commercialize or exploit it for business benefit. The situation is even tougher when the trade secret is required to be licensed to third parties along with the know-how. Prevention is always better than cure and hence one has to be extremely careful while dealing with trade secrets.

Misappropriation of a trade secret is a severe offence and hence employees shall take precautions to safeguard it. Reading the confidentiality clause of the agreement(s) signed with the employer is important in understanding obligations and duties. Take all precautions not to incorporate the prior employer's confidential information in your new company projects. Be careful while including work experience in the resume and talking to people about your work experience. As a responsible employee, refrain from unauthorized acts while handling confidential information and be careful about any disclosure during interviews as well.

18.6 SOME CASES

1. American Express Bank Ltd. vs. Ms. Priya Puri ((2006)III LLJ 540(Del))

The plaintiff in this case is a banking company and the defendant was working as a wealth management head covering the Northern Region. It was alleged by the plaintiff that the defendant quit her job and joined the competitor's bank, also using the confidential data of the plaintiff company to attract clients to the competitor's bank. Thereby, the plaintiff approached the court seeking a permanent injunction to stop the defendant from disclosing and using the confidential data of the plaintiff's bank. The plaintiff also submitted that this action of the defendant constituted breach of confidentiality obligations present in the employment letter and violated the code of conduct in handling the sensitive client details.

The defendant argued that while working as a relationship manager, she had gotten to know the details of the clients, such as names and phone numbers. Moreover, the details such as these names and phone numbers are freely available in the public domain. Thereby, this detail of the client is not confidential in nature and hence will not amount to breach of confidentiality.

The Delhi High Court held that just by merely possessing the names and phone numbers of clients the defendant could not make the clients leave the plaintiff's company. The court also held that the decision to switch banks lies with the customer and observed the following:

> The option of the customer/clients to bank with anyone cannot be curtailed on the plea of confidentiality of their details with any particular bank. Creating a database of the clients/customers and then claiming confidentiality about it will not permit such bank to create a monopoly about such customers that even such customers cannot be approached.

2. Mr. Anil Gupta and Anr. sv. Mr. Kunal Dasgupta and Ors (97 (2002) DLT 257)

The plaintiff in this case came up with the idea of producing a reality television program involving the selection of a spouse, with the participants in this television show being the real persons before a TV audience. The plaintiff has named the concept as "Swayamvar" and the same was registered under the Copyright Act. The plaintiff developed the concept under his production house itself and communicated the same to the defendants explaining various details of the program.

After a few months the plaintiff was shocked to find that the defendants promoting a similar program named "Shubh Vivah" that is similar to the concept of the reality show explained by the plaintiff to the defendants.

The plaintiff filed a case, contending that this resulted in breach of confidentiality and copyright Infringement.

The defendants argued that there were no breach of confidence and copyright since the idea behind the reality show is in the public domain. It was also contended that the information disclosed by the plaintiff is a just a preliminary detail with no references to the actual reality show program.

The court kept in mind the fact that a concept note was seen by the defendants and held that defendants cannot claim that there is no confidentiality obligation involved. It restrained the defendants from broadcasting the show for a period of four months from the date of the order, and the court also said that if within these four months no television program is telecasted in the name of "Swayamvar" then the defendants were free to broadcast.

3. John Richard Brady and Ors vs. Chemical Process Equipments P. Ltd. and Anr (AIR 1987 Delhi 372)

The plaintiff in this case disclosed technical material, detailed know-how, drawings and specifications concerning his Fodder Production Unit (FPU) under strict confidentiality to the defendants to manufacture all the components of FPU precisely. Since the plaintiff was not happy with the quality of the components and the inability of them to meet the requirements, he canceled the orders. During the process the defendants had visited the plaintiff's facility to understand the FPU, at some time without the permission of the plaintiff.

The plaintiff later, after terminating the order, found that the defendants were developing their own version of the FPU by false disclaimers that the invention originated from them. The plaintiff then filed a suit in Delhi High Court seeking for an injunction restraining the defendants from manufacturing and selling the FPU machines, which substantially resembled the plaintiff's product.

The defendants contended that they had not infringed the copyright of plaintiff nor were they liable for confidentiality breach. They further contended that there are several companies in the market who are involved in making the same products and the FPU machine is based on a long-known concept of the Hydroponic System.

The Single Judge Bench of the Delhi High Court held that in the interest of justice, the defendants had to be restrained from using the know-how, specifications, drawings, and other technical information regarding plaintiff's FPU. Otherwise, the plaintiff would be put to irreparable injury and loss by the actions of the defendant.

CHAPTER

DATA EXCLUSIVITY

19

Gursharan Singh
Life Sciences, SmartAnalyst India Private Limited, Gurgaon, Haryana, India

19.1 INTRODUCTION

Apart from the patents discussed in earlier chapters, there are two additional types of protection that serve to provide incentives to compensate the innovator company for the investment it has made in the development and approval of innovative pharmaceutical products. These are data exclusivity and market exclusivity.

Data exclusivity is the period of nonreliance and nondisclosure during which the preclinical and clinical data generated by the innovator company as part of their New Drug Application/Marketing Authorization Application, (also known as pharmaceutical registration data), cannot be referenced in the regulatory filings of another company for the same drug substance. Unlike patents, data exclusivity does not prevent another company from generating the data. While some drugs have both patent and exclusivity protections, there are others which will have either or neither type of protection. Patents and exclusivity do not always run concurrently as patent application is filed much earlier than marketing authorization applications. Moreover, patents and exclusivity do not always encompass the same claims [1,2].

Market exclusivity is the additional period of time beyond data exclusivity during which a generic company may not market an equivalent generic version of the originator's pharmaceutical product, although during this period their application for authorization may be processed, so as to enable them to market their product on the expiry of this additional period.

19.2 TRIPS AND DATA EXCLUSIVITY

The legal basis of data exclusivity originates from the Trade-Related Aspects of Intellectual Property Rights (TRIPs) Agreement, which was negotiated as part of the Uruguay round of trade negotiations in the General Agreement on Tariffs and Trade (GATT). Under the Article 39.3 of section 7 (protection of undisclosed information) of the TRIPs Agreement, all World Trade Organization (WTO) countries are required to provide registration data exclusivity periods. The article specifically recognizes the "protection of undisclosed information" as being a category of intellectual property subject to protection [3].

Pharmaceutical Medicine and Translational Clinical Research. DOI: http://dx.doi.org/10.1016/B978-0-12-802103-3.00020-1

> Members, when requiring, as a condition of approving the marketing of pharmaceutical or of agricultural chemical entities, the submission of undisclosed test or other data, the origination of which involves a considerable effort, shall protect such data against unfair commercial use. In addition, Members shall protect such data against disclosure, except where necessary to protect the public, or unless steps are taken to ensure that the data are protected against unfair commercial use [3].

Article 39 of the TRIPs Agreement itself has its origin in Article 10bis of the Paris Convention (1967) aimed at ensuring effective protection against unfair competition [3].

Despite their obligations under Article 39.3 of the TRIPS Agreement, as to date many countries do not provide adequate protection to proprietary registration data. Even amongst those countries which provide this protection, there are significant differences with regards to the implementation and the length of the period of data exclusivity. The United States (US), European Union (EU), and Switzerland are aligned that the governments should prevent regulatory authorities or third parties from relying on proprietary registration data without the originator's consent [1,3]. The data exclusivity provisions in different geographies are described below.

19.3 DATA EXCLUSIVITY IN THE UNITED STATES

In the US, exclusivity prevents the submission or effective approval of Abbreviated New Drug Applications (ANDAs) or applications described in Section 505(b)(2) of the Act. The period for which the Food and Drug Administration (FDA) protection to the proprietary registration data varies depends on the type of exclusivity. There are six types of exclusivities provided in the US [2].

Orphan Drug Exclusivity (ODE) is granted to drugs designated and approved to treat diseases or conditions affecting fewer than 200,000 in the U.S. (or more than 200,000 and no hope of recovering costs [2,4].

New Chemical Exclusivity (NCE) is granted to drugs that contain no active moiety that has been approved by FDA under section 505(b) [2,5].

New Clinical Investigation Exclusivity is granted to drugs when application or supplement contains reports of new clinical investigations (other than bioavailability studies) conducted or sponsored by an applicant and essential for approval [2,5].

Pediatric Exclusivity is granted when the sponsor has conducted and submitted pediatric studies on the active moiety in response to a written request from FDA [2,6].

Generic Drug or 180-Day Exclusivity is granted to the first company which submits an ANDA with the FDA has the exclusive right to market the generic drug for 180 days [2,7].

Qualified Infectious Disease Product Exclusivity is granted to certain exclusivity periods for products that have been granted a Qualified Infectious Disease Product (QIDP) designation (with some exceptions) [8] (Table 19.1).

Table 19.1 Data Exclusivity in the United States

S. No.	Type of Exclusivity	Period	Regulation/Statute
1	NCE exclusivity	5 years (or 4 years if para. IV)	21 CFR 314.108(b)(2)
2	New clinical study/ investigation exclusivity	3 years	21 CFR 314.108(b)(4) and (5)
3	Orphan drug exclusivity	7 years	21 CFR 316
4	Pediatric exclusivity	6 months beyond any existing marketing or patent exclusivity	21 USC 355 A/ Best Pharmaceuticals for Children Act (BCPA) and Section 505(A) of the Food and Drug Administration Modernization Act of 1997
5	Generic drug exclusivity	180 days	21 USC 355(j)(5)(B)(iv)
6	Qualified infectious disease product exclusivity	5 years beyond certain exclusivity periods	Generating Antibiotic Incentives Now (GAIN) Title VIII of the FDA Safety and Innovation Act (FDASIA)

Sources: *[2,4–8].*

The relevant data exclusivity periods for each approved drug product, as well as any applicable patents on the product or its use, are listed in the 'Orange Book', which is published by the FDA and also available electronically on the FDA's website.

19.4 DATA EXCLUSIVITY IN EUROPEAN UNION AND EUROPEAN ECONOMIC AREA

In the European Union, data exclusivity is provided per the provisions of Article 14.11 of Regulation 726/2004, which applies to products submitted for approval via the Centralized Authorization Procedure (CAP) after November 20, 2005 [9,11].

On the other hand, data exclusivity for products submitted for approval via the Mutual Recognition Procedure (MRP), National Approval Procedure (NAP), or Decentralized Approval Procedure (DCP) after October 30, 2005 is provided per the provisions of Article 10.1 & 10.5 of Directive 2001/83 (last amended on 31st March, 2004) [10,11].

The European Union follows the 8 + 2 + 1 formula for data exclusivity. Under this formula, eight years of data exclusivity is granted to protect against filing of a generic application. Then an additional two-year exclusivity is granted to protect against marketing of the generic. This effective ten-year exclusivity period can be further extended by an additional one year of market protection if, during the first eight years of those ten years, the marketing authorization holder obtains an authorization for one or more new therapeutic indications for the drug and where the drug is shown to offer a significant clinical benefit in comparison with existing therapies [9,10,11] (Fig. 19.1).

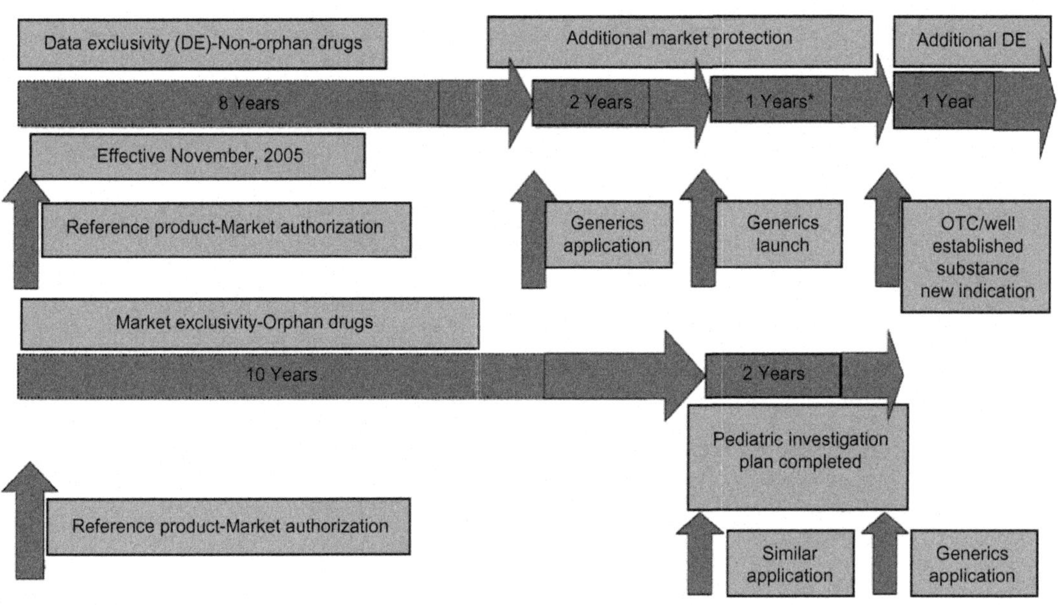

FIGURE 19.1

Data exclusivity and market protection in European Union-Orphan versus Non-Orphan Drugs. *If new indication is registered in first 8 years and brings significant clinical benefit over existing therapies.

Source: *Regulation (EC) No 726/2004 of the European Parliament and of the Council of 31 March 2004. http://ec.europa.eu/health/files/eudralex/vol-1/reg_2004_726/reg_2004_726_en.pdf [accessed 4.09.16]; Directive 2001/83/EC of the European Parliament and of the Council of 6 November 2001 on the Community Code Relating to Medicinal Products for Human Use. Last amended on March 31, 2004, http://www.ema.europa.eu/docs/en_GB/document_library/Regulatory_and_procedural_guideline/2009/10/WC500004481.pdf [accessed 4.09.16]; EMA. 2013 Data exclusivity, market protection and paediatric rewards. Workshop for Micro, Small and Medium Sized Enterprises. accessed online at http://www.ema.europa.eu/docs/en_GB/document_library/Presentation/2013/05/WC500143122.pdf on August 4, 2016.*

CASE STUDY: ADDITIONAL ONE YEAR MARKET PROTECTION IN EU

TORISEL (temsirolimus) was originally approved in Europe for Renal cell carcinoma. Later, marketing authorization for an additional indication "Treatment of adult patients with relapsed and/or refractory mantle cell lymphoma (MCL)" was obtained. Since at that time, there were no approved treatments for relapsed MCL in EU, additional one year market protection was granted

Source: EMA. 2013 Data exclusivity, market protection and paediatric rewards. Workshop for Micro, Small and Medium Sized Enterprises. accessed online at http://www.ema.europa.eu/docs/en_GB/document_library/Presentation/2013/05/WC500143122.pdf on August 4, 2016.

Apart from above, under Art. 10(5) Dir. 2001/83/EC, one year of data exclusivity which adds to 8 + 2 + 1 formula can be granted for a new therapeutic indication for a well-established substance, provided that significant preclinical or clinical studies were carried out in relation to the new indication. Similarly, additional one-year data exclusivity can be granted for a change in classification of a medicinal product (e.g., Over the Counter switch) on the basis of significant preclinical tests or clinical trials under Art. 74(a) Dir. 2001/83/EC [10,11].

Orphan Drug Exclusivity: The EU Orphan Regulation No 141/2000 provides for 10 years' market exclusivity for orphan drugs. This applies not only to the generic version of the reference product but also to a similar medicinal product which has been granted orphan designation for the same therapeutic indication [12].

Further, Regulation (EC) No 1901/2006 on medicinal products for pediatric use provides for extension of market exclusivity in the case of orphan drugs to twelve years if the requirement for data on use in the pediatric population is fully met [13].

Pediatric Exclusivity: While pediatric exclusivity for orphan drugs has been described above, in case of nonorphan drugs, Article 36 of Regulation (EC) No 1901/2006 provides for 6 months of SPC (Supplementary Protection Certificate) resulting in patent extension by 6 months. However, in case the new pediatric indication brings a significant clinical benefit in comparison with existing therapies, then in accordance with Article 14(11) of Regulation (EC) No 726/2004 or the fourth subparagraph of Article 10(1) of Directive 2001/83/EC, the applicant can apply for a one-year extension of the period of marketing protection for the medicinal product concerned. In this case, the SPC is not granted [13].

19.5 DATA EXCLUSIVITY IN JAPAN

Japan does not have an affirmative system for the protection of test data. Data exclusivity in Japan originates from a passive re-examination system (Article 14–4 and 23–29 of the Japanese Pharmaceutical Affairs Law). The reexamination system is aimed at reconfirmation of the clinical usefulness of drugs, through collecting information on the efficacy and safety of the drug during a specified period of time after approval. This system commenced in April 1980. Based on the revision of October 1993, the reexamination period for orphan drugs was extended to a maximum of

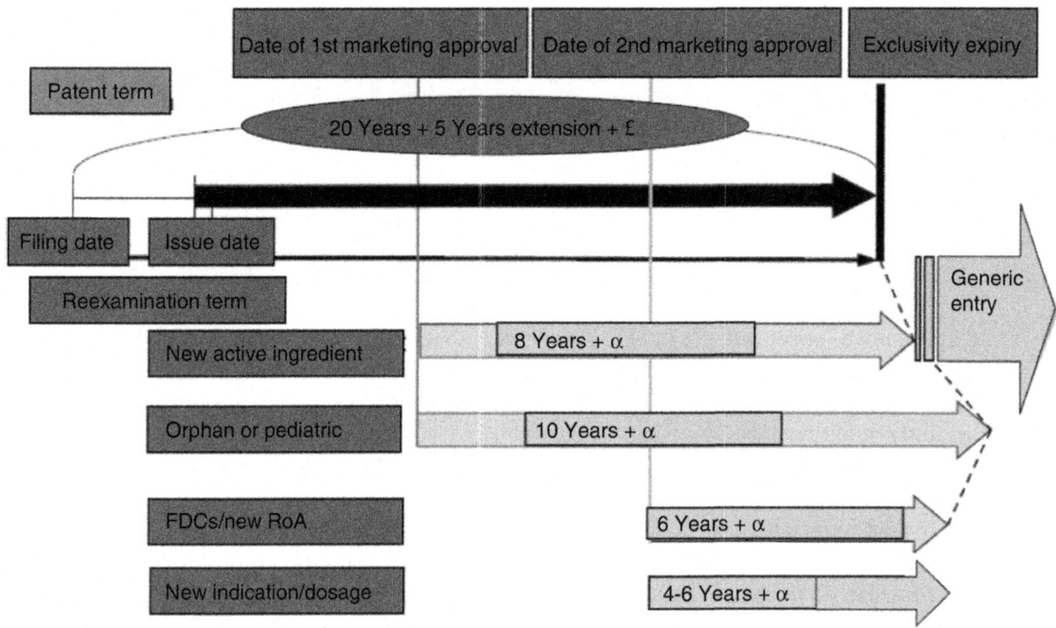

FIGURE 19.2

Data exclusivity and market protection in Japan.

Sources: *JPMA. Pharmaceutical Administration and Regulations in Japan. http://www.jpma.or.jp/english/parj/pdf/2015.pdf; 2015 [accessed 4.08.16]; JPMA. Drug Re-Examination/Data Exclusivity in JAPAN and Neighboring Countries. https://www.aippi.org/download/helsinki13/presentations/Pres_Pharma_4_YOkumura_300813.pdf; 2013 [accessed 4.08.16].*

ten years. The reexamination period for drugs with new active ingredients based on Notification No. 0401001 of the (Pharmaceutical and Food Safety Bureau) PFSB dated April 1, 2007, is eight years [14,15].

Applications for generic drugs cannot be filed until completion of the reexamination. In other cases reexamination period ranges from six years (Fixed dose combinations, new routes of administration) to four to six years (new indications, new dosages). When pharmacoepidemiological surveys or clinical studies for setting pediatric doses are performed, the reexamination period may be extended to ten years [14,15] (Fig. 19.2).

19.6 DATA EXCLUSIVITY IN OTHER GEOGRAPHIES

The data exclusivity in geographies other than those described above varies. Table 19.2 below captures the applicable law and the data exclusivity period for many of these geographies [16,17,18].

Table 19.2 Data Exclusivity Worldwide

S. No.	Country	Period of Data Exclusivity	Applicable Law/Comments
1.	Antigua & Barbuda	Not specified	Article 9(4), Protection Against Unfair Competition Act of 2006
2.	Argentina	Not specified	Articles 4 and 5, Law on the Confidentiality of Information and Products, No. 24,766
3.	Australia	5 Years	Data Exclusivity Provision of the Therapeutic Goods Act (Cth) 1989
4.	Bahrain	5 Years	Article 1–2, Law No. (7) for the Year 2003 on Trade Secrets
5.	Bolivia	Not specified	Article 266, Andean Community Decision 486 (2000)
6.	Brazil	Not specified	Article 195, Law 9.279 on Industrial Property; Title V, Crimes Against Industrial Property; Chapter VI, Protection Against Unfair Competition
7.	Canada	8 Years + 6 months[a]	Food and Drug Regulations, Section C.08.004.1 (as amended)
8.	Chile	5 Years	Articles 89–91, Law 19,039 on Industrial Property (as amended by Law 19,996), (implemented in part by Decree No. 153 by the Ministry of Health (2005) that is not reproduced)
9.	China	6 Years	Section 284, Report of Working Party on the Accession of China to the WTO (WT/ACC /CHN/ 49); Article 35, Regulations for Implementation of the Drug Administration Law of the People's Republic of China (Decree of the State Council No. 360)
10.	Colombia	5 Years	Article 1–5, Data Protection Decree No. 2085—September 19, 2002; Article 18–22, Data Protection of Chemicals used for Pharmaceutical/Purposes (Farmoquímicos) or Agroquemicals; Andean Community Decision 486 (2000); Article 266, Andean Community Decision 632 (2006)
11.	Costa Rica	5 Years	Article 9. Abbreviated Documentation for the Authorization of a Pharmaceutical Preparation; Article 10. Mandatory Licenses after the Expiration of the Period; Article 11. Definition of an Essentially Similar Medication; Article 12. Documentation of Data Exclusivity Compliance
12.	Croatia	6 Years	Article 15, Law on Medicines and Medical Products
13.	Dominican Republic	5 Years	Law 20–00 on Industrial Property; Article 181. Information and data protection for marketing approval DR-CA FTA – Art. 15.10(1)(b):
14.	Ecuador	Not specified	Article 286, Law No. 2006–13 on Industrial Property and Rights of the Author; Article 266, Andean Community Decision 486 (2000); Andean Community Decision 632 (2006)
15.	Egypt	5 Years	Undisclosed Information—Articles 55–62, Intellectual Property Law No. 82 (2002); Prime Ministerial Decree No. 2211

(*Continued*)

Table 19.2 Data Exclusivity Worldwide *Continued*

S. No.	Country	Period of Data Exclusivity	Applicable Law/Comments
16.	El Salvador	5 Years	Articles 181-A -E, Law on the Promotion and Protection of Intellectual Property; Decree No. 604 (1993) as amended by Article 103 of Legislative Decree No. 912 (2005)
17.	Guatemala	5 Years	Decree 57-2000, Law on Industrial Property, Articles 177 through 177(fifth), as amended.
18.	Honduras	5 Years	Article 19–24, Chapter: Protection of Undisclosed Data or Information, Title IV: Measures Related to Certain Regulated Products Sole, Decree No. 16-2006, Law on the Application of the Free Trade Treaty between the Dominican Republic, Central America and the United States
19.	Hong Kong	Not specified	Pharmacy and Poisons Regulations, Ordinance Cap 138
20.	India	NA	No Provision
21.	Iraq	5 Years	Patent, Industrial Design, Undisclosed Information, Integrated Circuits and Plant Variety Law. Applies only to approvals that were obtained on or after April 26, 2004 31) Chapter Three bis, Article 2
22.	Israel	5 Years/5½ Years	Chapter I: Health Amendment of the Pharmacists' Ordinance Commencement and Effect 5 years with regards to date of approval in Israel and 5 1/2 years with regards to date of approval elsewhere
23.	Jordan	5 Years	Article (8)—Trade Secrets and Unfair Competition Law, No. 15 (2000)
24.	Malaysia	5 Years	Directive of Data Exclusivity, Regulation 29 of the Control of Drugs and Cosmetics Regulations 1984 5 years for New Chemical Entity and 3 years for New Indication Drug product needs to be manufactured within 18 months (in case of NCE) and 12 months (in case of new indication) of approval
25.	Mexico	5 Years	Industrial Property Law (as amended) Article 86bis; Article 1711. Trade Secrets. NAFTA; Article 18–22 Data Protection of Chemicals used for Pharmaceutical Purposes (Farmoquímicos) or Agrochemicals, Treaty of Group of Three (Colombia, Mexico and Venezuela)
26.	Morocco	5 Years	Article 15.10: Measures Related To Certain Regulated Products
27.	New Zealand	5 Years	23B/ 23 C of Medicines Act 1981 No. 118
28.	Nicaragua	5 Years	Health Ministry Resolution No. 115-2006
	Oman	5 Years	Article 34, Sultanic Decree No. 38/2000, The Law of Trademarks, Trade Data, Undisclosed Trade Information and Protection from Unfair Competition; Law on Industrial Property and their Enforcement, Royal Decree 67/2008; Article 89, Regulations under the Law on Industrial Property Rights and their Enforcement for the Sultanate of Oman
29.	Panama	5 Years	Legislative Assembly Law No. 23, 2009 Section 7: Protection of Undisclosed Information Article 39

30.	Peru	5 Years	Legislative Decree 1072 Protection of Undisclosed Test Data or Other Undisclosed Data Related to Pharmaceutical Products; Article 1–3, Andean Community Decision 486 (2000); Article 266, Andean Community Decision 632 (2006)
31.	Republic Of Korea	4–6 Years	Article 32, Pharmaceutical Affairs Law (PAL); Article 25/ 27, KFDA Regulations regarding the Licensing, Report and Examination of Drug Products; Article 35, Ministerial Decree to PAL 6 years for NCE, new active ingredient, new composition ratio, new route of administration 4 years for new indication
32.	Russia	DE: 4 years (generics)/ 3 years (biosimilars) + 2 years (generics)/3 years (biosimilars) of marketing exclusivity	Paragraph 6 of the article 18 to the FZ-61 amended in December, 2014 w.e.f. 2016. 4 years data exclusivity for generics and 3 years data exclusivity for biosimilars. However, market launch of generic/biosimilar permitted only after 6 years thus giving additional 2 years/3 years of marketing exclusivity
33.	Saudi Arabia	5 Years	Article 5,6,8 of Decision No. 3218: Regulations for the protection of Confidential Commercial Information, later amended by Decision No. 4319 of 2005
34.	Singapore	5 Years	Medicines Act (Chapter 176)
	Switzerland	10 Years	Decree on Medications—Section 3, Article 17
35.	Taiwan	5 Years	Pharmaceutical Affairs Law, Article 40-1/40-2 February 5, 2005
36.	Trinidad And Tobago	Not specified	Protection Against Unfair Competition Act 27/1996
37.	Turkey	6 Years	Regulations on Licensing the Human Medical Products; Article 9—Abbreviated License
38.	Ukraine	5 Years	The Law of Ukraine—On Amending Article 9 of the Law of Ukraine "On Medicines"
39.	United Arab Emirates	Not specified	Ministers of Health and of Finance & Industry Decree No. 404/2000
40.	Venezuela	5 Years	Article 18–22, Data Protection of Chemicals used for Pharmaceutical Purposes (Farmoquímicos) or Agroquemicals, Treaty of Group of Three (Colombia, Mexico and Venezuela)
41.	Vietnam	5 Years	Intellectual Property Law Article 128—Obligations to maintain secrecy of data of tests

[a] *Pediatric Exclusivity.*

Sources: *IFPMA. Data Exclusivity: Encouraging Development of New Medicines. http://www.ifpma.org/wp-content/uploads/2016/01/IFPMA_2011_Data_Exclusivity__En_Web.pdf; 2011 [accessed 29.07.16]; BRIC WALL. Current State of Data Protection and Exclusivity in Russia. https://bricwallblog.com/2015/12/30/current-state-of-data-protection-and-exclusivity-in-russia/; 2015 [accessed 29.07.16]; Linking Patent and Data Exclusivity. http://www.lexology.com/library/detail.aspx?g = c6ec53ab-c22a-467e-b981-0b36f00611a4 [accessed 4.08.16]*

REFERENCES

[1] IFPMA. Encouragement of New Clinical Drug Development: The Role of Data Exclusivity. <http://www.who.int/intellectualproperty/topics/ip/en/DataExclusivity_2000.pdf>; 2000 [accessed 29.07.16].

[2] CDER/SBIA. Patents and Exclusivity. FDA/CDER SBIA Chronicles. accessed online at http://www.fda.gov/downloads/Drugs/DevelopmentApprovalProcess/SmallBusinessAssistance/UCM447307.pdf>; 2015 [accessed 29.07.16].

[3] Article 39, Section 7: Protection of Undisclosed Information, Agreement on Trade-Related Aspects of Intellectual Property Rights. <https://www.wto.org/english/docs_e/legal_e/27-trips.pdf>; [accessed 29.07.16].

[4] Orphan Drugs Part 316, Title 21, Subchapter D Electronic Code of Federal Regulations. <http://www.ecfr.gov/cgi-bin/text-idx?SID = b57e93376bfff0f6663e07e243d901f5&mc=true&node=pt21.5.316&rgn=div5> [accessed 04.08.16].

[5] New Drug Product Exclusivity, Part 314.108, Title 21, Subchapter D Electronic Code of Federal Regulations. <http://www.ecfr.gov/cgi-bin/retrieveECFR?gp=&SID=11996b27cfbe7f4cf3cdc3f62b010b04&mc=true&n=pt21.5.314&r=PART&ty=HTML#se21.5.314_1108> [accessed 04.08.16].

[6] 21 USC 355a: Pediatric studies of drugs. <http://uscode.house.gov/view.xhtml?req = (title:21 section:355a edition:prelim) OR (granuleid:USC-prelim-title21-section355a)&f=treesort&edition=prelim&num=0&jumpTo=true > [accessed 04.08.16].

[7] CDER. Guidance for Industry "180-Day Generic Drug Exclusivity Under the Hatch-Waxman Amendments to the Federal Food, Drug, and Cosmetic Act". <http://www.fda.gov/downloads/Drugs/.../Guidances/ucm079342.pdf>; 1998 [accessed 04.08.16].

[8] FDA Safety and Innovation Act. Title VIII— Generating Antibiotics Incentives Now (GAIN). <http://www.fda.gov/downloads/Drugs/DevelopmentApprovalProcess/SmallBusinessAssistance/UCM361320.pdf> [accesssed 04.08.16].

[9] Regulation (EC) No 726/2004 of the European Parliament and of the Council of 31 March 2004. <http://ec.europa.eu/health/files/eudralex/vol-1/reg_2004_726/reg_2004_726_en.pdf> [accessed 04.08.16].

[10] Directive 2001/83/EC of the European Parliament and of the Council of 6 November 2001 on the Community Code Relating to Medicinal Products for Human Use. Last amended on March, 31, 2004. <http://www.ema.europa.eu/docs/en_GB/document_library/Regulatory_and_procedural_guideline/2009/10/WC500004481.pdf> [accessed 04.08.16].

[11] EMA. Data exclusivity, market protection and paediatric rewards. Workshop for Micro, Small and Medium Sized Enterprises, accessed online at <http://www.ema.europa.eu/docs/en_GB/document_library/Presentation/2013/05/WC500143122.pdf>; 2013 [accessed 04.08.16].

[12] Regulation (EC) No 141/2000 of the European Parliament and of the Council of 16 December 1999 on orphan medicinal products. <http://ec.europa.eu/health/files/eudralex/vol-1/reg_2000_141/reg_2000_141_en.pdf> [accessed 04.08.16].

[13] Regulation (EC) No 1901/2006 of the European Parliament and of the Council of 12 December 2006 on medicinal products for paediatric use and amending Regulation (EEC) No 1768/92, Directive 2001/20/EC, Directive 2001/83/EC and Regulation (EC) No 726/2004. <http://ec.europa.eu/health/files/eudralex/vol-1/reg_2006_1901/reg_2006_1901_en.pdf> [accessed 04.08.16].

[14] JPMA. Pharmaceutical Administration and Regulations in Japan. accessed online at http://www.jpma.or.jp/english/parj/pdf/2015.pdf>; 2015 [accessed 04.08.16].

[15] JPMA. Drug Re-Examination/Data Exclusivity in JAPAN and Neighboring Countries. <https://www.aippi.org/download/helsinki13/presentations/Pres_Pharma_4_YOkumura_300813.pdf>; 2013 [accessed 04.08.16].

[16] IFPMA. Data Exclusivity: Encouraging Development of New Medicines. <http://www.ifpma.org/wp-content/uploads/2016/01/IFPMA_2011_Data_Exclusivity__En_Web.pdf>; 2011 [accessed 29.07.16].
[17] BRIC WALL. Current State of Data Protection and Exclusivity in Russia. <https://bricwallblog.com/2015/12/30/current-state-of-data-protection-and-exclusivity-in-russia/>; 2015 [accessed 29.07.16].
[18] Linking Patent and Data Exclusivity. <http://www.lexology.com/library/detail.aspx?g = c6ec53ab-c22a-467e-b981-0b36f00611a4> [accessed 04.08.16].

SECTION

GENERICS, SUPERGENERICS, BIOLOGICS, BIOSIMILARS, AND BIOBETTERS

CHAPTER

GENERIC DRUG AND BIOEQUIVALENCE STUDIES

20

Rajinder K. Jalali[1] and Deepa Rasaily[2]

[1]*Sun Pharmaceutical Industries Ltd, Gurgaon, Haryana, India*

[2]*Sun Pharmaceutical Industries Ltd, New Delhi, India*

20.1 GENERIC DRUG

20.1.1 INTRODUCTION

A drug is any substance that is intended for use in the diagnosis, cure, relief, treatment, and prevention of disease. According to the US Food and Drug Administration (US FDA), a generic drug is identical—or bioequivalent—to a brand name drug in dosage form, safety, strength, route of administration, quality, performance characteristics, and intended use. According to regulatory definitions, generic drug products need to be identical to their reference with respect to the active substance, the route of administration, as well as quality standards [1]. Generic drugs are therefore equally effective and safe as branded drugs. Generic drugs are produced and distributed without patent protection. Generic drugs are available once the patent protections afforded to the original branded drug have expired. Innovator drugs have to demonstrate their clinical efficacy and safety, whereas generics are considered therapeutically equivalent based on simple bioequivalence testing.

The rising cost of medication has been contributing to the rising cost of healthcare. As a result of the huge economic burden on healthcare cost, the use of generic drugs is steadily increasing. With the introduction of generic drugs, the market competition leads to lower prices for both the branded innovator drug and the generic drugs. The generic drug therefore costs less than their original brand equivalents. One of the strategies for lowering the cost of medication, and thereby reducing its contribution to total health care costs, has been the introduction of generic equivalents of brand name drugs/innovator drugs [2]. Generic drugs, therefore, provide the opportunity for savings in healthcare cost [3]. Generics account for more than 80% of prescription drugs in the US, and still continue to grow [4].

20.1.2 HISTORICAL PERSPECTIVE

Generics have existed throughout the history of the pharmaceutical industry. The 1984 the US Drug Price Competition and Patent Restoration (Hatch-Waxman) Act was the decisive moment in the development of the generics industry in the pharmaceutical market. The Hatch-Waxman

Pharmaceutical Medicine and Translational Clinical Research. DOI: http://dx.doi.org/10.1016/B978-0-12-802103-3.00021-3

Amendments created section 505(j) of the Act (21 U.S.C. 355(j)). Section 505(j) established the abbreviated new drug application (ANDA) approval process, which allows generic versions of previously approved innovator drugs to be approved and brought on to the market. The Hatch-Waxman Act provided for facilitated market entry for generic versions of all post-1962 approved products without having to prove safety and efficacy. With this, started the generic competition of pharmaceutical products, creating the modern generic pharmaceutical industry [5]. In the first year after the introduction of the Hatch-Waxman Act, the FDA received more than 1000 applications for approval of new generic drugs [6]. Since then, using the BE as the basis for approving generic drugs was established.

20.1.3 PREREQUISITES FOR GENERIC DRUGS

Generic drugs must meet the same standards as the innovator drug to gain FDA approval. Generic drugs, however, do not need to contain the same inactive ingredients as the brand name drug [7].

A generic drug must:

- Contain the same active ingredients as the innovator drug (inactive ingredients may vary)
- Be identical in strength, dosage form, and route of administration
- Have the same indications of use
- Be bioequivalent
- Meet the same batch requirements for identity, strength, purity, and quality
- Be manufactured under the same strict standards of FDA's good manufacturing practice regulations required for innovator products.

20.1.4 APPROVAL AND REGULATION

As compared to the innovator drug, which has to go through long and expensive process of demonstrating safety and efficacy in animals and human clinical trials, the generic drug manufacturers are required to prove that their generic drug products are bioequivalent to the innovator product.

Generic drugs are usually approved via an Abbreviated New Drug Application (ANDA) process. Generic drug applications are termed "abbreviated" because they are generally not required to include preclinical (animal) and clinical (human) data to establish safety and effectiveness.

An ANDA requires that the generic drug contains the same active ingredient and has the same dosage and route of administration as the reference drug. It must have the same indication for use. The applicant needs to show that the generic product is bioequivalent to the reference product, in terms of pharmacokinetic and pharmacodynamic properties.

Requirements of ANDAs include:

- An approved reference product (RLD) and a patent certification
- Must be therapeutically equivalent (pharmaceutically equivalent and bioequivalent) to a reference product
- All related facilities should be cGMP-compliant
- Must meet the quality standards for chemistry and/or microbiology

20.1.5 GENERIC DRUG EXCLUSIVITY

In certain circumstances, an ANDA applicant whose ANDA contains a paragraph IV certification is protected from competition from subsequent generic versions of the same drug product for 180 days. During this 180-day period, only one manufacturer (or sometimes a few) can produce the generic version of a drug. This marketing protection is commonly known as "180-day exclusivity." This provides an incentive of 180 days of market exclusivity to the "first" generic applicant who challenges a listed patent by filing a paragraph IV certification and running the risk of having to defend a patent infringement suit [4].

20.2 BIOEQUIVALENCE STUDY

20.2.1 INTRODUCTION

In the last three decades, the concepts of bioavailability (BA) and bioequivalence (BE) have gained considerable importance and play a major role in the drug development phase for both new drug products and their generic equivalents. BA and BE studies are also important in the post approval period when there are certain manufacturing changes.

BA data provide an estimate of the amount of the drug that is absorbed, as well as provide information related to the pharmacokinetic properties of the drug. In BE studies, the exposure profile of a test drug product is compared to that of a reference drug product. Thus, both BA and BE focus on the release of a drug substance from a drug product and subsequent absorption into the systemic circulation [8]. BE is a strategy to introduce generic equivalents of branded- innovator drugs through proper assessment as required by the various regulatory authorities. BA and BE have therefore become the cornerstone for the approval of generic drugs globally [9].

21 CFR 320.1(a) defines BA as the rate and extent to which the active ingredient or active moiety is absorbed from a drug product and becomes available at the site of action. For drug products that are not intended to be absorbed into the blood stream, BA may be assessed by measurements intended to reflect the rate and extent to which the active ingredient or active moiety becomes available at the site of action.

BE means the absence of a significant difference in the rate and extent to which the active ingredient or active moiety in pharmaceutical equivalents or pharmaceutical alternatives becomes available at the site of drug action when administered at the same molar dose under similar conditions in an appropriately designed study (21 CFR 320.1(e)).

20.2.2 WHY IS BIOEQUIVALENCE NEEDED?

Pharmaceutical equivalence does not necessarily mean therapeutic equivalence. Multisource drug products should conform to the same standards of quality, safety, and efficacy required as those of the reference product, and must be interchangeable. Since the generic drugs would be interchanged with innovator drugs in the market, it is therefore required to demonstrate that the safety and efficacy of both the innovator and generic drugs are comparable. Assessment of this "interchangeability" between the generic and innovator drug product is carried out by demonstrating bioequivalence. BA and BE are required by regulations to ensure therapeutic equivalence between a pharmaceutically equivalent test product (generic) and a reference product.

BE studies are particularly needed for pharmaceutical products for systemic action such as [10]:

1. Oral immediate release drug formulations when one or more of the following criteria apply:
   a. Indicated for serious conditions requiring assured therapeutic response.
   b. Narrow therapeutic window/safety margin; steep dose-response curve.
   c. Complicated pharmacokinetics.
   d. Unfavorable physicochemical properties, e.g., low solubility, instability, meta-stable modifications, poor permeability, etc.
   e. Documented evidence for bioavailability problems related to the drug or drugs of similar chemical structure or formulations.
   f. Where a high ratio of excipients to active ingredients exists.
2. Nonoral and nonparenteral drug formulations designed to act by systemic absorption, such as transdermal patches, suppositories, etc.
3. Sustained or otherwise modified release drug formulations.
4. Fixed dose combination products with systemic action.
5. Comparative clinical or pharmacodynamic studies are required to prove equivalence for nonsolution pharmaceutical products that are for nonsystemic use (oral, nasal, ocular, dermal, rectal, vaginal application, etc.) and are intended to act without systemic absorption.

20.2.3 ASSESSMENT OF BIOEQUIVALENCE

In BE studies, the plasma concentration time curve is generally used to assess the rate and extent of absorption. AUC, the area under the concentration time curve, reflects the extent of exposure. Cmax, the maximum plasma concentration or peak exposure, and the time to maximum plasma concentration, tmax, are parameters that are influenced by absorption rate [11].

The assessment of BE of two drug products is based on the fundamental assumption that two drug products are equivalent when the rate and extent of absorption of the test product is not significantly different from that of the reference drug when the two drugs are administered at the same molar dose of the therapeutic ingredient under similar experimental conditions in either a single dose or multiple doses.

BE studies are generally recommended by using the following endpoints, listed in order of preference [8,10]:

1. Pharmacokinetic studies
2. Pharmacodynamic studies
3. Comparative Clinical studies
4. In vitro dissolution studies

20.2.3.1 Pharmacokinetic studies

The therapeutic effect of a solid drug product is the function of the concentration of the active ingredient in the systemic circulation and is therefore related to its bioavailability. The principles and methodology involved in assessing bioequivalence of drug products involves the measurement of the concentration of the active ingredient in the systemic circulation, either directly in blood or indirectly through urinary excretion studies.

The rate and extent of absorption of the active moiety or ingredient to the site of action,emphasizes the use of pharmacokinetic measures to indicate release of the drug substance from the drug product into the systemic circulation. BA and BE uses pharmacokinetic measures such as AUC to assess the extent of systemic exposure and Cmax and Tmax to assess the rate of systemic absorption. This approach is based on the understanding that some predetermined relationship exists between the drug concentrations at the site of action and that in the systemic circulation.

Pharmacokinetic studies are most widely preferred to assess BE for drug products, where drug levels can be easily determined in an accessible biological fluid (such as plasma, blood, urine). Regulatory guidance recommends that measures of systemic exposure be used to reflect clinically important differences between test and reference products in BA and BE studies.

The two major pharmacokinetic methods used to assess BE are:

1. Plasma level-time studies
2. Urinary excretion studies

20.2.3.2 Pharmacodynamic studies

For orally administered drug products when the drug is absorbed into systemic circulation and a PK approach can be used to assess systemic exposure and evaluate BA or BE, PD studies are not recommended. PK endpoints are generally a more accurate, sensitive, and reproducible approach and are therefore preferred. However, a well justified PD studies can be used to demonstrate BA or BE if measurements of drug concentrations cannot be used as surrogate endpoints for the demonstration of efficacy and safety of the particular pharmaceutical product, e.g., for topical products without an intended absorption of the drug into the systemic circulation [10].

An essential component of a BA or BE study based on a PD response is documentation of a dose-response relationship. Pharmacodynamic evaluation measures the effect on a pathophysiological process after administration of two different products to serve as a basis for BE assessment.

These studies generally become necessary under two conditions:

1. If the drug and/or metabolite(s) in plasma or urine cannot be analyzed quantitatively with sufficient accuracy and sensitivity.
2. If drug concentration measurement cannot be used as surrogate endpoints for the demonstration of efficacy and safety of the particular pharmaceutical product.

The two pharmacodynamic methods involve determination of bioavailability from:

1. Acute pharmacologic response
2. Therapeutic (clinical) response

20.2.3.3 Comparative clinical studies

In the absence of pharmacokinetic and pharmacodynamic approaches, adequate and well-controlled clinical studies may be used to establish BA and BE. The use of BE studies with clinical trial endpoints can be appropriate to demonstrate BE for orally administered drug products when measurement of the active ingredients or active moieties in an accessible biological fluid (pharmacokinetic approach) or pharmacodynamic approach is not possible.

20.2.3.4 In vitro dissolution studies

Under certain circumstances, product BA and BE can be evaluated using in vitro approaches (21 CFR 320.24:b). For highly soluble, highly permeable, rapidly dissolving, and orally administered drug products, documentation of BE using an in vitro approach (dissolution/drug release studies) is appropriate based on the biopharmaceutics classification system.

20.2.4 GENERAL CONSIDERATIONS FOR BIOEQUIVALENCE STUDIES

As the pharmacokinetic approach is often used, most of the advances in assessment of bioequivalence have been made in this approach. Important areas in such studies include [8]:

1. Study design
2. Bioanalytical methodology
3. Selection of appropriate analyte(s)—Moieties to be measured
4. Pharmacokinetic measures
5. Statistical approaches
6. Criteria for bioequivalence

20.2.4.1 Study design

If two formulations are to be compared, a two period, two sequence crossover study is the design of choice and the two periods of the treatment should be separated by an adequate washout period which should ideally be equal to or more than five half-life's of the moieties to be measured [10].

Single-dose PK studies are generally recommended to assess BA and BE because they are generally more sensitive than steady-state studies in assessing rate and extent of release of the drug substance from the drug product into the systemic circulation [8].

Other study designs include:

1. Parallel design study used for drugs with very long half-life
2. Replicate design study for drugs with high variability

20.2.4.2 Bioanalytical methodology

The bioanalytical methods used to determine the drug and/or its metabolites in any suitable matrix such as plasma, serum, blood, or urine in a BA/BE study must be well characterized, standardized, and fully validated [8].

Validating the bioanalytical method includes performing all of the procedures to demonstrate that a particular method to be used for quantitative measurement of analytes in a given biological matrix is reliable and reproducible for the intended use. The acceptability of analytical data corresponds directly to the criteria used to validate the method used.

Fundamental parameters for this validation include the following:

- Accuracy
- Precision
- Selectivity
- Sensitivity
- Reproducibility
- Stability

20.2.4.3 Selection of appropriate analyte(s)—moieties to be measured

In biological fluid collected in BE studies, the moieties to be measured are either the active drug ingredient or its active moiety in the administered dosage form (parent drug) and when appropriate, its active metabolite [9].

In BE studies, measurement of the active ingredient or the active moiety, rather than metabolites, is generally recommended because the concentration-time profile of the active ingredient or the active moiety is more sensitive to changes in formulation performance than that of the metabolite.

In some situations, as mentioned below, the measurements of an active or inactive metabolite may be necessary.

1. The concentrations of the drug(s) may be too low to accurately measure in the biological matrix
2. Limitations of the analytical method
3. Unstable drug
4. Drug(s) with a very short half life
5. In cases of prodrugs

20.2.4.4 Pharmacokinetic measures

Systemic exposure measures are used to evaluate BA and BE. Exposure measures are defined relative to peak, partial, and total portions of the plasma, serum, or blood concentration-time profile [8].

- Peak Exposure:

  Peak exposure is assessed by measuring the Cmax obtained directly from the systemic drug concentration data without interpolation. The Tmax can provide important information about the rate of absorption.
- Total Exposure (Extent of Absorption):
  - For single-dose studies, the measurements of total exposure are:
    - Area under the plasma, serum, or blood concentration time curve from time zero to time t (AUC0 $-$ t), where t is the last time point with a measurable concentration.
    - Area under the plasma, serum, or blood concentration time curve from time zero to time infinity (AUC0 $- \infty$), where AUC0 $- \infty =$ AUC $0 - t + Ct/\lambda z$. Ct is the last measurable drug concentration and λz is the terminal or elimination rate constant calculated according to an appropriate method.

    - For drugs with a long half-life, Cmax and a suitably truncated AUC (e.g., AUC0-72 h) can be used to characterize peak and total drug exposure, respectively.
  - For steady-state studies, the measurement of total exposure is the area under the plasma, serum, or blood concentration time curve from time zero to time tau over a dosing interval at steady state (AUC0-tau), where tau is the length of the dosing interval.
- Partial Exposure:

  For orally administered drug products, BA and BE can generally be demonstrated by measurements of peak and total exposure. For certain classes of drugs and under certain circumstances (e.g., to assess onset of an analgesic effect), an evaluation of the partial exposure could be used to support the performance of different formulations by providing further evidence of therapeutic effect. The use of partial AUC can be used as a partial exposure measure. The time to truncate the partial area should be related to a clinically relevant PD measure. Sufficient quantifiable samples should be collected to allow adequate estimation of the partial area.

20.2.4.5 Statistical approaches

A general objective in assessing BE is to compare the log-transformed BA measure after administration of the test and reference products [10].

The 90% confidence interval for the ratio of the population means (test/reference) or two one sided-t tests with the null hypothesis of non-bioequivalence at the 5% significance level for the parameter under consideration are considered for testing bioequivalence [10].

To meet the assumption of normality of data underlying the statistical analysis, the logarithmic transformation should be carried out for the pharmacokinetic parameters Cmax and AUC before performing statistical analysis [10].

The parameter Tmax should be analyzed using non-parametric methods. In addition to the above, summary statistics such as minimum, maximum, and ratio should be given.

20.2.4.6 Criteria for bioequivalence

To establish bioequivalence, the calculated 90% confidence interval for AUC and Cmax should fall within the bioequivalence range, usually 80%–125% [10]. However, tighter limits for the permissible difference in bioavailability may be required for drugs that have:

1. A narrow therapeutic index
2. A serious, dose-related toxicity
3. A steep dose/effect curve
4. A nonlinear pharmacokinetics within the therapeutic dose range

20.2.5 CONDUCT OF BIOEQUIVALENCE STUDY

This can broadly be classified as:

1. Clinical Phase
2. Bioanalytical Phase (PK and statistical analyses)[1]

Clinical Phase of BA and BE studies:

1. *Required Documents*
   a. Study Protocol
   b. Informed Consent form
   c. Case Record form
   d. Investigator's Brochure, if applicable
   e. Subject Recruitment Procedures such as Advertisement
2. *Ethical Considerations*
   An Institutional Review Board (IRB)/Independent Ethics Committee (IEC) is responsible to ensure and safeguard the rights, safety, and wellbeing of trial-related subjects. An IRB/IEC reviews and approves the clinical trial protocol/amendments and related documents such as informed consent form(s), subject recruitment procedures (e.g., advertisements), written information to be provided to subjects, or any other documents as applicable.
3. *Selection of Subjects*
   Healthy adult volunteers, as determined through medical examination and necessary laboratory tests, are generally enrolled based on the inclusion and exclusion criteria as mentioned in the trial protocol. Male and female volunteers who are 18 years or older should be enrolled unless there is a specific reason to exclude one sex. Volunteers should be capable of providing written informed consent. Selection of volunteers should be standardized so as to minimize interindividual variability. In some instances, BA and BE may be necessary to be evaluated in patients for whom the drug product is intended to be used if the safety considerations preclude the use of healthy subjects.
4. *Sample Size*
   In crossover design studies, required sample size depends on the intrasubject variability either known through literature or a pilot study. Possible dropouts and withdrawals should be considered while determining sample size. In cases when parallel design is necessary, intersubject and intrasubject variability should be considered.
5. *Standardization*
   The test conditions should be standardized in order to minimize the variability of all factors involved except that of the products being tested [11]. Procedure of drug intake, time of administration (fasting or fed), fluid and food intake, prohibition of alcohol, posture, and restriction of activities,should be standardized as per the regulatory requirement.

[1] Refer Chapter 6 on Pharmacokinetics for PK Analyses.

20.2.6 REGULATORY REQUIREMENTS FOR BE STUDIES

	USFDA (USA)	EMEA (Europe)	CDSCO (India)	ANVISA[12] (Brazil)	ASEAN[a13]	Canada[14]	TGA[15] (Australia)
Age	18 years of age or older	≥ 18	Healthy adult volunteers (age not specified)	18–50	18–55	18–55	18–55
BMI (kg/m^2)	Not specified	18.5–30	Not specified	Not specified	18.0–25.0	18.5–30	Normal values
Sex	Both sexes	Both sexes	Both sexes	Both sexes	Both sexes	Both sexes	Both sexes
Sample Size	12	Not less than 12	Not less than 16 volunteers unless justified for ethical reasons	Not less than 12. 24 in case of nonavailability of inter subject variation	Not less than 12	Not less than 12	Not less than 12 unless justified
Fasting Requirement	Following an overnight fast of at least 10 h, with a subsequent fast of 4 h post dose	Should fast for at least 8 h prior to dosing, unless otherwise justified and no food is allowed for at least 4 h post dose	Overnight fast (at least 10 h), with a subsequent fast of 4 h following dosing. For multiple-dose fasting studies, when an evening dose must be given, 2 h before and after the dosing	Overnight fast (at least 10 h), with a subsequent fast of 4 h following dosing	Overnight fast, with a subsequent fast of 4 h following dosing	Normally, subjects should fast for 8 hours before drug administration	Should fast for at least 8 h prior to dosing, unless otherwise justified and no food is allowed for at least 4 h post dose
Fed study requirement	High fat high calorie breakfast comprising of 800–1000 calories. Following an overnight fast	Composition of the meal should be according to the SPC of the originator product, If no specific recommendation	High fat breakfast comprising of 950–1000 kcal to be consumed approximately 15 min before dosing	High fat high calorie breakfast comprising of 800–1000 calories	Composition of the meal should be according to the SPC of the originator product, If no specific recommendation	A high-fat (approximately 50% of total caloric content of the meal) and high calorie (approximately 800 to 1000	Composition of the meal should be according to the SPC of the originator product, If no specific recommendation

	of at least 10 h, subjects should start the recommended meal 30 min prior to dosing. Study subjects should eat this meal in 30 min or less; however, the drug product should be administered 30 min after start of the meal	is given in the originator SPC, the meal should be a high-fat high calorie (comprising of 800–1000Kcal) & should be started 30 min prior to administration of the drug product and eat this meal within 30 min			is given in the originator SPC, the meal should be a high-fat high calorie (comprising of 800–1000 kcal)	kcal) meal should derive approximately 150, 250, and 500–600 kilocalories from protein, carbohydrate, and fat, respectively	is given in the originator SPC, the meal should be a high-fat high calorie (comprising of 800–1000Kcal) & should be administered 30 min prior to dosing
Fluid intake	Subjects should be administered the drug product with 240 mL (8 fluid ounces) of water; water is not allowed as desired except for 1 h before and after the drug administration	Drug should be administered with at least 150 mL of fluid, water is allowed as desired except for one hour before and one hour after drug administration	Standardization of fluid intake	200 mL	Drug intake with at least 150 mL	The dose should be taken with water of a standard volume (e.g., 150–250 mL)	Drug should be administered with at least 150 mL of fluid, water is allowed as desired except for one hour before and one hour after drug administration
Sampling criteria	Blood samples should be drawn at appropriate times to describe the absorption,	*For Single dose sampling* Sufficient sampling is required; frequent sampling around	Should extend to at least 3 elimination half-lives; at least 3 sampling points during	Sampling should be carried out according to the pharmacokinetic profile of the drug to be studied and	*For Single dose sampling* Sufficient sampling is required; frequent sampling around	Minimum of 12–18 samples should be collected per each subject per dose	Sampling should be planned to provide an adequate estimation of Cmax & to cover the plasma

(*Continued*)

Continued

	USFDA (USA)	EMEA (Europe)	CDSCO (India)	ANVISA[12] (Brazil)	ASEAN[a13]	Canada[14]	TGA[15] (Australia)
	distribution, and elimination phases of the drug; 12–18 samples, including a predose sample be collected per subject per dose. This sampling can continue for at least three or more terminal half-lives of the drug	predicted Tmax, avoid Cmax be the first point; accommodate reliable estimate AUC(0-t) covers at least 80% of AUC(0-∞). At least three to four samples are needed during the terminal log-linear phase	absorption phase, 3–4 at the projected Tmax, and 4 points during elimination phase	research protocol, the first collection (baseline collection) being mandatorily performed before the drug is administered. sample collection schedule shall consider a time equal to or higher than 3–5 times the elimination half-life of the drug or metabolite, when this is active	predicted Tmax, avoid Cmax be the first point; accommodate reliable estimate $AUC(0-t)$ covers at least 80% of AUC $(0-\infty)$. At least three to four samples are needed during the terminal log-linear phase		concentration time curve long enough to provide a reliable estimate of the extent of absorption
Washout criteria	More than 5 half-lives of the moieties to be measured	Sufficient washout period (usually at least 5 terminal half-lives)	≥5 half-lives of the moieties to be measured	Interval between the periods shall be at least seven elimination half-lives of the drug or metabolite, when this is active	Subsequent treatments should be separated by periods long enough to eliminate the previous dose before the next one(adequate washout periods)	Should not be Normally, subjects should fast for 8 h before drug administration. less than 10 times the mean terminal half-life of the drug	At least three times the terminal half lives

[a]ASEAN, *Association of Southeast Nations.*

REFERENCES

[1] Meredith P. Bioequivalence and other unresolved issues in generic drug substitution. Clin Ther. 2003;25:2875−90.

[2] Midhal KK, McKay G. Bioequivalence: its history, practice, and future. AAPS J. 2009;11:664−70.

[3] King DR, Kanavos P. Encouraging the use of generic medicines: Implications for transition economies. Croat Med J 2002;43:462−9.

[4] FDA/CDER Small Business and Industry Assistance Chronicles. May 19, 2015.

[5] Barr Laboratories. History of Generic Industry; 2002. Available at: <http://www.barrlabs.com/pages/induhist.html>.

[6] Harnden D. Can generics survive under Hatch-Waxman? Scrip Mag. 1998;1998(December):32−4.

[7] <http://www.fda.gov/Drugs/ResourcesForYou/Consumers/BuyingUsingMedicineSafely/UnderstandingGenericDrugs/ucm144456.htm>.

[8] FDA Guidance for Industry: "Bioavailability and Bioequivalence Studies Submitted in NDAs or INDs — General Considerations" (March 2014).

[9] Study on requirements of bioequivalence for registration of pharmaceutical products in USA, Europe and Canada. Saudi Pharmaceut. J. 2014; 22, 391−402.

[10] Guidelines for bioavailability & bioequivalence studies. Central Drugs Standard Control Organization, Directorate General of Health Services, Ministry of Health and Family Welfare, Government of India, New Delhi (March, 2005). Available on <www.cdsco.nic.in/html/BE> Guidelines.

[11] CHMP (Committee for Medicinal products for Human use) Guideline on the Investigation of Bioequivalence (CHMP/ QWP/ EWP/ 1401/ 98 Rev 1/ Corr**, 20 January 2010).

FURTHER READING

ASEAN Guidelines for the conduct of Bioavailability and Bioequivalence and Bioequivalence Studies, July 21, 2004.

Guidance Document, health Canada: Conduct and Analysis of Comparative Bioavailability Studies, effective date May 22, 2012.

Guidelines for Bioavailability/Bioequivalence of Medicines, Resolution – RE 1170, dated April 19, 2006.

Therapeutic Goods Administration. CPMP Guideline –Note for Guidance on the Investigation of Bioavailability and Bioequivalence (CPMP/EWP/QWP/1401/98). Available at: <http://www.tga.gov.au/docs/pdf/euguide/ewp/140198entga.pdf>.

CHAPTER 21

VACCINES

Subodh Bhardwaj
Consultant Biopharmaceuticals, New York, NY, United States

The introduction of an antigen into the body for stimulating an immune response is termed a "vaccination"; it leads to immunization—the process of protecting individuals from disease by acquiring immunity, through the production of antibodies against the invading organism. The terms are often used synonymously, though passive immunization can be achieved by immunoglobulin administration for initial protection, while vaccines act through T cell responses or B cell immune responses to induce long lasting protection.

21.1 CLASSIFICATION OF VACCINES [1]

Vaccines are classified as "live attenuated" or "inactivated." Inactivated vaccines contain killed organisms. Examples of live vaccines include: measles vaccine, Bacillus Calmette–Guérin (BCG), oral polio vaccine, live attenuated influenza vaccine, rotavirus vaccine, oral typhoid vaccine, and oral cholera vaccine. Examples of inactivated or killed vaccines include: inactivated trivalent split virion Influenza vaccine, diphtheria-tetanus-pertussis vaccines, hemophilus influenza b polysaccharide vaccine, Hibconjugate vaccines, antirabies vaccines, meningococcal vaccine.

21.2 ACTIVE IMMUNIZATION

Live vaccines replicate in the host causing a mild or undetected disease and bring about an immune response which mimics natural infection. These vaccines are attenuated or weakened by serial animal or human cell culture, e.g., the polio virus oral vaccine or OPV. Attenuation can be achieved by altering the natural route of vaccination, as with adenovirus vaccines given by the gastrointestinal route. An example of a live bacterial vaccine is the afore-mentioned Bacillus Calmette–Guerin (BCG), derived from Mycobacterium bovis, which is used for the prevention of disseminated tuberculosis including tubercular meningitis. Examples of engineered live vaccines include tetravalent rhesus rotavirus vaccine (RRV-TV), oral typhoid 21a vaccine, and intranasal influenza vaccine.

Pharmaceutical Medicine and Translational Clinical Research. DOI: http://dx.doi.org/10.1016/B978-0-12-802103-3.00022-5

21.3 INACTIVATED OR KILLED VACCINES

Inactivated or killed vaccines may consist of whole, inactivated agents like whole cell pertussis vaccine and inactivated polio vaccine [IPV], or specific components derived from the pathogen through physical, chemical, or biological means. These also include toxoids, which are protein toxins chemically modified to reduce pathogenicity. Polysaccharide vaccines like meningococcal and pneumococcal vaccines are physically purified from bacterial capsules. Components can be engineered to improve immunogenicity or to alter the nature of immune response. E.g., conjugation of capsular polysaccharides to the protein carrier (PRP-T Hib vaccine) can change a T cell independent response to a T cell dependent response, resulting in immunological memory and inducing protection in children below 2 years of age, in whom polysaccharide vaccines are ineffective [2]. Vaccines can also be produced by recombinant DNA technology. The Hepatitis B surface antigen DNA sequence was isolated from virus (Hbs Ag) in 1978. The gene coding for Hbs Ag was inserted into bacterium *Escherichia coli* (E. coli) for growth. In 1981 this cloned DNA was cultured successfully in "*Saccharomyces cerevisiae*" (yeast cells). The Hbs Ag produced was separated from yeast cells and purified before being processed in to vaccine, which was licensed for use in 1986. This was the first human vaccine produced by recombinant DNA technology, which proved to be economical and efficient for producing Hepatitis B vaccine in unlimited quantities [3].

21.4 PASSIVE IMMUNIZATION

Passive immunization is achieved by administration of antibodies, resulting in short term protection. This process occurs naturally during the last two months of pregnancy, when large amounts of immunoglobulin's IgG class are transferred from the placenta to the baby to protect them from infections till their immune system can protect them through vaccine-induced immunity. Passive immunization is required by patients with immune defects like agammaglobulinemia, who receive intravenous polyclonal immunoglobulin (IVIG) to prevent infections. Hyper immunoglobulin derived from donors with high antibody titers to pathogens can be used in varicella zooster immunoglobulin (VZIG) for prevention of chickenpox in exposed immune-compromised individuals. Hepatitis B immunoglobulin (HBIG) is used in neonates born to Hepatitis B carrier mothers or individuals exposed to acute infection. Hyper immunoglobulin also contains antibodies to other pathogens. Donor screening of blood and processing of antibodies reduce risk of blood borne pathogens like Hepatitis and HIV. Passive immunization can also be through administration of toxoids or anti-sera. Passively acquired antibodies can inactivate live attenuated viral vaccines like varicella, measles, OPV, and rotavirus vaccines. The yellow fever vaccine is not affected by this phenomenon.

Another important example is the administration of rabies immunoglobulins [HRIG or ERIG] derived from human and equine sources. The equine rabies immunoglobulin is widely used in India and other developing countries with a high burden of animal bites. The dose is 20IU per kg body weight (HRIG) or 40IU per kg body weight (ERIG). Rabies immunoglobulin [RIG] is used to provide passive immunity against rabies, as a part of a postexposure prophylaxis regimen in individuals exposed to the disease or virus and not previously vaccinated against rabies. The manufacturers and the US Public Health Service Advisory Committee on Immunization Practices (ACIP) currently recommend that postexposure prophylaxis following rabies exposure includes immediate and thorough wound treatment (cleansing with soap and water) and passive immunization with rabies immunoglobulin (RIG), utilizing WHO intramuscular or intradermal schedules. RIG provides immediate, temporary antibodies until the patient responds, to achieve immunization with vaccine. Rabies immunoglobulin is administered preferably by local wound infiltration and, when local infiltration accounts only for a portion of the dose or is not anatomically feasible, by intramuscular injection at a site distant from vaccine administration, which could be the gluteal muscle or anterolateral aspect of the thigh in children. RIG and vaccine should never be given at the same site [4].

21.5 GENERAL CONSIDERATIONS

The immunization process depends on a number of factors which include vaccine characteristics, biology of immunization, epidemiology of disease, and host environments. Public health officials and clinicians play a key role in developing recommendations that maximize the benefits of vaccines, while minimizing the risks and cost of vaccines. The following are important guidelines: complete and accurate records listing date of vaccination, product, manufacture lot number, and expiry date must be recorded. Site and route of administration, name and address, and title of Health care Provider (HCP) are essential. Vaccines must be administered by sterile technique with proper hand washing. The use of gloves is not necessary unless HCP have contact with infectious body fluids or with open lesions on hands. Isopropyl alcohol (70%) can be used to clean the skin to prevent bacterial infection or abscess formation. A sterile disposal syringe must be used on each occasion and discarded after the session in a puncture-proof container to avoid needle stick injuries or reuse [4].

21.6 ROUTES OF ADMINISTRATION

The recommended route of administration, i.e., intramuscular (DTP vaccine), subcutaneous (MMR), intradermal (BCG), Intranasal (seasonal flu vaccine), or oral (typhoid/cholera vaccine) are as mentioned in the manufactures package insert or product information leaflet (PIL). Persons administering the vaccine should not deviate from the recommended route or anatomic site as this could result in reduced or inadequate immune response, e.g., anti-rabies vaccines given in the gluteal region instead of the deltoid region. The deep intramuscular route is recommended for adjuvanted vaccines in order to reduce local reactions. Aspiration prior to injection of vaccine is not recommended as there are no large blood vessels at the recommended injection sites. Techniques of vaccine administration and appropriate position of the needle are depicted in Fig. 21.1.

21.7 ORAL

The oral route is best accomplished with a child in feeding position using a needleless syringe to deliver the vaccine in the mouth. Vomiting requires repeating the dose after 10 minutes [5]. Oral

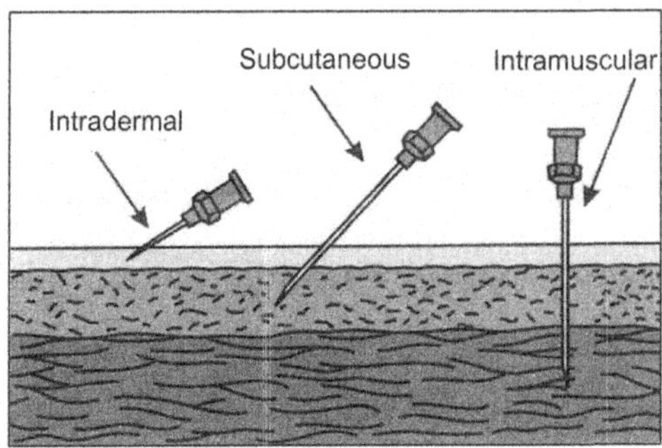

FIGURE 21.1

Routes of administration of injections (position of needle).

polio, rotavirus vaccine, and oral typhoid vaccines are given by this route. This does not apply to rotavirus vaccine; no readministration is done if the dose is spat out or regurgitated [6]. Intramuscular: many vaccines are given by the intramuscular route. The needle size and site is based on the age, subcutaneous tissue thickness, size of muscles, and volume and depth of the muscle. For majority of injections the sites are the vastus lateralis muscles (anterolateral) aspect of the upper arm just below the acromion process. The needle is introduced at 90° to penetrate muscle depth (Fig. 21.1) The buttocks should not be used because they contain fat which can lead to erratic absorption and damage to the sciatic nerve can occur. For large volume passive immunization the upper, outer mass of gluteus maximum is recommended. As a rule the anterolateral aspect of the thigh is preferred in infants and the deltoid in older children and adults. Intramuscular injections in 18-month old children have been reported to cause transient limping [7]. A 22–25 needle gauge size is suitable for intramuscular administration of most vaccines. A 5/8 inch (16mm) needle or 7/8 inch (22–25 mm) needle can be used. Needle length can vary from 1 inch (25mm) needle for men, and for women <60 kg (16–25 mm) needle, 60–90 kgs (25 mm) and 1.5inch (38 mm) for women >90 kgs and men >118 kgs [8]. The deltoid fat pad is thicker in women [9].

21.8 SUBCUTANEOUS

For subcutaneous injections the skin and subcutaneous tissue should be pinched up and needle directed in at 45 degrees as shown in Fig. 21.1.

21.9 INTRADERMAL

The BCG vaccine is the only product administered by the intradermal route in the deltoid region of the upper arm or volar surface of forearm. The needle (25–27 mm) is inserted into the epidermis at an angle parallel to the long axis of the forearm. A bleb 0.5–0.7 mm in diameter depicts proper intradermal and not subcutaneous administration. The usual site is the middle of the forearm near insertion of the deltoid [10].

21.10 INTRANASAL ROUTE

The intranasal route of administration is used only for the live attenuated influenza vaccine for healthy non-pregnant persons aged 2–49 years. The nasal sprayer with dose divider chip allows introduction of 0.1 mL spray into each naris. The tip should be inserted into the naris before administration. The dose needs no repetition even if subject sneezes or coughs. The risk for acquiring virus from the environment is unknown, but low as Live attenuated influenza virus (LAIV) vaccines are cold adapted viruses and attenuated, hence unlikely to cause symptomatic influenza. Not indicated for severely immunosuppressed individuals. High-risk age groups are those above 50 years, pregnant women and asthmatic patients who can recieve the vaccine [10].

21.11 BLEEDING DISORDERS

Patients of hemophilia and those receiving anticoagulant therapy are at a risk of bleeding with intramuscular injections. When indicated, the vaccine should be administered after clotting factor replacement. A 23 gauge needle should be used and pressure applied to injection site for 2 minutes without rubbing; alternatively the subcutaneous route can be used for vaccines if recommended, e.g., Hib vaccine [10].

21.12 VACCINE STORAGE AND COLD CHAIN MAINTENANCE

Vaccines must be stored at recommended temperatures to avoid loss of effectiveness. Recommended storage and handling requirements for each vaccine are given in each manufacturers product label. ACIP, CDC, AAP, and WHO publish correct shipping and storage handling practices [11].

21.13 STORAGE RECOMMENDATION FOR VACCINES [12]

Hepatitis A & B vaccines (no diluent), Pneumococcal Conjugate Vaccines (PCV), HPV vaccines, DT containing vaccines (DT, Td), and pertussis containing vaccines (Dtap, Tdap) are stored at 2°C–8°C (35°F–46°F). They should not be frozen as irreversible loss of potency occurs. Hib vaccine (PRP-OMP), IPV, quadrivalent meningococcal conjugate vaccine (MCV4), and trivalent inactivated influenza vaccine (TIV) are also stored at 2–8 degrees(35°F–46°F). Data on thermostability are lacking. PRP-T Hib should not be frozen. MMR vaccine should not be exposed to the light and should be stored at room-temperature or refrigerated 2°C–25°C (35°F–77°F). Varicella-containing vaccine MMRV, Varicella and Herpes zoster are stored at −50°C to −15°C (−58°F to

−50°F) and may be refrigerated or stored at room temperature while diluents at 2°C−25°C (35°F−77°F).

Live reassortant pentavalent (RV5) rota virus vaccine, LAIV, and monovalent rotavirus vaccines should be stored at 2°C−8°C or 35°F−46°F. Any vaccine should not be exposed to a temperature exceeding the recommended range. Live virus vaccines OPV and MMRV, should be kept frozen until just before administration. Yellow fever vaccine should be stored at 2°C−8°C (34°F−46°F). Hepatitis A, Hepatitis B, Hib, and Human pappilloma virus vaccines should never be frozen as they lose their potency. Maintenance of cold chain from production to end-user helps ensure vaccine potency. Temperature monitoring and control are of extreme importance during transportation and field usage. Temperature logs, twice daily at the same times every day, vaccine vial monitors, cold boxes, and temperature tails (temp tails) are devices used to ensure proper cold chain maintenance. Vaccines should not be reconstituted until immediately before use when all subjects are there to ensure optimal usage of multidose vials, which are to be discarded at the end of the immunization session. Only diluents provided by the manufacturer should be used. With the exception of oral polio vaccine (OPV), vaccines should not be refrozen after thawing. Vaccines from multidose vials when opened can be reused according to expiry date of properly stored vaccines with no visible contamination. It is advisable to have a generator back-up to ensure the efficient maintenance of cold chain [13].

21.14 IMMUNE RESPONSE TO VACCINES

The specific reactivity induced in a host by an antigenic stimulus is termed the "immune response." It can be of two types, either humoral (antibody mediated) or cellular (cell mediated). The two types usually develop together, and at different times one or the other may be predominant or exclusive. These are classified as active immunities. A subject is considered to be immune when they have specific protective antibodies or cellular immunity due to previous infection or immunization and so conditioned by previous experience to respond adequately to prevent infection and/or clinical illness following exposure to a specific infectious agent. Active immunity may be acquired following clinical infection, e.g., chicken pox, rubella, or measles, or following subclinical or inapparent infection, e.g., polio and diphtheria, or following immunization with live attenuated and inactivated vaccines or toxoids. Passive immunization is induced through the administration of immuniglobulins or antisera, e.g., rabies immunoglobulin.

21.15 HUMORAL IMMUNE RESPONSE

In a humoral immune response, the immunity conferred is specific to a particular disease and the individual in most cases is "immune" to further infection with the same organism.

Production of antibodies has three steps. These are the entry of the antigen, and its distribution and fate in the tissue and contact with immune-competent cells (afferent limb), the processing of the antigen by the cells and control of antibody process formation (central function) and the secretion of antibody and distribution in tissue, and the body fluids exerting its effects (efferent limb).

21.16 PRIMARY IMMUNE RESPONSE

Antibody response to initial antigenic stimulus differs qualitatively and quantitatively from response to subsequent stimuli with the same antigen. The former is called primary response and the latter secondary response, which are together illustrated in Figs. 21.2 and 21.3.

It has a short lag period and is prompt, powerful, and prolonged, producing antibodies that last for longer periods, and the avidity is high, i.e., the capacity to bind to the antigen is high. It

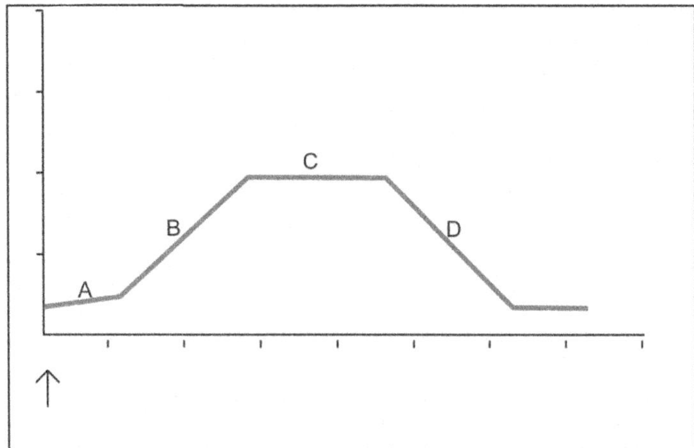

FIGURE 21.2

Primary immune response.

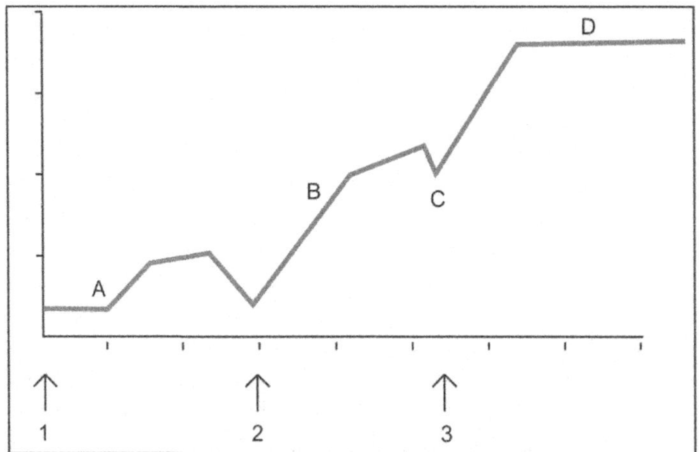

FIGURE 21.3

Secondary immune response to antigen.

involves the production of the IgM and IgG antibodies and the collaboration between B and T cells. There is a brief production of IgM antibody followed by much a larger and prolonged production of IgG antibody. This accelerated response is attributed to immunological memory and forms the basis of vaccination and revaccination. It is for this reason that inactivated or killed vaccines are given in multiple doses for active immunization, e.g., diphtheria, pertussis, and tetanus (DTP) vaccines as compared to live vaccines, where one dose is sufficient due to multiplication of organism in the body, which provides a continuous antigenic stimulus that acts as a priming and booster dose, e.g., measles, MMR vaccine. When an antigen is administered for the first time to an unexposed person there is a latent period of 3–10 days before antibodies appear in the blood. The primary response is sluggish and short-lived with a long lag phase and low titer of antibodies which do not persist for long. The IgM type of antibody predominates. However, if antigenic stimulus is sufficient, IgG antibodies appear in a few days, reach a peak by 4–6 weeks, and then decline over weeks or months. Primary response depends on a number of factors like dose, nature, route of administration, adjuvants, and avidity. An important outcome of the primary antigenic challenge is education of the recticulo-endothelial system where there is production of memory cells, or primed cells by B and T lymphocytes are responsible for immunological memory, which is established after immunization. The antigenic dose required for induction of IgG antibodies is 50 times more than that required for the IgM antibodies [14].

21.17 SECONDARY IMMUNE RESPONSE

The response to a booster dose is characterized by a shorter lag period and is also prompt, powerful, and prolonged—producing antibodies which last for longer periods with high avidity (capacity to bind to antigen is high). Both IgM and IgG antibodies are produced due to collaboration between B cells and T cells. This accelerated response is attributed to immunological memory and forms the basis of vaccination and revaccination. It is for this reason that inactivated vaccines are given in multiple doses for active immunization, e.g., diphtheria, pertussis and tetanus (DTP) vaccine as compared to live vaccines where one dose is sufficient as multiplication of the organism in the body provides a continuous antigenic stimulus that acts as both a priming and booster dose, for instance in Measles and MMR vaccines.

Live viral vaccines trigger innate immunity through multiple signals via RNA, allowing recognition by pattern recognition receptors (PRR) [15]. Following infection, viral particles spread throughout the body vascular network and reach target tissues. Dendritic cells are activated at multiple sites and migrate to draining lymph nodes, launching multiple foci of T and B cell activation. This explains higher immunogenicity induced by live vaccines compared to inactivated vaccines. The route of administration of live viral vaccines is of minor importance, e.g., the measles vaccine by the IM or SC route produces similar immunogenicity and reactogenicity responses [16]. BCG vaccines (live) multiply at the site of injection (rich in dendritic cells) bringing about a prolonged inflammatory reaction. Factors determining antibody response include the vaccine type (live or inactivated), protein or polysaccharide, adjuvants, avidity of antibody, dose, age at immunization, interval between dose at four weeks—or at least three weeks, extremes of age, and immune status normal or compromised.

Inactivated or non-live vaccines have limited potential for vaccine induced activation and active innate response at their site of infection, and unlike live vaccines their site and route are important. High content of dendritic cells in dermis allows a 10-fold reaction of antigen dose with intradermal route, a fact well utilized in the post exposure intradermal route of anti-rabies vaccines approved for use by WHO. Needle-free devices maybe available in the near future [17].

Patrolling dendritic cells are abundant in well-vascularized muscles, which are the preferred sites for intramuscular non-live vaccines. Adipose tissue has less dendritic cells, and subcutaneous vaccines may be less effective than the intramuscular site in conditions of limited immunogenicity, i.e., hepatitis B vaccine administered to adults by this route [18].

Mucosal immunization is limited to LAIV only. As the primary immune responses to non-live vaccines is focal and unilateral, this contributes to the fact that several vaccines can be given at a distance from one another by the intramuscular route, without interference draining into respective axillary, inguinal lymph nodes following injection into the deltoid or quadriceps muscles [19]. Adjuvants do not trigger immune responses comparable to live vaccines. This was observed with the AH1N1/09 pandemic influenza vaccine with AS03 adjuvant, which elicited immune responses comparable to convalescents [20]. Polysaccharide vaccines do not produce sustained immunity as they do not stimulate germinal centers.

21.18 B CELL RESPONSE

B cells do not produce antibodies and do not offer protection; however their participation in vaccine efficacy requires antigen driven proliferation and differentiation [21]. This reactivation may occur due to endemic or frequent pathogens or cross-reactions with organisms (natural boosters) or to booster immunization. The stimulus to B cells results in their rapid proliferation and differentiation into plasma cells, which produce high quantities of antibodies with high affinity resulting in strong immune responses [22]. As affinity of surface immunoglobulin from memory B cells is increased, their requirements for reactivation are lower than naïve B cells, hence they can be recalled by lower amounts of antigen and without CD4 + T cell help. Antigen specific memory cells are more numerous than naïve B cells initially and capable of antigen recognition [23]. This process is completed faster than primary responses as evidenced by the 4–7 day period for appearance of Polysaccharide specific antibodies in Hib vaccines in blood of previously primed infants [24]. Rapidity of appearance of antibodies is another hallmark of secondary responses. Post booster responses are higher in subjects with a strong primary response as evidenced by the increased Hbs Ag antibodies level observed with high (>100IU/L) than intermediate (>10–99 IU/L) anti-Hbs Ag observed post primary responses [25]. Persistence of memory B cells is important for long-term efficacy and antigenic persistence [26].

From neonatal tolerance to recognition of the fact that the neonatal and early life immune system is in contrast specifically adapted to unique challenges of early postnatal life and developing over time through a highly regulated process. Many factors determine quality and quantity of immune responses, e.g., state of prenatal and postnatal development of immune system, type of vaccine and immunogenicity potential, number of doses, their spacing, and influence of maternal antibodies [27]. Antibodies to polysaccharide (PS) antigens are not induced in the first two years of age, which is likely to reflect many factors including slow maturation of the splenic marginal zone and non- stimulation of germinal centers. Inhibitory influence of maternal antibodies is antibody titer dependent or reflects the ratio of maternal antibodies to vaccine antigen [27]. Increasing the dose of antigen may circumvent inhibitory effect of maternal antibodies as for hepatitis A and measles vaccine [28].

21.19 T CELL RESPONSE

Immune response to an antigen is brought about by three types of cells: antigen processing cells (APCs)—mainly the macrophages and dendritic cells, T cells, and B cells. The first step is the capture and processing of antigen by APCs and their presentation with the appropriate MHC molecule to T cells. T cell and B cell activation occurs when the processed antigen is presented on the surface of the APC; and with the appropriate MHC molecule to the T cell carrying the receptor (TCR) for the epitope the T cell is able to recognize it. In case of CD4 (Helper T/Th) cells the antigen has to be presented complexed with MHC class II, and for CD8 (Cytotoxic T/Tc) cells with MHC class I molecules. B cells which possess surface Ig and MHC class II molecules can also present antigens to T cells, particularly during secondary response. Helper T cells require two signals for activation. The first is combination of T cell receptor with MHC II complexed antigen and second signal is interleukin 1 (IL1), produced by the APC. The activated TH cell forms interleukin 2 and other cytokines required for B cell stimulation. These are IL4, IL5, and IL6, which act as B cell activating factor and B cell differentiating factor. They activate B cells which have combined with their respective antigens to proliferate and differentiate into antibody secreting plasma cells. A small number of activated B cells become long-lived memory cells responsible for antigen recognition on subsequent contact with antigen. Cytotoxic T cells (CD8/TC) are activated when they come into contact with the antigen presented with the MHC I class molecule and through IL2 as the second signal, secreted by Helper T cells, the activated T cells release cytotoxins that destroy the target, which may be a virus or tumor cells. Treg cells are CD4+ cells which produce cytokine TGF beta. These cells regulate immune response and tolerance to self-reacting cells [29].

21.20 ACIP RECOMMENDATIONS 2015 [30]

The Advisory Committee on Immunization Practices (ACIP) schedule, effective 1st January 2015, recommends various agees at immunization with vaccines intended for use in the age group 0–18 years in USA. It mentions primary schedules, boosters, and catch-up immunization. Any dose of vaccine not administered at the recommended age should be given at the subsequent visit, when indicated and feasible. This schedule is approved by the American academy of Pediatrics, American Academy of Family Physicians, and the American College of Obstetricians and Gynecologists. The adult immunization schedule relates to vaccine recommendations for persons aged 19 years or older. For contraindications and precautions to use of a vaccine, or additional information, users should consult the relevant ACIP statement online. Ages at immunization are discussed in different vaccine heads of this chapter based on ACIP, WHO, and other national schedules of developing countries. Vaccine doses administered 4 days or less before the minimum interval are considered valid. Clinically significant adverse events following vaccines must be reported to the vaccine adverse event reporting system by phone and online [31]. It is important that appropriate vaccines be administered as per recommendation to have an optimal effect in preventing disease.

21.21 MANAGEMENT OF EMERGENCIES

Acute emergencies are rare after vaccine administration, however it is important to be prepared to manage them and be prepared for reactions such as anaphylaxis and syncope. The AAP recommends 15–20 minutes observation period, although it is not recommended by the ACIP. The decision rests with the treating doctor. Vasovagal reactions like syncope are reported in 60%, especially females, young adults, and adolescents, within 5 minutes of vaccine receipt. Anaphylaxis is also sudden and fast. For management, emergency medicines like Adrenaline are administered (at the rate of 1 in 1000) at 0.1 mL/kg subcutaneously, and can be repeated every 10–20 minutes. Emergency supplies like oxygen, adult and pediatric airway, laryngoscope, antihistaminics, corticosteroids, and beta 2 agonists should be at hand [32]. In vaccination camps it is advisable to have a fully fledged ambulance with equipment and staff to deal with any emergency adverse event or to rush the patient to hospital if an untoward emergency occurs. Recently, subjects receiving HPV vaccines are advised to lie down for 15 minutes as syncope can occur. Do not give prophylactic acetaminophen as it can reduce immune responses to several vaccines.

REFERENCES

[1] Immunizing agents. In: Park K, editor. Parks textbook of preventive and social medicine. 23rd ed. M/S Banarsidas Bhanot Publishers, Jabalpur, India, vol. 103; 2015.

[2] American Academy of Paediatrics. Active Immunization. In: Pickering, Baker CJ, Long SS, McMillan J, editors. Red Book 2006; Report of the Committee on Infectious diseases, 27th ed., American Academy of Pediatrics, Elk Grove Village, IL; 2006, p. 9–54.

[3] Valenzuela P, Medina A, Rutter WJ. Synthesis and assembly of Hepatitis B Virus surface antigen particle in yeast. Nature 1982;298:347–50.

[4] Centers for Disease Control General recommendations on immunization. Recommendations of immunization. Recommendations of the Advisory Committee on Immunization practices (ACIP) and the American Academy of Family physicians (ACIP) and the American Academy of family Physicians (AAFP). MMWR 2002;51(RR-2):7.

[5] Centers for Disease control and prevention. Poliomyelitis prevention in the United States: Introduction of a sequential vaccination schedule of inactivated poliovirus vaccine followed by oral polio virus vaccine. MMWR 1977;46(RR-3):1–25.

[6] Center for Disease Control and Prevention. Prevention of rotavirus gastroenteritis among infants and children: recommendations of the Advisory Committee on immunization Practices (ACIP). MMWR 2006;55 (RR-12):1–13.

[7] Scheifele D, Bjornson G, Barreto L. Controlled Trial of Haemophilus influenzae type B diptheria toxoid conjugate combined with diphtheria tetanus and pertussis vaccines in 18 months old children including comparison of arm versus thigh injection. Vaccine 1992;10:455–60.

[8] Ipp MM, Gold R, Goldback M. Adverse reactions to diptheria, tetanus, pertussis-polio vaccination at 18 months of age: effect of injection site and needle length. Paediatrics 1989;83:679–82.

[9] Poland GA, Borrad A, Jacobson RM. Determination of deltoid fat pad thickness: implications for needle length in adult immunization. JAMA 1997;277:1709–11.

[10] Advisory Committee on Immunization Practices (ACIP) MMWR 2011; 60 (NRR2):1–61.

[11] World Health Organizations (WHO). Temperature sensitivity of vaccines (WHO/IVB/06.10). Geneva Switzerland: WHO; 2006.

[12] Centers for disease control and prevention. Vaccine storage and handling guide. <http://www.cdc.gov/vaccines/recs/storage/guide/vaccines-storage-handling guide.pdf> [accessed 25.07.15].

[13] Centers for Disease Control and prevention. Prevention and control of influenza with vaccines: Recommendations of Advisory Committee on Immunization practices (ACIP). MMWR Recomm Rep 2010;59(RR-8):1–25.

[14] Host defences. In: Park K, editor. Parks textbook of preventive and social medicine. 23rd ed. M/S Banarsidas Bhanot Publishers, Jabalpur, India; 2015, p. 101–2.

[15] Palm NW, Medzhitov R. Pattern recognition receptors and control of adaptive immunity. Immunol Rev 2009;227:221–33.

[16] Hong Kong Measles vaccination Committee. Comparative trial of live attenuated measles vaccine by intramuscular and intradermal injection. Bull World Health Organ 1967;36:375–84.

[17] Praunitz MR, Mikszta JA, Cornier M. Microneedle-based vaccines. Curr Trop Microbial Immunol 2009;333:369–93.

[18] De Lalla F, Rinaldi E, Santoro D. Immune response to hepatitis B given at different injection sites and by different routes: a controlled randomized study. Eur J Epidermis 1988;4:256–8.

[19] Spreafico R, Ricciardi-Castagnoli P, Mortellaro A. The controversial relationship between NLRP3, alum, danger signals and the next generation adjuvants. Eur J Immunol 2010;40:638–42.

[20] Meier S, Bel M, L'Huillier A. Antibody response to natural influenza 29 A/H1N1/09 disease or following immunization with adjuvanted vaccines in immunization with adjuvanted vaccines in immunocompetent and immunocompromised children. Vaccine 2011;29:3548–57.

[21] Kurosaki T, Aiba Y, Kometani K. Unique properties of memory B cells of different iso types. Immunol Rev 2010;237:104–16.

[22] Good-Jacobson KL, Shiomchik MJ. Plasticity and heterogenicity in the generation of memory B cells and long lived plasma cells: theinfluence of germinal center interactions and dynamics. J Immunol 2010;185:3117–25.

[23] Goodnow CC, Vinuesa CG, Randall KL. Control systems and decisions making for antibody production. Nat Immunol 2010;11:681–8.

[24] Pichichero ME, Voloshen T, Passador S. Kinetics of booster responces to Haemophilus influenza type B conjugate after combined diphtheria-tetanus acethular pertussis-Hamophilus influenza type B vaccination in infants. Pediatr Infect Dis J 1999;18:1106–8.

[25] Zanettiar MA. Romanol l. Long term immunogenicity of hepatitis B Vaccination and policy for booster: an Italian multicenter study. Lancet 2005;366:1379–84.

[26] Gray D, Skarvall H. B cell memory is short lived in the absence of antigen. Nature. 1988;336: 70–3.

[27] Siegrist CA, Aspinall. B cell responses to vaccination at extremes of age. Nat Rev Immunol 2009185–94.

[28] Cutts FT, Nyandu B, Markowitz LE. Immunogenicity of high titre AIK-C or Edmonston Zagreb vaccines in 3.5 month old infants, and of medium or high titre Edmonston Zagreb Vaccine in 6 month old infants in Kinhasa, Zaire. Vaccine 1994;12:1311–16.

[29] Kapil A, editor. Text book of microbiology: immune response, 9th ed. reprint. Universities Press(India) Private Limited, Hyderabad; 2014, vol. 15, p. 147.
[30] Advisory Committee on Immunization practices Recommendations; 2015. <http://www.cdc.gov/vaccines/hcp/acip-recs/index.html>.
[31] Vaccine adverse event reporting system. <http://www.vaers.hhs.gov>.
[32] Sapien R, Hodge D. Equipping and preparing for office emergencies. Pediatr Ann 1990;19:659–67.

FURTHER READING

Ausellio CM, Cassone A. Acellular pertussis Vaccines and pertussis resurgence: Revise or replace? mBio 2014;5(3); e01339–14.

CHAPTER

BIOSIMILARS

22

Gursharan Singh

Life Sciences, SmartAnalyst India Private Limited, Gurgaon, Haryana, India

22.1 INTRODUCTION

Biopharmaceutical agents are medicinal products derived from a biological source. These include vaccines, blood and blood components, allergenics, somatic cells, gene therapy, tissues, recombinant therapeutic proteins, etc. [1,2,3]. The increasing role of biologics in the treatment of various diseases can be appreciated from the fact that in the year 2015, out of the 45 novel products approved by the US FDA, 13 were biologics. These included antibodies, antibody fragments, peptides, and enzymes [4].

Though novel biologics offer a ray of hope for the treatment of many serious diseases, these are much more expensive than conventional small molecule drugs. For example, the annual cost of therapy with Lumizyme (recombinant alglucosidase alfa enzyme) for the treatment of Pompe disease, Soliris (eculizumab) for the treatment of paroxysmal nocturnal hemoglobinuria, and Avastin (bevacizumab) for the treatment of metastatic colo-rectal cancer, are in the range of USD 100,000 to USD 600,000 [5]. Hence, there is a need for low-cost versions of original biologics to reduce healthcare costs [6].

As in the case of conventional small molecule drugs, following expiry of patent and other exclusivity rights, the market opens to noninnovator versions of biological medicines, so-called biosimilar medicines or more commonly biosimilars, which tend to be less expensive due to a lower research and development (R&D) cost and possible impact of competition [7].

Biosimilars are similar in structure, function, activity, immunogenicity, and safety to the innovator biologic product which is considered the "Reference Biologic" [8]. Compared to small molecules, the biopharmaceutical products are highly complex and usually more difficult to characterize. Various parameters of biopharmaceutical products like three-dimensional structure, the amount of acido-basic variants or posttranslational modifications (e.g., glycosylation profile etc.) can be significantly altered by trivial changes in the manufacturing process. Therefore, the standard approach for approval of small molecule generic products based on demonstration of bioequivalence with a Reference Medicinal Product by appropriate bioavailability studies is not considered appropriate for biosimilars. The approval of biosimilars is based on a step-wise comparability exercise. Despite this comparability exercise, there may be subtle differences between biosimilars from different manufacturers or compared with "Reference Biologics" [9].

The above factors give rise to additional regulatory, clinical, exclusivity, and commercialization considerations for biosimilars, which are briefly discussed in this chapter.

Pharmaceutical Medicine and Translational Clinical Research. DOI: http://dx.doi.org/10.1016/B978-0-12-802103-3.00023-7

22.2 BIOSIMILARS: TERMINOLOGY

A wide variety of terms like "Follow-on biologic," "Subsequent Entry Biologic," "Biosimilar Biologic," "Similar Biologic," "Similar Biotherapeutic," etc. have been used to refer to biosimilars. However, the term biosimilar is now being increasingly used and emerging as a consensus term. The regulatory terminology and definition of biosimilars differs across geographies, as presented in Table 22.1.

22.3 BIOSIMILARS: REGULATORY FRAMEWORK

The approval of biosimilars is based on a comparability exercise with the "Reference Biologic" product. In this section, we discuss the "Reference Biologic," comparability exercise, regulatory approved biosimilars, as well as considerations with regards to indication extrapolation, pharmacovigilance, and substitution and interchangeability.

22.3.1 REFERENCE BIOLOGIC

While there are certain differences with regards to how regulatory authorities define a "Reference Biologic," in principle a "Reference Biologic" is one which is to be used as the comparator for head-to-head comparability studies with the biosimilar in order to demonstrate similarity in terms of quality, safety, and efficacy. Most regulatory authorities consider only the innovator product as the "Reference Biologic" and the same should itself have been approved on the basis of a full registration dossier. Table 22.2 provides the regulatory considerations for the "Reference Biologic" in various geographies.

22.3.2 COMPARABILITY EXERCISE

The approval of biosimilars based on comparability exercises follows a sequential approach, starting with a comprehensive physicochemical and biological characterization. The extent and nature of the required nonclinical in vivo studies and clinical studies depend on the level of evidence obtained from the physicochemical, biological, and nonclinical in vitro data. Before conducting any preclinical studies, consistency of the manufacturing process, product characterization, and product specifications needs to be demonstrated [10–18].

22.3.2.1 Manufacturing process

Regulatory authorities require a well-defined manufacturing process with appropriate process controls for biosimilars to assure that an acceptable product in terms of identity, purity, and potency is produced on a consistent basis. Data requirements pertaining to manufacturing process include description of characteristics of host cells, vector system, cell banks, stability of clone, cell culture/fermentation, harvest, dosage form, excipients, formulation, purification, filling into bulk or final containers, storage, primary packaging interactions, container closure system, and usage

Table 22.1 Regulatory Terminology and Definition of Biosimilars in Various Geographies

S. No.	Country/ Geography	Regulatory Guideline	Regulatory Term	Regulatory Definition
1.	European Union (EU)/ Australia	Guideline on Similar Biological Medicinal Products (CHMP/437/04 Rev 1) [10]	Biosimilar	A biological medicinal product that contains a version of the active substance of an already authorized original biological medicinal product (Reference Medicinal Product) in the European Economic Area (EEA)
2.	United States (US)	Biologics Price Competition and Innovation Act of 2009 [11]	Biosimilar	The biological product is biosimilar to a reference product based upon data that demonstrates that the biological product is highly similar to the reference product notwithstanding minor differences between the biological product and the reference product in terms of the safety, purity, and potency of the product
3.	Japan	Guideline for the quality, safety, and efficacy assurance of follow-on biologics [12]	Follow-on biologic	A biotechnological drug product developed to be comparable in regard to quality, safety, and efficacy to an already approved biotechnology-derived product ("original biologic") of a different company
4.	Canada	Guidance Document: Information and Submission Requirements for Biosimilar Biologic Drugs [13]	Biosimilar Biologic Drug	A biologic drug that obtains market authorization subsequent to a version previously authorized in Canada, and with demonstrated similarity to a "Reference Biologic" drug
5.	India	Guidelines on similar biologics [14]	Similar biologic	A Similar Biologic product is that which is similar in terms of quality, safety, and efficacy to an approved Reference Biological product based on comparability
6.	China	Technical Guideline for the Research, Development and Evaluation of Biosimilars [15]	Biosimilar	A biosimilar is defined as a therapeutic biologic that is similar to a reference product approved in China or elsewhere in quality, safety, and efficacy; the biosimilar should in principle have the same amino acid sequence as the reference product's
7.	Korea	Guidelines on the evaluation of biosimilar products [16]	Biosimilar	A "biosimilar product" is a biological product that is comparable to already marketed reference products in terms of quality, safety, and efficacy
8.	WHO	Guidelines on evaluation of similar biotherapeutic products [17]	Similar Biotherapeutic Products	A biotherapeutic product which is similar in terms of quality, safety, and efficacy to an already licensed reference biotherapeutic product

From CHMP. Guideline on Similar Biological Medicinal Products. http://www.ema.europa.eu/docs/en_GB/document_library/Scientific_guideline/2014/10/WC500176768.pdf; 2015 [accessed 22.12.16]; Title VII—Improving Access to Innovative Medical Therapies, Subtitle A—Biologics Price Competition and Innovation Act of 2009. http://www.fda.gov/downloads/Drugs/GuidanceComplianceRegulatoryInformation/UCM216146.pdf [accessed 22.12.16]; MHLW. Guideline for the Quality, Safety, and Efficacy Assurance of Follow-on Biologics. https://www.pmda.go.jp/files/000153851.pdf; 2009 [accessed 30.12.16]. Health Canada. Guidance Document- Information and Submission Requirements for Biosimilar Biologic Drugs. http://hc-sc.gc.ca/dhp-mps/alt_formats/pdf/brgtherap/applic-demande/guides/seb-pbu/seb-pbu-2016-eng.pdf; 2016 [accessed 30.12.16]; CDSCO. Guidelines on Similar Biologics: Regulatory Requirements for Marketing Authorization in India. http://cdsco.nic.in/writereaddata/CDSCO-DBT2016.pdf; 2016 [accessed 30.12.16]; Ropes and Gray. China Announces Final Biosimilars Guideline. http://www.mondaq.com/x/379908/food + drugs + law/China + Announces + Final + Biosimilars + Guideline; 2015 [accessed 30.12.16]; Joung J. Korean regulations for biosimilarsGenerics and Biosimilars Initiative Journal (GaBI Journal) 2015; 4 (2):93–94; WHO. Guidelines on Evaluation of Similar Biotherapeutic Products. http://www.who.int/biologicals/areas/biological_therapeutics/BIOTHERAPEUTICS_FOR_WEB_22APRIL2010.pdf; 2009 [accessed 30.12.16].

Table 22.2 Regulatory Considerations for the "Reference Biologic" in Various Geographies for Approval of Biosimilars

S. No.	Country/ Geography	Regulatory Considerations for the "Reference Biologic"
1	EU/Australia	The reference medicinal product is a product authorized in the EEA, on the basis of a complete dossier in accordance with the provisions of Article 8 of Directive 2001/83/EC, as amended. A single reference medicinal product, defined on the basis of its marketing authorization in the EEA, is recommended to be used as the comparator throughout the comparability program [10] However, with the aim of facilitating the global development of biosimilars, flexibility has been provided to allow for comparison of the biosimilar in certain clinical studies and in in vivo nonclinical studies (where needed) with a non-EEA authorized product approved in other ICH countries provided it is demonstrated that the comparator authorized outside the EEA is representative of the reference product authorized in the EEA [10] For Australia, additionally, the reference product has to be approved in Australia and marketed for a suitable duration and must have a volume of marketed use to serve as sufficient evidence regarding safety and efficacy [18]
2	US	The single biological product licensed under subsection (a) of section 351 of the Public Health Service Act, against which a biological product is evaluated in an application submitted under subsection [11]
3	Japan	The original biologic should be already approved in Japan and be the same product throughout the development period of the follow-on biologic (i.e., during the entire period from characterization of quality attributes through nonclinical and clinical studies [12]
4	Canada	A biologic drug authorized on the basis of a complete quality, nonclinical, and clinical data package, to which a biosimilar is compared to demonstrate similarity A non-Canadian "Reference Biologic" drug is permitted provided it has the same medicinal ingredient(s), dosage form, and route of administration as the version authorized in Canada and is approved in an ICH region. For use in clinical studies, a non-Canadian "Reference Biologic" drug must satisfy chemistry and manufacturing (quality) information [13]
4	India	The "Reference Biologic" should be licensed/approved in India or ICH countries and should be the innovator's product. The "Reference Biologic" should be licensed based on full safety, efficacy, and quality data [14]
5	China	A reference product (including active ingredients used for production or extracted from finished products) to which a biosimilar product is compared in analytical and preclinical studies must be approved in China or elsewhere. The same reference product, however, must be approved in China at the time when clinical studies are initiated. The reference product is usually, but not necessarily, the originator's product. Nevertheless, approved biosimilars cannot themselves act as reference products [15]
6	Korea	A "reference product" is a drug product already authorized by a regulatory authority on the basis of full regulatory submissions. The reference product is used in demonstrating the comparability of a biosimilar product through quality, nonclinical studies, and clinical studies [16]
7	WHO	A reference biotherapeutic product (RBP) is one which is to be used as the comparator for head-to-head comparability studies with the similar biotherapeutic product in order to show similarity in terms of quality, safety, and efficacy. Only an originator product licensed on the basis of a full registration dossier can serve as a RBP [17]

From CHMP. Guideline on Similar Biological Medicinal Products. http://www.ema.europa.eu/docs/en_GB/document_library/Scientific_guideline/2014/10/WC500176768.pdf; 2015 [accessed 22.12.16]; Title VII—Improving Access to Innovative Medical Therapies, Subtitle A—Biologics Price Competition and Innovation Act of 2009. http://www.fda.gov/downloads/Drugs/GuidanceComplianceRegulatoryInformation/UCM216146.pdf [accessed 22.12.16]; MHLW. Guideline for the Quality, Safety, and Efficacy Assurance of Follow-on Biologics. https://www.pmda.go.jp/files/000153851.pdf; 2009 [accessed 30.12.16]. Health Canada. Guidance Document- Information and Submission Requirements for Biosimilar Biologic Drugs. http://hc-sc.gc.ca/dhp-mps/alt_formats/pdf/brgtherap/applic-demande/guides/seb-pbu/seb-pbu-2016-eng.pdf; 2016 [accessed 30.12.16]; CDSCO. Guidelines on Similar Biologics: Regulatory Requirements for Marketing Authorization in India. http://cdsco.nic.in/writereaddata/CDSCO-DBT2016.pdf; 2016 [accessed 30.12.16]; Ropes and Gray. China Announces Final Biosimilars Guideline. http://www.mondaq.com/x/379908/food + drugs + law/China + Announces + Final + Biosimilars + Guideline; 2015 [accessed 30.12.16]; Joung J. Korean regulations for biosimilarsGenerics and Biosimilars Initiative Journal (GaBI Journal) 2015; 4 (2):93–94; WHO. Guidelines on Evaluation of Similar Biotherapeutic Products. http://www.who.int/biologicals/areas/biological_therapeutics/BIOTHERAPEUTICS_FOR_WEB_22APRIL2010.pdf; 2009 [accessed 30.12.16]; TGA. Evaluation of Biosimilars. https://www.tga.gov.au/sites/default/files/pm-argpm-biosimilars-150420_1.pdf; 2013 [accessed 30.12.16].

instructions, etc. The manufacturing process is optimized to minimize differences between the biosimilar and "Reference Biologic" [10–17].

22.3.2.2 Product characterization

This involves studies to characterize the variety of quality attributes of biosimilars, such as (1) structure and composition, (2) physicochemical properties, (3) bioactivity, (4) immunochemical properties, and (5) purity, impurities, and contaminants. Product-related as well as process-related impurities are also characterized and evaluated, taking into consideration their elimination data during the purification process [10–17].

The dosage form and administration route of a biosimilar is same as those of the "Reference Biologic." As long as there is no adverse effect on efficacy and safety, it is not necessary for the formulation of the biosimilar to be the same as that of the "Reference Biologic." In certain cases, it may be justified to select different excipients. Long-term (minimum of six months), real-time, real-condition stability studies are done to determine the recommended storage period [10–17].

The quality attributes of biosimilar are compared with "Reference Biologic" through appropriate comparative studies of (1) the structural analyses and physicochemical properties, (2) bioactivity, and (3) immunologic responses, etc. From the perspective of establishing similarity, Quality Attributes of a biosimilar may be considered in two categories:

1. Critical Quality Attributes (CQA) which have direct impact on the clinical safety or efficacy. These include attributes that directly impact the known mechanism(s) of action of the molecule. CQAs must be controlled within limits established based on the "Reference Biologic."
2. Key Quality Attributes (KQA) which are not known to impact clinical safety and efficacy but are considered relevant from a product and process consistency perspective. It may be acceptable to have slight differences in comparison to the "Reference Biologic." The acceptable criteria for differences in quality attributes vary depending on the characteristics of the product and the intended use and dosing regimen in clinical practice [10–17].

22.3.2.3 Specifications and test procedures

Specifications are employed to verify the routine quality of the drug substance and drug product rather than to fully characterize them. Specifications are set as described in established guidelines and monographs (where applicable) with additional test parameters as appropriate. All analytical methods referenced in the specification are required be validated with documentation of the corresponding validation as well as acceptable limits. It may be noted that methods used for setting specifications may or may not be the same as the analytical methods used for product characterization and for establishing product comparability.

While specifications for the "Reference Biologic" generally differ from the biosimilar, still the regulations require the same to be captured to the extent known [10–17].

22.3.2.4 Nonclinical/preclinical data

The non clinical studies include in vitro assays as well as in vivo pharmacodynamic, toxicological, and immunological studies [10–17].

In vitro assays: These include receptor-binding studies or cell-based assays (e.g., cell-proliferation or cytotoxicity assays). These are required to establish comparability of the biological/pharmacodynamic activity of the biosimilar and "Reference Biologic" [10–17].

In vivo studies: Comparative animal studies are performed in a relevant species to evaluate pharmacodynamic activity, toxicity, and immunogenicity. Pharmacodynamic activity relevant to the clinical application is assessed. If feasible, this may be evaluated as part of the nonclinical repeat dose toxicity study. These studies may not be required if validated and reliable in vitro assays are available, which reflect the clinically relevant pharmacodynamic activity. Toxicity is evaluated in at least one repeat dose toxicity study in relevant species (including toxicokinetics). These also include immunologic evaluation by determination and characterization of antibody responses, including anti-product antibody titres, cross reactivity with homologous endogenous proteins, and product neutralizing capacity. The duration of the studies is such as to allow detection of potential differences in toxicity and antibody responses between the biosimilar and "Reference Biologic." The studies, when performed with the final formulation intended for clinical use, allow detection of potential toxicity associated with both the drug substance and product- and process-related impurities. Antibody measurements help assess differences in structure or immunogenic impurities between the biosimilar and "Reference Biologic" [10–17].

Specialized toxicological studies, including safety pharmacology, reproductive toxicology, mutagenicity and carcinogenicity studies, are generally not required for a biosimilar submission unless triggered by results of the repeat dose toxicity study or the local tolerance study and/or by other known toxicological properties of the biosimilar [10–17].

22.3.2.5 Clinical data (Efficacy, Safety and Immunogenicity)

The aim of clinical data is to address slight differences detected at previous steps and to confirm comparable clinical performance of the biosimilar and the "Reference Biologic." While clinical studies are generally required, they may not be needed if the following criteria are fulfilled:

1. Similar efficacy and safety can be clearly deduced from the similarity of physicochemical characteristics, biological activity/potency, and PK and/or PD profiles of the biosimilar and the "Reference Biologic."
2. No additional concerns arise due to the impurity profile and the nature of excipients of the biosimilar itself.

In all other cases, comparative clinical trials, including PK/PD, to show the similarity in efficacy, purity, potency, and safety (including immunogenicity) of the biosimilar and "Reference Biologic," are required [10–17].

The biosimilar product used in clinical studies (investigational product) should be produced through well-established manufacturing processes. The type and contents of necessary clinical studies vary widely according to available information and the properties of the original biologics and is determined on a case-by-case basis. Important considerations in the conduct of clinical studies include determination of appropriate sample size and the comparability margin using clinically established endpoints. In general, equivalence designs are preferred over noninferiority design for the comparison of efficacy and safety of the biosimilar with the "Reference Biologic" [10–17].

While safety can be assessed as part of clinical efficacy studies, dedicated safety studies may be required in cases where the requirement of clinical efficacy studies has been waived off due to demonstration of comparable efficacy in confirmatory PK/PD studies. Safety data from a sufficient number of patients and adequate duration of exposure are required to be generated to compare the nature, severity, and frequency of adverse reactions before approval. Apart from safety, immunogenicity is an important concern with biologic products, particularly because certain antibodies may impact the safety and/or efficacy of the biosimilar product by altering pharmacokinetics, inducing anaphylaxis, or by neutralizing the product and/or its endogenous protein counterpart. The purpose of the comparative immunogenicity study(ies) is to rule out clinically meaningful differences in immunogenicity between the biosimilar and the "Reference Biologic." Immunogenicity studies are designed considering factors such as immunocompetence, prior or concomitant use of immunosuppressant therapies, and historical data with respect to the immunogenicity of the "Reference Biologic." For example, for two soluble insulins, the euglycaemic clamp study is considered the most sensitive method to detect differences in efficacy. However, immunogenicity and local tolerance of a subcutaneously administered biosimilar cannot be assessed in such a study and is therefore required to be evaluated in the target population [10–17].

22.3.3 REGULATORY APPROVED BIOSIMILARS IN US AND EU

Subsequent to the development of regulatory framework for biosimilars, many such products have been approved across the world. While the EU took the lead in approval of biosimilars with more than two-dozen products approved to date, four biosimilars have been approved in the United States to date. Biosimilars have also been approved in Japan and many emerging markets. Table 22.3 provides a list of biosimilars approved by the US FDA and European Medicines Agency (EMA).

22.3.4 INDICATION EXTRAPOLATION

Regulations require that comparative clinical studies for biosimilar approval should be carried out in at least one approved indication of the "Reference Biologic." Extrapolation of the safety and efficacy data to other clinical indications may be possible based on a scientific rationale considering same mechanism of action or receptor (s)/pathophysiologic mechanisms involved in other clinical indications, e.g., growth hormone action in different conditions of short stature in children, erythropoiesis-stimulating action of epoetins in different conditions associated with anaemia, etc. [10–17].

Other considerations for extrapolation of data to other indications include availability of sufficient data on safety and immunogenicity of the biosimilar so that no unique/additional safety issues arise for the extrapolated indication(s) e.g., immunogenicity data in immunosuppressed patients would not allow extrapolation to patients with autoimmune diseases. Also, if the efficacy trial used a noninferiority study design then results from this trial where a low dose is used may be difficult to extrapolate to an indication where a higher dose is used, from both an efficacy and safety perspective [10–17].

Table 22.3 Biosimilars Approved by US FDA and EMA

S. No.	Active Substance	Product Name	Approved Indication(s)	Approval Year	Company Name
US FDA					
1.	Adalimumab	Adalimumab-atto (Amjevita)	Ankylosing spondylitis, Adult Crohn's disease, Psoriatic arthritis, Plaque Psoriasis, Rheumatoid arthritis, Ulcerative colitis, Juvenile Idiopathic Arthritis	2016	Amgen
2.	Etanercept	Etanercept-szzs (Erelzi)	Rheumatoid Arthritis, Polyarticular Juvenile Idiopathic Arthritis, Psoriatic Arthritis, Ankylosing Spondylitis, Plaque Psoriasis	2016	Sandoz
3.	Filgrastim	Filgrastim-sndz (Zarxio)filgrastim/ Neupogen	Patients with Cancer Receiving Myelosuppressive Chemotherapy, Patients with Acute Myeloid Leukemia Receiving Induction or Consolidation Chemotherapy, Patients with Cancer Undergoing Bone Marrow Transplantation, Patients Undergoing Autologous Peripheral Blood Progenitor Cell Collection and Therapy, Patients with Severe Chronic Neutropenia	2015	Sandoz
4.	Infliximab	Infliximab-dyyb (Inflectra)	Crohn's Disease, Pediatric Crohn's Disease, Ulcerative Colitis, Rheumatoid Arthritis, Ankylosing Spondylitis, Psoriatic Arthritis, Plaque Psoriasis	2016	Celltrion/ Pfizer
EMA					
5.	Epoetin alfa	Abseamed	Anaemia, Cancer, Chronic kidney failure	2007	Medice Arzneimittel Pütter
6.		Binocrit			Sandoz
7.		Epoetin alfa Hexal			Hexal
8.	Epoetin zeta	Retacrit	Anaemia, Autologous blood transfusion, Cancer,	2007	Hospira
9.		Silapo	Chronic kidney failure		Stada R & D
10.	Etanercept	Benepali	Axial spondyloarthritis, Psoriatic arthritis, Plaque psoriasis, Rheumatoid arthritis	2016	Samsung Bioepis
11.	Filgrastim	Accofil	Neutropenia	2014	Accord Healthcare
12.		Grastofil		2013	Apotex
13.		Biograstim	Cancer, Haematopoietic stem cell transplantation, Neutropenia	2008	CT Arzneimittel
14.		Tevagrastim			Teva Generics

15.		Filgrastim Hexal		2009	Hexal
16.		Zarzio			Sandoz
17.		Filgrastim ratiopharm		2008 [2011][a]	Ratiopharm
18.		Nivestim		2010	Hospira
19.	Follitropin alfa	Bemfola	Anovulation (IVF)	2014	Finox Biotech
20.		Ovaleap		2013	Teva Pharma
21.	Infliximab	Flixabi	Ankylosing spondylitis, Crohn's disease, Psoriatic arthritis, Psoriasis	2016	Samsung Bioepis
22.		Inflectra	Rheumatoid arthritis Ulcerative colitis	2013	Hospira
23.		Remsima			Celltrion
24.	Insulin glargine	Lusunda	Diabetes	2016	Merck (MSD)
25.		Abasaglar		2014	Eli Lilly/ Boehringer Ingelheim
26.	Rituximab	Truxima	Non-Hodgkin's lymphoma, Chronic Lymphocytic Leukemia, Rheumatoid Arthritis, Granulomatosis with Polyangiitis, and Microscopic Polyangiitis	2016	Celltrion
27.	Somatropin	Omnitrope	Pituitary dwarfism Prader-Willi syndrome Turner syndrome	2006	Sandoz
28.		Valtropin	Pituitary dwarfism Turner syndrome	2006 [2012][a]	BioPartners
29.	Teriparatide	Movymia	Osteoporosis	2016	Stada Arzneimittel
30.		Terrosa			Gedeon Richter

[a] *Year of withdrawal.*

FDA DHHS. "Amjevita BLA Approval Letter". <http://www.accessdata.fda.gov/drugsatfda_docs/appletter/2016/761024Orig1s000ltr.pdf>; 2016 [accessed 30.12.16]; FDA DHHS. "Erelzi BLA Approval Letter". <http://www.accessdata.fda.gov/drugsatfda_docs/appletter/2016/761042Orig1s000ltr.pdf>; 2016 [accessed 30.12.16]; FDA DHHS. "Zarzio BLA Approval Letter". Accessed online at: http://www.accessdata.fda.gov/drugsatfda_docs/appletter/2015/125553Orig1s000ltr.pdf>; 2015; US Prescribing Information of Inflectra. <http://www.accessdata.fda.gov/drugsatfda_docs/label/2016/125544s000lbl.pdf>; 2016 [accessed 30.12.16]; GaBI. Biosimilars approved in Europe. <http://www.gabionline.net/Biosimilars/General/Biosimilars-approved-in-Europe>; 2011 [accessed 30.12.16]; EMA Truxima Key Facts. <http://www.ema.europa.eu/ema/index.jsp?curl = pages/medicines/human/medicines/004112/smops/Positive/human_smop_001068.jsp&mid = WC0b01ac058001d127> [accessed 30.12.16] [19–24].

22.3.5 PHARMACOVIGILANCE

Regulations generally require a risk-management pharmacovigilance plan for biosimilars with close monitoring of safety after marketing authorization, with periodic safety update reports and/or post-marketing safety and immunogenicity study [10–17].

22.3.6 SUBSTITUTION AND INTERCHANGEABILITY

There is a wide variation in regulations with regards to substitution and interchangeability. In the EU, EMA does not evaluate and make assessments in this regard and these issues are to be addressed by the Individual Member State. The EMA considers this matter to be addressed between patients and their doctor and pharmacist Substitution of biologics (including between an innovator and a biosimilar) by a pharmacist without the permission of the prescribing doctor either is not allowed or is advised against in the vast majority of EU countries (including Italy, Spain, the UK, the Netherlands, Sweden, and Germany). Some of these countries do allow generic substitution in the case of conventional small molecule drugs, indicating that a different approach is being adopted in the case of biosimilars. However, France is an exception as it has allowed substitution of biosimilars as part of a law concerning the social security budget (Article 47 of the Law of 23 December 2013). However, even under this law, substitution of biosimilars is allowed only when initiating a course of treatment if the biosimilar belongs to the same group as the prescribed product (known as a "similar biologic group") and if the prescribing physician does not explicitly prohibit substitution of the prescribed biological by indicating "*non substiuable*" (nonsubstitutable) in hand-written characters on the prescription [25–27].

In the US, achieving a favorable FDA interchangeability opinion (refers to substitutability at the pharmacy level) requires manufacturers to undertake additional studies to demonstrate that for a biosimilar that is administered more than once to an individual, the risk in terms of safety or diminished efficacy of alternating or switching between its use and the use of "Reference Biologic" is not greater than the risk of using the "Reference Biologic" without such alternation or switch [11].

Japanese regulation does not give a definition of interchangeability for biosimilars. The naming system does not allow for substitution of biosimilars at the pharmacy level. Switching to a biosimilar is therefore dependent on the decision of the physician. Health Canada does not support automatic substitution of a biosimilar for its "Reference Biologic" and recommends that physicians make only well-informed decisions regarding therapeutic interchange. Similarly, automatic substitution of biosimilars at the pharmacy level is not allowed in South Korea [16,28,29].

22.4 BIOSIMILARS: EXCLUSIVITY

The regulations provide exclusivity for the innovator biologics which prevent filing and approval of an application for a biosimilar for a specified duration of time. The application for a biosimilar cannot be submitted for at least 4 years and cannot be approved for at least 12 years after the date on which the innovator biologic was first licensed. Thus, the innovator biologic is provided a total of 12 years of exclusivity. However, no additional exclusivity is provided for a supplement for the innovator biologic. Similarly, no additional exclusivity is provided in case of a subsequent application filed by the same sponsor or manufacturer of the innovator biologic (or a licensor, predecessor in interest, or other related entity) for a change in a new indication, route of administration, dosing schedule, dosage form, delivery system, delivery device, or strength which does not include a modification to the structure of the biological product or for a modification to the structure of the biological product that does not result in a change in safety, purity, or potency [11].

The regulations also provide exclusivity for first interchangeable biosimilar product. The period of such exclusivity varies and is the least of the following:

1. One year after first commercial marketing of the interchangeable product.
2. 18 months after the resolution of patent litigation with the sponsor of the reference product.
3. 42 months after initial approval of the interchangeable product if patent infringement litigation is ongoing.
4. 18 months after approval of the first interchangeable biosimilar if the biosimilar applicant has not been sued for patent infringement by the sponsor of the reference product.

However, there is no provision for exclusivity for a subsequent interchangeable biosimilar or for a noninterchangeable biosimilar [11].

Unlike in the US, there are no separate provisions for exclusivity for biologics other than those available for small molecule drugs. Thus, exclusivity for biologics in the EU is around 11 years, comprising 10 years for new biologics (8-year data exclusivity and 2-year market exclusivity) and a 1-year extension for a new indication [30].

22.5 BIOSIMLARS: MARKET ACCESS AND COMMERCIALIZATION

As discussed earlier, the advancement in technology has made it possible to develop low cost versions of biologics in the form of biosimilars. Also, with increasing pressure on healthcare budgets, more and more countries are developing regulatory mechanisms for approval of biosimilars [7]. However, the market access and commercial success of biosimilars are likely to be influenced by perceptions of payers, providers, and patients [31].

Compared to innovator biologics, biosimilars offer a 15%–30% price discount, which is significantly less than that offered by small molecule generics. The obvious reason for this is the more complex and costly development of biosimilars compared to small molecule generics. Given the high cost of biologics, even this price discount is significant from a payer perspective and most payers are likely to include biosimilars on their formularies. Payers are even likely to recommend the use of a biosimilar for an off-label indication approved for an innovator biologic even if the biosimilar lacks clinical data for that indication. Payors may even agree to switch existing patients to biosimilars once there is sufficient experience with the biosimilar, particularly if the biosimilar offers a greater price discount [32,33].

However, due to lower price differential, innovator biologics themselves are still likely to compete with biosimilars due to unique concerns associated with them. Even though regulators may approve biosimilars based on abbreviated data package, clinicians would still require substantial evidence directly comparing safety and efficacy between the innovator biologic and the biosimilar. The biosimilars are more likely to be prescribed to treatment-naïve patients than to existing patients. Just like clinicians, patients loyal to the reference brand may not have the same level of confidence with regards to the safety and efficacy of the biosimilars, and may actually consider them as inferior products [32,34].

Apart from above, the introduction of second- or next-generation biologics or biobetters by the innovator may help innovators maintain a greater market share against biosimilars [31].

REFERENCES

[1] Schellekens H. Follow-on biologics: challenges of the 'next generation'. Nephrol Dial Transplant (May 2005);20 (suppl 4): iv31–iv36.

[2] Tsang L, Cortez N. Biopharmaceuticals: definition and regulation. Pharmaceut Sci Encyclopedia 2010;1:1–18.

[3] US FDA. What Are "Biologics" Questions and Answers. <http://www.fda.gov/AboutFDA/CentersOffices/OfficeofMedicalProductsandTobacco/CBER/ucm133077.htm>; 2015 [accessed 22.12.16].

[4] CDER. Novel Drugs 2015 Summary. <http://www.fda.gov/downloads/Drugs/DevelopmentApprovalProcess/DrugInnovation/UCM485053.pdf>; 2016 [accessed 22.12.16].

[5] AHIP. High-Priced Drugs: Estimates of Annual Per-Patient Expenditures for 150 Specialty Medications. <https://www.ahip.org/wp-content/uploads/2016/04/HighPriceDrugsReport.pdf>; 2016 [accessed 22.12.16].

[6] Kelly C, Mir F. Biological therapies: how can we afford them? BMJ|Volume 339. <http://www.bmj.com/bmj/section-pdf/186337?path = /bmj/339/7722/Analysis.full.pdf>; 2009 [accessed 22.12.16].

[7] Moorkens E, Jonker-Exler C, Huys I, Declerck P, Simoens S, Vulto AG. Overcoming barriers to the market access of biosimilars in the european union: the case of biosimilar monoclonal antibodies. Front Pharmacol 2016;7:193.

[8] Bennett CL, Chen B, Hermanson T, Wyatt MD, Schulz RM, Georgantopoulos P, et al. Regulatory and clinical considerations for biosimilar oncology drugs. Lancet Oncol 2014;15(13):e594–605.

[9] CHMP. Guideline on Similar Biological Medicinal Products. <http://www.ema.europa.eu/docs/en_GB/document_library/Scientific_guideline/2009/09/WC500003517.pdf>; 2005 [accessed 22.12.16].

[10] CHMP. Guideline on Similar Biological Medicinal Products. <http://www.ema.europa.eu/docs/en_GB/document_library/Scientific_guideline/2014/10/WC500176768.pdf>; 2015 [accessed 22.12.16].

[11] Title VII—Improving Access to Innovative Medical Therapies, Subtitle A—Biologics Price Competition and Innovation Act of 2009. <http://www.fda.gov/downloads/Drugs/GuidanceComplianceRegulatoryInformation/UCM216146.pdf>; [accessed 22.12.16].

[12] MHLW. Guideline for the Quality, Safety, and Efficacy Assurance of Follow-on Biologics. <https://www.pmda.go.jp/files/000153851.pdf>; 2009 [accessed 30.12.16].

[13] Health Canada. Guidance Document- Information and Submission Requirements for Biosimilar Biologic Drugs. <http://hc-sc.gc.ca/dhp-mps/alt_formats/pdf/brgtherap/applic-demande/guides/seb-pbu/seb-pbu-2016-eng.pdf>; 2016 [accessed 30.12.16].

[14] CDSCO. Guidelines on Similar Biologics: Regulatory Requirements for Marketing Authorization in India. <http://cdsco.nic.in/writereaddata/CDSCO-DBT2016.pdf>; 2016 [accessed 30.12.16].

[15] Ropes and Gray. China Announces Final Biosimilars Guideline. <http://www.mondaq.com/x/379908/food + drugs + law/China + Announces + Final + Biosimilars + Guideline>; 2015 [accessed 30.12.16].

[16] Joung J. Korean regulations for biosimilars.. Generics Biosimilars Initiative J. (GaBI J.) 2015;4(2): 93–4.

[17] WHO. Guidelines on Evaluation of Similar Biotherapeutic Products. <http://www.who.int/biologicals/areas/biological_therapeutics/BIOTHERAPEUTICS_FOR_WEB_22APRIL2010.pdf>; 2009 [accessed 30.12.16].

[18] TGA. Evaluation of Biosimilars. <https://www.tga.gov.au/sites/default/files/pm-argpm-biosimilars-150420_1.pdf>; 2013 [accessed 30.12.16].

[19] FDA DHHS. "Amjevita BLA Approval Letter". <http://www.accessdata.fda.gov/drugsatfda_docs/appletter/2016/761024Orig1s000ltr.pdf>; 2016 [accessed 30.12.16].

[20] FDA DHHS. "Erelzi BLA Approval Letter". <http://www.accessdata.fda.gov/drugsatfda_docs/appletter/2016/761042Orig1s000ltr.pdf>; 2016 [accessed 30.12.16].

[21] FDA DHHS. "Zarxio BLA Approval Letter". Accessed online at: http://www.accessdata.fda.gov/drugsatfda_docs/appletter/2015/125553Orig1s000ltr.pdf>; 2015.

[22] US Prescribing Information of Inflectra. <http://www.accessdata.fda.gov/drugsatfda_docs/label/2016/125544s000lbl.pdf>; 2016 [accessed 30.12.16].

[23] GaBI. Biosimilars approved in Europe. <http://www.gabionline.net/Biosimilars/General/Biosimilars-approved-in-Europe>; 2011 [accessed 30.12.16].

[24] EMA Truxima Key Facts. <http://www.ema.europa.eu/ema/index.jsp?curl = pages/medicines/human/medicines/004112/smops/Positive/human_smop_001068.jsp&mid = WC0b01ac058001d127> [accessed 30.12.16].

[25] European biosimilars conference highlights extrapolation as key issue, Generics and Biosimilars Initiative Journal (GaBI Journal) 2015;4(3):148–9. <http://gabi-journal.net/european-biosimilars-conference-highlights-extrapolation-as-key-issue.html#R3> [accessed 30.12.16].

[26] Mestre-Ferrandiz J, Towse A, Berdud M. Biosimilars: how can payers get long-term savings?. Pharmacoeconomics 2016;34:609–16.

[27] EMA. Questions and answers on biosimilar medicines (similar biological medicinal products). <http://www.ema.europa.eu/docs/en_GB/document_library/Medicine_QA/2009/12/WC500020062.pdf>; 2012 [accessed 30.12.16].

[28] Health Canada. Questions & Answers To Accompany the Final Guidance for Sponsors: Information and Submission Requirements for Subsequent Entry Biologics (SEBs). <http://www.hc-sc.gc.ca/dhp-mps/brgtherap/applic-demande/guides/seb-pbu/01-2010-seb-pbu-qa-qr-eng.php#q15>; 2010 [accessed 30.12.16].

[29] GaBI. Naming and interchangeability for biosimilars in Japan. <http://www.gabionline.net/Reports/Naming-and-interchangeability-for-biosimilars-in-Japan>; 2016 [accessed 30.12.16].

[30] Regulation (EC) no 726/2004 of the European Parliament and of the Council of 31 March 2004. <http://ec.europa.eu/health/files/eudralex/vol-1/reg_2004_726/reg_2004_726_en.pdf> [accessed 12.16].

[31] Singh SC, Bagnato KM. The economic implications of biosimilars. Am J Manage Care 2015;16 Suppl:21.

[32] Cohen JP, Felix AE, Riggs K, et al. Barriers to market uptake of biosimilars in the US: generics and biosimilars initiative. GaBI J 2014;3(3):108–15.

[33] Payer perspectives on biosimilars. <http://www.xcenda.com/Insights-Library/Payer-Perspectives/PayerPerspectives-on-Biosimilars/>; [accessed 30.12.16].

[34] Rompas S, Goss T, Amanuel S, Coutinho V, Lai Z, Antonini P, Murphy MF. Demonstrating value for biosimilars: a conceptual framework. Am Health Drug Benefits 2015;8(3):129–39.

CHAPTER

RE-INNOVATION IN PHARMACEUTICAL INDUSTRY: SUPERGENERICS AND BIOBETTERS

23

Gursharan Singh

Life Sciences, SmartAnalyst India Private Limited, Gurgaon, Haryana, India

23.1 INTRODUCTION

The classical R&D model of developing new chemical or biologic entities (radical innovation) involves a high risk of failure (due to toxicity, pharmacokinetics, lack of clinical efficacy, safety concerns, etc.), requires a huge investment (up to $2.6 billion per product reaching the market), and is time consuming (8–12 years). The declining return on investment in the NCE/NBE R&D has resulted in pharmaceutical and biotechnology companies exploring novel models of innovation. These include incremental innovation and re-innovation [1,2].

Incremental innovation involves creating minor improvements or changes to an existing chemical entity. This strategy has been pursued in the past primarily by innovator companies for the ever-greening of patents and of late by generic companies to avoid patent infringement. This usually offers no or little value-addition [1].

Re-innovation involves creating changes to an existing chemical/biologic entity which result in value-addition through application of breakthrough technology or new technology platforms, new components, and new configurations [1].

The focus of this chapter is on re-innovation of small molecule drugs and biologics in the form of supergenerics and biobetters respectively. A closely related term "Next generation biotherapeutics" is also discussed.

23.2 SUPERGENERICS: TERMINOLOGY

The term Supergenerics is used for novel treatments and treatment modalities, derived from existing off-patent small molecule drugs. Many generics companies have developed specialty R&D capabilities in an effort to circumvent the patents' innovator companies. This is evident from the increasing number of patents being issued to generic companies. Many big generic companies are now focusing on hybrid models of pharmaceutical R&D and developing re-innovated products in

Pharmaceutical Medicine and Translational Clinical Research. DOI: http://dx.doi.org/10.1016/B978-0-12-802103-3.00024-9

Table 23.1 Commonly Used Terms in Literature to Refer to "Supergenerics"

1	Value added generics
2	New therapeutic entities
3	Improved therapeutic entities
4	Hybrid pharmaceuticals
5	Enhanced therapeutics
6	Re-innovated generics
7	Branded generics
8	Premium generics

From Barei F, Ross M. The refinement of the super generic concept: semantic challenge for product re-innovation? Generics Biosimilars Initiative J (GaBI Journal) 2015;4(1):25–32 [5].

the form of supergenerics. Since supergenerics are derived from existing small molecule drugs with proven efficacy and safety, the supergeneric R&D involves lower risk of failure. Moreover regulations allow approval of supergenerics based on bridging studies. Therefore supergenerics require lower investment in terms of time and money [3,4]. Many other terms given in Table 23.1 are commonly used to refer to supergenerics.

There is a trend towards avoiding the term generic in nomenclature of re-innovated products as pharmaceutical companies search for a "nongeneric" identity and want to project themselves as research based pharmaceutical companies. Moreover, using the "generics" designation for re-innovated products seems to present an obstacle to generating higher revenues [5].

23.3 SUPERGENERICS: REGULATORY AND CLINICAL DEVELOPMENT FRAMEWORK

The United States (US) and European Union (EU) regulatory framework provides for NDA (New Drug Application) 505 B(2) and hybrid application route, respectively, for products involving re-innovation of already approved small molecules [6,7,8].

For NDA 505 B(2) application, the applicant is not required to conduct all studies as for a full NDA [referred to as NDA 505 B(1)] application and can rely on studies conducted by others from whom the applicant has not obtained a right of reference. The applicant can rely on literature and/or upon the FDA's finding of safety and effectiveness for a previously approved drug product (Reference Listed Drug). The purpose of the NDA 505 B(2) provision is to avoid the need of duplication of studies to demonstrate what is already known about a drug. Therefore, the applicants with a product based on incremental innovation or re-innovation need to conduct only bridging studies [7]. Table 23.2 provides a list of changes to approved drugs which can be submitted as a 505(b)(2) application [7].

The Centre for Drug Evaluation and Research (CDER) Manual of Policies and Procedures (MAPP) 5018.2 describes the NDA classification code assigned to an NDA based on characteristics of the product in the application. Table 23.3 provides NDA submission classes applicable for 505 (B)(2) application [8].

Table 23.2 Changes to Approved Drugs Which Can be Submitted as a 505(b)(2) Application

1	Changes in dosage form, strength, route of administration, formulation, dosing regimen
2	New indication
3	New combination product
4	Change to an active ingredient (e.g., different salt, ester complex, chelate, etc.)
5	New molecular entity when studies have been conducted by other sponsors and published information is pertinent to the application (e.g., a pro-drug or active metabolite of an approved drug)
6	Change from an Rx indication to an OTC indication
7	Change to an OTC monograph drug (e.g., nonmonograph indication, new dosage form)
8	Bioinequivalence for drug products otherwise intended to be filed as generics

From CDER. Draft Guidance for Industry, Applications Covered by Section 505(b)(2). http://www.fda.gov/downloads/Drugs/.../Guidances/ucm079345.pdf%; 1999 [accessed 15.12.16].

Table 23.3 NDA Submission Classes Applicable for NDA 505(B)(2) Route

Type of Change to Approved Drug	NDA Submission Classification
New active ingredient	Type 2
New dosage form	Type 3
New combination	Type 4
New formulation or other differences	Type 5
Previously marketed but without an approved NDA	Type 7
Rx to OTC	Type 8
New indication or claim, drug to be marketed under type 10 NDA after approval	Type 10
Type 2, New active ingredient, and Type 3, New dosage form	Type 2/3
Type 2, New active ingredient and Type 4, New combination	Type 2/4
Type 3, New Dosage Form, and Type 4, New combination	Type 3/4

From CDER. MAPP 5018.2 (NDA Classification Codes). http://www.fda.gov/downloads/aboutfda/centersoffices/officeofmedicalproductsandtobacco/cder/manualofpoliciesprocedures/ucm470773.pdf; 2015 [accessed 15.12.16].

Analogous to the NDA 505 B(2) route in the US, Article 10 (3) of EU DIRECTIVE 2001/83/EC lays down the requirement of appropriate preclinical or clinical studies in case of changes in the active substance(s), therapeutic indications, strength, pharmaceutical form, or route of administration—compared to the reference medicinal product [9].

The type of bridging studies required for approval of products filed under NDA 505 B(2) or Article 10(3) depend on the extent of modification made to the existing drug product or its label. While a change in dosage form (e.g., tablet to suspension) may be approved based on comparative bioavailability study(ies), extensive clinical and preclinical studies may be required for approval of new indication(s) for existing drug products (also known as drug repurposing). The NDA 505 B(2) application for Epaned Kit (enalapril oral suspension) consisted of three relative bioavailability studies, the results of which were used to bridge to the finding of efficacy and safety of the reference listed drug Vasotec® Tablets [10]. The NDA 505 B(2) application for Horizant (Gabapentin enacarbil, a prodrug of Gabapentin) for moderate-to-severe primary restless legs syndrome (RLS) in adults consisted of several clinical and preclinical studies [11].

23.4 SUPERGENERICS: INTELLECTUAL PROPERTY CONSIDERATIONS

Approval or filing of a 505(B)(2) application may be delayed because of patent and exclusivity rights that apply to the listed drug. Therefore, the 505(B)(2) products can be launched only after the loss of data exclusivity and if the patent rights of the innovator are not infringed [7].

On the other hand, a 505(B)(2) application may itself be granted 3 years of Waxman-Hatch exclusivity if one or more of the clinical studies, other than bioavailability/bioequivalence (BA/BE) studies, was essential to approval of the application and was conducted or sponsored by the applicant. A 505(B)(2) application may also be granted 5 years of exclusivity if it is for a new chemical entity, and 7 years exclusivity if it is repurposed in an orphan indication or pediatric exclusivity. A 505(B)(2) application also contains information on patents claiming the drug or its method of use [7].

23.5 SUPERGENERICS: VALUE ADDITION

Supergenerics may offer improvement in terms of patient convenience (once daily dosing etc.), new dosage form (suspension instead of tablet or capsule for children), route of delivery (topical administration for localized pain), pharmacokinetic profile (e.g., no food effect, etc.), safety, efficacy, stability, manufacturing process, etc. [3,5]. This improvement results in value added products for patients, physicians, and healthcare systems by addressing specific, unmet needs. Table 23.4 gives examples of differentiated products which have been approved by NDA 505 B(2) route in the USA.

Unlike in the US, fewer products have been approved in the EU by the analogous Article 10 (3) route. Table 23.5 gives examples of differentiated products which have been approved by the Article 10 (3) route in EU.

From the perspective of pharmaceutical companies, supergenerics offer higher returns compared to conventional generics as the end product may gain a significant price premium depending on the value-addition offered. The price of Vimovo Capsules is almost four times the price of individual mono-components (corresponding strengths of esomeprazole and naproxen purchased separately) [24].

23.6 FOLLOW ON BIOLOGICS

Biologics can hit targets that a small molecule cannot, and are relatively safer. Therefore, biologics offer safe and effective treatment options for diseases which cannot be treated with small-molecules. Pharmaceutical companies are shifting their focus to biologics as they offer a higher

Table 23.4 List of Products Approved by NDA 505 B(2) Route

S. No.	Product (Brand Name/Generic Name)	Company	Year of Approval	Value-Addition Offered	Original Product/ Company/Year of Approval
1	Carnexiv/ Carbamazepine Injection	Lundbeck LLC	2016	New Dosage Form	Tegretol/Novartis/ 1968
2	Simvastatin Oral Suspension	Rosemont Pharmaceuticals Limited	2016	New Dosage Form	Zocor/Merck/1991
3	Mitigare/Colchicine Capsules	Hikma Pharmaceuticals	2014	New Dosage Form	Colchicine/No NDA for original product*
4	Epaned Kit/Enalapril Oral Solution	Silvergate Pharms	2013	New Dosage Form	Vasotec/Valeant Pharms North/ 1985
5	Nuedexta/ Dextromethorphan hydrobromide and quinidine sulfate Capsules	Avanir Pharmaceuticals	2010	New combination for a novel indication	Dextromethorphan
6	Silenor/Doxepin Tablets	Pernix Therapeutics	2010	New Dosage Form for new indication	Sinequan/Pfizer/ 1969
7	Zegerid/Omeprazole and Sodium Bicarbonate Capsules	Santarus	2006	New Combination with $NaHCO_3$ to prevent acid degradation of omeprazole; avoids the need of enteric coating	Prilosec/ Astrazeneca /1989
8	Altocor/Levostatin Extended Release Tablets	Aura Laboratories, Inc./Covis Pharma SARL	2002	Longer Tmax and lower Cmax	Mevacor/Merck/ 1987
9	Canasa/Mesalamine FIV-ASA Suppositories	Axcan Scandipharma Inc.	2001	New Formulation with improved dissolution profile	Rowasa Suppositories/ Solvay/1990
10	Luxiq/Betamethasone Valerate Foam	Delcor Asset Corp	1999	New Dosage Form	Valisone/Schering /1969

From Refs. [12–21].

Table 23.5 List of Products Approved by EMA Through Article 10 (3) Route

S. No.	Product (Brand Name/Generic Name)	Company	Year of Approval	Value-Addition Offered	Original Product/Company/ Year of Approval
1	PecFent/Fentanyl Nasal Spray	Archimedes Development Ltd	2010	New Dosage Form	Actiq, lozenge/Cephalon/ 2000
2	Controloc control/Pantoprazole 20 mg gastro-resistant tablets	Nycomed	2009	OTC switch	Pantozol 20 mg gastro-resistant tablets/Nycomed /1998

From EMA. Assessment report for PecFent. <http://www.ema.europa.eu/docs/en_GB/document_library/EPAR_-_Public_assessment_report/human/001164/WC500096495.pdf>; 2009 [accessed 15.12.16] [22]; EMA. Assessment report for Controloc Control. <http://www.ema.europa.eu/docs/en_GB/document_library/EPAR_-_Public_assessment_report/human/001097/WC500034153.pdf>; 2009 [accessed 15.12.16] [23].

return on investment compared to small molecules. A higher percentage of biologics have turned out as blockbusters. The increasing role of biologics in the treatment of various diseases can be further appreciated from the fact that in 2014, out of the 41 novel products approved the FDA, 11 were biologics. The year also saw the maximum number of BLAs approved in last 20 years. As biologics are very complex molecules, this constitutes a major hurdle in imitating them both from a scientific and a regulatory perspective [25,26].

With the passage of time, biopharmaceutical companies focus on development of follow on products. These include biosimilars, biobetters, and the next-generation biotherapeutics. Biosimilars have the same amino acid sequence and molecular profile as the reference biologic and represent a relatively low cost alternative version of originator biologic without any clinical advantage [27]. Biobetters and next generation biotherapeutics as the term itself suggests represent value-added versions of original biologic. Here we discuss the basic concept of biobetters and next-generation biotherapeutics and the value-addition they offer over existing biologics. We also discuss various strategies which are adopted to modify the existing biologics.

23.7 BIOBETTERS AND NEXT GENERATION BIOTHERAPEUTICS: TERMINOLOGY

Biobetter or biosuperior is a generic term used to refer to value-added biologics. However, in the case of monoclonal antibodies, which are one of the most common classes of biologics, various terminologies are used for value-added products depending on the type of modification made to the original biologic. Biobetter antibodies are those that target the same validated epitope as an existing antibody, but have improved properties. Second-generation antibodies have improved variable domains (e.g., humanized or human variable domains, etc.) to decrease immunogenicity. Third-generation antibodies are those which target different epitopes or trigger other mechanisms of action. These are often engineered for improved Fc-associated immune functions or half-life [27,28]. Compared to developing a new biologic, the biobetter benefits from an established mechanism, safety, and efficacy profile of a known biologic. This reduces the risk of failure. Further for certain biobetters, a possibility to demonstrate pharmacologic comparability with the original biologic may help accelerate development programs. Experience with the original biologic may help in the selection of dose and biomarkers during both nonclinical and clinical development, and the regulatory agencies might agree to reduce the sample size and duration of Phase 2 trials and to focus safety monitoring on the known side effects of the target pathway. An example is the approval of Oncaspar for acute leukemia patients hypersensitive to asparaginase. The safety and effectiveness of Oncaspar was evaluated in 4 open-label studies enrolling a total of 42 patients with multiple-relapsed, acute leukemia with a history of a prior clinical allergic reaction to asparaginase [29,30].

23.8 APPROACHES FOR VALUE ADDITION TO ORIGINAL BIOLOGICS

Various strategies have been employed by protein engineering companies to modify the existing biologic to create a product with superior characteristics. These approaches include chemical modification (pegylation), protein fusion [albumin fusion, carboxy-terminal peptide (CTP fusion)], antibody-drug conjugates, affinity maturation, and a variety of antibody engineering techniques/ altered amino acid sequence (engineered fc domain/glyco-engineering, humanization of the glycosylation pattern), etc. [31,32].

Pegylation is one of the most common protein engineering tools. This involves covalent addition of polyethylene glycol which increases the molecular size of the biologic thus preventing the glomerular filtration and resulting in prolongation of serum half-life. Pegylation also offers the advantage of reduced immunogenicity [33].

In view of the long half-life of human serum albumin, it has been possible to prolong the serum half-life of many biologics employing both covalent and non-covalent binding. Apart from prolongation of half-life, site-specific conjugation of albumin to a permissive site of a target protein may even help extend the therapeutic activity [34].

CTP is a naturally occurring and hence non-immunogenic peptide contained in hCG (human chorionic gonadotropin), which is responsible for its long lifespan. It can be attached to other protein therapeutics to prolong their serum half-life [35].

Antibody drug conjugates combine the targeted anticancer property of a monoclonal antibody with a non-specific cytotoxic agent. This approach results in improved efficacy of the monoclonal antibody by preventing up-regulation of alternative mechanisms normally responsible for diminishing the activity of targeted therapy. In contrast to separate administration of a cytotoxic agent, the conjugation approach achieves targeted delivery and hence is associated with fewer adverse reactions [36].

Affinity maturation involves introducing random mutations and filter assays to improve the binding affinity of the variable fragment of the antibody (which binds to the antigen) and/or modulating the target selectivity [37].

Antibody engineering techniques involve alteration of amino acid sequences. The constant fragment (Fc) of an antibody is responsible for interactions with immune cells, and the associated properties of the Fc can be modulated by engineering at several levels. These include modifying the anti- and pro-inflammatory properties by altering the glycosylation status, site-directed mutagenesis to alter binding to Fc receptors to modulate the antibody-dependent cellular cytotoxicity (ADCC), Fc engineering to increase binding to the neonatal Fc receptor, to increasing the serum half-life, etc. Humanization of the glycosylation pattern of the variable fragment helps to reduce immunogenicity [37].

23.9 BIOBETTERS AND NEXT GENERATION BIOTHERAPEUTICS: VALUE ADDITION

Biobetters and next generation biotherapeutics have optimized characteristics compared to the original biologic in terms of improved pharmacokinetics (duration of action), route of delivery (subcutaneous versus intravenous), efficacy/potency, safety, immunogenicity, patient convenience, stability, longer shelf-life, etc. They may find usage in different patient segment with higher unmet needs [35]. Table 23.6 provides a list of biobetter and next generation biotherapeutics products.

Table 23.6 Value Added Follow on Biologics: Biobetters and Next Generation Biotherapeutics

S. No.	Type	Product	Original Biologic	Modification and Value Addition
1	Biobetter antibodies	Margetuximab	Trastuzumab	Fc-optimization resulting in improved cell-killing properties by increased binding to activating receptors and decreased binding to the inhibitory receptor on immune effector cells
2		Trastuzumab emtansine (Kadcyla)		Trastuzumab emtansine is an antibody-drug conjugate and it connects two anti-cancer properties: the HER2 inhibition of trastuzumab (the active ingredient found in Herceptin) and the microtubule inhibition of DM1. It is indicated as a single agent for the treatment of adults with HER2-positive, unresectable locally advanced or metastatic breast cancer who would otherwise need the combination
3		TrasGex		Glycooptimization of trastuzumab; claimed to yield a manifold higher anti-tumor activity
4	2nd generation antibodies	Panitumumab	Cetuximab	Human versus chimeric mAb with reduced dosing frequency to every 2 weeks as well as reduced incidence of infusion reactions.
5		Ofatumumab (Arzerra)	Rituximab	Human versus chimeric mAb efficacy shown in rituximab-refractory follicular lymphoma
6		Veltuzumab		Humanized versus chimeric mAb with efficacy claimed in CD20-positive-B cell NHL patients who had received more than two rituximab regimens
7	3rd generation antibodies	Obinutuzumab (Gazyva)	Rituximab	A novel type II anti-CD20 mAb that demonstrated an overall survival advantage when combined with chemotherapy in previously untreated older patients with CLL and comorbidities. Obinutuzumab was superior to rituximab in terms of response rates and progression-free survival
8		Ocaratuzumab		Humanized mAb optimized through protein engineering for both increased affinity to CD20 and enhanced effector function in antibody-dependent cell-mediated cytotoxicity (ADCC) assays. Efficacy claimed in CD20 + follicular non-Hodgkin's lymphoma (NHL) patients with poor response to a prior rituximab treatment
9	Non mAb Value Added Biologics/ Biobetters	Pegaspargase (Oncaspar)	L-asparaginase	*E. coli*-derived L-asparaginase covalently conjugated to monomethoxypolyethylene glycol (mPEG) to reduce dosing frequency to every 2 weeks as well as reduced incidence of hypersensitivity reactions

Table 23.6 Value Added Follow on Biologics: Biobetters and Next Generation Biotherapeutics *Continued*

S. No.	Type	Product	Original Biologic	Modification and Value Addition
10		Darbepoetin Alfa (Aranesp)	Erythropoetin	Produced in Chinese hamster ovary (CHO) cell instead of mammalian cells offer the advantage of reducing dosing frequency to every 2 weeks
11		Methoxy polyethylene glycol-epoetin beta (Mircera)		Chemical bonding between either the N-terminal amino group or the ε-amino group of any lysine present in erythropoietin and methoxy polyethylene glycol (PEG) butanoic acid helps in reducing dosing frequency to every 4 weeks
12		Pegfilgrastim (Neulesta)	Filgrastim (Neupogen)	Pegylated to reduce dosing frequency to every 3 weeks

From Refs. [30,38–52].

23.10 BIOBETTERS: REGULATORY AND CLINICAL DEVELOPMENT FRAMEWORK

Biobetters are different from existing biologics and are regarded as new biologic entities. Therefore, the requirement is to file a new BLA (Biologic License Application) for approval. Therefore, like any new biologic, a full development package would be required for approval including manufacturing, preclinical, and clinical data [29].

From a clinical development perspective, the requirement would be to demonstrate the superiority of biobetter over existing biologic. The superiority may be demonstrated in the form of efficacy in a population resistant to original biologic, or reduced dosing frequency or lower incidence of hypersensitivity reactions, etc., compared to the original biologic. Accordingly, approval of Aranesp was based on clinical studies which demonstrated that Aranesp administered once weekly was non-inferior in efficacy to epoetin alfa administered twice or thrice weekly [51]. A Phase 2 Study of margetuximab in patients with advanced breast cancer who have either relapsed or are refractory to available therapies including trastuzumab is currently ongoing [53].

23.11 BIOBETTERS: INTELLECTUAL PROPERTY CONSIDERATIONS

From an I.P. (intellectual property) protection perspective, in the US, the biologics price competition and innovation act, 2009 provides 12 years of exclusivity to innovator biologics. However, biobetters need to fulfill certain requirements to be eligible for this exclusivity. A new period of exclusivity is not available for a biological product if the licensure is for a subsequent application filed by the same sponsor or manufacturer of the biological product (or a licensor, predecessor in interest, or other related entity) for a change (not including a modification to the structure of the biological product) that results in a new indication, route of administration, dosing schedule, dosage form, delivery system, delivery device, or strength; or a modification to the structure of the biological product that does not result in a change in safety, purity, or potency. The term "predecessor in interest" refers to an entity (e.g., a corporation) that the sponsor has taken over, merged with, or purchased, or from which the sponsor has purchased all rights to the drug [reference product]. This also includes an entity which has granted to the applicant exclusive rights to a new drug application or the data upon which exclusivity is based, which may include licensors, assignors, and joint venture partners, depending on the circumstances of the case. For protein products, structural differences include, as appropriate, any differences in amino acid sequence, glycosylation patterns, tertiary structures, post-translational events (including any chemical modifications of the molecular structure such as pegylation), and infidelity of translation or transcription, among others. To be eligible for exclusivity, sponsors need to demonstrate change in safety, purity, or potency for the structurally modified biologic products [54].

REFERENCES

[1] Barei F, Le Pen C, Simoens S. The generic pharmaceutical industry: moving beyond incremental innovation towards re-innovation. Generics Biosimilars Initiative J (GaBI J) 2014;2(1):13–19.

[2] DiMasi JA, Grabowski HG, Hansen RA. Innovation in the pharmaceutical industry: new estimates of R&D costs. J Health Econ 2016;47:20–33.

[3] Teva Pharmaceuticals Industries Limited. NTEs: A New Pipeline ....Through Integration. <http://www.tevapharm.com/research_development/rd_integrated/new_therapeutic_entities/>; 2015.

[4] Ross MS. Innovation strategies for generic drug companies: moving into supergenerics. IDrugs. 2010 Apr;13(4):243–7.

[5] Barei F, Ross M. The refinement of the super generic concept: semantic challenge for product re-innovation? Generics Biosimilars Initiative J (GaBI J) 2015;4(1):25–32.

[6] Stegemann S, Klebovich I, Antal I, Blume HH, Magyar K, Németh G, et al. Improved therapeutic entities derived from known generics as an unexplored source of innovative drug products. Eur J Pharm Sci 2011 Nov 20;44(4):447–54.

[7] CDER. Draft Guidance for Industry, Applications Covered by Section 505(b)(2). <http://www.fda.gov/downloads/Drugs/.../Guidances/ucm079345.pdf%>; 1999 [accessed 15.12.16].

[8] CDER. MAPP 5018.2 (NDA Classification Codes). <http://www.fda.gov/downloads/aboutfda/centersoffices/officeofmedicalproductsandtobacco/cder/manualofpoliciesprocedures/ucm470773.pdf>; 2015 [accessed 15.12.16].

[9] Directive 2001/83/EC of the European Parliament and of the Council of 6 November 2001 on the Community Code Relating to Medicinal Products for Human Use last amended on 31st March, 2004. <http://www.ema.europa.eu/docs/en_GB/document_library/Regulatory_and_procedural_guideline/2009/10/WC500004481.pdf> [accessed 15.12.16].

[10] CDER. Epaned Kit Summary Review. <http://www.accessdata.fda.gov/drugsatfda_docs/nda/2013/204308Orig1s000SumR.pdf>; 2013 [accessed 15.12.16].

[11] CDER. Horizant Medical Review. <http://www.accessdata.fda.gov/drugsatfda_docs/nda/2011/022399Orig1s000MedR.pdf>; 2011 [accessed 15.12.16].

[12] FDA DHHS. Carnexiv NDA approval letter. <http://www.accessdata.fda.gov/drugsatfda_docs/appletter/2016/206030Orig1s000ltr.pdf>; 2016 [accessed 15.12.16].

[13] FDA DHHS. Simvastatin Oral Suspension approval letter. <http://www.accessdata.fda.gov/drugsatfda_-docs/appletter/2016/206679Orig1s000ltr.pdf>; 2016 [accessed 15.12.16].
[14] FDA DHHS. Mitigare Capsules approval letter. <http://www.accessdata.fda.gov/drugsatfda_docs/appletter/2014/204820Orig1s000ltr.pdf>; 2014 [accessed 15.12.16].
[15] FDA DHHS. Epaned Kit approval letter. <http://www.accessdata.fda.gov/drugsatfda_docs/appletter/2013/204308Orig1s000ltr.pdf>; 2013 [accessed 15.12.16].
[16] FDA DHHS. Nuedexta approval letter. <http://www.accessdata.fda.gov/drugsatfda_docs/appletter/2010/021879s000ltr.pdf>; 2013 [accessed 15.12.16].
[17] FDA DHHS. Silenor Tablets approval letter. <http://www.accessdata.fda.gov/drugsatfda_docs/appletter/2010/022036s000ltr.pdf>; 2010 [accessed 15.12.16].
[18] FDA DHHS. Zegerid Capsules approval letter. <http://www.accessdata.fda.gov/drugsatfda_docs/appletter/2006/021849s000ltr.pdf>; 2006 [accessed 15.12.16].
[19] FDA DHHS. Altocor Extended Release Tablets approval letter. <http://www.accessdata.fda.gov/drugsatfda_docs/appletter/2002/21316ltr.pdf>; 2002 [accessed 15.12.16].
[20] FDA DHHS. Canasa Suppositories approval letter. <http://www.accessdata.fda.gov/drugsatfda_docs/appletter/2001/21252ltr.pdf>; 2001 [accessed 15.12.16].
[21] FDA DHHS. Luxiq Foam approval letter. <http://www.accessdata.fda.gov/drugsatfda_docs/appletter/1999/20934ltr.pdf>; 2001 [accessed 15.12.16].
[22] EMA. Assessment report for PecFent. <http://www.ema.europa.eu/docs/en_GB/document_library/EPAR_-_Public_assessment_report/human/001164/WC500096495.pdf>; 2009 [accessed 15.12.16].
[23] EMA. Assessment report for Controloc Control. <http://www.ema.europa.eu/docs/en_GB/document_library/EPAR_-_Public_assessment_report/human/001097/WC500034153.pdf>; 2009 [accessed 15.12.16].
[24] Drug Price Database. Vimovo Price. <https://www.drugs.com/price-guide/>; 2016 [accessed 15.12.16].
[25] Wong G. Biotech scientists bank on big pharma's biologics push. Nat Biotechnol 2009;27:293–5.
[26] Mullard A. 2014 FDA drug approvals. Nat Rev Drug Discov 2015;14:77–81.
[27] Beck, A., editor. Biosimilar, biobetter and next generation therapeutic antibodies. mAbs 2011;3(2), 107–110.
[28] Oflazoglu E, Audoly LP. Evolution of anti-CD20 monoclonal antibody therapeutics in oncology. mAbs 2010;2(1):14–19.
[29] Anour R.) Biosimilars versus 'biobetters'—a regulator's perspective. Generics Biosimilars Initiative J (GaBI J) 2014;3(4):166–7.
[30] US Prescribing Information of Oncaspar. <http://www.accessdata.fda.gov/drugsatfda_docs/label/2016/103411s5188lbl.pdf>; 2014 [accessed 18.12.16].
[31] Carter PJ. Introduction to current and future protein therapeutics: a protein engineering perspective. Exp Cell Res May 15 2011;317(9):1261–9.
[32] Li J, Zhu Z. Research and development of next generation of antibody-based therapeutics. Acta Pharmacol Sin 2010;31(9):1198–207.
[33] McDonnell T, Ioannou Y, Rahman A. PEGylated drugs in rheumatology—why develop them and do they work? Rheumatology (Oxford, England) 2014;53(3):391–6.
[34] Lim SI, Hahn YS, Kwon I. Site-specific albumination of a therapeutic protein with multi-subunit to prolong activity in vivo. J Controlled Release Offic J Controlled Release Soc 2015;207:93–100.
[35] Strohl WR. Fusion proteins for half-life extension of biologics as a strategy to make biobetters. Biodrugs 2015;29(4):215–39.
[36] Peters C, Brown S. Antibody–drug conjugates as novel anti-cancer chemotherapeutics. Biosci Rep 2015;35(4):e00225.
[37] Beck, et al. Strategies and challenges for the next generation of therapeutic antibodies. Nat Rev Immunol 2010;10:345–52.
[38] Margetuximab. <https://www.macrogenics.com/margetuximab-anti-her2/> [accessed 18.12.16].

[39] US Prescribing Information of Kadcyla. <http://www.accessdata.fda.gov/drugsatfda_docs/label/2016/125427s096lbl.pdf>; 2016 [accessed 18.12.16].

[40] TrasGex. <http://www.glycotope.com/pipeline/trasgex> [accessed 18.12.16].

[41] US Prescribing Information of Vectibix (panitumumab). <http://www.accessdata.fda.gov/drugsatfda_-docs/label/2015/125147s200lbl.pdf>; 2016 [accessed 18.12.16].

[42] US Prescribing Information of Erbitux (cetuximab). <http://www.accessdata.fda.gov/drugsatfda_docs/label/2015/125084s262lbl.pdf>; 2015 [accessed 18.12.16].

[43] Graham CN, et al. Economic analysis of panitumumab compared with cetuximab in patients with wild-type KRAS metastatic colorectal cancer that progressed after standard chemotherapy. Clin Ther June 2016;38(6):1376–91.

[44] US Prescribing Information of Arzerra (ofatumumab). <http://www.accessdata.fda.gov/drugsatfda_docs/label/2016/125326s063lbl.pdf>; 2016 [accessed 18.12.16].

[45] US Prescribing Information of Rituxan (rituximab). <http://www.accessdata.fda.gov/drugsatfda_docs/label/2014/103705s5432lbl.pdf>; 2014 [accessed 18.12.16].

[46] Czuczman MS, et al. Ofatumumab monotherapy in rituximab-refractory follicular lymphoma: results from a multicenter study. Blood 2012;119(16):3698–704 Apr 19.

[47] Morschhauser F, et al. Humanized anti-CD20 antibody, Veltuzumab, in refractory or recurrent non-Hodgkin's lymphoma: Phase I/II results. J Clin Oncol 2009;27:3346–53.

[48] Owen & Stewart. Obinutuzumab for the treatment of patients with previously untreated chronic lymphocytic leukemia: overview and perspective. Therapeutic Adv Hematol 2015;6(4):161–70.

[49] Ocaratuzumab (AME-133v). <http://www.mentrik.com/oncology.html>.

[50] US Prescribing Information of Aranesp. <http://www.accessdata.fda.gov/drugsatfda_docs/label/2015/103951s5363lbl.pdf>; 2015 [accessed 18.12.16].

[51] US Prescribing Information of Mircera. <http://www.accessdata.fda.gov/drugsatfda_docs/label/2016/125164s071s072s073lbl.pdf>; 2016 [accessed 18.12.16].

[52] US Prescribing Information of Neulasta. <http://www.accessdata.fda.gov/drugsatfda_docs/label/2016/125031s184lbl.pdf>; 2016 [accessed 18.12.16].

[53] NCT01828021. A Phase 2 Study of the Monoclonal Antibody MGAH22 (Margetuximab) in Patients With Relapsed or Refractory Advanced Breast Cancer. <https://clinicaltrials.gov/ct2/show/NCT01828021?term = margetuximab&rank = 3> [accessed 18.12.16].

[54] CDER/CBER. DRAFT Guidance for Industry. Reference Product Exclusivity for Biological Products Filed Under Section 351(a) of the PHS Act. <http://www.fda.gov/downloads/Drugs/GuidanceCompliance RegulatoryInformation/Guidances/UCM407844.pdf>; 2014 [accessed 18.12.16].

SECTION

VI

MEDICAL SERVICES

CHAPTER

PHASE IV STUDIES AND LIFECYCLE MANAGEMENT

24

Gerfried K.H. Nell[1,2]

[1]NPC Nell Pharma Connect Ltd, Vienna, Austria

[2]Rokitan Ltd, Vienna, Austria

24.1 INTRODUCTION

Lifecycle Management is the overarching term for all activities aiming to develop, register, market, and promote a drug from the beginning of development till the final withdrawal from the market. It is immediately clear that this term goes far beyond the common understanding of the period during which a drug can be marketed under exclusivity because of patent protection or other regulations. In fact medical profiling and planning of the commercial success of a drug has to be started based on the progress of the development of the medicinal product years before the registration of a drug. It will reach its climax just before and during the first years of commercialization and then abate during the period when patent protection expires and the generic competition is going to start. As already stated, the common perception of Lifecycle Management is focused on the period of marketing the drug under patent protected exclusivity. This view is too narrow since the success of many drugs is already decided during the development period, and at the other extreme of the lifecycle it is possible to extend the commercial viability far into the period of generic competition. In order to facilitate rethinking, some propose to introduce the term Drug Life Optimization instead of Lifecycle Management in order to shift the focus from the time of market exclusivity to the whole lifecycle of a drug (see e.g., [1]).

It is clear from the aforementioned remarks that Lifecycle Management is a comprehensive term and comprises developmental and regulatory affairs activities (e.g., new indications, new dosage schedules, new galenic formulations, new patent applications), improvements in manufacturing, safety studies, pharmacovigilance, and a whole array of commercial measures (pricing, changes in prescription status, strategic alliances with other companies). For the purpose of this chapter we will focus on the area of measures which fall into the field of Pharmaceutical Medicine, i.e., mainly all activities which are connected to clinical trials of the Phase IV type, observational studies—especially regarding safety, and obtaining as much valid information on the evolving benefit risk balance of the drug as possible. Other activities such as exchange of information and intelligence of the therapeutic area are described in Chapter 25 (e.g., Medical Science Liaison).

Pharmaceutical Medicine and Translational Clinical Research. DOI: http://dx.doi.org/10.1016/B978-0-12-802103-3.00025-0

24.2 OBJECTS AND EXECUTION OF PHASE IV STUDIES

For the purpose of this chapter we will subsume all systematic data collection methods in clinical research to the Phase IV activities since strictly speaking trials are interventional studies which have to follow the rules of Good Clinical Practice (GCP, see [2]), whereas observational studies, which are equally important especially for the evaluation of safety issues, do not fall into this category but are regulated ICH E2 [3]. It is also clear by definition that Phase IV activities can commence only after the (first) approval of the drug.

The objectives of Phase IV activities are to extend the database on efficacy and safety, to collect comparative data to other treatment modalities (e.g., comparator drugs), and to create a database of outcome assessments in order to establish a clinical basis for health economics and HTA. The generation of data on new indications or new formulations or modes of administration are not Phase IV activities by definition because phase IV is defined as designing clinical studies within the approved label of the drug. Thus, exploration of new indications does constitute a new clinical development plan starting, e.g., with phase II again. Often health authorities request the Phase IV activities in order to address specific safety issues. It is further obvious that there cannot exist a generic plan for Phase IV activities in the sense of "one size fits it all," because of the divergent properties and therapeutic uses of drugs (influenza vaccine vs lipid lowering drug which had to be taken for a lifetime).

Which departments are involved in executing Phase IV activities? This depends on the size and organization of the sponsoring company. A big player in the field may have a fully capable medical affairs department which is well equipped to perform the required phase IV activities. Another way to organize this is that the groups originally involved in the Phase I–III program move on to execute the Phase IV activities. The latter approach has the advantage that the group has already accumulated an appreciable amount of expertise with the drug but requires an understanding of marketing considerations. Another possibility is to outsource such a program to a Clinical Research Organization (CRO), which may conduct the total program or parts of it in case there is no sufficient internal capacity available.

24.3 TYPES OF PHASE IV STUDIES

24.3.1 EXTENSION OF THE DATA ON SAFETY

The typical question is the detection of relatively rare but potentially serious adverse reactions. These effects might not be detected with a database of a few thousand patients only. Often the health authorities request this type of investigation as a condition of approval.

24.3.2 COMPETITIVE EFFICACY CLAIMS

This is an important area of differentiation of treatment options in a relatively crowded therapeutic field, e.g., hypertension, hyperlipidemia. Head to head results are considered more convincing then meta-analyses of several placebo-controlled clinical trials. Sometimes there exists a "gold standard" in a specific therapeutic area which will be selected as comparator, e.g., omeprazole as the paradigmatic gastric proton pump inhibitor in ulcer disease (see e.g., [4]).

24.3.3 NEW INDICATIONS

This is a particularly important field since this is the typical way to explore the full potential of a medicinal product. One of many examples in this field is the history of development of valsartan. This drug, an ARB (angiotensin receptor blocker), was first developed as an antihypertensive medication. Because of the already-known beneficial effects of ACE (angiotensin converting enzyme) inhibitors to reduce mortality and morbidity in patients after myocardial infarction, subsequently a whole trial program was set up which demonstrated the efficacy of valsartan in reducing mortality and morbidity in high-risk post myocardial infarction patients with left ventricular systolic dysfunction and/or heart failure and in heart failure patients, respectively, in two major trials (VALIANT and Val-Heft [5]). Both indications were approved by the health authorities. As already mentioned, from the regulatory point of view these indications represented new clinical developments and are thus not considered as mere Phase IV activities. However, they are typical examples of a successful lifecycle management.

24.3.4 NEW DOSAGE FORMS

The initial dosage form is often the most stable and most suitable form for the majority of patients. In many cases it will be considered desirable to develop specific formulations for, e.g., linctuses for children instead of tablets, and fast or slow release formulations depending on which time course of efficacy is considered favorable. In some cases a single bioequivalence test in humans will be sufficient for registration, in cases of deliberately different pharmacokinetic properties additional clinical trials might be necessary, e.g., to prove fast onset of action of an analgesic. In case the new formulation includes a technical innovation it might be patentable and used as a means to prolong at least partially the market exclusivity of a drug.

24.3.5 ETHNOPHARMACOLOGY, GENDER SPECIFIC MEDICINE AND SPECIAL POPULATIONS

Particular examples are additional studies in populations of different racial origin, studies in subgroups like the elderly, and in patients with comorbidities. Separate development in children is already mandatory globally which can be waived only in case there is certainly no therapy in children intended (e.g., Alzheimer's dementia). Gender-specific medicine also represents a special attention to differences in presentation and treatment of diseases in male and female patients.

These commitments show on the one hand the engagement of the company to fully explore the potential of its drug, and on the other hand often lead to expansion of the therapeutic armamentariums to the benefit of the patients.

24.3.6 DRUG INTERACTIONS

This area can be interpreted as a particular kind of clinical investigation in specific populations, namely those who are likely to be treated with several drugs simultaneously because of the peculiarities of the disease(s) from which these patients suffer. If there is a known potential for interaction regarding drugs which are most likely to be coadministered or if e.g., in vitro investigations on the cytchrome P450 system are pointing in this direction, studies on interactions are already part the clinical development plan. However, potential interactions may be detected only after drug approval. In this case retrospective case-control analyses of the clinical trial data might be requested. Proving a lower potential for drug interactions might be an important competitive advantage for a particular drug.

24.3.7 OUTCOMES RESEARCH

In this context outcome research aims at providing data on pharmacoepidemiology and pharmacoeconomics in order to create a database for decisions on which interventions work best for the patients, and under what circumstances [6]. Methodologically, outcomes research can take advantage of the whole armamentarium of Phase IV activities from interventional clinical trials to observational studies, meta-analyses, and systematic reviews.

24.3.8 PHARMACOVIGILANCE

This research serves to broaden the safety database for a specific drug and does apply a full range of study designs, in particular of the observational type.

24.4 DESIGNS AND METHODS APPLIED IN PHASE IV ACTIVITIES

24.4.1 CLINICAL TRIALS

The features of clinical trial designs applied in Phase I–III studies, i.e., randomization, blinding, placebo, or active comparator controlled and proper statistical evaluation planning, are also exercised in Phase IV clinical trials. Trials designed to evaluate efficacy and safety with an active comparator are in fact very similar to developmental phase III studies and follow the same methodology. This can be either the conventional Randomized Clinical Trial (RCT) approach or the sponsor may take into consideration more modern techniques such as adaptive trial design. While this methodology has been developed to increase the flexibility for identifying the optimal clinical benefit of the test treatment without undermining the validity and integrity of phase I to III studies in clinical development, adaptive designs might also help to improve Phase IV comparative studies. Since adaptive designs serve to plan pre-specified modifications in the design or statistical procedures of an ongoing trial, depending on the data generated it renders the trial more flexible [7].

This may allow terminating an inferior treatment earlier, allowing for a change of the comparator drug or leading to a recalculation of the sample size. Thus, adaptive designs should be considered also in Phase IV studies.

Phase IV trials are often designed to provide data under conditions which are closer to the ordinary clinical situation as compared to the complex Phase I–III trials. This allows for larger numbers, which may lead to mega-trials. The prerequisite is that the statistical evaluation does focus on few well-defined efficiency or safety criteria. Large patient numbers are often necessary for comparison of active treatments. In the VALIANT trial efficacy and safety of long-term treatment with valsartan, captopril, and their combination in high-risk patients with myocardial infarction and left ventricular systolic dysfunction and/or heart failure was compared. The number of patients amounted to almost 15,000 [5].

In order to align more with real world conditions and to facilitate study management trials may be designed single blind or open. Especially in open trials attention has to be given to confounding factors, since there might be a tendency—especially in patients—to consider one treatment as preferable to others. In case there is an imbalance in withdrawals which are not related to intolerability then interpretation difficulties may arise [4].

The statistical approach of comparative trials centers on the test of equivalence, noninferiority, and superiority. The problem with the evaluation of the statistical significance of differences of e.g., efficacy of drug treatments, is that this analysis is often based on the rejection of the null hypothesis of no treatment difference based on a statistical significance test, resulting in a the calculation of a P-value (usually $P < 0.05$). However, in order to draw clinically meaningful conclusions regarding equivalence, noninferiority, and superiority, the power of the statistical evaluation and the clinical significance of differences to be detected is also of paramount importance. An example for this line of statistical analysis is the methods that are required to show bioequivalence of two drugs [8]. In order to show bioequivalence of the test product after a single dose the 90% confidence interval of the ratio of $AUC_{(0-t)}$ and C_{max} (Area under the plasma concentration curve from administration to last observed concentration at time t; maximum plasma concentration) of the test and the reference product should be contained within the acceptance interval of 80.00%–125.00%. In the same manner the difference of treatment effects which are considered as clinically significant should be specified in advance, then the power and the sample size should be calculated and the interpretation of the significance, whether it is equivalence, noninferiority, or superiority should be based on confidence intervals (see for discussion e.g., [9])

24.4.2 META-ANALYSES

Meta-analyses have proven an invaluable scientific activity. They establish whether scientific findings are consistent and to what extent they can be generalized across populations and treatment variations. Methods have been developed to limit bias from inhomogenous data, multivariate effects, nonlinear regressions and flawed data transformations in order to allow for comparability. Today meta-analyses are increasingly applied in Phase III programs and Phase IV activities instead of unsystematic literature searching and as the basis of treatment guidelines (e.g., Cochrane collaborators). Based on the guidelines which are used by the Cochrane Collaborators, meta-analyses produce probability statements that are almost as valid as they are in randomized controlled clinical trials (see for discussion [10]).

Closely connected with the completeness of data of clinical trials in a particular area is the question how can the publication bias be minimized. This point is also addressed in Chapter 25—Medical Affairs. In addition to the discussion there we should like to go further into detail regarding the publicly accessible databases on clinical trials in order to evaluate to what extent these databases help to minimize the publication bias, i.e., the probability that positive results of trials are much more likely to be published than negative ones.

The Declaration of Helsinki (October 2013, [11]) defines the ethical responsibility for research registration and publication and dissemination of results:

35. Every research study involving human subjects must be registered in a publicly accessible database before recruitment of the first subject.
36. Researchers, authors, sponsors, editors, and publishers all have ethical obligations with regard to the publication and dissemination of the results of research (abbreviated).

This request led to the creation of public portals with a global (e.g., FDA) regional (e.g., EU) and national level for registration of clinical trials. In addition, in some cases publication of summary of the results is mandatory (EudraCT, clinical trials.gov) or encouraged on a voluntary basis (India, Japan). Several companies have set up portals where researchers can request data on clinical trials drilling down to the level of single anonymized patients (see for discussion [11]). These developments in the direction of openness and publication of results with dissemination of data are certainly encouraging but still in their infancy and there is a long way to go until we will have a global standard of public access to trial results.

24.4.3 METHODS FOR POSTAUTHORIZATION SAFETY STUDIES

In this section mainly methods of surveillance and observational studies will be described since clinical trials have already been covered under point 4.1. Summaries of these methods are given by several health authorities based on ICH E2E (see EMA Guideline, [12]).

Firstly, methods of active surveillance of adverse events are described. Active surveillance seeks to ascertain the number of adverse events in a given population via a continuous organized process. One example is intensive monitoring schemes. This might be done in hospitals or institutional settings where the infrastructure is appropriate for this task. One possibility of intensive monitoring is done by reviewing medical records or interviewing patients or center personnel in order to ensure complete and accurate data on reported adverse events. The downsides of the sentinel approach are small patient numbers and potential selection bias.

Another method of active surveillance is prescription event monitoring. Patients may be identified by e.g., electronic prescription data. Then a follow up questionnaire is sent for data collection.

A patient registry is an organized system that uses observational methods to collect uniform data on specified outcomes in a population defined by a particular disease, condition, or exposure. An example for an exposure registry is the registry of rheumatoid arthritis patients exposed to biological therapies which is designed to determine if a medicinal product has a special impact on this group of patients.

Secondly, observational studies are a key component in the evaluation of adverse effects. Several major types of observational studies can be performed. In a cross-sectional study data on a population are collected at a single point in time. These studies are best used to examine the prevalence of a disease or condition at one point in time. Trends can only be detected by repeated data collection.

In a cohort study a population at risk for an event of interest is followed over time for the occurrence of that event. In this study type incidence rates can be calculated. Patients may be identified from large automated databases. Cohort studies can also be used to examine safety concerns in special populations.

Another common type of observational studies is case-control studies. In this type of study, cases of events are identified and patients without the event of interest are selected as controls from the source population and the exposure status of the two groups is then used to calculate the relative risk among the exposed as compared to the nonexposed. In a nested case control study the control group represents the person–time distribution of exposure in the source population.

Drug utilization studies are a type of observational study aiming at "real world evidence." They may be used to monitor use in everyday practice and thus examine the relationship between recommended and actual clinical practice.

Summarizing this point it is fair to state that observational designs are indispensable in drug safety research and collecting data on real world evidence as basis of HTA evaluations.

24.5 INTEGRATION OF PHASE IV ACTIVITIES INTO THE STRATEGY OF LIFECYCLE MANAGEMENT

All Phase IV activities have to be planned to fit into the overall strategy of Lifecycle Management. This can be achieved only by a close cooperation of all functions involved in Lifecycle Management. Phase IV activities are an excellent tool to promote a drug and to differentiate it from the competitors. All action regarding further development, whether it be new formulations, new indications, special studies in specific population groups, new data on pharmcoeconomics, or safety studies, has to be part of an integrated medical/marketing plan in which market and medicinal aspects are carefully balanced. Medical Phase IV activities should be checked on their potential return on investment and value in regard to the likelihood to achieve the desired results for promotional activities or reimbursement issues. These considerations fit into the general picture of a flexibilization of the Lifecycle Management. Up till now drugs were mainly approved at a single point in time based on the results of the clinical development program, and decisions on reimbursement were made essentially on the same body of clinical data. During the last years a tendency for a more flexible approach has been adopted in the United States ("accelerated approval" and "conditional approval" in the EU and Japan) in order to allow for more rapid patient access to promising new therapies. These developments led to the concept of adaptive licensing, which is defined as a "prospectively planned, flexible approach to regulation of drugs and biologics." During the development of this concept it became clear that the evaluation of financing the new treatments have also to follow a flexible approach in order to lay the groundwork for a better coordination for the actions of regulators and payers on the overall market access pathway (see for discussion [13]). These developments will render the pathway to registration and reimbursement a stepwise approach and most likely add more flexibility to the exploitation of the full potential of a drug during its whole lifecycle.

24.6 LEGAL ASPECTS

Obviously all Phase IV activities do produce new data which may have an impact on the labeling of efficacy and safety of the drug. Therefore medical personnel have to work closely with drug regulatory authorities in order to allow for an ongoing updating of the label of the drug and also to guarantee that necessary actions are taken when changes in the labeling regarding safety information are necessary. The spectrum of actions might range from update at next routine printing to withdrawal of the product. The specific ways to accomplish this depend on national legislation but close cooperation with drug regulatory and legal departments is mandatory [4].

24.7 CONCLUSION

Phase IV activities comprise the whole array of medical endeavors in order to fully evaluate and exploit the potential of a drug during its whole lifecycle. A classic example is the development of new indications of acetyl salicylic acid (®Aspirin) during a lifecycle of 100 years duration up till now. It is a challenging and rewarding range of activities which require thorough experience with the whole armamentarium of study designs whether clinical trials or observational studies. The medically responsible personnel work on the interface with marketing, drug regulatory affairs, pharmacovigilance, health, economic functions and the legal department and may also cooperate regarding new formulations development. Being engaged in Phase IV activities is a stimulating and rewarding experience which requires a thorough knowledge of all medical aspects of the therapeutic area, one's own and the competitors' drugs properties, an attitude of openness for innovative approaches and ideas, and at the same time a strong team orientation. Based on the present tendencies of building in more flexibility in drug registration and reimbursement, working in a Phase IV unit will become even more attractive.

REFERENCES

[1] Bernard S. Rethinking Product Lifecycle Management. Pharmaceutical Executive; Feb 01, 2013. <http://www.pharmexec.com/rethinking-product-lifecycle-management> [accessed 27.11.15].

[2] ICH E6 (R2): Integrated Addendum to ICH E6(R1) Guideline to Good clinical Practice ICH. <http://www.ich.org/fileadmin/Public_Web_Site/ICH_Products/Guidelines/Efficacy/E6/E6_R2__Addendum_Step2.pdf> (ICH = International conference on Harmonization [accessed 27.11.15].

[3] ICH E2. Pharmacovigilance Planning. <http://www.ich.org/products/guidelines/efficacy/article/efficacy-guidelines.html> [accessed 27.11.15].

[4] Johnson-Pratt LR. In: Edwards LD, Fletcher AJ, Fox AW, Stonier PD, editors. Phase IV Drug Development: Post-Marketing Studies. Principles and Practice of Pharmaceutical Medicine. 2nd ed. John Wiley & Sons, Ltd; 2007. p. 119–25.

[5] Jugdutt BI. Valsartan in the treatment of heart attack survivors. Vascular Health Risk Manage 2006;2(2): 125–38.

[6] Krumholz HM. Outcomes research: myths and realities. Circ Cardiovasc Qual Outcomes 2009;2:1–3.

[7] Mahajan R, Gupta K. Adaptive design clinical trials: methodology, challenges, and prospect. Indian J Pharmacol 2010;42(4):201–7.

[8] EMA. Guideline on the investigation of bioequivalence. <http://www.ema.europa.eu/docs/en_GB/document_library/Scientific_guideline/2010/01/WC500070039.pdf> [accessed 30.11.15].

[9] Cleophas TJ, Zwinderman AH. Statistics applied to clinical studies. Chapter 62: Clinical trials: superiority testing. 5th ed. Springer; 2012. p. 665–73.

[10] Cleophas TJ, Zwinderman AH. Statistics Applied to Clinical Studies. Chapter 32: Meta-analysis, basic approach. 5th ed. Springer; 2012. p. 365–77.

[11] Declaration of Helsinki. <http://www.wma.net/en/30publications/10policies/b3/>; October 2013 [accessed 30.11.15].

[12] Guideline on good pharmacovigilance practices (GVP) Module VIII (Rev 1) EMA/813938/2011 Rev1. <http://www.ema.europa.eu/docs/en_GB/document_library/Scientific_guideline/2012/06/WC500129137.pdf> [accessed 30.11.15].

[13] Eichler H-G, Baird LG, Barker R, Bloechl-Daum B, Børlum-Kristensen F, Brown J, et al. From adaptive licensing to adaptive pathways: delivering a flexible life-span approach to bring new drugs to patients. Clin Pharmacol Therapeutics 2015;97(3):234–46.

FURTHER READING

Bragman K., Cottam B. Global Clinical Trials: A Bridge Too Far. FPM e-newsletter No. 46. p. 5–8. <https://www.fpm.org.uk/> [accessed 25 Nov DY].

CHAPTER

MEDICAL AFFAIRS

25

Gerfried K.H. Nell[1,2]
[1]NPC Nell Pharma Connect Ltd, Vienna, Austria
[2]Rokitan Ltd, Vienna, Austria

25.1 INTRODUCTION

Medical Affairs describes a complex combination of a series of functions starting with the interpretation of clinical results, providing a balanced view of the results of clinical research based on first-hand experience of the interface with clinical development, phase IV trials, observational studies, medical information including medical science liaison and contribution to drug regulatory affairs, the market access function, drug safety, epidemiology and pharmacovigilance, medical writing, advertising and promotion as well as training functions inside and outside the company. From this enumeration it is easily seen that Medical Affairs is a central platform for collecting and providing medical information and conferring medical input to an array of functions in charge of interaction with drug authorities, health technology assessment bodies, patient organizations, and the public at large. Activities involved are liaising with clinicians regarding participation in trials relating to company sponsored clinical development or investigator sponsored research, providing medical input in data collection under "real world conditions," involvement in the work of advisory boards as attendee or presenter, proving medical input regarding training of marketing colleagues and sales personnel, medical support of marketing teams, cooperating regarding medical marketing which is especially important regarding medical information to healthcare professionals, patient groups, and the public at large, since promotional material has to be scientifically correct and appropriate for the target audience. A particular task includes contact to health technology assessment (HTA) providers and committees deciding on reimbursement issues on international, national, or local levels whereby medical input regarding efficiency under real world conditions can play a crucial role.

This multitude of activities targeted at other functions within the company requires carefully designed interfaces and a flexible and efficient organization.

The multitude of interactions within the company and the array of contacts outside from authorities to customers require careful planning of the organization. Equally important properties in employees working in Medical Affairs are well developed managerial, organizational, and leadership skills. Regarding organization no generalization is possible since clearly the kind of organization chosen reflects the size, the degree of internationalization and the local conditions under which a company is operating. Large international companies have most often a central department of Medical Affairs which has the task to streamline and coordinate the worldwide activities in this

Pharmaceutical Medicine and Translational Clinical Research. DOI: http://dx.doi.org/10.1016/B978-0-12-802103-3.00026-2

field. In addition, regional managers might be applied for reasons of synergy and in order to accommodate to national or local conditions, which is of utmost importance since usually the medical prestige on a national level is represented by the perceived competencies of the Medical Affairs department.

25.2 CONTRIBUTION TO CLINICAL TRIALS

A member of the Medical Affairs department has to be able to plan, execute, evaluate, and report a clinical trial which represents one of the base competencies of a specialist in Medicines Development whether they are by education a physician or they come from another background in life sciences. Regarding day-to-day work it is useful to discriminate between two main directions in clinical research as conducted or supported by pharmaceutical companies.

One of these pathways is the whole line of clinical development from Phase I to Phase III, leading to submission of a dossier of the new drug to the health authorities. Work of this type is usually not performed by Medical Affairs but by the Clinical Development department. Depending on the size of the company Clinical Development might be totally independent of Medical Affairs and, e.g., organized on an international level or outsourced to Clinical Research Organizations (CROs) whereas Medical Affairs necessarily has a strong national affiliation. However, even in this case members of Medical Affairs advice might be sought, e.g., concerning differences in local therapeutic attitudes or eligibility of clinical investigators.

The other pathway comprises all clinical trial activities which are not directly included into the clinical development plan leading to a registration dossier. This might comprise clinical trials of phase II and III, sometimes labeled IIIb in case they are conducted during the dossier review by the health authorities. Medical Affairs departments should be able to perform clinical trials according to GCP. In case they are too small they should either cooperate on a regional or global basis or have the means to outsource the necessary activities in order to guarantee the required quality and the adherence to legal standards. The purpose of these studies is to provide additional data regarding safety and efficacy. These additional data might sometimes be required by health authorities.

The particular domain of clinical trials conducted by Medical Affairs is the Phase IV studies. These trials can be commenced after approval of the drug only and have to comply with the drug label approved by the health authorities. However, they are clinical trials by definition and have to be performed according to GCP. They may be requested by the health authorities or are initiated by the company in order to obtain more data on efficacy and safety especially in particular populations (e.g., elderly patients, particular co-morbidities).

Another category of trial activities are non interventional studies which differ from the clinical trials described above in that they have to strictly follow the usual medical routine. These studies can only be initiated after approval of the drug. The most important point is that any influence on the treatment decision has to be avoided. Only after that decision has been taken the patient can be included into the trial.

Other kinds of observational studies comprise epidemiological investigations, which are especially relevant to phamacovigilance. Types of these studies are patient registries, cohort studies, and nested control studies [1].

25.3 DATA MINING AND REAL WORLD EVIDENCE

Data mining is the computational process of discovering patterns in large data sets involving methods at the intersection of artificial intelligence, machine learning, statistics, and database systems [2]. The goal is to extract "nontrivial, implicit, previously unknown and potentially useful information" especially from large data sets [3]. As applied to Pharmaceutical Medicine these methods are usually aiming at retrospective analyses of pooled clinical trial databases. These methods are by no means inferior to the analyses of the results of a single trial. On the contrary the advantage is the increase in patient numbers. However, in order to avoid biased results such an approach needs careful formulation of research questions and method planning. It can only be done by a multidisciplinary team which includes medical expertise. In the described case the type of information sought is mainly connected to safety questions. Other targets pursued by data mining methodology are the detection of signals raising safety concerns either in available pharmacoepidemiologic data sets or in databases collecting spontaneous reports by pharmaceutical companies or health authorities or other organizations collecting data on drug safety (e.g., World Health Organization).

In the pharmaceutical industry data mining is also used to categorize potential customers in order to focus on the most efficient approach of sales and marketing. Because this kind of data mining has also medical implications, e.g., exploring therapeutic trends, it may need medical input.

Clinical development of new products and then Phase IV data can only provide selective evidence on the safety and efficacy of a drug. The "real" efficiency of a therapeutic measure can be derived only from data collecting in patients in day-to-day practice. This "real world evidence" is sought by industry who want to obtain information on the performance of their products in day-to-day practice, firstly, because this is the ultimate test of efficiency and safety and, secondly, the health authorities and especially the reimbursement bodies need this material in order to obtain data on which to base reimbursement modalities in the real world. Thus, obtaining "real world evidence" on the efficacy and safety of therapeutic measures is one of the main goals of HTA providers, health authorities, and the pharmaceutical industry.

25.4 MEDICAL INFORMATION AND COMMUNICATION

One of the most important functions of a Medical Affairs department is to provide information on the company products to customers, patients, and the public at large but also to internal clients. The handling of these requests is usually organized taking into account the level of complexity of the questions. Routine questions can be answered by nurses or pharmacists. Common practice is to prepare documents containing the answers to frequently asked questions. If the inquiry goes beyond this document further research is required by a medical information specialist. Since questions can come in from physicians, pharmacists, nursing staff, patients, or any other persons like journalists, the legal framework of answering those questions has to be observed carefully. Information on prescription drugs can only be given to health professionals in most countries. Questions regarding over-the-counter products can also be answered to patients as well as inquiries related to the patient leaflet.

In addition, promotional activities should be avoided. However, unsolicited questions on efficacy and safety of drugs which refer to not approved ("off label") use can be answered if the information has been described in a journal or at a medical meeting. Considering the necessity to avoid any confusion of medical information with promotional activities which are confined to the approved indications, legal advice should be solicited when in doubt [1].

25.5 DRUG REGULATORY AFFAIRS AND MARKET ACCESS

Preparing and submitting a registration dossier is usually the task of Clinical Development and Drug Regulatory Affairs. However, depending on company size and organization as well as on regional factors on the side of health authorities, medical input in regulatory questions might be sought from Medical Affairs.

During the last decade mounting hurdles to market access of new drugs have caused a shift in the paradigm of how to launch a new drug optimally. For the purpose of handling the new approach Market Access teams were created. In order to explain the functions of these teams we need to look back to the approach in the past, in which the same data showing evidence on efficacy and safety of new drugs from randomized clinical trials, which led to drug approval, were also sufficient to achieve reimbursement of the drug. After having obtained approval and reimbursement no further reassessments were usually required.

This paradigm has changed fundamentally in the last years due to the growing role of payers and economic evaluations at national/sub-national level. The reason for this development is the worldwide trend to cost containment in healthcare in industrialized as well as developing countries. Thus, companies have to provide evidence for an incremental benefit versus standard of care. This cannot be based only on the results of randomized clinical trials since payers want to pay only for drugs with proven efficiency in the real world and therefore request real world evidence which can only be provided after drug approval. This in turn creates the need for re-assessment for comparative effectiveness and safety after the commercialization of a drug in principle during its whole lifecycle. The process is started by preparing value assumptions during the R&D on an international level then creating comprehensive launch value packages on a national or regional level, which should be implemented locally. During the whole commercialization period the evidence for successful reassessment should be collected on a regional or national level as appropriate. In order to achieve this goal multifunctional teams have to be created collecting input from business strategy, health economy, marketing, and Medical Affairs. The result is an integral approach combining medical, economic, regulatory, and legal aspects during the whole lifecycle of a drug.

25.6 MEDICAL WRITING

Depending on the size and organization of the company—either a group or a department of medical writers work within Medical Affairs. Usually they come from a scientific background but they may also have a medical degree.

Their activities comprise input in documents needed for the execution of clinical trials. They provide assistance to the composition of investigator's brochures and are involved in the drafting of clinical trial reports and the publication of clinical trial results. The latter is a particularly sensitive point in light of the widespread suspicion of bias regarding clinical trial results in that sense that favorable results only are published whereas equivocal or negative findings are less likely to be publicly available. This is the so-called publication bias, which is quite understandable since it is certainly easier to publish positive findings like an improvement in a therapeutic modality in a renowned journal than to expand on a report on a refuted hypothesis. However, in the development of new therapeutic agents it is also valuable to learn about failures since it may draw attention to other lines of research and prevent unnecessary duplication of work which has already been proven not to lead to successful new treatments and thus may also raise ethical concerns. The need to avoid these complications and provide more transparency to clinical research cooperation between government sectors, academia, and industry led to the development of databases providing access to ongoing and concluded clinical trails. Information on ongoing trials is necessary because it allows an overview on clinical trials in progress which helps to avoid duplication of work. Another feature of these repositories is that they allow patients to check whether they might be eligible for certain clinical trials ([4], see for more information Chapter 24). An additional domain of activity is drafting Phase IV publications, booklets, and pamphlets.

25.7 DRUG SAFETY AND EPIDEMIOLOGY UNIT

Dependent on the size of the company this is either a standalone unit or does report into Medical Affairs, especially in smaller companies or local affiliates. Pharmacovigilance is one of the most important tasks of a pharmaceutical company since it is mandatory to follow the benefit risk balance during the whole lifecycle of a drug.

There are several important tasks in which Medical Affairs is involved. Firstly, signal detection in an ongoing phase IV program. This is usually done by an adjudication committee—mostly consisting of external experts—which evaluates the reports on adverse events observed during the trial. Medical Affairs is in charge of setting up and working together with these bodies. In addition, Medical Affairs is involved in assessment and evaluation of causality of incoming spontaneous reports of adverse effects. Also, regular literature searches for reports on adverse reaction in connection with the company's medicinal products or medical devices which have to be monitored. At regular time intervals medical expertise is needed for contribution to safety reports according to the legal requests, e.g., Periodic benefit-risk evaluation report (PBRER[5]).

25.8 MEDICAL INPUT INTO ADVERTISING AND PROMOTIONAL ACTIVITIES

Promotional activities of pharmaceutical companies are supervised by committees composed of legal, drug regulatory, and medical personnel. The input from the side of Medical Affairs is one of the most important tasks of this function and may be well described as the "medical and social conscience" of a company. The committee is in charge of all promotional material and information that can be considered as promotional. Promotional material may be distributed by print, radio, TV, Internet, or provided in person, e.g., sales reps. In addition, material which may contain medical/

scientific content such as sale reps instructions, press releases, financial statements, and communications to authorities and health insurances should be checked for appropriateness [1].

The principles of appropriate promotional practice are laid down in guidelines of industry associations (e.g., IFPMA Code of Practice, [6]) and are embedded in general guidelines on ethical conduct and promotion as well as in numerous national and regional legislations. These principles stipulate that pharmaceutical companies are responsible for providing accurate, balanced, and scientifically valid data on their products. Promotion must be ethical, accurate, and balanced, and must not be misleading, and information in promotional material must support proper assessment of the risks and benefits of the product and its appropriate use.

25.9 MEDICAL SCIENCE LIAISON

The medical science liaison function (MSL) was created already 50 years ago firstly by Upjohn and was initially conceived as an education service. The concept has evolved over the years, and is now defined as a key driver of knowledge exchange and cooperation with health professionals, especially of higher-level physicians and researchers [1]. The MSL is designed as the main interface in order to exchange scientific and medical information during the whole lifecycle of a drug. In order to enable this a strict distinction has to be made between medical/scientific exchange and promotional activities. This distinction necessitates a careful definition of the role of MSL versus other functions such as, e.g., sales personnel. Since the activities of MSL are not considered promotional by definition their action is not confined to the post launch period of the lifecycle of a drug but MSL is engaged in contacts with health professionals in the early development phase of a new drug. MSLs focus on interaction with key opinion leaders in therapeutic areas in which a company is active regarding support of the clinical trial program, acting as the company's interface with the medical/scientific world and providing insight into clinical practice in certain therapeutic areas either globally or regionally/locally within the company, and they are the messengers for outside views of healthcare professionals on the company's developmental strategy and development plans. The focus of the MSL activities depends on the therapeutic area, and the lifecycle of the drug. It is considered to be of utmost importance that their facilitating action on the exchange of knowledge and information is not seen as a promotional task but as a medical/scientific service. That means that their success cannot be measured in commercial metrics. Usually MSL personnel are specialized in specific areas of research and are scientists with MD, PharmD, or PhD qualifications [1].

25.10 SUMMARY AND OUTLOOK

As seen in this chapter, the function of Medical Affairs is one of the most versatile parts of the organization of a pharmaceutical company. It is also a vital one since it is the interface between seen the internal departments of a company and the customers, authorities, and the public. In Medical Affairs people thrive who love diversity and interaction with a lot different personalities inside and outside the company. It is not possible to expand on how to best organize such a department because it depends on the size of the company, the therapeutic area, and on whether we are considering headquarters or some regional or local function. However, one thing that is for sure is the growing importance of exchange of information between the company and the outside world, because this determines which value is assigned to the medicinal products in the public especially by the payers and the patients which will increasingly shape the financial success of the enterprises.

REFERENCES

[1] Geba PG. In: Edwards LD, Fletcher AJ, Fox AW, Stonier PD, editors. Medical Affairs. Principles and Practice of Pharmaceutical Medicine. 2nd ed. John Wiley & Sons, Ltd; 2007. p. 519–27.

[2] <https://en.wikipedia.org/wiki/Data_mining> [accessed 20.11.15].

[3] Rahman MI, Dabbous OH. In: Edwards LD, Fletcher AJ, Fox AW, Stonier PD, editors. Data Mining. Principles and Practice of Pharmaceutical Medicine. 2nd ed. John Wiley & Sons, Ltd; 2007. p. 545–55.

[4] Bragman K, Cottam B. Global Clinical Trials: A Bridge Too Far. FPM e-newsletter No. 46, pp. 5–8. <https://www.fpm.org.uk/> [accessed 25.11.2015].

[5] <https://www.ema.europa.eu/docs/en_GB/document_library/Regulatory_and_procedural_guideline/2012/12/WC500136402.pdf> [accessed 25.11.15].

[6] IFPMA Code of Practice (International Federation of Pharmaceutical Manufacturers &Associations). <https://www.ifpma.org/ethics/ifpma-code-of-practice/ifpma-code-of-practice.html> [accessed 25.11.15].

FURTHER READING

Careers in Pharmaceutical Medicine: Medical Affairs Physician. <https://www.fpm.org.uk/> p. 9 [accessed 20.11.15].

SECTION

PHARMACOVIGILANCE VII

CHAPTER

PHARMACOVIGILANCE AND DRUG SAFETY

26

Rajinder K. Jalali
Sun Pharmaceutical Industries Ltd, Gurgaon, Haryana, India

Pharmacovigilance is also known as drug safety and is defined by the World Health Organization as the science and activities relating to the detection, assessment, understanding and prevention of adverse effects or any other drug-related problems [1,2]. Pharmacovigilance is the science of collecting, monitoring, researching, assessing, and evaluating information from healthcare providers and patients on the adverse effects of the prescription medicines, over the counter medicines, vaccines and biologicals, blood products, medical devices, herbals and other traditional and complimentary medicines, with a view to identify new information about side effects associated with their use, thereby taking necessary measures and preventing harm to patients. Many other issues are also of relevance to pharmacovigilance such as substandard or spurious medicines and medication errors. Medication errors such as misuse, abuse, or overdose of a drug, as well as drug exposure during pregnancy and lactation, are also important because they may result in an adverse drug reaction.

Medicines improve quality of life by controlling diseases, however, despite all their benefits, medicines are associated with adverse reactions that may result in significant morbidity and mortality. If more than one medicine is prescribed, there is always a risk of negative interactions. Further, due to differences in genetic pattern, a patient may respond differently to a medication and develop different adverse effects. A typical example would be acute hemolysis developed in patients with glucose-6-phosphate dehydrogenase (G6PD) deficiency on exposure to a commonly prescribed anti-malarial drug—primaquine. Adverse drug reactions (ADRs) rank among the top 10 leading causes of mortality. The ultimate goal of pharmacovigilance is to foster the rational and safe use of medicines, minimize the risks related to the drug use, and maximize the benefits.

It is a known fact that a drug has to go through a series of tests which include animal testing and various phases of clinical trials in humans to establish its safety and efficacy before permission to market is granted. However, clinical trials being conducted in controlled conditions have various limitations such as small patient population tested, ranging to a few thousand patients, and also exclusion of special populations, e.g., children, pregnant and lactating women, and geriatric patients. Further, they often consist of a highly restricted group of patients enrolled through stringent inclusion and exclusion criteria and do not necessarily reflect the general population who will use the drug once marketed. In addition, some other factors causing adverse drug reactions such as genetic factors, environmental factors, and complete drug-drug interactions may not have been studied during the clinical trials. Data generated during clinical trials will provide information about the common adverse events but more rare adverse events may not be encountered at all, e.g., fatal

Pharmaceutical Medicine and Translational Clinical Research. DOI: http://dx.doi.org/10.1016/B978-0-12-802103-3.00027-4

dyscrasia occurring in 1 in 5000 patients treated with a new drug is only likely to be recognized after 15000 patients have been treated and observed [3]. In short, during the preclinical and clinical development of a drug, minimal safety information of the product is known to us. Continuous observation and evaluation is required under the real use of drugs in clinical practice and a strong pharmacovigilance system is essential to monitor it. Both clinical trial safety and postmarketing pharmacovigilance are critical throughout the product lifecycle [4]. Several postauthorization safety studies (PASS) are conducted, either at the time of authorization or postauthorization phase for collecting data for the assessment of the safety and/or efficacy of the medicinal product in clinical practice. Data from long term controlled clinical studies, observational epidemiological studies such as case-control, cohort, and cross sectional studies, drug use surveys, automated databases linking drugs and disease, and drug registries, will provide information about rare adverse events associated with drugs [5–7]. Drug safety monitoring and risk management are vitally important to safeguard public health. To promote safe use of the medicines, a thorough assessment of the new evidence generated through pharmacovigilance activity is required. Pharmaceutical companies should ensure that they maintain efficient systems and processes for drug safety monitoring. Pharmacovigilance is a critical component for determining the benefit to risk ratio of medicines. A close coordination and collaboration is required between healthcare professionals, patients, pharmaceutical companies, and regulatory agencies to have an effective Pharmacovigilance system globally.

The first committee set up to monitor drug safety, after a report was published in Lancet, was in reaction to the death of a 15-year old girl in 1848 who received chloroform anesthesia for the removal of ingrown toe-nail [8]. Cases of aplastic anemia associated with the use of chloramphenicol were reported in the United States of America in 1950 [9]. As a consequence, the Council on Drugs of the American Medical Association set up a Blood Dyscrasia Registry [10]. Food and Drug Administration (FDA) began the systematic collection of reports of all types of ADRs in 1961. Thalidomide, because of its association with congenital malformation in newborn infants, was withdrawn from the market in 1961 [11]. Ten countries, having national drug monitoring centers, joined the World Health Organization (WHO) Pilot Research Project for International Drug Monitoring in 1968 [12]. These countries were Australia, Canada, Czechoslovakia, Germany, Netherlands, Ireland, New Zealand, Sweden, United Kingdom, and USA. At the end of 2010, one hundred and thirty-four countries including India were part of the WHO pharmacovigilance program.

Adverse Event (AE) reporting is a key pharmacovigilance activity for assessing the risk-benefit profile of a drug and involves receipt, triage, data entry, assessment, distribution, and reporting of AEs to regulatory agencies. The source of AE reports may include: spontaneous reports from healthcare professionals (HCP) or patients, their relatives or lawyers; solicited reports from patient support programs; reports from clinical or postauthorization studies; reports from medical literature; reports from the Internet; and reports reported to drug regulatory authorities directly by HCPs or consumers.

Aggregate reporting or periodic safety reporting also plays a key role in the safety assessment of drugs. It involves the compilation of safety data for a drug over long periods of time constituting quarterly, 6 monthly, yearly, 3 or 5 year periodic safety reports and provides a broader view of the safety profile of a drug seen over a longer duration of time, allowing comprehensive assessment of worldwide safety data of a marketed drug. Different types of aggregate reports are the Periodic Safety Update Reports (PSURs) or Periodic Adverse Drug Experience Reports (PADERs), submitted to drug regulatory agencies in the European Economic Area, the US, and other countries around the world. After having released Good Pharmacovigilance Practices guidance documents in Europe in 2012, PSUR is now referred as the Periodic Benefit Risk Evaluation Report (PBRER) with focus on analysis of cumulative data and benefit-risk profile of the drug. Other forms of aggregate reports are the Addendum to Clinical Safety and Overview (ADCOs) submitted along with product renewal registration, Psychiatric Reports for regulatory submission, Safety Evaluation Reports (SERs) used by companies for internal monitoring of drug safety, Annual Safety Reports, and Development Safety Update Report (DSUR) generated to report safety in clinical trials.

A medicinal product is authorized for the treatment of a specific indication on the basis that risk-benefit at the time of authorization is judged to be positive. However, not all risks are identified and not all target population has been treated with the medicinal product at the time of authorization. Therefore, it is important to monitor risks during postauthorization and throughout the entire lifecycle of the medicinal product to arrive at an actual risk-benefit profile of the medicinal product and take necessary measures to minimize risks. This is called risk management and encompasses identifying signals and monitoring risk-benefit profile of drugs.

The WHO defines a safety signal as reported information on a possible causal relationship between an adverse event and a drug, the relationship being unknown or incompletely documented previously [13]. Usually more than a single report is required to generate a signal.

Data mining pharmacovigilance databases is a common and easy approach to study adverse event profile associated with use of a medicinal product. Drug regulatory authorities and pharmaceutical companies perform data mining of the huge safety data available in their databases to identify hidden patterns of associations or unexpected occurrences. The current method of signal detection is primarily based on spontaneous reporting. Individual Case Safety Reports (ICSRs) are extracted from the databases, drug-event pairs created and statistical methods applied to calculate statistical measures of association. If the statistical measures cross arbitrarily set threshold criteria, a signal is declared for a given drug associated with a given adverse event. These are called signals of disproportionate reporting (SDR). All SDRs are subjected to further clinical analysis to confirm or refute a signal. Other sources of signal detection are prescription event monitoring, case control surveillance, and record linkage approaches. Signal detection and its assessment is the most important aspect of pharmacovigilance to monitor drug use and safety surveillance. The aim is to identify ADRs that were previously considered unexpected and provide guidance in the product's label so as to minimize the risk and promote safe use of medicine.

A risk management plan (RMP) is a complete depiction of a risk management system that specifies the risks and gives a description of both adverse drug reactions and potential adverse reactions that may arise during the use of a medicinal product. It also describes how the risks can be minimized by including a warning on the product label of possible adverse reactions which if observed should alert a patient to seek medical advice. The aim of RMP is to promote safe use of medicine and assure a positive risk-benefit balance once the drug is marketed. RMP is required to be submitted in a specified format prescribed by the regulatory agencies where the drug is to be marketed. The risks described in an RMP are categorized into important identified risks and potential risks. It also identifies the missing information and describes measures that the Marketing Authorization Holder (MAH) will undertake to minimize the risks associated with the use of the medicinal product. These measures focus on the product's labeling and information for healthcare professionals and patients. Risks described in a preauthorization RMP become part of the product label. A drug, once authorized, if used in indications other than approved by regulatory authority, is called off-label use. This potential off-label use and the risks associated with it are also described within the RMP. These RMPs are usually submitted within EU or outside EU at the time of request for authorization or when a regulatory authority demands same in addition to routine one, particularly if there are reports suggesting a change in the risk-benefit profile of the drug.

In US, FDA requires Risk Evaluation and Mitigation Strategies (REMS) for a drug that has a specific risk and needs mitigation. A few examples are isotretinoin pregnancy prevention program, antiretroviral pregnancy registry or REMS submitted for use of controlled substances. REMS requires a sponsor to perform certain activities or to follow a protocol, to assure that a positive risk-benefit balance for the drug is maintained during marketing [12].

Pharmacoepidemiology refers to the incidence of adverse drug reactions in a large number of patient populations using medicinal products. It is the study of populations and not a study of individual patients. Pharmacoepidemiology is used to answer questions about adverse events/serious adverse events in populations after a signal is identified on the basis of spontaneous reports. The aim is to confirm or reject a signal.

Pharmacogenetics and pharmacogenomics refers to the study of genetic variation that results in differing responses to drugs and is thereby likely to provide information as to which patient with a particular genetic profile will benefit or is at risk from using a particular drug.

REFERENCES

[1] World Health Organization. Safety of medicines. A guide to detecting and reporting adverse drug reactions—why health professionals need to take action. Geneva: World Health Organization; 2002.

[2] World Health Organization. Safety monitoring of medicinal products: guideline for setting up and running a Pharmacovigilance Centre. Uppsala: Uppsala Monitoring Centre, World Health Organization; 2000.

[3] World Health Organization. The importance of pharmacovigilance. Safety monitoring of medicinal products. Uppsala Monitoring Centre, World Health Organization; 2002.

[4] The Uppsala monitoring system, WHO collaborating Centre for International Drug Monitoring World Health Organization; 2002.

[5] Edlavitch SA. Post marketing surveillance methodologies. Drug Intell Clin Pharm 1988;22:68—78.

[6] Shapiro S. The epidemiological evaluation of drugs. Acta Med Scand Suppl 1984;683:23—7.

[7] Strom BL. Overview of different logistical approaches to post marketing surveillance. J Rheumatol Suppl 1988;17:9—13.

[8] Routledge P. 150 years of Pharmacovigilance. Lancet 1998;351:1200—1.

[9] Rich ML. Fatal case of aplastic anemia following chloramphenicol therapy. Ann Intern Med 1950;33:1459.

[10] Wallerstein RO, Condit PK, Kasper CK, Brown JW, Morrison FR. Statewide study of chloramphenicol therapy and fatal aplastic anemia. JAMA 1969;208:2045—50.

[11] Randall T. Thalidomide is back in the news, but in more favorable circumstances. JAMA 1990;263:1467—8.

[12] Van Grootheest K. The dawn of Pharmacovigilance. Intern J Pharm Med 2003;17:195—200.

[13] World Health Organization (WHO). Guidelines on safety monitoring of herbal medicines in Pharmacovigilance systems. World Health Organization, Geneva: WHO 2004.

CHAPTER

CLINICAL AND POST APPROVAL SAFETY DATA MANAGEMENT 27

Rajinder K. Jalali
Sun Pharmaceutical Industries Ltd, Gurgaon, Haryana, India

A medicinal product goes through preclinical, clinical, and post approval safety surveillance throughout its lifecycle. Before testing a new medicine in humans, various in vitro, in vivo, and toxicological studies are conducted to understand the effects of medicinal products in animals. This forms the basis of potential risk expected in humans, which is further investigated in various phases of clinical trials before a medicine is granted permission to be marketed. The limitations of testing safety in humans in clinical trials is fraught with small size of patient population studied, narrow populations tested, insufficient data on special population groups, and short duration of these studies. On the other side, benefits of post marketing monitoring is the ability to identify rare adverse effects, adverse effects in special population groups, long term effects and effects of interaction with other drugs. Further, it is important to understand that a medicinal product will be under various stages of development and/or marketing in different countries at some point in time. Both clinical safety and post approval safety data need to be processed in the company's safety database and reported to various regulatory agencies as per regulatory requirements in various countries [1,2].

Definitions and terminology [1–4] used in clinical and post approval safety reporting are in general as follows:

1. Adverse Event (AE)
   An adverse event is any untoward medical occurrence (unfavorable or unintended sign including an abnormal laboratory finding, symptom or disease) temporally associated with the use of a medicinal product administered to a clinical investigation subject or patient and which does not necessarily have a causal relationship with the medicinal product.
2. Adverse Drug Reaction (ADR)
   a. During clinical phase, all noxious and unintended responses to a medicinal product related to any dose are considered ADRs.
   b. During post approval phase, ADR is a response to a drug which is noxious and unintended and which occurs at doses normally used in humans for prophylaxis, diagnosis or treatment of disease or modification of physiologic function.
3. Unexpected Adverse Drug Reaction
   An adverse reaction, the nature, severity, specificity or outcome of which is not consistent with the applicable product information (Investigator Brochure during clinical safety reporting and Prescribing Information/Summary of Product Characteristics during post approval safety

Pharmaceutical Medicine and Translational Clinical Research. DOI: http://dx.doi.org/10.1016/B978-0-12-802103-3.00028-6

reporting). This includes class-related reactions, which are mentioned in the regulatory authority approved product information but which are not specifically described as occurring with the product.

4. Serious adverse Event or Adverse Drug Reaction
   Serious adverse event is a key component that drug regulatory authorities consider in the decision-making as to whether to grant or deny marketing authorization for a drug.
   A serious adverse event or serious adverse drug reaction is any untoward medical occurrence at any dose and results in death, is life-threatening, requires inpatient hospitalization or prolongation of existing hospitalization, results in persistent or significant disability/incapacity, is a congenital anomaly/birth defect or is a medically important event or reaction. Other Important Medical Events have also been listed in European Medicine Agency release and are considered serious in post approval safety reporting for adverse events.
5. Expectedness of an Adverse Drug Reaction
   The following documents are used to determine whether an adverse event/reaction is expected:
   For a medicinal product not yet approved for marketing in a country, a company's Investigator Brochure is used as a source document. During postapproval phase Prescribing Information or Summary of Product Characteristic is used as a source document.
   a. Reports which add significant information on specificity or severity of a known, already documented serious ADR constitute unexpected events. Some of the examples are (1) acute renal failure as a labeled ADR with a subsequent report of interstitial nephritis, (2) hepatitis with subsequent report of hepatic necrosis, and (3) cerebral thrombosis with first report of cerebral vascular accident.

During clinical trial phase, clinical trial or epidemiological investigation is the source of ICSRs and during post approval phase, it is divided into unsolicited or solicited sources [1–7].

1. Unsolicited Source
   a. Spontaneous Reports
      A spontaneous report is an unsolicited report by a healthcare professional (medically qualified person such as physician, dentist, pharmacist, nurse) or consumer (patient, friend, relative, or lawyer of a patient) to a pharmaceutical company, Regulatory Agency, or other organization such as World Health Organization or a Regional ADR Monitoring Centre, that describes one or more adverse drug reactions in a patient who was given one or more medicinal products and that does not derive from a clinical study or any organized data collection scheme. Other forms of spontaneous reports are stimulated reports. Stimulated reports are those that may have been motivated, prompted, or induced such as "Dear Doctor letter," public advisory issued by regulatory agencies, media reports, or questioning of healthcare professional by company representatives.
   b. Literature
      Each Marketing Authorization Holder (MAH) is expected to undertake global literature monitoring by accessing widely used systematic literature reviews or reference databases such as PubMed or Embase or outsourcing literature monitoring activity to an outside contract organization. The frequency of literature searches depends upon any local regulatory requirements or should be done at least weekly. Cases of ADRs from the

scientific and medical literature, including relevant published abstracts from conferences and draft manuscripts from meetings, might qualify for expedited reporting. A copy of the article/abstract is required to accompany the report. The regulatory reporting time clock starts as soon as the MAH, its Affiliates, or outsourced contract organization has knowledge that the case meets minimum criteria for reporting. If the product source, brand or trade name, or batch number is not specified, the MAH should assume that it was its product. If a number of products have been mentioned in the article, a report should be submitted only by the MAH whose product is considered as a suspect by the author of the article.

c. Websites
MAH should regularly screen websites under its management for any potential ADR case reports. MAH is not expected to screen external websites for ADR information. However, if MAH becomes aware of ADR from such external websites, the MAH should assess whether the report needs to be sent to regulatory authority. ADR data collection should be facilitated by pharmaceutical companies by providing ADR forms on company website.

d. Other sources of unsolicited reports
If MAH becomes aware of ADR from one of its products from press or other media, it should be handled as a spontaneous report.

2. Solicited Sources
Reports derived from organized data collection systems are known as solicited reports. Examples include reports derived from clinical trials, patient registries, post approval patient programs, other patient support and disease management programs, surveys of patients or healthcare providers or information gathering on efficacy and patient compliance. Adverse event reports obtained from any of these sources are classified as study reports for safety reporting purpose, and should have an appropriate causality assessment by a healthcare professional (HCP) or an MAH.

3. Safety Data Exchange Agreements
The marketing of many products takes place through contractual agreements between two or more companies, which may market the same product in the same or different regions/countries. It is very important to have Safety Data Exchange Agreements in place defining the process for exchange of safety information, including timelines and regulatory reporting responsibilities. Processes should be in place to avoid duplicate reporting.

4. Regulatory Authority Sources
Individual serious unexpected adverse drug reaction reports originating from foreign regulatory authorities are subject to expedited reporting to other authorities by each MAH. Such reports usually need not be sent to originating regulatory authority unless otherwise stated.

An individual case safety report (ICSR) is considered to be valid if it meets at least four minimum criteria, i.e., it has: (1) an identifiable reporter, (2) an identifiable patient, (3) an adverse reaction, and (4) a suspect product. If one or more of these four criteria is missing, the case is not a valid AE report, i.e., reportable to a regulatory agency. However, if there is information regarding drug and adverse event, it needs to be databased, and the information may be used in assessment of risk-benefit balance and signal generation.

All ADRs from clinical or epidemiological investigation, spontaneous reports, and reports from regulatory agencies or literature that are both serious and unexpected are subject to expedited reporting [5,6]. The source of report (investigation, spontaneous, other) should always be specified. Causality assessment is required for clinical trial cases. All cases judged by either clinical investigator or sponsor as having a reasonable suspected causal relationship to the investigational product are subjected to expedite reporting. For purposes of safety reporting, adverse event reports associated with marketed products (spontaneous reports) are considered to have implied causality. Serious adverse events from clinical trials which are not related to study medication are not subjected to expedite reporting. Reports which are serious but expected are not usually required to be expedited. However, expected serious ADRs showing significant increase in frequency will require to be processed on an expedite basis.

Lack of efficacy for life-threatening disease, vaccines, or contraceptives are considered as expedited reports. Reports of overdose, dosing error with no adverse event outcome need not be reported as adverse reactions.

In a blinded clinical study, when a serious adverse reaction is judged reportable on an expedited basis, it is recommended that the blind be broken only for that specific patient by the sponsor even if the investigator has not broken the blind. It is also recommended that the blind be maintained for those personnel involved in analysis and preparation of clinical trial reports. The blind can be broken independently by the pharmacovigilance team involved in processing and reporting of adverse events for MAH to the regulatory agencies. However, when a fatal or other serious outcome is the primary efficacy endpoint in a clinical investigation, the integrity of the clinical investigation may be compromised if the blind is broken. Under these circumstances, it may be appropriate to reach agreement with regulatory authorities in advance regarding serious events that would be treated as disease-related and not subject to routine expedited reporting.

Serious adverse events associated with an active comparator should be reported by sponsor to the manufacturer of the active control or to appropriate regulatory agencies. Adverse events associated with placebo will usually not satisfy the criteria for an ADR and, therefore, for expedited reporting. For poststudy events, a causality and expectedness assessment is needed for a decision on whether or not expedited reporting is required.

All information related to SAEs should be shared with other Investigators and Ethics Committees and Institutional Review Boards as per ICH GCP Guidelines.

In addition to ICSRs, any safety information that could change the risk-benefit evaluation for the product should be communicated to regulatory authorities such as safety findings from an in vitro animal or epidemiological, preclinical, or clinical study that suggests a significant human risk, such as evidence of mutagenicity, teratogenicity, carcinogenicity, or lack of efficacy with a drug used in a life-threatening or other serious disease.

Fatal or life-threatening, unexpected ADRs from clinical trials need to be reported to regulatory agencies no later than 7 calendar days from the day of knowledge by the sponsor followed by complete report with assessment within additional 8 calendar days. Other serious, unexpected ADRs need to be reported within 15 calendar days after first knowledge by sponsor/company. When additional medically relevant information is received for a previously reported case (called initial case), the reporting time clock is considered to begin again for submission of the follow-up report. In addition, a case initially classified as a nonexpedited report would qualify for expedited reporting upon receipt of follow-up information that suggests that the case should be re-classified from nonserious to serious.

Cases of nonserious ADRs need not be reported on an expedited basis. These nonserious ADRs are included in periodic safety update reports. However, in the European Union and some African and Latin American countries these are reported within 90 calendar days.

The following elements need to be considered while processing ICSRs:

1. Assessing Patient and Reporter Identifiability
   Patient and reporter identifiability is important in order to avoid case duplication and facilitate follow-up of appropriate cases. One or more of the following should automatically qualify a patient as identifiable: age (or age category, e.g., adolescent, adult, elderly), gender, initials, date of birth, name, patient identification number. Further, a case should have minimum four criteria for case reporting.
2. The Role of Narratives
   The objective of the narrative should be to summarize all relevant clinical and related information in a logical sequence. Normally it should be presented in the chronology of patient experience. The narrative should summarize all relevant clinical and related information, including patient demography, medical and concomitant disease history, concomitant and other treatment details, investigations performed, clinical course of disease, diagnosis and ADRs with their outcome. Any relevant autopsy or postmortem findings—if any—should be included in the narrative summary. In follow-up reports, new information should be clearly identified.
3. Clinical Case Evaluation
   The report should include the verbatim term as used by the reporter or an accurate translation of it. Additional information should be sought from HCP, if required.
4. Follow-up Information
   Follow-up information should be sought for incomplete cases and the priority should be as follows: (1) serious and unexpected, (2) serious and expected, and (3) nonserious and unexpected. Follow-up information can be obtained via a telephone call and/or site visit and/or written request preferably using a targeted questionnaire. For serious ADRs, it is important to continue follow-up and report new information until the outcome has been established or the condition is stabilized. The duration for follow up of such cases is a matter of judgment.
5. Pregnancy Exposure
   All cases of pregnancy exposure received from HCPs or consumers where the embryo/fetus could have been exposed to one of its medicinal products, should be appropriately followed up. Patient may become pregnant either during the treatment or after the completion of treatment with a medicinal product, and then all such cases of exposed pregnancy shall be monitored for an adverse outcome or adverse drug reaction in the mother, fetus, infant, or child. Reports of pregnancy exposure are considered serious if they are associated with serious ADRs and/or with abnormal pregnancy outcome. All pregnancy exposure cases reported spontaneously shall be followed up to obtain pregnancy exposure outcome. Reports of induced termination of pregnancy without information on congenital malformation, reports of pregnancy exposure without outcome data, or reports which have a normal outcome, are not required to be reported as individual case reports since there is no adverse drug reaction. These reports form part of periodic safety update reports. However, all cases of use of a medicinal product during

pregnancy are required to be submitted as individual case reports on expedited basis if there is a risk management program for one or more products to which the embryo/fetus has been exposed.

The CIOMS-I form has been widely accepted standard for expedited adverse event reporting. Med Watch is used for reporting to US FDA, if E2B submission through electronic gateway runs into any trouble or is not possible. The Medical Dictionary for Regulatory Activities (MedDRA) should be used for coding adverse events. WHO DDE should be used for coding noncompany products and ICH E2B/M2 guidelines should be followed for electronic reporting of ICSRs.

REFERENCES

[1] ICH E 2A: Clinical Safety data Management–Definitions and Standards for Expedited Reporting (CPMP/ICH/377/95), June 1995, European Medicines Agency.

[2] ICH E2D: Post Approval Safety Data Management–Note for Guidance on Definitions and Standards for expedited Reporting (CPMP/ICH/3945/03), November 1993, European Medicines Agency.

[3] The Uppsala monitoring system, WHO collaborating Centre for International Drug Monitoring World Health Organization; 2002, pp. 28–34.

[4] World Health organization. Safety of medicines. A guide to detecting and reporting adverse drug reactions–why health professionals need to take action. Geneva: World Health Organization; 2002.

[5] Guidance for Industry: Postmarketing Safety Reporting for Human Drug and Biological Products Including Vaccines, Food and Drug Administration, March 2001 (draft). <http://www.fda.gov/cder/guidance/4153dft.pdf>.

[6] Safety Reporting Requirements for Human Drug and Biological Products, Proposed Rule, Food and drug Administration, March 2003.

[7] <http://www.ema.europa.eu>.

CHAPTER

INDIVIDUAL CASE SAFETY REPORTS

28

Rajinder K. Jalali
Sun Pharmaceutical Industries Ltd, Gurgaon, Haryana, India

Individual Case Safety Report (ICSR) captures information necessary to support reporting of suspected adverse reaction(s) due to use of a medicinal product by a patient at a specific point of time. ICSR content and format requirements for drug and biologics reporting to various regulatory authorities is based upon the International Conference on Harmonization ICH E2B(R2), ICH E2B(M), and ICH E2B(R3) specifications and the Guideline on Good Pharmacovigilance practices (GVP): Module VI for direct database-to-database transmission of information using standardized data elements [1–4].

ICSR reporting is one of the key activities performed by drug safety departments and encompasses the receipt, triage, data entry, quality, and medical review, distribution, reporting, and archiving of adverse event (AE) data. This is necessary for assessing the risk-benefit profile of a given medicinal product.

ICSR is considered to be valid for reporting to a regulatory authority if it has at least: (1) one single identifiable patient, (2) one identifiable reporter, (3) one or more suspect adverse drug reaction, and (4) one or more suspect identifiable product [5–7]. If one or more of these four basic criteria is missing, the case is not a valid AE report. An identifiable patient usually means that one or more identifiers are available: age, age group, and patient number, date of birth, gender, initials, or name. Health Canada recognizes case reports referring to "a patient" as having a valid patient identifier, even if other patient identifiers are missing [8]. Cases received from a regulatory authority without a patient identifier can be considered valid if it has other minimum reportable information available. This approach that the existence of the patient holds true is considered since the case is being reported to the company by the regulatory authority itself. An identifiable reporter is usually a healthcare professional, pharmacist, patient, or a patient's relative or lawyer. An identifiable reporter usually means that one or more identifiers are available: name or initials, qualification, contact details, e.g., address, e-mail address, phone number, fax number. The suspect drug may be known by its generic name or brand name. Identifiers are essential to prevent duplicate reporting of the same case, and also allow reporting follow-up if additional information is received.

A suspected adverse drug reaction initially considered valid after being reported by a patient cannot be downgraded to nonvalid even if the patient's healthcare professional disagrees with the patient's suspicion. In such a scenario, the opinions of both patient and healthcare professional are included in the ICSR.

Pharmaceutical Medicine and Translational Clinical Research. DOI: http://dx.doi.org/10.1016/B978-0-12-802103-3.00029-8

Adverse event coding is the process by which adverse event information received from a reporter/patient, called the "verbatim," is coded using standardized terminology from a medical coding dictionary, such as Medical Dictionary for Regulatory Activities (MedDRA). MedDRA is a clinically validated international medical terminology dictionary used to convert adverse event information into terminology that can be readily identified, retrieved, and analyzed.

Seriousness assessment is done separately for every reported adverse event within a case taking overall case assessment into consideration. An adverse event is considered to be serious [5,6] if it meets one or more of the following criteria:

1. Results in death: death is usually the outcome of an event that causes it and the cause of death is regarded as the AE. However, "sudden death" is considered as AE.
2. Is life threatening: the term "life-threatening" refers to an event where the patient was at immediate risk of death at the time of event, not an event that might have caused death had it been more severe.
3. Requires inpatient hospitalization or prolongation of existing hospitalization: the term "hospitalization" refers to the situation when an AE is associated with an unscheduled admission to the hospital, for the purpose of investigating and/or treating the AE.
4. Results in persistent or significant disability/incapacity: the AE has a clinically important effect on the patient's physical or psychological wellbeing, to the extent that the patient is unable to function normally.
5. Is a congenital anomaly/birth defect: any congenital anomaly in a child should be regarded as a serious AE when the mother or father was exposed to a medicinal product at any stage during conception or pregnancy.
6. Is a medically important event or reaction: medical and scientific judgment should be exercised in deciding whether expedited reporting is appropriate in other situations such as medical events that may jeopardize the patient or may require intervention to prevent one of the other outcomes listed above. A few examples of such events are intensive treatment in an emergency room or at home for allergic bronchospasm, and blood dyscrasias or convulsions that do not result in hospitalization. Other Important Medical Events have also been listed in the European Medicine Agency release and are considered serious in postapproval safety reporting for adverse events.

Seriousness of a case can be 'upgraded' or 'downgraded' based on information received in the follow-up report. A follow-up report may be required to be submitted as an expedited report if the seriousness of the case changes even if the initial report was not serious and, therefore, not submitted as an expedited report.

Expectedness (listedness) of the adverse event terms using appropriate reference safety information (current product label) of the country for which expedited report is scheduled for submission and expedited reporting to the country regulatory authority is based on expectedness assessment. An ADR whose nature, severity, specificity or outcome is not consistent with the term or description used in the product label (Prescribing Information or Summary of Product Characteristic) should be considered as 'unexpected'.

The narrative should summarize all relevant clinical information, including patient demography details, medication received, medical and concomitant history, investigations performed, clinical course of the events, diagnosis, ADRs and their outcome. The narrative should describe information in the chronology of patient experience. Abbreviations and acronyms should be avoided with the possible exception of laboratory parameters and their units. Postmortem/autopsy findings should also be summarized in the narrative. The narrative should include the verbatim term as used by the patient/reporter or an accurate translation of it. Patient/reporter confidentiality should be maintained. The narrative should also include the medical reviewer's evaluation and comment. The purpose is to present the company's opinion on the case assessment including causality assessment [9]. If the reporter clearly states that there is no causal relationship with the medicinal product and the company medical reviewer agrees with it, the case can be downgraded to unrelated case otherwise for postapproval ADR cases implied causality is considered. Any follow-up information added should be presented in a separate paragraph. Follow-up information should be considered significant when new information received in the follow-up has significant impact on the medical assessment of the case.

Expedited reports refer to ICSRs that involve a serious event which is related to the use of the drug and the ADRs are unexpected. Spontaneous reports are typically considered to have an implied causality, whereas a clinical trial case is assessed for causality both by the investigator and sponsor of the trial. The timeframe for reporting expedited cases is generally 15 calendar days from the time a pharmaceutical company receives information (referred to as "Day 0") about such a case. Within clinical trials, such a case is referred to as a SUSAR (a Suspected Unexpected Serious Adverse Reaction). With a SUSAR that is life-threatening or fatal, the initial report should be sent to regulatory authority by 7 calendar days and follow-up report by additional 8 calendar days. Cases with a nonserious event are not sent on expedited basis and form part of periodic safety reports.

Safety information from clinical trials is used to establish a drug's safety profile in humans and is a key component that drug regulatory authorities consider in the decision-making as to whether to grant or deny market authorization (market approval) for a drug to a pharmaceutical company. Serious Adverse Event (SAE) reporting from clinical trials occurs as a result of study subjects experiencing SAEs during the conduct of clinical trials. SAE information, which may also include relevant information from the patient's medical background, are reviewed and assessed for causality by the study investigator. This information is forwarded to the company sponsoring the trial that is responsible for the reporting of this information, as appropriate, to drug regulatory authorities.

The origin of ADRs is usually from two sources: Unsolicited and Solicited.

Unsolicited Reports would be of the following:

A spontaneous report is an unsolicited communication by healthcare professionals or consumers to a company, regulatory authority, or other organization (e.g., WHO, Regional Monitoring Centers, Poison Control Center) that describes one or more adverse drug reaction in a patient who was given one or more medicinal product and that does not derive from a study or any organized data collection scheme [4–7]. Stimulated reporting may occur in certain situations, such as a notification by a "Dear Healthcare Professional" letter, a publication in the press, or questioning of healthcare professionals by company representatives. These reports are considered spontaneous.

Spontaneous reporting system relies on reporting of adverse events by physicians and other healthcare professionals such as pharmacists, nurses, and patients to pharmaceutical companies, regulatory authorities, and regional and national pharmacovigilance centers. It is an important source of regulatory authority actions such as ordering pharmaceutical companies to take a drug off the market or to change a label due to safety concerns. Under-reporting is one of the major weaknesses of spontaneous reporting with AE reporting behaviors varying from country to country, and in general probably about 5% of all adverse events that occur are actually reported. Adverse reaction reports from consumers should be handled as spontaneous reports, irrespective of any subsequent "medical confirmation"—a process required by some regulatory authorities for reporting. The Marketing Authorization Holder (MAH) is expected to regularly screen the worldwide scientific literature, by accessing widely used systematic literature reviews or reference databases or outsourcing this activity to an outside organization. Cases of ADRs from the scientific and medical literature, including relevant published abstracts from meetings and draft manuscripts, might qualify for expedited reporting. If an article or abstract is in a different language, then the article should be translated to determine whether there is a valid case. The publication reference(s) should be given as the report source; a copy of the abstract or article should accompany the report when it is being submitted to a regulatory authority. In addition to the global literature monitoring, local literature monitoring is also required in some countries. The regulatory reporting time clock starts once it is determined that the case meets minimum of four criteria for reporting. MAHs should search the literature according to local regulation, or at least weekly. If the product source, brand, or trade name is not specified, the MAH should assume that it was its product and report to regulatory authorities.

MAHs should regularly screen the websites under their management, for any potential ADR case reports. MAHs should also facilitate ADR data collection by providing ADR forms on the websites for direct reporting and also provide contact details for any direct communication. MAHs are not expected to screen external websites for ADR information. However, if an MAH becomes aware of an adverse reaction on a website that it does not manage, the MAH should review the adverse reaction and determine whether it should be reported. Unsolicited cases from the Internet should be handled as spontaneous reports. For the determination of regulatory reporting, the same criteria should be applied as for other spontaneous cases.

Solicited reports are derived from organized data collection systems such as a patient registry, and also include clinical trials, postapproval named patient use programs, other patient support and disease management programs, surveys of patients or healthcare providers, or information gathering on efficacy or patient compliance. Adverse event reports obtained from any of these should not be considered spontaneous. For the purposes of safety reporting, solicited reports should be handled as if they were study reports, and therefore should have an appropriate causality assessment done by a company medical reviewer.

Initial reports received may have incomplete information to make the complete clinical assessment and scientific evaluation. Follow-up of all serious AEs should be done if additional information is required. At least two follow-ups may be adequate but for more critical cases additional attempts may be required. Regulatory agencies do not expect follow-up for nonserious expected cases. However, for nonserious unexpected cases, follow-ups can be attempted. Any attempt to obtain follow-up information should be documented.

Cases of exposure of fetus or embryo to a medicinal product through maternal or paternal exposure need to be followed to know pregnancy outcome. Such cases should preferably be followed for at least 6 months after birth. Individual cases with an abnormal outcome associated with a medicinal product following exposure during pregnancy are classified as serious reports and should be reported as expedited reports. Other cases may be reports of induced termination of pregnancy without information on congenital malformation or reports with normal outcome or without any outcome data should not be reported as there is no adverse reaction. Such reports should be included in periodic update safety reports. However, in certain circumstances even with no abnormal outcome, pregnancy exposure reports need to be reported. This may be part of risk management plans. Examples could be a medicinal product contraindicated during pregnancy or requiring surveillance because of high teratogenicity potential, e.g., thalidomide, isotretinoin.

Suspected adverse reactions occurring in infants following exposure to medicinal product from breast milk should be reported in accordance with the criteria, whether serious or nonserious. The collection of safety information in the pediatric or elderly population is important. Reasonable attempts should be made to obtain complete information about age or age group of patient when a case is reported by a HCP or consumer in order to be able to identify potential safety signal to a particular population.

Reports of overdose, abuse, off-label use, misuse, medication error, or occupational exposure without any adverse reactions form part of periodic safety update reports and not as ICSRs. However, if they are associated with adverse reactions or tilt risk-benefit balance, they are reported in accordance with criteria outlined. Necessary follow-ups are required to obtain complete information to make a complete clinical and scientific evaluation.

Reports of lack of efficacy (LOE) are discussed in periodic safety update reports. However, if the case is deemed serious and LOE is unexpected, the ICSR would qualify for 15 day submission. Also, LOE associated with medicinal products used in critical conditions or for the treatment of life-threatening diseases, vaccines, and contraceptives are reported within a 15-day timeframe. If the reporter relates LOE to progression of disease, they are not reported within 15 calendar days. Follow-ups are required if information is not complete.

Only valid ICSRs are reported. The clock for reporting ICSRs starts from the time when any company employee, affiliate, or any business partner becomes aware of a valid ICSR. Both initial and follow-up serious and unexpected cases are reported within 15 calendar days. Significant follow-up information is one that corresponds to new medical or administrative information that could impact the assessment or management of a case or could change seriousness criteria; nonsignificant information constitutes updated comments on the case assessment or typographical errors in the previous case version.

REFERENCES

[1] ICH Harmonized Tripartite Guideline. Maintenance of the ICH Guideline on Clinical Safety Data Management: Data Elements for Transmission of Individual Case Safety Reports E2B (R2), February 2001.

[2] Guidance for Industry E2BM. Data Elements for Transmission of Individual Case Safety Reports, April 2002.

[3] ICH E2B Expert Working Group. Implementation Guide for Electronic Transmission of Individual Case Safety Reports (ICSRs). E2B (R3) Data Elements and Message Specification, April 2013.

[4] Guideline on Good Pharmacovigilance practices (GVP): Module VI – Management and Reporting of adverse reactions to medicinal products, European Medicines Agency (Rev 1), Sep 2014.

[5] ICH E2A: Clinical Safety Data Management – Definitions and Standards for Expedited Reporting, October 1994.

[6] ICH E2D: Post Approval Safety Data Management – Definitions and Standards for Expedited Reporting, November 2003.

[7] Guidance for Industry. Post Marketing Safety Reporting for Human Drug and Biological Products including Vaccines. Draft Guidance U.S. Department of Health and Human Services, Food and Drug Administration, Centre for Drug Evaluation and Research (CDER) and Centre for Biologics Evaluation and Research (CBER), March 2001.

[8] Guidance Document for Industry – Reporting Adverse Reactions to Marketed Health Products. Health Canada, March 2011.

[9] The use of WHO-UMC system for standardized case causality assessment. <http://Who-umc.org/Graphics/26649.pdf> 17 April 2012.

CHAPTER

DEVELOPMENT AND PERIODIC SAFETY REPORTS

29

Rajinder K. Jalali

Sun Pharmaceutical Industries Ltd, Gurgaon, Haryana, India

During the medicinal product development and after the medicinal product is on the market, the sponsor or marketing authorization holder (MAH) has obligations to submit the cumulative safety information to regulatory agencies periodically. The Development Safety Update Report (DSUR) is used to provide periodic safety information during preapproval and for marketed drugs that are under further study and the Periodic Safety Update Report (PSUR) is intended to provide an evaluation of the risk-benefit balance of a medicinal product for submission by MAHs at defined time points during the postapproval phase [1–6]. The Development Safety Update Report (DSUR) is prepared in accordance with ICH E2F Guideline and is intended to be a common standard for periodic reporting on drugs under development among the ICH regions. The term PSUR comes from the previous ICH E2C (R1) which has been subsequently revised to ICH E2C (R2) and renames PSUR as Periodic Benefit-Risk Evaluation Report [3,4]. The Periodic Benefit-Risk Evaluation Report (PBRER) is used to report periodic safety information to European agencies and some countries replacing PSUR whereas PSUR is required in other regions [5,6]. Periodic Adverse Drug Experience Report (PADER) is used to report periodic safety information to US FDA. All adverse drug experiences not submitted as 15-day calendar reports must be submitted in the periodic report.

- The US FDA requires PADERs quarterly during the first 3 years after the medicine is approved, and annual reports thereafter.
- The EMA requires PSURs every 6 months for 2 years, annually for the 3 following years, and then every 5 years (at the time of renewal of registration) if a specific frequency is not defined within the European Union Reference Dates (EURD) list. The EURD list consists of a list of active substances and combinations of actives that are subject to PSUR single assessment procedures.
- In Japan, the authorities require periodic safety reports with the date designated by the Ministry at the time of approval, which is established as the base date. The frequency of reports is every 6 months during the first 2 years from this base date. Thereafter, reports are to be submitted once each year during the remaining period of reexamination.
- Development Safety Update Report (DSUR), replace IND Annual Report and Annual Safety Report, for drugs under developments and Periodic Safety Update report (PSUR) has been updated to Periodic Benefit-Risk Evaluation Report (PBRER) for drugs already on the market.

Pharmaceutical Medicine and Translational Clinical Research. DOI: http://dx.doi.org/10.1016/B978-0-12-802103-3.00030-4

The Periodic Benefit-Risk Evaluation Report (PBRER) is prepared in accordance with ICH E2C (R2) and is intended to be a common standard for periodic benefit-risk evaluation reporting on marketed products and replaces PSUR prepared in accordance with ICH E2C (R1) in many markets [3,4]. However, the European Medicinal Agency has not fully adopted the PBRER and continues to use the term PSUR, as defined in its recent guidelines.

The Development Safety Update Report is prepared in accordance with the ICH E2F [1,2] and is intended to be a common standard for periodic reporting on drugs under development. US and EU regulators consider that the DSUR, submitted annually, would meet national and regional requirements currently met by the US IND Annual Report and the EU Annual Safety Report, respectively, and can therefore take the place of these existing reports.

DSURs are new, internationally-harmonized safety documents (which became mandatory in European Union member states in September 2011) covering the safety summary of medicinal products during their development or clinical trial phase.

The new DSUR (defined in guideline ICH E2F) is based heavily on the PSUR format already used for updating the safety record of drugs in their marketing phase. The new DSUR format replaces the previous European Union ASR (Annual Safety Report) and in due course is expected also to replace the United States IND Annual Report.

Sponsors are required to submit a DSUR within one year of the Development International Birth Date (DIBD—the date of first authorization of a clinical trial in any country worldwide) and provide annual DSUR submissions until all open clinical studies have been completed (the final clinical study is completed and its study report has been submitted).

1. A DSUR should be submitted until the last visit of the last patient in the Member States concerned, as specified within the protocol.
2. The data lock point (DLP) for a DSUR reporting period is the last day (or the last day of the month, ICH E2F section 2.2) before the anniversary of the DIBD (Development International Birth Date), the date on which the product was first authorized for testing in humans anywhere in the world. IBD (International Birth Date), the date on which the product was first approved for market anywhere in the world, can also be acceptable.
3. The first DSUR period should not be longer than 1 year. The DSUR is always submitted on a yearly basis.
4. For transitional period: The DIBD and the European Birth Date (EBD) of the previous DSUR should be considered.
5. ASR should be aligned in such a way that DSUR periods that are substantially longer than 12 months, as well as overlapping DUSR periods, are avoided.
6. A DSUR is required for Phase IV clinical trials, if only such trials are conducted.
7. Submission of one single DSUR is strongly recommended if the same IMP is used in the clinical trials.
8. A separate DSUR for a comparator, placebo, or non-IMP is not required. However, relevant safety information of the above-mentioned drug types should be addressed in the DSURs of the investigational drugs.
9. The Investigator Brochure (IB) or Summary of Product Characteristics can be updated during the DSUR reporting periods.

During the clinical development of an investigational drug, periodic analysis of safety information is crucial to the ongoing assessment of risk of trial subjects. It is also important to inform regulators, ethics committees, and other investigators involved in the conduct of study, at regular intervals of the evolving safety profile of the investigational drug and the actions proposed or being taken to address safety concerns. The main objective of a DSUR is to present an annual review and evaluation of pertinent safety information collected during the reporting period to (1) examine whether the information obtained by the sponsor during the reporting period is in accordance with previous knowledge of the product's safety; (2) summarize the current understanding and management of identified and potential risks; (3) describe the new safety issues that could have an impact on the protection of clinical trial subjects; and (4) to provide an update on the status of the clinical investigation. A DSUR should be concise and provide information to assure regulators that sponsors are adequately monitoring and evaluating the safety profile of the investigational drug. The main focus of the DSUR is data and findings from clinical trials of the drugs and biologicals under investigation, whether or not they have a marketing approval. The DSUR should concentrate primarily on the investigational drug, providing information on comparators only where relevant to the safety of trial subjects.

The DSUR should provide safety information from all ongoing clinical trials and other studies that the sponsor is conducting or has completed during the review period, including: (1) Phase I to Phase IV clinical trials; (2) therapeutic use of an investigational drug, e.g., expanded access programs (IND based program in US that allows sponsor to supply an investigational drug to patients with an indication for which benefit has been demonstrated), compassionate use programs (a program in the EU to supply an investigational drug to patients with an indication for which benefit has been demonstrated), single patient INDs (within the United States, preapproval demand is generally met through single-patient INDs), treatment INDs (allows the supply of an investigational drug to treat patients with serious or immediately life-threatening disease for which satisfactory alternative treatment is not available) and particular patient use (outside the US, preapproval use of drugs in response to a request by physicians on behalf of specific patients before those medicines are licensed in the patient's home country); and (3) clinical trials conducted to support changes in the manufacturing process of medicinal products. The DSUR should also include significant safety findings from observational and epidemiological studies, nonclinical studies, other related DSURs, manufacturing or microbiological changes, studies recently published in literature, clinical trials with results indicating lack of efficacy that could have direct impact on patient safety, e.g., worsening of an underlying serious or life-threatening condition.

A single DSUR should be prepared for all ongoing clinical trials with different formulations and indications with a single data lock point. The DSUR is intended to be an annual report that should be submitted as long as clinical trial(s) are ongoing. The development International Birth Date, i.e., date of first approval of conduct of a clinical trial in any country, is used to determine the start of the annual period for the DSUR. The DSUR should be submitted no later than 60 calendar days from the DSUR data lock point. A single DSUR should be prepared in case of clinical trials on fixed dose combination or combination therapies. Once a drug has received a marketing approval in any country or region, and clinical trials continue or are initiated in other countries, both PSUR and DSUR need to be prepared, and in such a situation, DSUR data lock point should be changed to International Birth Date so that both PSUR and DSUR are synchronized. When an annual report of clinical trials is no longer required in an individual country, the final DSUR should be submitted and the agency should be notified about it. Investigator Brochure should be referred to in order to assess whether the safety information received is consistent with the previous knowledge of the safety profile of the investigational drug or there is any new safety information.

The content of DSUR [1] is shown below:

- Title page
- Executive summary

Table of contents

- Introduction
- Worldwide marketing authorization status
- Actions taken in the reporting period for safety reasons
- Changes to reference safety information
- Clinical trials ongoing and completed during the reporting period
- Estimated cumulative exposure
  - Cumulative exposure in the development program
  - Patient exposure from marketing experience
- Line listings and summary tabulations
  - Reference information
  - Line listings of serious adverse reactions during the reporting period
  - Cumulative summary tabulations of serious adverse events
- Significant findings from clinical trials during the reporting period
  - Completed clinical trials
  - Ongoing clinical trials
  - Long term follow-up
  - Other therapeutic uses of investigational drug
  - New safety data related to combinations therapies
- Safety findings from noninterventional studies
- Other clinical trial/study safety information
- Safety findings from marketing experience
- Nonclinical data
- Literature
- Other DSURs
- Lack of efficacy
- Region-specific information
- Late-breaking information
- Overall safety assessment
  - Evaluation of the risks
  - Benefit-risk considerations
- Summary of important risks
- Conclusion
- Appendices to the DSUR

The Periodic Safety Update Report for marketed drugs (PSUR) is a standalone document that allows a periodic and comprehensive assessment of the worldwide safety data of a marketed drug [6]. It is an important source for the identification of new safety signals, a means of determining changes in the benefit-risk profile, an effective means of risk communication to regulatory authorities, and an indicator for the need for risk management initiatives as well as a tracking mechanism monitoring the effectiveness of such initiatives. The main objective of a PSUR is to present a comprehensive, concise, and critical analysis of risk-benefit balance of a medicinal product and taking into account new or emerging information in the context of cumulative information on risks and benefits. The scope of the risk benefit information should include both clinical trial and postmarketing data. The evaluation should include examining the information which has emerged during the interval period to determine whether it has generated new signals, led to the identification of new potential, or identified risks or contributed to knowledge of previously identified risks; summarize relevant new safety, efficacy, and effectiveness information that could have an impact on the risk-benefit balance of the medicinal product; integrated risk-benefit analysis taking into account cumulative data available from development to international birth date; summarize any risk minimization actions that may have been taken/implemented during reporting interval or that are planned to be implemented; and outlining plans for signal or risk evaluations including timelines and/or proposals for additional Pharmacovigilance activities [7,8].

In summary, the aim of cumulative reports of safety is to:

- Report all the relevant new information from appropriate sources.
- Relate these data to patient exposure to the medicine.
- Summarize the medicine's approval status in different countries and any significant variations related to safety.
- Create periodically the opportunity for an overall reevaluation of safety.
- Indicate whether changes should be made to an approved medicine's label in order to optimize the use of the product.

PSUR is to be submitted to the European Medicines Agency (through a PSUR repository) according to following timelines:

1. Within 70 calendar days of data lock point (DLP), considered as Day 0, for PSUR covering intervals up to 12 months.
2. Within 90 calendar days of DLP for PSURs more than 12 months.
3. The ad hoc PSURs requested by the agency should be submitted within 90 calendar days unless otherwise specified.

A single PSUR needs to be prepared for all medicinal products containing the same active substance with information covering all authorized indications, route of administration, dosage forms, and dosing regiments, irrespective of whether authorized under different names and through separate procedures. Where relevant, data relating to particular indication, dosage form, route of administration, or dosing regimen shall be presented in a separate section of the PSUR and any safety concerns shall be addressed accordingly.

The PSUR must present an explicit assessment on the status of the benefit-risk balance. The potential benefit of any medicinal product must be balanced against the potential risk and MAH should provide an integral benefit-risk balance based on scientific assessment of new safety information in order to update the benefit-risk balance on interval data. The PBRER format of the PSUR is designed to simplify the presentation of safety, efficacy, and benefit information collected during the reporting period and to present an explicit risk-benefit evaluation in the context of what was known at the start of the interval [5,6]. A PBRER is a comprehensive, concise, and critical analysis of new or emerging information on the risks of the medicinal product, and on its benefits in approved indications, to enable an appraisal of the product's overall benefit risk profile. The evaluation includes:

- Presentation of data received during the reporting period. This includes a brief summary of data relevant to the benefit and risks of the medicinal product from all sources including case reports, clinical and nonclinical safety studies, the published reports, and an overview of signals, tabulated as new, ongoing, or closed.
- Critical assessment of relevant new safety, efficacy, and benefit related information. This assessment includes what impact the new knowledge has on the risk-benefit balance of the medicinal product. This evaluation also considers data related to unintended use of drug including medication error, misuse, abuse, overdose, off-label use and use in special populations such as children and pregnant women. Patient exposure should be presented based on the region, patient details, and formulations, among others, as the risk assessment should be based on populations exposed.
- Summary of any risk minimization activity that may have been taken or planned as stated in the risk management plan (RMP).
- Outline of plans for signal or risk evaluation including timelines and/or proposal for additional Pharmacovigilance activities.

The PSUR conclusion should provide a preliminary proposal to optimize or further evaluate the risk-balance. Conclusions of further evaluation or optimization should be incorporated into the Pharmacovigilance plan and risk minimization plan of the RMP.

Case narrative shall be provided in the relevant risk evaluation section of the PSUR, integral to the scientific analysis of a signal or safety concern.

The PSUR is formatted [4,5] into following sections:

- Title page including signature
- Executive summary

Table of contents

- Introduction
- Worldwide marketing authorization status
- Actions taken in the reporting interval for safety reasons
- Changes to reference safety information
- Estimated exposure and use patterns
  - Cumulative subject exposure in clinical trials
  - Cumulative and interval patient exposure from marketing experience
- Data in summary tabulations
  - Reference information
  - Cumulative summary tabulations of serious adverse events from clinical trials
  - Cumulative and interval summary tabulations from postmarketing data sources

- Summaries of significant findings from clinical trials during the reporting interval
  - Completed clinical trials
  - Ongoing clinical trials
  - Long-term follow-up
  - Other therapeutic use of medicinal product
  - New safety data related to fixed combination therapies
- Findings from noninterventional studies
- Information from other clinical trials and sources
  - Other clinical trials
  - Medication errors
- Nonclinical Data
- Literature
- Other periodic reports
- Lack of efficacy in controlled clinical trials
- Late-breaking information
- Overview of signals: new, ongoing, or closed
- Signal and risk evaluation
  - Summaries of safety concerns
  - Signal evaluation
  - Evaluation of risks and new information
  - Characterization of risks
  - Effectiveness of risk minimization (if applicable)
- Benefit evaluation
  - Important baseline efficacy and effectiveness information
  - Newly identified information on efficacy and effectiveness
  - Characterization of benefits
- Integrated benefit-risk analysis for authorized indications
  - Benefit-risk context—Medical need and important alternatives
  - Benefit-risk analysis evaluation
- Conclusions and actions
- Appendices to the PSUR

The content and format of a PBRER is specified in the guidelines from the International Conference on Harmonization (ICH) E2C (R2), which aims to harmonize regulatory requirements globally [4]. The format is as below:

Part I: Title page including signature
Part II: Executive summary

List of abbreviations should be placed immediately after the executive summary

Part III: Table of contents—it has following elements:

- Introduction carrying the brief description of the product, international birth date, and reporting interval.
- Worldwide Marketing Authorization Status containing cumulative worldwide marketing authorization status of the medicinal product.

- Actions taken in reporting interval for safety reasons. This should include update on regulatory agency or pharmaceutical company's actions taken for safety reasons, such as: approval/withdrawal/suspension; restrictions on distribution; failure to obtain renewal of approval; suspension of a clinical study; dosage modification; changes in target population and indications.
- Changes to reference safety information (e.g., any significant changes made within the reporting interval including changes relating to contraindications, precautions, warnings, serious adverse drug reactions, adverse events of special interest, or drug–drug interactions).
- Data on patient exposure derived from an estimate of the cumulative number of patients exposed in clinical trials and cumulative and interval patient exposure from marketing experience along with the method used to calculate the estimate.
- Data in summary tabulations of all events received from clinical trials, noninterventional trials, postapproval spontaneous reports from healthcare professionals, patients/consumers, literature, and regulatory authorities and noninterventional solicited sources.
  - Reference information stating version of coding dictionary used for analysis of adverse events/adverse drug reactions.
  - Cumulative summary tabulations of serious adverse events from clinical trials.
  - Cumulative and interval summary tabulations from postmarketing data sources.
- Summaries of significant findings from clinical trials during the reporting interval describing any significant safety findings that had impact on conduct of:
  - Completed clinical trials
  - Ongoing clinical trials
  - Long-term follow-up
  - Other therapeutic use of medicinal product in expanded access programs, compassionate use programs, particular patient use, and other organized data collection.
  - New safety data related to fixed dose combination therapies
- Findings from noninterventional studies, e.g., observational studies, epidemiological studies, registries, and active surveillance programs.
- Information from other clinical trials and sources:
  - Other clinical trials such as results from pool analysis or meta-analysis of randomized clinical trials, safety information provided by codevelopment partners or from investigator initiated trials.
  - Medication errors by patients and healthcare professionals.
- Nonclinical data from in vivo and in vitro studies, e.g., summary of carcinogenicity, reproduction or immunotoxicity studies that are ongoing or completed during the reporting interval.
- Literature review to identify any relevant safety concern arising out of publications from reference databases, e.g., Embase, Pubmed.
- Other periodic reports in case of fixed dose combination products or products with multiple indications or formulations where multiple PSURs are prepared.
- Lack of efficacy in controlled clinical trials, e.g., for products intended to treat life-threatening infections, vaccine or contraceptives having direct impact on patient safety.
- Late breaking information for any safety or efficacy concern that has arisen after DLP but during preparation of PSUR.

- Overview of signals: new, ongoing, or closed signals during the reporting period.
- Signal and risk evaluation:
  - Summaries of safety concerns
  - Signal evaluation
  - Evaluation of risks and new information
  - Characterization of risks
  - Effectiveness of risk minimization
- Benefit Evaluation:
  - Important baseline efficacy and effectiveness information
  - Newly identified information on efficacy and effectiveness
  - Characterization of benefits
- Integrated benefit-risk analysis for authorized indications:
  - Benefit-risk context—medical need and important alternatives
  - Benefit-risk analysis evaluation
- Conclusions and actions
- Appendices

29.1 PERIODIC ADVERSE DRUG EXPERIENCE REPORTS (PADERS)

All adverse drug experiences not submitted as 15 day reports must be submitted in the periodic report. A periodic report for each NDA/ANDA/BLA must be submitted quarterly for the first three years and thence yearly. Each report must be submitted 30 calendar days after the close of the quarter for quarterly reports and 60 calendar days after the close of the annual date for yearly reports [9,10].

Each PADER is required to contain:

1. Descriptive information.
   - a. A narrative summary and analysis of the information in the report.
   - b. An analysis of the 15-day Alert reports submitted during the reporting interval (all 15-day Alert reports being appropriately referenced by the applicant's patient identification code, adverse reaction term(s), and date of submission to FDA).
   - c. Tables of frequency of occurrence of adverse events from the reporting period, organized by body system.
   - d. A history of actions taken since the last report because of adverse drug experiences (for example, labeling changes or studies initiated).
   - e. An index consisting of a line listing of manufacturer control numbers and adverse event terms are also to be included in the PADER.
   - f. Formerly, a MedWatch form for each adverse event not previously submitted as a 15-day report was required to be submitted to USFDA. However, post June 2015 and as per Federal Register/Vol. 79, No. 111/Tuesday, June 10, 2014: Postmarketing Safety Reports for Human Drug and Biological Products; Electronic Submission Requirements, non-15 day reports are electronically submitted to FDA and are captured in the body of the PADER.

2. ICSRs for serious, expected, and nonserious adverse drug experiences.

PADERs do not include adverse events occurring outside the U.S. or those obtained from the scientific literature or from postapproval studies, except for 15-day reports. Follow-up information for adverse events submitted in a previous periodic report as received or submitted during the reporting interval of PADER are also to be included. Further, actions taken for safety reasons since the last report form part of PADER.

A PADER is formatted as under:

- Title page
- Introduction
- Narrative summary and analysis
- Narrative discussion of actions taken
- Index line listing—NDA line listing extracted from Argus safety
- A copy of most recent US product label effective during the reporting period of a PADER

REFERENCES

[1] ICH E2F: Guidance For Industry: Development Safety Update Report. U.S. Department of Health and Human Services. Food and Drug Administration–Centre for Drug Evaluation and Research (CDER), Centre for Biologics Evaluation and Research (CBER), August 2011.

[2] ICH E2F: ICH guideline E2F on development safety update report (Step 5). European Medicines Agency, September 2011.

[3] ICH E2C (R1) ICH Topic E 2 C (R1). Note for Guidance on Clinical Safety Data Management: Periodic Safety Update Reports for Marketed Drugs. European Medicines Agency, June 1997.

[4] ICH E2C (R2). ICH Harmonised Tripartite Guideline: Periodic Benefit-Risk Evaluation Report (PBRER), December 2012.

[5] Guideline on Good Pharmacovigilance Practices (GVP). Module VII–Periodic Safety Update Report (Rev 1). European Medicines Agency, December 2013.

[6] VOLUME 9A of The Rules Governing Medicinal Products in the European Union –Guidelines on Pharmacovigilance for Medicinal Products for Human Use. EuropeanMedicines Agency September 2008.

[7] Guideline on Good Pharmacovigilance Practices (GVP). Module IX–Signal Management. European Medicines Agency, June 2012.

[8] Guideline on Good Pharmacovigilance Practices (GVP). Module V–Risk Management Systems. European Medicines Agency, April 2014.

[9] 21 CFR 314.80, revised as of April 1, 2015.

[10] Draft FDA Guideline Post-marketing safety reporting for Human Drug and Biological Products including Vaccines, March 2001.

CHAPTER

RISK MANAGEMENT IN PHARMACOVIGILANCE

30

Rajinder K. Jalali

Sun Pharmaceutical Industries Ltd, Gurgaon, Haryana, India

Risk management in Pharmacovigilance is a global activity to safeguard health of patients. A medicinal product is authorized on the basis of results of preclinical and clinical studies. These studies are usually conducted on a small number of patients in controlled environments, e.g., restricted age, comorbidity, comedication, and excluding special populations such as elderly population, children, pregnant and lactating women. At the time of authorization, risk-benefit is judged to be positive. However, not all actual or potential risks have been identified at the time of authorization. Risk management is a set of activities performed for identification of risk, risk assessment, risk minimization or prevention, and risk communication [1,2]. Risk Management Plan (RMP) is developed in accordance with applicable regulations and guidelines. However, in absence of guidelines for a country, the plan is prepared in line with ICH E2E guideline on pharmacovigilance planning [3].

The FDA identifies risk management as an iterative process designed to optimize the benefit-risk balance for regulated products throughout the product lifecycle. In March 2005, the FDA issued three guidance documents that defined the formal basis of risk management. These were Premarketing Risk Assessment, Good Pharmacovigilance Practices and Pharmacoepidemiologic Assessment, and the Development and Use of Risk Minimization Action Plans (RiskMAPs). These three documents subsequently provided the building blocks for the more recent risk evaluation and mitigation strategies (REMS). The final content of a product's REMS, however, reflects the unique mix of product attributes as well as the intended prescriber and patient populations [4]. The REMS program seeks to manage known or potential serious risks, and the content must have a timetable for submission of assessments. Additional components for a particular REMS program vary according to the severity of identified risks, the population likely to be exposed, and other factors.

The EU RMP is an engagement of wider scope than the US REMS, and is binding on a larger set of medicines. The US REMS is compulsory only for some medicines, and can be limited to two years' post product launch. The REMS concerns itself with communication of risk, with the prescriber information—the package insert (PI), being central to risk minimization. Components of a typical FDA REMS are a communication plan; patient selection; web-based materials and a medical scientific liaison; elements to assure safe use; an implementation system; a patient or physician survey and patient understanding of risk. The EU RMP is a more comprehensive, more extensive safety package that the sponsor is obligated to follow throughout the lifecycle of all new drugs or biologics [5].

Pharmaceutical Medicine and Translational Clinical Research. DOI: http://dx.doi.org/10.1016/B978-0-12-802103-3.00031-6

European Union (EU) legislation necessitates that, when required, a description of the risk management system should be submitted in the form of an EU-RMP. Accordingly, risk management has the following stages: (1) identification and characterization of the safety profile of the medicinal product including what is known or not known (with emphasis on important identified and important potential risks and missing information) about the product and, importantly, which risks need to be further characterized or managed proactively (the "safety specification"); (2) planning of pharmacovigilance activities to characterize and quantify clinically relevant risks and to identify new adverse reactions and to increase the knowledge in general about the safety profile of the medicinal product (the "pharmacovigilance plan"), (3) planning and implementation of risk minimization measures and mitigation and assessment of the effectiveness of these activities (the "risk minimization plan"), and (4) document postapproval obligations that have been imposed as a condition of the marketing authorization [2]. All these activities together constitute Risk Management Plan (RMP), which is required to be submitted during the authorization of the drug. An updated RMP is required when there are significant changes to the safety profile of the drug. The overall aim of risk management is to ensure that the benefits of medicinal product outweigh the risks by a wide margin for the treatment of a particular indication both at individual level and for the target population as a whole.

RMP is structured [2] with the following elements:

- *Part I*: Product overview
- *Part II*: Safety specification
  - *Module SI*: Epidemiology of the indication(s) and target population
  - *Module SII*: Nonclinical part of safety specification
  - *Module SIII*: Clinical trial exposure
  - *Module SIV*: Populations not studied in the clinical trials
  - *Module SV*: Postauthorization experience
  - *Module SVI*: Additional requirements for the safety specification
  - *Module SVII*: Identified and potential risks
  - *Module SVIII*: Summary of the safety concerns
- *Part III*: Pharmacovigilance plan (including postauthorization safety studies)
- *Part IV*: Plans for postauthorization efficacy studies
- *Part V*: Risk minimization measures and evaluation of the effectiveness of the risk minimization activities
- *Part VI*: Summary of the risk management plan
- *Part VII*: Annexes

The RMP is part of the scientific dossier of a product and as such should be science-based and not promotional. The submitted RMP should follow the RMP template. The amount of information—particularly in RMP Part II—to be provided will depend on the type of medicinal product, its risks, and where it is in its lifecycle.

RMP part I "Product(s) overview" is a complete overview of the product. This includes complete information on the active substance(s) and all related administrative information on RMP.

RMP part II "Safety specification" consists of a synopsis of complete safety profile of the medicinal product and summarizes important identified risks (an untoward occurrence for which there is adequate evidence of an association with the medicinal product of interest), important potential risks (an untoward occurrence for which there is some basis for suspicion of an association with the medicinal product of interest but where the association has not been confirmed) and missing information, i.e., gaps in knowledge about a medicinal product, related to safety or use in particular patient populations, which could have clinical significance [2,6]. It also addresses populations potentially at risk and identifies the need for specific data collection to answer outstanding safety information that could have relevance to patient safety. The safety specifications facilitate understanding of the risk-benefit profile during postapproval period and construction of pharmacovigilance plan and basis of risk minimization activities.

The safety specification consists of eight RMP modules of which RMP modules SI–SV, SVII, and SVIII correspond to safety specification headings in ICH-E2E. RMP module SVI includes additional elements required to be submitted in the EU. For generic medicinal products the expectation is that the safety specification is the same as that of the reference product or of other generic products for which an RMP is in place. RMP summaries for most recently approved centrally authorized medicinal products (CAPs) are published on EMA website [7].

RMP part III "Pharmacovigilance plan": The purpose of the pharmacovigilance plan is to discuss how the marketing authorization holder plans to identify and/or characterize the risks identified in the safety specification. It provides structured plan for: (1) the identification of new safety concerns; (2) further characterization of known safety concerns including elucidation of risk factors; (3) the investigation of whether a potential safety concern is real or not; and (4) how missing information will be sought [2]. In short, pharmacovigilance plan describes routine and additional pharmacovigilance activities and, thereof, action plans for each safety concern elucidated. Further, it specifies actions to be taken, in addition to procedures in place to detect safety signals, to address identified safety concerns.

The pharmacovigilance plan focuses on the safety concerns summarized in RMP module SVIII of the safety specifications and should be proportionate to the benefits and risks of the product. The action plan for each safety concern is presented and justified according to the following structure: safety concern, objective of proposed actions, actions proposed, rationale for proposed actions, monitoring by the MAH for safety concerns and proposed actions, milestone for evaluation and reporting [6].

RMP part IV "Plans for postauthorization efficacy studies" includes a list of postauthorization efficacy studies (PAES) imposed as conditions of the marketing authorization (MA) or when included as specific obligations in the context of a conditional MA or a MA under exceptional circumstances. Not all medicines require postauthorization studies. However, there may be situations when efficacy of products may vary over time and longer-term efficacy data postauthorization is necessary. In addition, regulations on pediatric medicinal products [8] and advanced therapy medicinal products [9] require long-term efficacy as part of postauthorization surveillance for these types of medicinal products.

RMP part V "Risk minimization measures (including evaluation of the effectiveness of risk minimization activities)" provides details of the risk minimization measures which will be taken to reduce the risks associated with respective safety concerns. Some safety concerns may be adequately addressed by the proposed actions in the pharmacovigilance plan, but for others the risk may be of a particular nature and seriousness so that risk minimization activities are required. It is possible that the risk minimization activities may be limited to ensuring that suitable warnings are included in the product information or by careful use of labeling and packaging, i.e., routine risk minimization activities. However, for some risks, routine risk minimization activities will not be sufficient and additional risk minimization activities will have to be performed. If these are required, they should be described in the risk minimization plan. Additional risk minimization methods include education training material or training programs for medical practitioners, pharmacists, and patients and controlled access programs, and other risk minimization measures [2,6,10]. Consideration must be given to the risk proportionality of the risk minimization activity proposed, the feasibility of implementing any additional risk minimization activity, whether the proposed measures are necessary for the safe and effective use of the product in all patients, and the possibility to adapt distribution modalities [6,10].

For each safety concern, the following information should be provided: (1) objective of the risk minimization activities; (2) routine risk minimization activities; (3) additional risk minimization activities (if any), individual objectives, and justification of why needed; (4) how the effectiveness of each risk minimization activity will be evaluated in terms of attainment of their stated objectives; (5) criteria for judging success; and (6) milestones for evaluation and reporting [2]. This helps to improve risk communication with healthcare professionals and patients and reduced harm caused by new drugs [11]. Evaluation of the effectiveness of risk minimization activities is required beyond the pharmacovigilance plans. Assessment is performed for each safety concern whether any risk minimization activities are needed. The success of risk minimization activities needs to be evaluated throughout the lifecycle of a product to ensure that the burden of adverse reactions is minimized and hence the overall benefit-risk balance is optimized.

RMP part VI "Summary of the risk management plan" shall be made publicly available and shall include the key elements of the risk management plan. The audience of RMP summaries is very broad. To ensure that the summary can satisfy the different needs, it should be written and presented clearly, using a plain-language approach. The summary of the RMP part VI should be consistent with the information presented in RMP part II modules SVII, SVIII, and RMP parts III, IV, and V.

The summary must include key elements of RMP with a specific focus on risk minimization activities. Regarding safety specification of the medicinal product, it should contain important information on potential and identified risks as well as missing information [2].

The summary of the RMP should be formatted as below:

- Overview of disease epidemiology
- Summary of treatment benefits
- Unknowns relating to treatment benefits
- Summary of safety concerns
    - Important identified risks
    - Important potential risks
    - Missing information
- Summary of risk minimization activities by safety concern
- Planned post authorization development plan
- Studies which are a condition of the marketing authorization
- Major changes to the risk management plan over time

The information provided in each section should be brief and focused.

RMP part VII provides for the "Annexes to the risk management plan."

Both REMS and RMPs provide positive guidance for identification, monitoring, and minimization of risk to patient safety. The importance of risk management cannot be over-emphasized and the regulatory burden is increasing, and appropriately so. It is in the interests of patients, industry, and agencies that the least harm and maximum benefit results from taking a medicine; risk strategies such as the US REMS and the EU RMP contribute to this. They also channel drug developers to give greater consideration to how patients can avoid some adverse reactions to drugs and achieve better tolerance, by paying attention to criteria such as contraindications, warnings, and precautions. The EU RMP is an engagement of wider scope, and is binding on a wider set of medicines than the US REMS [5].

REFERENCES

[1] Dieck GS, Sharrar RG. Preparing for safety issues following drug approval: pre-approval risk management considerations. Ther Adv Drug Saf 2013;4:220–8.

[2] Guidelines on good Pharmacovigilance practices (GVP). Module V–Risk management systems (Rev 2). European Medicines Agency; March 2017.

[3] ICH E2E Pharmacovigilance Planning, November 2004.

[4] Yvonne Lis, Meissa H. Roberts, Shital Kamble, Jeff J. Guo, Dennis W. Raisch, December 2012 Volume 15, Issue 8, Pages 1108–1118// <http://www.valueinhealthjournal.com/article/S1098-3015(12)03797-7/pdf>.

[5] Dr. Hoss A. Dowlat, Bio Practice, Reference: Regulatory Reporteur – Vol 8, No 2 Feb 2011. <www.biopractice.com/biosimilar/EU_RMP_US_REMS.pdf>.

[6] Begona Calvo and Leyre Zuniga. Risk Management Plan and Pharmacovigilance System – Biopharmaceuticals: Biosimilars. <www.intechophen.com>.

[7] <http://www.ema.europa.eu>.

[8] Regulation (EC) No 1901/2006 of the European Parliament and of the Council of 12 December 2006 on medicinal products for pediatric use.

[9] Regulation (EC) No 1394/2007 of the European Parliament and of the Council of 13 November 2007 on advanced therapy medicinal products.

[10] Guideline on good pharmacovigilance practices (GVP): Module XVI – Risk minimization measures - selection of tools and effectiveness indicators (Rev 2), March 2017.

[11] Frau S, Font Pous M, Luppino MR, Conforti A. Risk Management Plans: are they a tool for improving drug safety? Eur J Clin Pharmacol 2010;66:785–90.

CHAPTER

RECENT DEVELOPMENTS IN PHARMACOVIGILANCE AT UMC 31

Sten Olsson
Uppsala Monitoring Centre, Uppsala, Sweden

31.1 INTRODUCTION

The Uppsala Monitoring Centre (UMC), with the formal name WHO Collaborating Centre for International Drug Monitoring, is the operational center of the WHO Programme for International Drug Monitoring (PIDM). This Programme consists of national pharmacovigilance centers in the participating member countries, coordinated by the Safety and Vigilance section at WHO in Geneva. Currently (September 2015) there are 122 full member countries and 29 associate member countries of the Programme. The associate members have not yet started contributing to the global WHO database of Individual Case Safety Reports (ICSRs). This database, VigiBase®, currently containing more than 11.2 million ICSRs, is managed and maintained by the UMC. The WHO Programme is supported also by newer and smaller WHO Collaborating Centers located in Ghana, Morocco, Netherlands and Norway.

The UMC is an independent, self-funding foundation created in 1978 with the mission of supporting the WHO PIDM. It has a multiprofessional, multicultural staff of approximately 100 people. This chapter describes the major functions of the UMC and recent developments in the professional area in which the UMC is operating.

31.2 THE SCOPE OF PHARMACOVIGILANCE

In 2002 WHO published a definition of pharmacovigilance [1], elaborated by its Advisory Committee on Safety of Medicinal Products (ACSoMP). The definition refers not only to adverse effects but also to "all other possible drug related problems." The definition is patient-centered and considers all drug-related harm, whether caused by the ingredients of the medicinal product taken, its quality, or the way it is being used. The definition led to a considerable expansion of the scope of work of the WHO Programme and UMC. As a result national pharmacovigilance centers participating in the Programme were encouraged to submit to VigiBase case reports not only regarding suspected adverse reactions but also on unexpected lack of effect, possibly related to quality problems or resistance, and medication errors.

Pharmaceutical Medicine and Translational Clinical Research. DOI: http://dx.doi.org/10.1016/B978-0-12-802103-3.00032-8

31.3 NEW STAKEHOLDERS

An expansion has also occurred of the stakeholders involved in pharmacovigilance activities. Development and harmonization of regulatory requirements, led by the ICH (International Conference on Harmonization) countries in America, Europe, and Japan, has required manufacturers to engage in lifecycle stewardship of the safety of their products [2]. Many national pharmacovigilance systems now receive a major part of their ICSRs from marketing authorization holders. In many low- and middle-income countries' public health programs, supported by many global donor organizations, have started to collect ICSRs and share them with the pharmacovigilance centers. As a consequence, anti-retroviral medicines are now the most commonly reported medicines associated with adverse reactions in Africa. In many countries however, national immunization programs prefer to operate vaccine safety surveillance activities in parallel with the national pharmacovigilance system, with the result that reports on Adverse Events Following Immunization (AEFI) to a large extent escape signal analysis on a global level. Activities of patient and consumer groups, development of social media, and research by pioneering pharmacovigilance scientists has led to the acceptance of direct patient reporting of drug-related harm as an essential element of modern pharmacovigilance. Acceptance of ICSR reports directly from patients and their carers is now to be considered as good pharmacovigilance practice [3].

31.4 THE WHO NETWORK AND ITS REPORTING BEHAVIOR

The global coverage of the WHO PIDM continues to grow and currently countries representing 95% of the global population are taking part in the collaboration. (Fig. 31.1) Although the majority

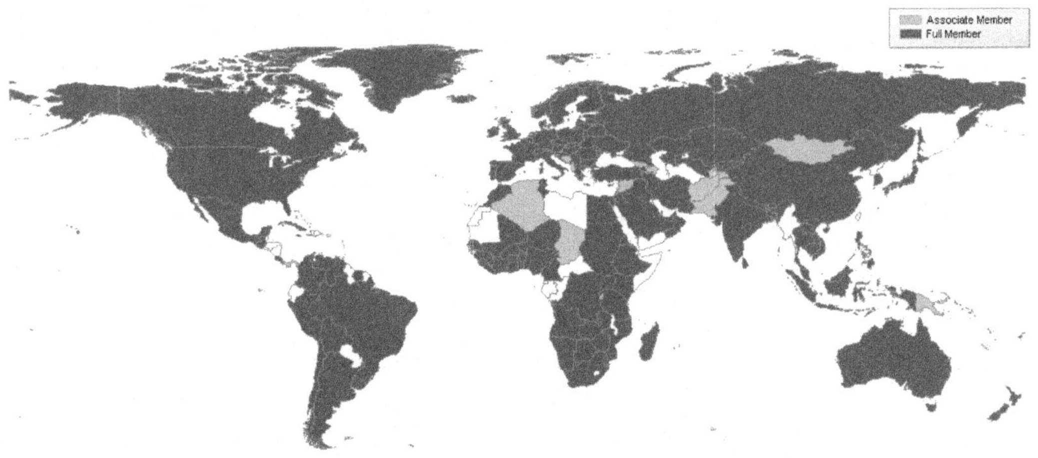

FIGURE 31.1

Member countries of the WHO Programme for International Drug Monitoring; full members dark, associate members shaded.

of ICSRs submitted are still coming from America and Europe, many Asian countries, notably P.R. China, India, Republic of Korea, and Singapore, have recently started to make huge contributions. Reporting from China has been made possible through a long-term collaborative project between the UMC and the Chinese pharmacovigilance center, solving the challenges of translating ICSR data in Chinese to the international E2b standard format. Many of the countries joining the Programme prefer using the VigiFlow™ data management tool developed by the UMC. More than 65 countries are now choosing this solution, which has automatically made them compliant with the international reporting standards. India started using VigiFlow in 2010, which enabled a rapid roll-out of the pharmacovigilance system to currently approximately 150 centers around the country and a fast increase in reporting. Recently the UMC developed a VigiFlow module designed to facilitate direct reporting from the general public. This e-reporting module is being offered to all VigiFlow users as free extra module. At least ten countries have included this facility as an integral part of the website of their pharmacovigilance center.

Reporting from Africa has also increased recently, although from a very low level, mainly through the involvement of public health programs, e.g., against HIV/AIDS and malaria. Although reporting from low- and middle-income countries has increased by 18% over the last year the total contribution from these countries to VigiBase is still only about 10% (Fig. 31.2).

Not only is the number of adverse reaction case reports of importance for the creation of new knowledge about drug-related harm, the frequency of reporting and the completeness of the documentation in each case is critical. The UMC continuously follows these parameters and provides feedback to reporting countries. According to the latest statistics 55% of countries report to VigiBase within a month and a further 15% on a three-monthly basis. UMC has developed a method to measure the completeness of documentation in each case—VigiGrade™ [4]. It is

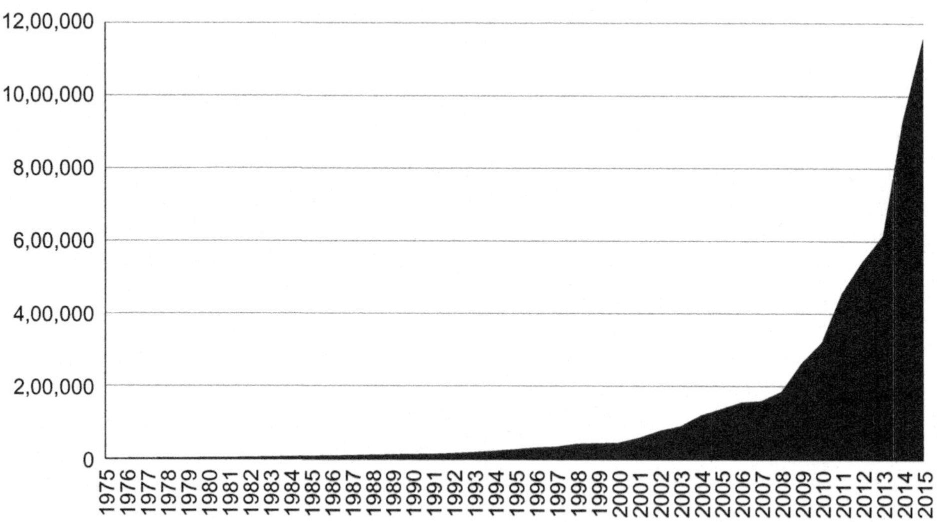

FIGURE 31.2

Cumulative number of ICSRs in VigiBase from low-and middle-income countries.

intended to help in the identification of high-quality case reports that are particularly useful when assessing the evidence for a drug–reaction signal. Countries are provided feedback regularly regarding the VigiGrade score of their contributions to VigiBase, thus stimulating the collection of high-quality data, not only high numbers.

The UMC maintains a support unit which is in frequent contact with all member countries and helps them with technical issues around data management, report submissions, and pharmacovigilance methodology in general. New pharmacovigilance centers in low- and middle-income countries receive special attention and support [5]. UMC staff members often visit countries and attend regional conferences.

31.5 DICTIONARIES FOR DRUG NAMES AND ADVERSE REACTIONS

Both the public and commercial sectors involved in pharmaceutical development and patient care have a need to use common nomenclature standards for drug names and adverse reaction descriptions to avoid misunderstandings and optimize efficiency. UMC has developed and maintains the world's most comprehensive and widely-applied drug coding reference, the WHO Drug Dictionary Enhanced™, WHO-DDE [5]. It is used to facilitate the collection, analysis, and communication of medication information in clinical trials and drug safety data. It is also the central point of an expanding family of related drug dictionaries and services offered in several languages, including Chinese and Japanese. WHO DDE is seamlessly connected to other UMC products and services. Today, more than a thousand pharmaceutical companies, CROs, and regulatory bodies around the world use UMC drug dictionaries in their day-to-day activities. Common examples include the coding, analysis, and reporting of concomitant medications found in clinical trials, as well as searching for drug names appearing on ICSRs. The UMC drug dictionaries also use a hierarchical system for Anatomical Therapeutic Chemical (ATC) classification and Standardized Drug Groupings (SDGs), facilitating aggregation of statistics in analysis and reporting. Recently the US-FDA and the Japanese drug regulatory authority, the PMDA, recommended the use of WHO DDE in regulatory submissions.

The WHO Adverse Reaction Terminology (WHO-ART), maintained by UMC, is also widely used globally in pharmacovigilance. It is kept as an easier-to-use subset of the larger MedDRA terminology, mandated for use by the pharmaceutical industry in many countries. The WHO database, VigiBase, is designed to manage input and output in either WHO-ART or MedDRA.

Revenues from provision of dictionaries, primarily the WHO DDE, to subscribers in the commercial pharmaceutical sector around the world, allow the UMC to maintain and develop VigiBase and to operate the services of the WHO PIDM free of charge for national pharmacovigilance centers.

31.6 ANALYZING DATA IN VIGIBASE

VigiBase is a common resource of the WHO PIDM, available to all participating national centers. The UMC has developed an interactive, web-based tool, VigiLyze™, through which national centers can have easy access to the collective pool of ICSR data. VigiLyze offers a variety of filters and drill-down functionalities to allow flexible searching. Graphical presentations, tabulations and free-text presentations are being offered as output options from the system. UMC has undertaken an ambitious webinar campaign to teach national center staff how to make use of the facility. If national center staff members do not feel confident using VigiLyze or if they have needs for very complicated searches, they can approach UMC staff who will assist in finding the relevant information in VigiBase.

Some consumer organizations and media representatives have in the past been critical of the WHO for not allowing anyone other than national center representatives to access information in VigiBase. Many regulatory authorities contributing information to VigiBase have recently made their national ICSR databases open to the public. In line with the recent trend of involving the public in reporting of their adverse experiences, the WHO has decided to also make summary data from VigiBase publicly available. A new search tool, VigiAccess™ [6], was launched at a WHO press briefing in Geneva in May 2015. Anyone with Internet access can now get a general overview of the medicine-related problems that have been reported to the WHO.

31.7 REGULAR SIGNAL ANALYSIS

One of the key functions of the UMC is to engage in signal analysis of data submitted to VigiBase. Over the years, many different methods and approaches have been attempted in the search for the optimal signal analysis process. It has to be realized that the most common drug-related problems are identified already in the clinical development phase of new medicines and also by the national pharmacovigilance centers receiving a large amount of ICSRs before sending them on to the global database. UMC will focus on rare but clinically significant problems that will be seen only in the global database. For more than 15 years the UMC has been employing an analytical approach combining statistical calculations of disproportionality in reporting, literature reviews, and clinical assessments of individual case reports. Clinical assessments are carried out either by UMC in-house staff or with the help of external experts in a signal review panel. Recently, groundbreaking innovations in the statistical analysis were introduced with the VigiRank™ [7] and VigiMatch™ tools [8]. VigiRank is a statistical analytical process that takes into account not only the number of reports of a drug–reaction combination but also other factors like completeness of documentation, number of countries reporting, timing of reporting, and access to a narrative in the statistical analysis. VigiMatch is a sophisticated statistical method for identification of suspected duplicates in the reporting. The introduction of VigiRank and VigiMatch has increased the efficiency of the signal analysis work dramatically. Most of the work is now also carried out in focused campaigns trying to optimize the identification of new risks in certain specific areas, e.g., pediatrics or vaccines [9]. Since VigiBase now contains more than a million ICSRs from low- and middle-income countries it is now feasible also to look for drug-related problems of regional significance.

Results of the signal analysis activities are fed back to national pharmacovigilance centers in an internal document called "Signals," in which marketing authorization holders are also invited to present their analysis of the newly identified problem. Since 2012 signals are also published in the WHO Pharmaceuticals Newsletter [10].

31.8 PHARMACOVIGILANCE COMPETENCE DEVELOPMENT

There are very few academic courses are available globally that cover all aspects of pharmacovigilance. Recognizing this fact UMC initiated its international pharmacovigilance training course in 1993. Over the years around 600 professionals from all over the world have been trained during an intense two-week period. Although pharmacovigilance courses are now being offered also by the WHO Collaborating Centres in Ghana, Morocco, and the Netherlands, these annual courses have failed to meet the global demand for pharmacovigilance training. UMC has systematically recorded course lectures and they are now made available on the UMC YouTube channel [11]. This initiative allowed the UMC to make the annual course more advanced, leaving out the basic principles covered in the videos.

UMC has also entered into several partnerships to increase the outreach of its competence development activities. Collaboration agreements were signed with the International Society of Pharmacovigilance (ISoP) and with the JSS University in Mysore, India. Joint pharmacovigilance training courses are now being organized with ISoP and at the JSS campus in Mysore [12]. A comprehensive WHO-ISoP core pharmacovigilance curriculum has been developed and published [13]. In collaboration with Uppsala University a distance-learning course in Pre-clinical Safety and Pharmacovigilance is also being offered [14].

The network of WHO Collaborating Centres in pharmacovigilance collaborate and regularly contribute to each other's training activities. UMC staff members frequently make presentations at regional and national training courses particularly in Latin America, Africa, and Asia, either onsite or remotely.

The UMC also offers e-learning courses and webinars for national center staff wanting to start using the various pharmacovigilance tools, e.g., VigiFlow, VigiLyze, or VigiGrade.

In order to assess progress made in the development of pharmacovigilance systems around the world and the remaining gaps, WHO embarked on a major project in 2008, together with national pharmacovigilance centers, to develop pharmacovigilance indicators, to which the UMC has contributed. The WHO set of pharmacovigilance indicators was recently published [15].

31.9 COMMUNICATIONS IN PHARMACOVIGILANCE

Effective communication with all stakeholders is an essential and integral part of pharmacovigilance. The UMC has published extensively on the subject and also on the topic of crises planning and management [16]. The crises management manual "Expecting the Worst" has been translated into Japanese, Chinese, and Spanish. With the development of social media and the increasing empowerment of patients to become involved in their own safety and in reporting, the UMC has developed new communication approaches, more directly oriented towards the public. In 2015 the "Take and Tell" campaign was launched [17]. It includes a simple reporting app, and the first pharmacovigilance song ever released. The UMC free quarterly newsletter "Uppsala Reports," distributed electronically and in print to over 5000 recipients globally, is an important vehicle in the campaign for medicine safety and in spreading information about the activities of the WHO PIDM.

31.10 RESEARCH AND DEVELOPMENT

In addition to the research carried out to support the signal analysis work in VigiBase, mentioned above, UMC is also involved in research to extend the borders of pharmacovigilance as a science and to employ other sources for safety information than ICSRs. A lot of this research has been carried out in collaboration with international consortia, e.g., the PROTECT, SALUS, Monitoring Medicines, and WEB-RADR consortia funded from the European Union and the OMOP project, chaired by the US FDA. Examples of methods developed through this collaborative research are:

- Analysis of medicine safety issues from longitudinal electronic healthcare records [18]
- Generation of ICSRs from within electronic health record systems [19]
- Analyzing social media for pharmacovigilance purposes [20]
- Methods for detection of substandard/spurious/falsely labeled/falsified/counterfeit medicines from ICSR databases [21]
- Early identification of dependence-producing medicines from ICSR data [22]
- Methods to support drug safety surveillance in public health programs [23]
- Reporting and learning systems for medication errors [24]

31.11 CONCLUSION

Global pharmacovigilance, the WHO PIDM and UMC have made important and rapid progress during the last decades, but many challenges remain before the vision of a world where all patients and health professionals take wise decisions in their use of medicines is attained. The UMC is committed to continuing its mission in global pharmacovigilance capacity building in a broad sense, strengthening human and institutional resources in identification, prevention, and treatment of medicine-related harm.

REFERENCES

[1] The importance of pharmacovigilance. WHO, Geneva, 2002 <http://apps.who.int/medicinedocs/pdf/s4893e/s4893e.pdf?ua = 1>.
[2] <http://www.ich.org/products/guidelines/efficacy/article/efficacy-guidelines.html>.
[3] Margraff F, Bertram D. Adverse drug reaction reporting by patients: an overview of fifty countries. Drug Saf 2014;37(6):409–19.
[4] Bergvall T, Norén GN, Lindquist M. vigiGrade: a tool to identify well-documented individual case reports and highligth systematic data quality issues. Drug Saf 2014;37(1):65–77.
[5] Olsson S, Pal S, Dodoo A. Pharmacovigilance in resource-limited countries. Expert Rev Clin Pharmacol 2015;8(4):449–60.
[6] <http://www.umc-products.com/DynPage.aspx?id = 73588&mn1 = 1107&mn2 = 1139>.
[7] <http://www.vigiaccess.org/>.
[8] Caster O, Juhlin K, Watson S, Norén GN. Improved statistical signal detection in pharmacovigilance by combining multiple strength-of-evidence aspects in vigiRank. Drug Saf 2014;37(8):617–28.

[9] Tregunno PM, Fink DB, Fernandez-Fernandez C, Lázaro-Bengoa E, Norén GN. Performance of probabilistic method to detect duplicate individual case safety reports. Drug Saf 2014;37(4):249–58.
[10] Caster O. Introducing vigiRank: promise for more effective signal detection. Uppsala Reports 2014;67:12–13 http://www.who-umc.org/graphics/28295.pdf.
[11] <http://www.who.int/medicines/publications/newsletter/en/>.
[12] <https://www.youtube.com/channel/UC1SmOUUe6noAWVY4P2JEljw>.
[13] <http://jsspharma.org/node/502>.
[14] Beckmann J, Hagemann U, Bahri P, Bate A, Boyd IW, Dal Pan GJ, et al. Teaching pharmacovigilance: the WHO-ISoP core elements of a comprehensive modular curriculum. Drug Saf 2014;37(10):743–59.
[15] <http://www.uu.se/en/admissions/master/selma/Kurser/?kKod = 3FX011&typ = 1&lasar = 15/16>.
[16] <http://www.who.int/medicines/areas/quality_safety/safety_efficacy/EMP_PV_Indicators_web_ready_v2.pdf?ua = 1>.
[17] <http://who-umc.org/DynPage.aspx?id=105833&mn1=7347&mn2=7259&mn3=7297&mn4 = 7501.
[18] <http://www.takeandtell.org/>.
[19] Norén N, Hopstadius J, Bate A, Edwards IR. Safety surveillance of longitudinal databases: methodological considerations. Pharmacoepidemiol Drug Saf 2011;20:714–17.
[20] <http://www.who-umc.org/DynPage.aspx?id = 119005&mn1 = 7349&mn2 = 7352&mn3 = 7697>.
[21] <http://web-radr.eu/>.
[22] Juhlin K, Karimi G, Andér M, Camilli S, Dheda M, Siew Har T, et al. Using VigiBase to identify substandard medicines: detection capacity and key prerequisites. Drug Saf 2015;38:373–82.
[23] Caster O, Edwards IR, Norén GN, Lindquist M. Earlier discovery of pregabalin's dependence potential might have been possible. Eur J Clin Pharmacol 2011;67(3):319–20.
[24] Pal S, Duncombe C, Falzon D, Olsson S. WHO strategy for collecting safety data in public health programmes: complementing spontaneous reporting systems. Drug Saf 2013;36(2):75–81.

FURTHER READING

<http://www.who.int/medicines/areas/quality_safety/safety_efficacy/emp_mes/en/>.

SECTION

DRUG UTILIZATION AND PHARMACOECONOMICS

CHAPTER

TOOLS FOR ASSESSING AND MONITORING MEDICINE USE

32

Sangeeta Sharma
Institute of Human Behaviour & Allied Sciences, New Delhi, India

Medicine use is influenced by many factors—characteristics of prescribers, patients, work place, drug regulation, supply, etc., and rationality is determined depending on the context. It is important that before any activities are initiated to improve rational use of medicines an effort should be made to describe and quantify the extent of the problem. In order to address the problems involving the use of medicines in a health facility it is important to initially:

1. Measure the problem of drug use
2. Analyze the same
3. Understand the causes of irrational use of medicines

In order to achieve the above objectives interventions may be planned in a stepwise manner:

STEP 1: General investigation to identify problem areas including:
a. Aggregate data methods
b. Indicator study methods
STEP 2: In-depth investigation of the specific problem identified
a. Prescription audit
b. Qualitative methods
c. Drug utilization review
STEP 3: Develop, implement, and evaluate strategies to correct the problem

32.1 STEP 1. GENERAL INVESTIGATIONS TO IDENTIFY THE PROBLEM AREAS

32.1.1 AGGREGATE DATA METHODS

Aggregate data methods involve data that do not relate to individual patients and can be collected relatively easily. Methods such as ABC analysis, VEN analysis and DDD methodology, etc., are used to identify broad problem areas in drug use.

For this an overview of drug use is collected by aggregating data available in the stock records of the health facility. These data may then be used for managing the formulary of the health

Pharmaceutical Medicine and Translational Clinical Research. DOI: http://dx.doi.org/10.1016/B978-0-12-802103-3.00033-X

facility. These data may also be analyzed according to the different department of the health facility, such as the surgical wards, medical wards, casualty department, etc.

Source of aggregate data. Data for the aggregate analysis may be obtained from the procurement records, warehouse drug records, pharmacy stock and dispensing records, medication error records, and adverse drug reaction (ADR) records.

Uses of aggregate data. The data thus available may be used for the following:

- To measure the quantities (in units—of tablets, injections, etc.) of drugs used by the department/health facility
- The most frequently and infrequently used drugs
- Actual drug consumption—and this may further be matched with the expected consumption, which in turn may be based on the morbidity records
- Per capita use of specific products
- Relative use of therapeutically substitutable products (generic substitution and therapeutic interchange)
- Incidence of adverse drug reactions and medication errors

Cost of drugs used. This may be calculated for individual drugs or for drug categories. Subsequently from this data one may be able to identify:

- The most expensive drugs
- The drugs for which most money is spent
- The most expensive therapeutic categories
- The percentage of the budget spent on certain drugs or drug classes

32.1.2 ANALYSIS OF AGGREGATE MEDICINE USE DATA

Aggregate data analyzed and collected may be used to conduct:

I. ABC analysis
II. Therapeutic category analysis
III. VEN analysis
IV. Calculation of DDD (defined daily dosage)

a. ABC analysis

ABC analysis is the analysis of annual medicine consumption and cost in order to determine which items account for the greatest proportion of the budget.

ABC classification is based on Pareto's Law, following the 80/20 Rule that 80% of problems are caused by 20% of the activities. The concept behind ABC classification is that some items are more important than others, and therefore deserve more managerial attention. Classifying inventory based on degree of importance determined by their consumption allows us to give priority to important inventory items and manage those with care. In other words, prevents us from wasting precious resources on managing items that are of less importance.

ABC classification states that a small percentage of items account for a large percentage of value. This value can be consumption, sales, profits, or another measure of importance. Roughly

10% to 20% of inventory items account for 70% to 80% of inventory value. These highly valuable items are classified as A inventory items. Moderate value items account for approximately 30% of inventory items and contribute to roughly 35% of the total. They are called B items. Finally, approximately 50% of the items only contribute to roughly 10% of total inventory value. These are called C items and are of least importance. Fig. 32.1 provides an example of ABC analysis (Fig. 32.1).

Applications of ABC analysis. It is an extremely powerful tool with uses in selection, procurement, management, distribution, and promotion of rational use of drugs. It reveals high usage items for which there are lower-cost alternatives on the list. Hence this information may be used to:

- Choose more cost-effective alternative medicines
- Identify opportunities for therapeutic substitution
- Negotiate lower prices with suppliers
- Measure the degree to which actual drug consumption reflects public health needs and so identify irrational drug use, through comparing drug consumption to morbidity patterns
- It can be used to critically review:
  - How drugs are utilized
  - How funds are spent in a drug system
- Identify purchases for items not on the hospital essential medicines list

The major advantage of ABC analysis is that it identifies those medicines on which most of the budget is spent. However, a major disadvantage is that it cannot provide information to compare medicines of differing efficacy.

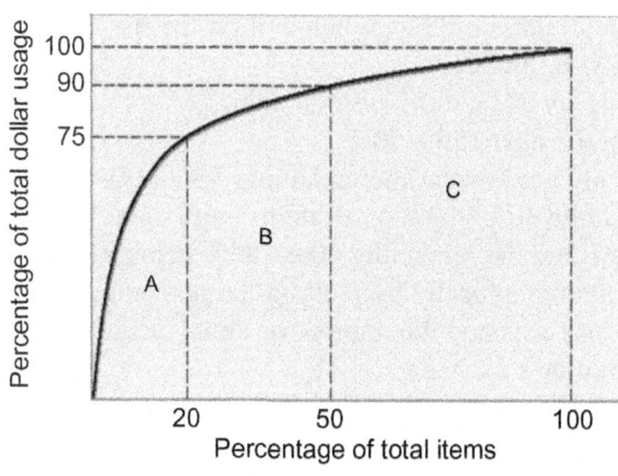

FIGURE 32.1

Classifying items as ABC.

A medicines: High percentage of funds spent on large-volume or high-cost items:

- Greatest potential for savings
- Greatest potential for identifying expensive medicines that are overused

B medicines: Moderate cost and moderate number of items; important items.

C medicines—Small amount of funds spent on the majority of the inventory.

Table 32.1 Applications of ABC Analysis

	A Type	B Type	C Type
Criteria	10%	20%	70%
Annual usage	70%	20%	10%
Control	Very strict	Moderate	Less
Ordering	Daily/weekly	Monthly	Yearly
Safety stock	Less	Moderate	High
Shelf life Monitoring	Priority	Regularly	Routine
Handled by	Senior officers	Middle management	Fully delegated

Applications of ABC in inventory control are shown in Table 32.1.

ABC analysis can be applied to drug consumption data over a one-year period. Shorter period snapshot of usage, i.e., one week or one month, is not desired as one could pick a month where an unusual demand from one customer could render entire analysis worthless. It can also be applied to a particular tender or set of tenders.

The steps to conduct an ABC analysis are as follows:

- *Step 1*: List all items purchased and enter the unit cost
- *Step 2*: Enter consumption quantities for each item
- *Step 3*: Calculate the value of consumption for each item. To do this the cost of each item is multiplied by its annual unit sales to get an annualized cost of usage
- *Step 4*: Sort the list, in descending order, by total value
- *Step 5*: Calculate the percentage of total value represented by each item
- *Step 6*: Calculate the cumulative percentage of total value for each item. Beginning with the first (top) item, add its percentage to that of the item below it on the list
- *Step 7*: Choose cutoff points for A, B, and C:
  - A—those few items accounting for 75%–80% of total value
  - B—those items which take up the next 15%–20%
  - C—the bulk of items which only account for the remaining 5%–10% of value
  - Typically, class A items constitute 10%–20% of all items, with class B items constituting another 10%–20% and the remaining 60%–80% being in category C.
- *Step 8*: The results may be presented graphically by plotting the percentage of total cumulative value on the vertical or y axis and the number of items (accounting for this cumulative value) on the horizontal or x axis (Fig. 32.1).

b. Therapeutic category analysis

Building on the ABC analysis the therapeutic category analysis can:

- Identify the therapeutic category of medicine that accounts for the highest consumption and greatest expenditure
- Indicate the potential inappropriate use of medicines when compared with the information on the morbidity pattern
- Identify medicines that are overused or whose consumption is not accounted for by the number of cases of a particular disease, e.g., chloroquin and malaria
- Help the authorities to choose the most cost-effective drug within a therapeutic class and to choose an alternative medicine for therapeutic substitution

c. Vital, essential, and nonessential (VEN) analysis

VEN analysis is a well-known method to help set-up priorities for purchasing medicines and keeping stock. Drugs are divided, according to their health impact, as vital, essential, and nonessential or desirable categories. VEN analysis allows medicines of differing efficacy and usefulness to be compared, whereas the ABC and therapeutic categories analyze and compare drugs of similar efficacy or action.

- *Vital drugs (V)*: potentially lifesaving or crucial to providing basic health services. These are medicines used for diseases of major public health importance.
- *Essential drugs (E)*: effective against less severe but significant forms of disease, but not absolutely vital to providing basic healthcare.
- *Nonessential drugs (N)*: used for minor or self-limited illnesses, these may or may not be formulary items and efficacious, but they are the least important items stocked.

Some systems divide drugs into two categories only—vital and essential. Sample guidelines for VEN classification are shown in Table 32.2.

Applications of VEN analysis

- System helps manager to set priorities for selection, procurement, and use of drugs according to potential health impact of individual drugs
- VEN classification should be done at regular basis as lists update regularly and public health priorities also change
- Drugs ordering and stock monitoring should be directed at vital and essential drugs
- Safety stock should be higher for vital and essential drugs
- Enough quantities of vital and essential drugs to be bought first

Steps in conducting VEN analysis

- Each DTC member should classify all the medicines as V, E, or N
- The results of each member's classification should be compiled and an overall classification agreed in the DTC

Table 32.2 Sample Guidelines for VEN Classification

Characteristics of Drug or Target Condition	Vital	Essential	Not Very Essential or Desirable
Persons affected (% of Population)	Over 5 %	1–5%	Less than 1%
Persons treated (No. per day at Average Health Center)	Over 5	1–5	Less than 1
Severity of target condition			
Life-threatening	Yes	Occasionally	Rarely
Disabling	Yes	Occasionally	Rarely
Therapeutic effect of drug			
Prevents serious disease	Yes	No	No
Cures serious disease	Yes	Yes	No
Treats minor, self-limited symptoms and conditions	No	Possibly	Yes
Has proven efficacy	Always	Usually	May or not
Has unproven efficacy	Never	Rarely	May or Not

The DTC should then:

- Classify all drugs on list V, E, and N
- Reduce quantities of 'N' categories or eliminate purchases entirely
- Identify and limit therapeutic duplication
- Reconsider proposed purchase quantities by building on the ABC analysis and VEN classification (Table 32.3)

d. Defined daily dose (DDD)

In order to measure drug use, it is important to have both a classification system and a unit of measurement. A common problem when comparing drugs is that different medications can be of different potency, e.g., 20 mg of the beta blocker propranolol are much less effective than 20 mg of the beta blocker bisoprolol. The main problem is that a single item can be used in any quantity or for any duration, e.g., six months or one week. There is a need for a system which measures drug volume more reliably. To deal with the objections against traditional units of measurement, a technical unit of measurement called the Defined Daily Dose (DDD) to be used in drug utilization studies was developed. The DDD is the assumed average maintenance dose per day for a drug used for its main indication in adults. It is defined globally for each medicine by the WHO Collaborating Centre for Drug Statistics in Oslo, Norway (http://www.whocc.no). A DDD will only be assigned for drugs that already have an ATC code. The DDD is based on the average maintenance dose for adults, but it can be adjusted for pediatric medicine use. Different DDDs may be assigned for different drug formulations (i.e., parenteral versus oral).

Structure and nomenclature Structure. In the Anatomical Therapeutic Chemical (ATC) classification system, the active substances are divided into different groups according to the organ or system on which they act, and their therapeutic, pharmacological, and chemical properties. Drugs are classified in groups at five different levels. The drugs are divided into fourteen main groups (1st level), with pharmacological/therapeutic subgroups (2nd level). The 3rd and 4th levels are chemical/pharmacological/therapeutic subgroups and the 5th level is the chemical substance. The 2nd, 3rd, and 4th levels are often used to identify pharmacological subgroups when that is considered more appropriate than therapeutic or chemical subgroups. The complete classification of metformin illustrates the structure of the code:

A Alimentary tract and metabolism (1st level, anatomical main group)
A10 Drugs used in diabetes (2nd level, therapeutic subgroup)
A10B Blood glucose lowering drugs, excluding insulin (3rd level, pharmacological subgroup)
A10BA Biguanides (4th level, chemical subgroup)
A10BA02 Metformin (5th level, chemical substance)

Table 32.3 ABC-VEN Matrix

Class	V items	E items	N items
A items	Constant control Regular follow-up Forecast carefully	Moderate stocks	Nil stocks Min service level
B Items	Moderate stocks	Moderate stocks	Very low stocks
C items	High stocks Max service level	Moderate stocks	Low stocks

Regarding nomenclature, International Nonproprietary Names (INN) are preferred. If INN names are not assigned, USAN (United States Adopted Name) or BAN (British Approved Name) names are usually chosen. The WHO's list of drug terms (Pharmacological action and therapeutic use of drugs—List of Terms) is used when naming the different ATC levels. Pharmaceutical forms with similar ingredients and strengths including immediate and slow release tablets have the same ATC code. However, more than one ATC code can be given if it is available in two or more strengths or routes of administration with clearly different therapeutic uses. For example, prednisolone in single ingredient products is given several ATC codes due to different therapeutic use and different local application formulations. Different combination products sharing the same main active ingredient are usually given the same ATC code. For details see guidelines on ATC classification and DDD assignment, WHO Collaborating Center for Drug Statistic Methodology 2015.

Below are particular groups of drugs where the concept of a DDD is unsuitable or difficult to apply:

- Skin preparations such as ointments or creams whose daily use is heavily influenced by the patients. Since DDDs have not been established, alternative consumption of dermatological preparations can be presented in grams of ointment, cream, etc., and antineoplastic agents in ATC group L01 can be presented in grams of active ingredient
- Vaccinations and other kinds of "one-off" treatments

DDD is expressed as follows:

- DDD per 1000 inhabitants per day for total medicine consumption
- DDD per 100 beds per day (100 bed-days) for hospital use

DDDs per 1000 inhabitants per day. Sales or prescription data presented in DDDs per 1000 inhabitants per day may provide a rough estimate of the proportion of the study population treated daily with a particular drug or group of drugs. As an example, the figure 10 DDDs per 1000 inhabitants per day indicates that 1% of the population on average might receive a certain drug or group of drugs daily. This estimate is most useful for chronically used drugs when there is good agreement between the average prescribed daily dose (see below) and the DDD. It may also be important to consider the size of the population used as the denominator. Usually the general utilization is calculated for the total population including all age groups, but some drug groups have very limited use among people below the age of 45 years. To correct for differences in utilization due to differing age structures between countries, simple age adjustments can be made by using the number of inhabitants in the relevant age group as the denominator.

For anti-infectives (or other drugs normally used in short periods) it is often considered most appropriate to present the figures as numbers of DDDs per inhabitant per year, which will give an estimate of the number of days for which each inhabitant is, on average, treated annually. For example, 5 DDDs/inhabitant/year indicates that the consumption is equivalent to the treatment of every inhabitant with a 5-day course during a certain year. Alternatively, if the standard treatment period is known, the total number of DDDs can be calculated as the number of treatment courses, and the number of treatment courses can then be related to the total population.

DDDs per 100 bed-days. The DDDs per 100 bed-days may be applied when drug use by inpatients is considered. The definition of a bed-day may differ between hospitals or countries, and bed-days should be adjusted for occupancy rate. The same definition should be used when performing comparative studies. As another example, 70 DDDs per 100 bed-days of hypnotics provide an estimate of the therapeutic intensity and suggest that 70% of the inpatients might receive a DDD of a hypnotic every day. This unit is quite useful for benchmarking in hospitals.

Prescribed Daily Dose (PDD). The prescribed daily dose (PDD) is defined as the average dose prescribed according to a representative sample of prescriptions. The PDD can be determined from studies of prescriptions or medical or pharmacy records. It is important to relate the PDD to the diagnosis on which the dosage is based. The PDD will give the average daily amount of a drug that is actually prescribed. The PDD can vary according to both the illness treated and national therapeutic traditions. For instance, for the antiinfectives, PDDs vary according to the severity of the infection treated. The PDDs also vary substantially between different countries, for example, PDDs are often lower in Asian than in Caucasian populations. The fact that PDDs may differ from one country to another should always be considered when making international comparisons.

The DDD does not necessarily reflect the Prescribed Daily Dose (PDD) or the recommended dose. Doses for individual patients and patient groups differ from the DDD and are based on individual characteristics (e.g., age and weight) and pharmacokinetic considerations. For the optimal use of drugs, one must recognize that genetic polymorphism due to ethnic differences can result in pharmacokinetic variations. However, only one single DDD is assigned per ATC code and route of administration (e.g., oral formulation). The DDD reflects global dosage irrespective of genetic variations. Further actual prescription to a patient can vary depending on both the illness and local guidelines. In such situations, the prescribed daily dose (PDD) is established by reviewing a sample of prescriptions and then converted into readily available aggregate data. When the prescribed dose actually differs significantly from the DDD, the reasons and implications must be understood for correct interpretation.

Uses of DDD. Drug consumption data presented in DDDs only give a rough estimate of consumption and not an exact picture of actual use. For example, the figure 10 DDDs/1000 inhabitants/day indicates that 1% of the population on average gets a certain treatment daily. This is only true if the prescribed dose corresponds to the DDD.

DDD provides a unit of measurement that is independent of price, package size, and formulation, making it possible to assess trends in consumption of medicines and to perform comparisons between population groups. By applying the DDD to a defined population, it is possible to:

- Compare consumption of different medicines within the same therapeutic class/group, having similar efficacy but different dose requirements
- Compare the consumptions of medicines belonging to different therapeutic class/groups
- Document and follow changes in the use of a class of drugs over a period between population groups and healthcare systems
- Compare medicine utilization over time for monitoring purposes
- Measure the impact of DTC interventions to improve the use of medicines or evaluate the effect of educational programs directed at different stakeholders
- Compare consumption in different hospitals or geographic areas (international)

- Using cost per DDD, compare the cost of different medicines within the same therapeutic category where the medicines have no treatment duration, such as analgesics and anti-hypertensives
- Evaluate the impact of regulatory interventions on prescribing patterns1:471321:47132

  The DDD method should only be used in settings where reliable procurement, inventory, or sales data have been recorded.

32.1.3 INDICATOR STUDY METHODS

I. *Drug use evaluation (DUE)* is a system of ongoing criteria-based evaluation of drug use that will help to ensure appropriate use at the individual patient level. This method involves the detailed analysis of individual patient data. A drug utilization study is therefore a study designed to describe—quantitatively and qualitatively—the population of users of a given drug (or class of drugs) and/or the conditions of use (indications, dosage, duration of treatment, previous or associated treatments and compliance).

   DUE is the same as drug utilization review (DUR) and these terms are used synonymously and are divided into descriptive and analytical studies. Drug use evaluation (DUE) helps ensure appropriate use of medicines (at the individual patient level). If therapy is deemed to be inappropriate, interventions with providers or patients will be necessary to optimize drug therapy. A DUE is drug- or disease-specific and can be structured and designed to assess the actual process of prescribing, dispensing, or administering a drug (indications, dose, drug interactions, etc.). The emphasis of the former has been to describe patterns of drug utilization and data on the extent and variability in usage and costs of drug therapy, and to identify problems deserving more detailed studies.

   Descriptive studies serve to profile the present situation and to pinpoint problems. Descriptive studies emphasize different aspects such as drug utilization of high-risk groups in the population (drug usage patterns in children, pregnant women, geriatric patients), drug utilization of specific drug groups (antibiotics, benzodiazepines), trends in drug expenditure (an overall change in volume due to changes in the morbidity pattern or a change in price), inter-prescriber variation in prescribing behavior.

   Drug utilization research may provide insights into the following aspects of drug use and drug prescribing:

   a. *Pattern of use*: This covers the extent and profiles of drug use and the trends in drug use and costs over time.
   b. *Quality of use*: This is determined using audits to compare actual use to national treatment guidelines or local drug formularies. An audit in drug use is an examination of the way in which drugs are used in clinical practice—carried out at intervals frequent enough to maintain a generally accepted standard of prescribing.
   c. Indices of quality of drug use may include the choice of drug (compliance with recommended choice of a drug), drug cost (compliance with budgetary recommendations), drug dosage (awareness of inter-individual variations in dose requirements and age-dependence), awareness of drug interactions and adverse drug reactions, and the proportion of patients who are aware of or unaware of the costs and benefits of the treatment.

d. *Determinants of use*: These include user characteristics (e.g., socio-demographic characteristics and beliefs towards drugs), prescriber characteristics (e.g., specialty, education, and other factors influencing therapeutic decisions), and drug characteristics (e.g., therapeutic properties and affordability).
e. *Outcomes of use*: These are the health outcomes (i.e., the benefits and adverse effects) and the economic consequences.
f. Efficiency of drug use, i.e., whether a certain drug therapy provides value for money and the results of such research can be used to help to set priorities for the rational allocation of healthcare budgets.
g. Rational use of drugs in populations. For the individual patient, the rational use of a drug implies the prescription of a well-documented drug at an optimal dose, together with the correct information, at an affordable price. Without knowledge of how drugs are being prescribed and used, it is difficult to initiate a discussion on rational drug use or to suggest measures to improve prescribing habits. Drug utilization research in itself does not necessarily provide answers, but it contributes to rationalize drug use in important ways.

II. Drug Indicators Study involves collection of data at the level of the individual patient (data collected from prescriptions or patient-provider interactions) and informs the investigator of the drug use pattern but does not usually include sufficient information to make judgments about drug appropriateness for diagnosis. Such data can be collected by non-prescribers trained for data collection. The results can be used to identify problem areas in medicine use and patient care, and evaluate impact of the interventions designed to correct the problems identified.

WHO/INRUD drug use indicators for health facilities are intended to measure aspects of health provider behavior in primary healthcare facilities in a reliable way, irrespective of who collects the data. The indicators provide information concerning medicine use, prescribing habits, and important aspects of patient care. The indicators are relevant, easily generated and measured, valid, consistent, reliable, representative, sensitive to change, and understandable.

The drug use indicator studies can be put to use to describe current treatment practices to determine the following:

- Problems in medicine use, in facilities or prescribers. When an indicator study shows unacceptable results a problem may be investigated in greater depth and action taken to improve the same
- To show trends over time providing a monitoring mechanism. Prescribers and facilities whose performance falls below a specific standard of quality can be subjected to intensive supervision
- To motivate healthcare providers and DTC members to improve and follow established healthcare standards
- To evaluate the impact of interventions designed to change prescribing behavior by measuring indicators in control and intervention facilities before and after the intervention

Since most of these indicators do not relate diagnosis to disease, they cannot tell us exactly what proportion of people were treated correctly or the exact nature of the drug use problem; they can only indicate that there is a drug use problem. Furthermore, different disease patterns and prescriber type will greatly affect the indicators, so analysis should be done by diagnosis or prescriber type if these vary between the facilities to be compared.

32.1.3.1 WHO/INRUD drug use indicators for primary healthcare facilities

Prescribing indicators

Average number of drugs per encounter
Percentage of drugs prescribed by generic name
Percentage of encounters with an antibiotic prescribed
Percentage of encounters with an injection prescribed
Percentage of drugs prescribed from essential medicines list or formulary.

Patient care indicators

Average consultation time
Average dispensing time
Percentage of drugs actually dispensed
Percentage of drugs adequately labeled
Patients' knowledge of correct doses

Facility indicators

Availability of essential medicines list or formulary to practitioners
Availability of standard treatment guidelines
Availability of key drugs

Complementary drug use indicators

Percentage of patients treated without drugs
Average drug cost per encounter
Percentage of drug cost spent on antibiotics
Percentage of drug cost spent on injections
Percentage of prescriptions in accordance with treatment guidelines
Percentage of patients satisfied with the care they receive
Percentage of health facilities with access to impartial drug information

Selected indicators used in hospitals

Average number of days per hospital admission
Percentage of drugs prescribed that are consistent with the hospital formulary list
Average number of drugs per inpatient-day
Average number of antibiotics per inpatient-day
Average number of injections per inpatient-day
Average drug cost per inpatient-day
Percentage of surgical patients who receive appropriate surgical prophylaxis
Number of antimicrobial sensitivity tests reported per hospital admission
Percentage of inpatients who experience morbidity as a result of a preventable ADR
Percentage of inpatient deaths as a result of a preventable ADR
Percentage of patients who report adequate postoperative pain control.

a. Data collection strategies

Data may be collected from patient charts and other records retrospectively or prospectively. A retrospective drug utilization study focuses on the evidence that has been provided in the past, i.e., is collected after completion of treatment/care to service users and used to analyze and interpret medication use data within specific a healthcare facility. Retrospective studies, since they use easily accessible data, are inexpensive. Retrospective data collection can be collected rapidly, without any distractions, away from the patient care areas. Selection of a data collection strategy depends on:

- Feasibility in terms of resources and timeframe to implement the data collection strategy
- What data collection strategy is most likely to result in complete and reliable data?

Concurrent reviews focuses on the evidence that will be provided at the present, i.e., concurrent data is collected while treatment/care is being provided, i.e., are conducted simultaneously (dispensing process). They are more expensive and time-consuming than retrospective studies, but offer advantages in terms of prevention of therapeutic misadventures, e.g., if a potential problem is discovered, the dispensing function stops until clarification or authorization is received for dosage correction or to initiate a modification or to continue as before. There is a potential for bias when data is collected concurrent with patient care—may influence clinical practice/behavior by the knowledge that an audit is ongoing. Concurrent reviews require a computer system or a well-organized manual record of medicine usage.

Prospective review focuses on the evidence that will be provided in the near future, i.e., prospective audit involves planning the recording of data on care which will be provided, and gathering specific data for the audit. Prospective review is the option closest to the ideal and is more comprehensive than a concurrent review. Use of prospective review, based upon a complete drug and medical history obtained either from an interview or from historical records, not only permits the practitioner to evaluate the patient's treatment on the basis of Standard Treatment protocols but also prevents interactions by discontinuing certain drugs with potential for drug-interaction for a patient. The reviewer can intervene if any errors in dosage, indications, interactions, or other mistakes are noticed, and prevent error from reaching the patient at the time the medicine is prescribed or dispensed. The ideal set-up for prospective reviews is where the prescriber captures the prescription electronically. The database program then informs the prescriber why they should select another medicine, and guides them through a series of steps/questions, after which some suggested medicines are displayed on the screen. These reviews are only possible if a computer and the required software for error proofing are available.

The terms concurrent and prospective are sometimes used interchangeably. Prospective or concurrent audits should be undertaken if the data required is not routinely collected.

Sample size. Data must be collected from a suitable random sample of prescription records or patient charts from the healthcare facility. The treatment of at least 30 patients, or 100 patients for common select clinical conditions, should be reviewed per clinical specialty or health facility/hospital. The larger the facility and the more the number of practitioners/specialists, the larger the number of prescriptions needed for review.

The number of prescribing encounters per facility and the number of facilities which should be examined would depend on the objective of the study.

At least 30 prescribing encounters from each of 20 health facilities (i.e., a total of 600 prescribing encounters) should be examined. If number of health facilities being examined is less than 20, then more prescribing encounters should be examined.

If the objective is to study prescribers in one health facility, as may be the case for a hospital, then at least 100 prescriptions should be examined at the single facility or specialty/department; if there is more than one prescriber in a given health facility, 100 prescriptions for each individual prescriber should be obtained.

32.2 STEP 2. IN-DEPTH INVESTIGATION OF SPECIFIC PROBLEMS

1. Prescription audit—aims to analyze the type of drugs prescribed, their dosing schedule, and the adequacy of the prescription for a specific diagnosis. By using data on prescriptions and prescribing, it is possible to relate prescribing patterns to many other aspects, for example, to:
   a. Analyze patterns of drug use among patient categories defined by age, gender, or diagnosis
   b. Study the relationship between the prescribed medicine and the apparent indication
   c. Identify the most common illnesses being treated
   d. Identify and examine prescription determinants, such as the extent of influence of particular information (from medical representatives) or publicity campaigns on prescribing (use of ORS)
   e. Examine specific safety problems related to drug use in the light of actual practice such as prescriptions for drugs which are contraindicated or banned drugs being prescribed
2. Qualitative methods to investigate causes of problems of medicine use.
   Qualitative methods include:
   a. Focus group discussion
   b. In-depth interview
   c. Structured observation
   d. Structured questionnaires

Table 32.4 depicts the advantages and disadvantages of the various qualitative methods.

Focus Group Discussion (FGD): The group consists of a small number of people who share similar characteristics such as age, gender, or type of work (a group of prescribers or a group of mothers). This group discussion may be conducted on a specific selected topic and may last for 1–2 hours. A trained moderator encourages participants to highlight their opinions regarding the reasons for the problem under discussion. The discussion is recorded, either by two observers or on tape, and analyzed systematically to identify key themes, issues, or new issues, if any arise during discussion. DTC can use FGDs to identify a range of beliefs, opinions, and motives of a healthcare provider.

The in-depth interview: For this about 10–30 topics may be selected. The discussion is usually flexible and often unstructured, to encourage the respondent to discuss at length a particular topic. The in-depth interview technique can be used to expand the results of a quantitative study for exploring the problems/behavior of a health facility/healthcare provider. This can also be used for evaluating the impact of an intervention that may have been introduced to promote rational medicine use.

Beside face-to-face in-depth interview and telephone interview the use of new communication forms such as e-mail and MSN messenger opens new ways for qualitative research workers for data collection. In e-mail and MSN messenger interview social cues of the interviewee are not

Table 32.4 Summary of Qualitative Methods

Method	Advantages	Disadvantages
Focus group discussion		
• 1–2 hours of discussion (may be recorded) • Homogenous not diverse: 6–10 respondents with similar characteristics, e.g., age, gender, social status • 2–4 discussions for each significant target population • Discussion topics pre-defined. Moderator leads the discussion • Flexible not standardized • Informal, relaxed, ambient • Beliefs, opinions and motives may be highlighted • Words not Numbers: rely upon words spoken by participants	• Inexpensive • Quick • Easy to organize • Identifies a range of beliefs • Identifies new issues • Expressions other than those in verbal form such as gestures and stimulated activities can provide researcher with useful insights	• Focus group output is not projectable. Groups may not represent the interest of the larger population • Successful outcome is dependent on the skills of the moderator. Inexperienced moderator may face problems in controlling some participants who try to dominate the group • Respondents may be reluctant to share some sensitive ideas and concerns publicly • Focus groups are a very artificial environment which can influence the responses that are generated. Use of recording devices may affect the participant's behavior. However, if the points are being recorded by a note-taker the chances of incomplete recording of facts may ensue
In-depth interviews		
• One-to-one extended interview • Questions may be predetermined and open-ended • Often covers up to 10–30 topics • Beliefs, attitudes, and knowledge as confounders may be identified	Provide detailed data Unsought but significant data may be identified and get deeper insights	• May generate data which are difficult to manage • Time-consuming and expensive • Bias due to respondent • Interview may be interviewer-dependent • Interviewer needs skills in effective interview technique like body language and friendly speaking • Findings not generalizable
Structured observation		
• Data collection instrument is structured • Observers are trained to blend into their surrounding • Observers are trained to record what they actually see • Useful for recording provider–patient interactions • Assesses actual behavior on site	Observes real life behavior as against the stated behavior; will be recorded	• Time-consuming and expensive • Observation may cause a temporary change in the behavior of health workers • Observer bias may be a confounder

Table 32.4 Summary of Qualitative Methods *Continued*

Method	Advantages	Disadvantages
Structured questionnaires		
• Useful for a large sample of respondents • Respondents are selected so as to represent the larger population • Questions are standardized with a fixed set of responses • Measures the frequency of attitudes, beliefs, and knowledge	Results may be generalized to include the wider population	• Interviewers may ask questions and interpret answers incorrectly (because of language or inclusion of illiterate population) • Questions may be ambiguous • Respondents may give answers to please the interviewer

important information sources for the interviewer (of course dependent on the research problem); there is a huge time difference, because interviewer and interviewee live in different parts of the world separated by several time zones, and synchronous interviewing means one party (interviewer or interviewee) interviewing at night. The type of interview technique chosen by the researcher can depend upon the advantages and disadvantages, which are linked to every interview technique.

Using face-to-face interviews for collecting information are preferred when:

- Social cues of the interviewee are very important information sources for the interviewer (of course dependent on the research problem)
- The interviewer has enough budget and time for travelling, or the interviewees live near the interviewer
- Standardization of the interview situation is important

32.2.1 THE STEPS OF A DUE

The steps of a DUE are as follows:

1. Establish responsibility: It is the responsibility of the DTC to establish procedures for the implementation of a DUE program starting from appointing a responsible member of the DTC or a subcommittee to monitor and supervise the DUE process in the hospital or health facility. Draw annual plans, outlining which medicines or clinical conditions are to be included in the DUE process.
2. Develop the scope of activities and define the objectives: The DTC should decide upon the objectives of the DUE and the scope of the activities necessary. The scope can be very extensive or focus on a single aspect of drug therapy depending upon the type of problem identified, for example:
   a. Overuse of a more expensive medicine when a cheaper equivalent is available, as revealed in aggregate data
   b. Incorrect use (indication, dosage, administration) of a particular drug, as revealed in patient charts, ADR reports, medication error reports

c. Inappropriate choices of antibiotic, as revealed in antibiotic sensitivity reports
d. Poor dispensing process, as revealed by dispensing error reports, patient complaints or feedback

Due to the large number of medicines available at a hospital or clinic, the DTC must prioritize and concentrate on those medicines with the highest potential for problems for optimum returns on the effort involved in conducting these studies. These high-priority areas include:

a. high-volume drugs
b. high-cost drugs
c. drugs with a high incidence of ADRs
d. drugs with high misuse potential
e. prophylactic, empirical, and therapeutic use of antimicrobial drugs
f. critically important therapeutic categories, such as cardiovascular, emergency, toxicology, intravenous drugs, and narcotic analgesics
g. drugs undergoing evaluation for addition to the formulary
h. drugs used for off-labeled indications
i. drugs with a narrow therapeutic index
j. drugs used in high-risk patients
k. common clinical conditions often inappropriately treated

3. Establish criteria for review of the medicine: DTC should establish DUE criteria. DUE criteria are statements that define correct drug usage with regard to various components, as shown in the box below. Criteria for the use of any medicine should be established using the hospital's STGs. In the absence of hospital STGs, criteria may be based on recommendations from national or other locally available satisfactory drug use protocols, other relevant literature sources, and/or recognized international and local experts. Credibility and acceptance of the results of DUE by staff depends on using criteria that have been developed from reading established evidence-based medicine information and that have been discussed and agreed upon with prescribers.

COMPONENTS OF DRUG USE FOR DRUG USE EVALUATION (DUE) CRITERIA

- *Uses*: appropriate indication for drug, absence of contraindications or high-risk situations
- *Selection*: appropriate drug for a particular clinical condition
- *Dosing*: indication-specific dosing, dose intervals, and duration of treatment
- *Interactions*: absence of drug–drug, drug–food, drug–disease interactions
- *Preparation*: steps involved with preparing a drug for administration
- *Administration*: steps involved in administration, quantity of medicines dispensed
- *Patient education*: patients knowledge on drug and disease-specific instructions and knowledge of red flag signs
- *Monitoring*: clinical and laboratory, frequency of monitoring, when to return in emergency, recognition of red flag or danger signs
- *Outcome*: decreased blood glucose, asthma attacks, blood pressure control

Reviewing many criteria makes the DUE process more difficult, and may hamper successful completion of the review. Therefore, the number of criteria established for each medicine should be small (usually between 3 and 5). After establishing the criteria, decide thresholds or benchmarks for each criterion in order to define the goals or expectations for compliance with the criteria. Ideally 100% compliance of all cases with the criteria is expected but in reality this may not be feasible. Therefore, DTC might decide to set a threshold of 90%–95%. Compliance below this standard should initiate investigation to ascertain reasons and corrective action accordingly.

4. Data collection: Data may be collected either prospectively or retrospectively, from patient charts and other records, at the time a medicine is prescribed or dispensed. Retrospective data collection has the advantage as it may be easier and quicker.

   Data Sources include patient charts, dispensing records, medication administration records, ADR reports, medication error reports, laboratory reports, antimicrobial sensitivity reports, documented staff and patient complaints, etc.

5. Data analysis: Collation and transferring of the data collected from the data collection tool onto summary sheets (manual data collection) or onto a spreadsheet or database (manual or electronic data collection) for interpretation. Calculate and summarize the percentages of cases that meet the threshold for each criterion followed by analysis of the data to find proportions of cases that met the criteria of good practice for presentation to the DTC. A quarterly report of all DUE programs should be prepared.

6. Feedback to the prescribers and develop a plan of action: After information is presented (on inappropriate drug use or unacceptable patient outcome), the DTC should determine the differences between actual and desired results and the reasons for these differences. The DTC should then decide the necessary follow-up action and whether to continue, discontinue, or revise the criteria or expand the functions of the DUE in question.

   After discussion with the prescribers, recommendations should be drawn and should include specific steps to correct any drug use problem revealed by the DUE results. For example, if a specific medicine is being prescribed at too often or too high a dose, the recommendations need to specify in detail how the prescription pattern or dosing of this medicine can be improved. Various interventions to improve drug use include:

   a. Feedback to the prescribers
   b. Education, e.g., in-service education, workshops, face-to-face discussions, letters, newsletters
   c. Revising the formulary list and use of non-formulary drugs
   d. Revising the standard treatment guidelines
   e. Introduction of a structured drug order forms
   f. Rotation of antibiotics
   g. Automatic stop orders such as for high end antibiotics
   h. Restriction on prescribing such as antibiotics with no restriction, for restricted use, and very restricted use

7. Follow-up: In every DUE, follow-up is critical to ensure success of the DUE program and to see effective resolution of the problem. Did an intervention achieve its objective? DUE will not serve any purpose if an intervention is not evaluated, or drug use problems are not resolved. The need to continue, modify, or discontinue the DUE must be assessed from time to time. Therefore, DUE activities should be evaluated regularly (at least annually), to review appropriateness of the criteria and identify ineffective interventions and redesign those which do not have a significant impact on drug use.

Common problems associated with DUEs include poor prioritization of problems, lack of good quality data, lack of documentation, unclear responsibilities for different activities, lack of personnel, and inadequate follow-up. With regular follow-up, performance of the prescribers is likely to improve in all areas knowing that their performance is being monitored.

32.3 STEP 3. DEVELOP, IMPLEMENT, AND EVALUATE INTERVENTIONS TO CORRECT THE PROBLEM

Drug utilization research undertaken in the following ways may enable us to assess the impact of interventions on drug use:

- Monitor and evaluate the effects of measures taken to ameliorate undesirable prescription patterns (e.g., provision of regional or local hospital formularies, information campaigns, and regulatory policies). Prescribers may switch to other equally undesirable prescribing practices. Therefore, to assess the full impact of the measure, these potential alternative medicines should be included in the study.
- Conduct broad surveys to assess the impact of regulatory changes or changes in reimbursement schemes or insurance. It is important to keep in mind that total cost to society may remain the same or may even increase if more expensive drugs are used as alternatives.
- To monitor the extent to which the promotional activities of the pharmaceutical industry and the educational activities of the community influence the patterns of drug use.

Drug use should be controlled according to a quality control cycle (PDCA) that offers a systematic framework for continuous quality improvement. The components of a PDCA cycle are illustrated as below:

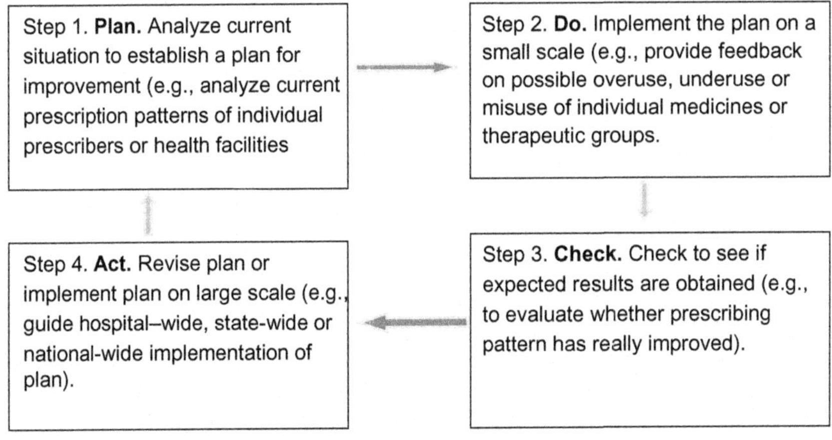

32.4 SUMMARY

1. No single tool is adequate to address irrational use of medicine. There is not yet enough evidence to recommend one method over another and choices should therefore be made according to the specific objectives of each study and the level at which it is performed depending upon availability of human resources, technical expertise, and resources. So one may choose to investigate one or more.
2. *Stepwise Approach*

   Step 1.
   a. Start with Aggregate Data Method viz., ABC, VEN, DDD analysis, to broadly identify problem.
   b. Results of Aggregate Data Method can be supported by Drug Use Indicator study.

   Step 2.
   a. In-depth investigation of specific problem by using "Qualitative Method," Prescription Audit, Drug Utilization Review.
   b. Above will help in understanding reason for problem in medicine use.

   Step 3.
   a. Develop and implement interventions. DTC/PTC can play vital role in developing and implementing interventions.

   Step 4.
   a. Evaluate impact of intervention by using same tools as used for pre-intervention investigation.
3. For drug utilization studies to be reliable, adhere to strict methodological standards and appropriate sample size and methodology.

FURTHER READING

Quick JD, Hogerzeil H, Rankin JR, Laing RO, Dukes MNG, O'Connor RW, et al. Managing Drug Supply: Management Science for Health in collaboration with World Health Organization. Second edition Kumarian Press; 1997. p. 137–49.

How to Investigate Drug Use in Health Facilities: Selected Drug Use Indicators - EDM Research Series No. 007 *World Health Organization* 1993.

Guidelines for ATC classification and DDD assignment. 2016. WHO Collaborating Centre for Drug Statistics Methodology, Available at https://www.whocc.no/atc_ddd_index/.

CHAPTER

PHARMACOECONOMICS IN HEALTHCARE

33

Mahendra Rai[1] and Richa Goyal[2]

[1]Tata Consultancy Services, Mumbai, Maharashtra, India [2]QuintilesIMS, Mumbai, Maharashtra, India

33.1 INTRODUCTION

Pharmacoeconomics can be considered as a branch of health economics which identifies, measures, and compares the costs and consequences of pharmaceutical products and services. It describes the economic relationship which combines the drug research, its production and distribution, storage, pricing, and further use by the people. Pharmacoeconomics can be classified as the field of study that evaluates the behavior or welfare of individuals, firms, and markets for the relevance of pharmaceutical products, services, and programs. Pharmacoeconomics can be summarized is a collection of descriptive and analytic techniques that evaluate pharmaceutical interventions, spanning from individual patients to the healthcare system as a whole, which can aid the policy makers and the healthcare providers in evaluating the affordability of and access to rational drug usage [1]. The pharmacoeconomic techniques include cost-minimization, cost-effectiveness, cost-utility, cost-benefit, cost-of-illness, cost-consequence, or other economic analytic techniques helpful in providing information to healthcare decision makers. It has been observed that cost-effectiveness evaluations of pharmaceutical options are becoming mandatory for attaining adequate reimbursement and payment for services. Further, pharmacoeconomic methods help in verifying the costs and benefits of various therapies and pharmaceutical services which may help in establishing significances for those options and help in appropriate resource allocation in ever-changing healthcare landscapes.

33.2 BASICS OF PHARMACOECONOMICS

33.2.1 PERSPECTIVES

Some key factors are of pivotal consideration in pharmacoeconomics, and one of those is the pharmacoeconomic perspective. Pharmacoeconomic perspectives take into consideration who pays the costs and who receives the benefits. Further, the values of saving money for society may be viewed differently by private third-party payers, administrators, health providers, governmental agencies, or even the individual patient (Fig. 33.1).

Pharmaceutical Medicine and Translational Clinical Research. DOI: http://dx.doi.org/10.1016/B978-0-12-802103-3.00034-1

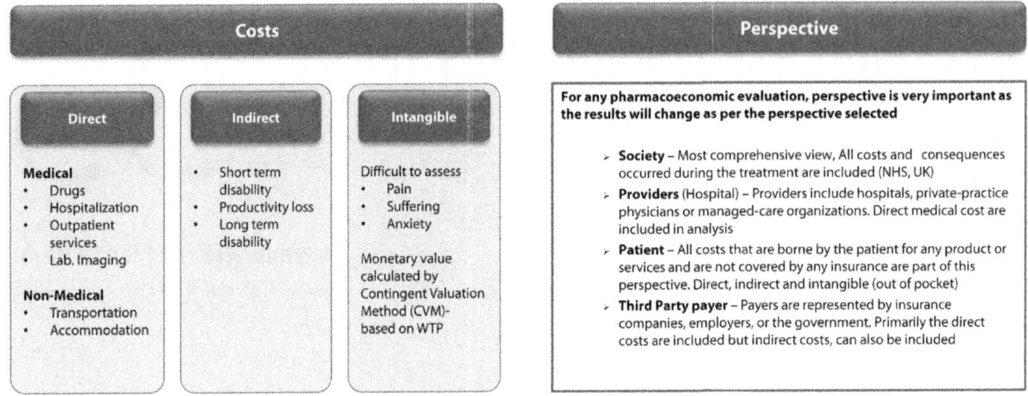

FIGURE 33.1

Various costs in Pharmacoeconomic analysis.

33.2.2 COST

While considering costs, a distinction must be made between financial and economic concepts. Financial costs relate to monetary payments associated with the price of a good or service traded in the marketplace. Economic costs relate to the wider concept of resource consumption, irrespective of whether such resources are traded in the marketplace. In pharmacoeconomics, costs can be classified as direct, indirect, and intangible costs [2] (Fig. 33.1).

33.3 TYPES OF PHARMACOECONOMIC ANALYSIS

The cost of the drugs is important as it plays a substantial part of the total cost of healthcare—typically 10%–15% in developed countries and up to 30%–40% in some developing countries. However, drug costs are generally inferred in the context of the overall (net) costs to the health system. Drugs cost money to buy, but their use may also save costs in other areas. For example, the purchase of one specific type of drug may lead to reductions in the following:

- Use of other drugs
- Number of patients requiring hospitalization or in the length of stay in hospital
- Number of doctor visits required
- Administration and laboratory costs compared with those incurred by using another drug to treat the same condition

Assessing the true cost to a health system of using a specific drug will therefore require the cost of acquisition of the drug to be adjusted against cost savings resulting from the use of the drug and the extra health benefits which may occur. However, costs may arise from adverse drug reactions both in the short-term and, particularly, the long-term. Assessing the value for money of using a drug requires the extra health benefits achieved to be weighed against the extra net cost. This comparison is usually expressed as an incremental cost-effectiveness ratio (ICER), which is the net incremental cost (costs minus cost offsets) of gaining an incremental health benefit over another therapy. Further, concerns about the cost of medical care in general, and pharmaceuticals in particular, are currently being expressed by all health systems. There is a focus on providing quality care within limited financial resources. Decision-makers are increasingly dependent on clinical economic data to guide policy formulation and implementation. Some of the concepts used in making such decisions include: cost-minimization, cost-effectiveness, cost-benefit, and cost-utility.

Using pharmacoeconomics and disease management concepts, health providers can produce more cost-effective outcomes in a number of ways. For example:

- Decrease drug–drug and drug–lab interactions
- Increase the percentage of patients in therapeutic control
- Reduce the overall costs of the treatment by utilizing more efficient modes of therapy
- Reduce the unnecessary use of emergency rooms and medical facilities
- Reduce the rate of hospitalization attributable to or affected by the improper use of drugs
- Contribute to better use of health manpower by utilizing automation, telemedicine, and technicians
- Decrease the incidence and intensity of iatrogenic disease, such as adverse drug reactions

33.3.1 COST MINIMIZATION ANALYSIS

Cost minimization analysis (CMA) comprises for the least costly alternatives when the outcomes of two or more therapies are virtually identical. CMA involves calculating drug costs to analyze the least costly drug or therapeutic modality. It also reflects the cost of preparing and administering a dose. This method of cost evaluation is the one used most often in evaluating the cost of a specific drug. This method can only be used to compare two products that have been shown to be equivalent in dose and therapeutic effect. Therefore, this method is most useful for comparing generic and therapeutic equivalents drugs. In many cases, there is no reliable equivalence between two products and if therapeutic equivalence cannot be demonstrated, then cost-minimization analysis is inappropriate. Many sources of clinical evidence can be used to support economic analyses; however, the "gold standard" is normally considered to be the randomized controlled trial (RCT), which holds everything constant with the exception of the drug being evaluated. However, as the results of clinical trials cannot be known in advance, it is impossible to plan to undertake a CMA alongside an RCT because it is not certain that the health outcomes being compared will be equivalent [3]. Therefore, no prospective economic evaluation starts out as a CMA; only when the health outcomes generated are empirically demonstrated to be "identical or similar" will the CMA be adopted as an appropriate methodology by the health economist.

The CMA is frequently portrayed as being the "poor relation" among health economic methodologies, with its apparent simplicity making it unworthy of being considered alongside more theoretically rigorous health economic methodologies. However, it is important that health economists recognize and acknowledge that the theoretical underpinnings of CMA are just as rigorous as those underpinning other methods of economic evaluation.

33.3.2 COST EFFECTIVENESS ANALYSIS

Cost effectiveness analysis (CEA) is used when two or more therapeutic approaches have differential effectiveness. The numerator states the costs of all direct medical utilization for the treatment of the condition and the indirect costs (work impact) and the denominator states the patient-level unit of benefit measured in temporal units (life-years saved or healthy days). Further, the incremental cost-effectiveness analyses assess the difference between the two therapies.

$$\text{Cost effectiveness ratio} = \text{Cost}/\text{Outcome}$$

Cost-effectiveness analysis (CEA) involves a broader look at drug costs. Cost is measured in monetary terms and effectiveness is measured independently and may be measured in terms of a clinical outcome for, e.g., number of lives saved or complications prevented or diseases cured [4]. CEA thus measures the incremental cost of achieving an incremental health benefit expressed as a particular health outcome that varies according to the indication for the drug. CEA provides a framework to compare two or more decision options by examining the ratio of the differences in costs and the differences in health effectiveness between options. The overall goal of CEA is to provide a single measure, the ICER, which relates the amount of benefit derived by making an alternative treatment choice to the differential cost of that option. When two options are being compared, the ICER is calculated by the formula [5]:

$$\text{Cost (treatment 2)} - \text{Cost (treatment 1)/Effectiveness (treatment 2)} - \text{Effectiveness (treatment 1)}$$

In medical or pharmacoeconomic cost-effectiveness analysis, health resource costs (the numerator) are in monetary terms, representing the difference in costs between choosing treatment 1 or treatment 2. In cost-effectiveness analysis, the differential benefits of the various options (the denominator) are non-monetary and represent the change in health effectiveness values implied by choosing treatment 1 over treatment 2. Typically, these health outcomes are measured as lives saved, life years gained, illness events avoided, or a variety of other clinical or health outcomes. Unlike CEA, cost-benefit analysis values both the costs and benefits of interventions in monetary terms.

CEA compares medical intervention strategies through the calculation of the ICER, a measure of the cost of changes in health outcomes. These analyses can be performed on clinical trial data when information on both costs and effectiveness is available or, more commonly, through the use of decision analysis models to synthesize data from many sources. Interpretation of CEA results can be challenging due to the variety of health outcomes that can be used as the effectiveness term in these analyses and to the absence of a definitive criterion for "cost-effective."

33.3.2.1 How is ICER Used?

Most countries do not use $/QALY for policy making. Some countries do use this measure as an aid in decisions, but the threshold varies:

- Relatively uniform ICER across disease
- ICER varies by disease
- Implementation varies by country

Threshold of ICER:

- USA: US $50,000/QALY
- Canada: CAN $20,000 to CAN $100,000/QALY
- Australia: AU $42,000 to AU $76,000/LYG
- NICE: £20,000 to £30,000/QALY

Plane depicting ICER is described in Fig. 33.2.

33.3.3 COST UTILITY ANALYSIS

Cost utility analysis (CUA) is an economic analysis in which the incremental cost of a program from a particular point of view is compared to the incremental health improvement expressed in the unit of quality adjusted life years (QALYs) [6]. CUA is used to determine cost in terms of utilities, to say in quantity and quality of life. Differing from cost-benefit analysis, cost-utility analysis is used to compare two different drugs or procedures whose benefits may be different. CUA expresses the value for money in terms of a single type of health outcome. The ICER in this case is usually expressed as the incremental cost to gain an extra QALY. This approach incorporates both increases in survival time and changes in quality of life into one measure. An increased quality of life (QoL) is expressed as a utility value on a scale of 0 (dead) to one (perfect quality of life). The use of incremental cost-utility ratios enables the cost of achieving a health benefit by treatment with a drug to be assessed against similar ratios calculated for other health interventions

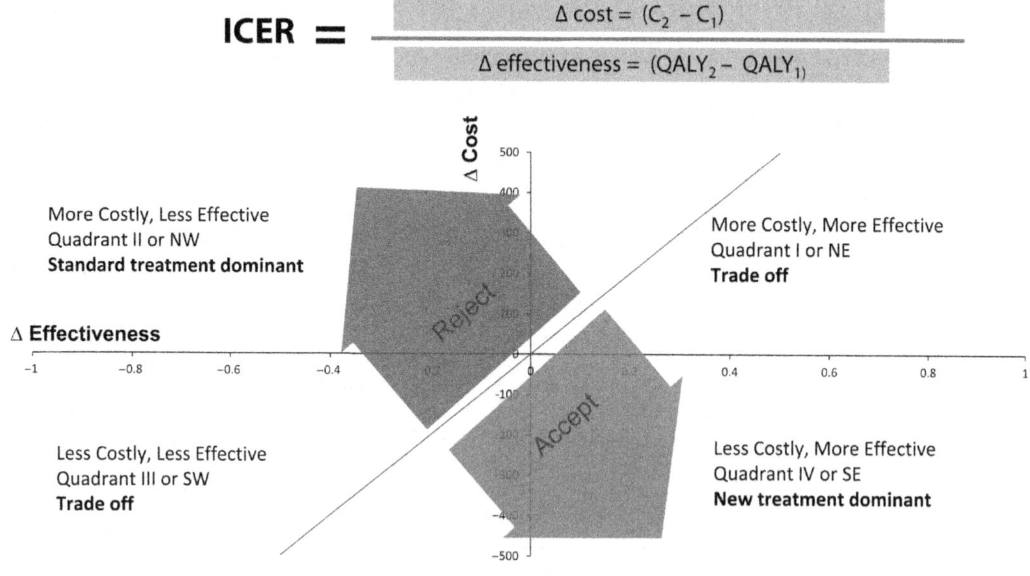

FIGURE 33.2

Plane depicting ICER.

(e.g., surgery or screening by mammography). It therefore provides a broader context in which to make judgments about the value for money of using a particular drug.

CUA is a special case of CEA, where the numerator of the ICER is a measure of cost and the denominator is measured typically using a metric called the QALY. A QALY accounts for both survival and QoL benefits associated with the use of a healthcare technology. The QoL component of the QALY is measured using a metric known as a health utility; hence, the term cost-utility analysis is used to describe this form of CEA. Given that the QALY can be used to measure the survival and QoL benefits of a healthcare technology, the QALY can serve as a common metric from which to compare the benefits of very different healthcare technologies.

33.3.4 COST BENEFIT ANALYSIS (CBA)

Cost benefit analysis (CBA) is an economic analysis in which both the costs and consequences are expressed in monetary terms. The numerator states the monetary benefit gained from the treatment and the denominator defines the monetary investment for the treatment. CBA comes to play so as to value both incremental costs and outcomes in monetary terms and therefore allowing a direct calculation of the net monetary cost of achieving a health outcome. A gain in life-years may be regarded as the cost of the productive value to society of that life-year. The methods for analyzing gains in quality of life include techniques such as *willingness-to-pay*, where the amount that individuals would be willing to pay for a quality-of-life benefit is assessed (Table 33.1).

33.4 IMPACT OF PHARMACOECONOMICS ON HEALTHCARE

Pharmacoeconomics began as an applied field. It was an urgent practical response to a number of new products mostly pharmaceuticals reaching the market at an unprecedented rate just before the

Table 33.1 Summary of Various Types of Pharmacoeconomic Analyses

Cost Minimization Analysis (CMA)	Cost Effectiveness Analysis (CEA)	Cost Utility Analysis (CUA)	Cost Benefit Analysis (CBA)
Compares the costs of two or more alternatives that produce equivalent health outcomes No value is given to benefits; only costs, measured in monetary units, are compared	• Most common analysis used to decide between different treatments for the same condition • Measures cost per unit of effect • Costs measured in monetary units, benefits measured as outcome measures or in natural units (e.g., life-years gained, symptom-free days)	• A type of CEA • Can compare treatments for different disorders, as the outcome measure is the same • Costs measured in monetary units, benefits measured as "utilities" • Best known utility measure is the QALY	• Not often used in modern health economics • Both costs and consequences are measured in monetary units • Can use willingness-to-pay to value benefits • Any treatment with benefits greater than costs is considered "worthwhile"

turn of the century, and to the growing perception that healthcare budgets were being strained as a consequence of pharmaceuticals expenditures outpacing those in other healthcare sectors. Manufacturers were suddenly facing requests from payers to justify the price of their products, and they looked up to clinical experts, decision analysts, and economists for help in providing answers. Initially, most of these justifications were not guided by theory. They mainly involved documentation of the clinical effects—often in broader, more patient-oriented terms—and some attempt at quantifying the expected costs.

Pharmacoeconomics can certainly help in decision making when evaluating the affordability of and access to the right medication to the right patient at the right time, comparing two drugs in the same therapeutic class or drugs with similar mechanism of action, and in establishing accountability that the claims by a manufacturer regarding a drug are justified. Proper application of pharmacoeconomics will allow the pharmacy practitioners and administrators to make better and more informed decisions regarding products and services they provide. Pharmacotherapy decisions traditionally depended solely on clinical outcomes like safety and efficacy, but pharmacoeconomics teaches us that there are three basic outcomes including clinical, economic, and humanistic, that should be considered in drug therapy. It is accepted by all that appropriate drug selection decisions could not be made today based on acquisition costs only. Hence, applied pharmacoeconomics can help in decision making, in assessing the affordability of medicines to the patients, access to the medicines when needed, and comparing various products for treatment of a disease. It will provide evidence contraindicating the promotion of certain types of high-cost medicines and services.

One theory suggests that the role of healthcare systems is to maximize collective health across society within a fixed budget and that the worth of any new intervention can be appraised by estimating the amount of additional cost that is required to produce an additional unit of health [7]. Furthermore, the practitioners proposed that health should be measured in QALYs which is a unit that conflates life expectancy with the expected quality of that life relative to some undefined "perfect" health. However, the concept did not have any empirical basis. It was observed that the actual decision makers were not seeking to maximize aggregate health—they were trying to deal with illness and its consequences. This led to requirement of a cost-effectiveness threshold. Thresholds that were and continue to be put forward have been arbitrary, inconstant, and out of line with exploratory research on society's evaluation of health outcomes.

33.5 GOING FORWARD

There are some areas which require more evolution in the field of pharmacoeconomics so as to deal with the actual problems occurring in decisions, including resource allocation. This might require estimating the total additional costs an intervention will accrue in the population of patients who will be affected but not the per-patient costs. The analyses will place much more emphasis on the near term, which is the time period of most relevance to actual decisions. This will also have to consider how quickly and which patients take up the intervention. The analyses will call for using real cost offsets that can be expected by the healthcare system, rather than hypothetical ones that might occur under ideal circumstances.

The expected advantages can be with the intervention for patients using it with realistic estimates of the uptake estimated based on real data. Also, it will be important to compare effect profile and cost impact with those of the other interventions available for the condition at issue. In most instances, this will mean that not only the new technologies are evaluated, but also the full array of available interventions. The analysis will need to take up the question of what existing interventions would need to be given up to cover the new one, or what level of budget increases will be required if nothing is removed. Most important, new interventions will need to fit into the relevant practice guidelines, and what effect the new intervention will have on the guideline. There will be several consequences of the maturation of pharmacoeconomics, not only methodological developments, but also effects on other research areas such as efficacy clinical trials, and even on the structure of pharmaceutical companies.

REFERENCES

[1] McGhan WF, editors. Introduction to Pharmacoeconomics. Pharmacoeconomics from theory to practice. CRC Press. Taylor and Francis Group; 2010, vol 1.

[2] Haycox A. What is Health Economics? What is? Series. Second edition. 2009, pp. 2–3.

[3] Donaldson C, Hundley V, McIntosh E. Using economics alongside clinical trials: Why we cannot choose the evaluation technique in advance. Health Econ Lett 1996;5:267–9.

[4] <https://http://apps.who.int/medicinedocs/en/d/Js4876e/5.3.html>. Essential medicines and health products information portal. A World Health Organization resource [accessed 25.09.15].

[5] Smith K.J. and Roberts M.S. CRC Press. Taylor and Francis Group, editors. Cost Effectiveness Analysis. Pharmacoeconomics from theory to practice; 2010, pp. 95–96.

[6] Phillips C. What is Health Economics? What is? Series. Second edition. 2009, p. 1.

[7] Gold MR, Russell LB, Siegel JE, editors. Cost-effectiveness in health and medicine. New York: Oxford University Press; 1996.

Index

Note: Page numbers followed by "*f*," "*t*," and "*b*" refer to figures, tables, and boxes, respectively.

B

C

D

E

H

J

K

L

N

O

P

S

U